ADVANCES IN PERFORMANCE-BASED EARTHQUAKE ENGINEERING

GEOTECHNICAL, GEOLOGICAL, AND EARTHQUAKE ENGINEERING

Volume 13

For further volumes:
http://www.springer.com/series/6011

Advances in Performance-Based Earthquake Engineering

ACES Workshop

edited by

MICHAEL N. FARDIS

Department of Civil Engineering,
University of Patras, Greece

Springer

Editor
Michael N. Fardis
University of Patras
Department of Civil Engineering
P.O. Box 1424
265 04 Patras
Greece
fardis@upatras.gr

ISBN 978-90-481-8745-4 e-ISBN 978-90-481-8746-1
DOI 10.1007/978-90-481-8746-1
Springer Dordrecht Heidelberg London New York

Library of Congress Control Number: 2010929682

Printed on acid-free paper

Springer is part of Springer Science+Business Media (www.springer.com)

Preface

Historically, introduction and enforcement of structural design codes and standards has been the responsibility of competent Authorities, with public safety as their overriding consideration. So, traditional seismic design codes or standards, especially for buildings, aim at protecting human life by preventing local or global collapse under a specific earthquake level with low probability of exceedance. However, in the 1960s the international earthquake engineering community was already aware of the importance of property loss and other economic consequences caused by more frequent seismic events. Recognizing that it is not feasible to avoid any damage under strong earthquakes, the Structural Engineers Association of California (SEAOC) adopted in its 1968 recommendations the following requirements for seismic design:

"Structures should, in general, be able to:

- Resist a minor level of earthquake ground motion without damage.
- Resist a moderate level of earthquake ground motion without structural damage, but possibly experience some nonstructural damage.
- Resist a major level of earthquake ground motion having an intensity equal to the strongest either experienced or forecast for the building site, without collapse, but possibly with some structural as well as nonstructural damage."

Major earthquakes that hit developed countries in the second half of the 1980s and the first half of the 1990s caused relatively few casualties but very large damage to property and other economic losses. In response to this, "Performance-based earthquake engineering" emerged in the SEAOC Vision 2000 document and developed into the single most important idea of recent years for seismic design or retrofitting of buildings.

"Performance-based engineering" focuses on the ends, notably on the ability of the engineered facility to fulfil its intended purpose, taking into account the consequences of its failure to meet it. Conventional structural design codes, by contrast, are process-oriented, emphasizing the means, i.e., the prescriptive, easy to apply, but often opaque rules that disguise the pursuit of satisfactory performance. These rules have been developed over time as a convenient means to provide safe-sided, yet economical solutions for common combinations of structural layout, dimensions and materials. They leave limited room for the designer to exercise judgement and creativity and do not provide a rational basis for innovative designs that benefit from recent advances in technology and structural materials.

"Performance-based earthquake engineering" in particular tries to maximize the utility from the use of a facility by minimising its expected total cost, including the short-term cost of the work and the expected value of the loss in future earthquakes (in terms of casualties, cost of repair or replacement, disruption of use, etc.). Ideally we should take into account all possible future seismic events with their annual probability of occurrence and carry out a convolution with the corresponding consequences during the design working life of the facility. However, this is not so practical. Therefore, at present "performance-based earthquake engineering" advocates replacing the traditional single-tier design against collapse and its prescriptive rules, with a transparent multi-tier seismic design, meeting several discrete "performance levels", each one under a different seismic event with its own annual probability of exceedance.

By the end of the first decade of the millennium the concept and the methodologies for Performance-based earthquake engineering have come of age. They have been introduced, be it cautiously, in codes of practice, especially for seismic assessment and retrofitting. Along with the further advancement of the methodologies, the international scientific and technical community of earthquake engineering has now the task to provide the full portfolio of tools and the techniques necessary for its implementation: from detailed knowledge of the full range of ground motions that a structure may conceivably experience and their damaging potential, including significant residual deformations, to techniques to limit such damage.

The Workshop on "Advances in Performance-Based Earthquake Engineering" that took place on the Greek island of Corfu in early July 2009 aspired to serve exactly that purpose. It attracted 54 renowned experts from around the globe to send and present contributions to its four parts:

- Ground motions for performance-based earthquake engineering.
- Performance-based seismic design and retrofitting – Methodologies.
- Performance-based seismic design and retrofitting – Implementation.
- Advanced seismic testing for performance-based earthquake engineering.

In addition to those invited to present papers, 10 scientists and engineers, mainly from Universities in Greece and Italy, attended its sessions and contributed to the discussions.

The idea for the subject of the Workshop came from the International Workshops organized by Professors Peter Fajfar and Helmut Krawinkler in Bled (Slovenia) in 1997 and 2004, on "Seismic Design Methodologies for the Next Generation of Codes" and "Performance-Based Seismic Design – Concepts and Implementation", respectively. These two events serve as milestones in the development and promotion of Performance-based earthquake engineering. Quite a few of the contributors to the Corfu Workshop have attended either one or both of these very successful past events. Among others, the 2009 Workshop in Corfu aspires to keep the Bled events vivid in their memory and to pave the way towards the 3rd Bled Workshop, hopefully in the near future.

Special thanks are due to Professors Peter Fajfar and Helmut Krawinkler for many reasons: for indirectly giving the idea for the theme of the Corfu Workshop; for the advice and support they generously provided to its organizers; for helping to attract some of the best experts on the subject; and, last but not least, for steering the discussions during the Workshop itself.

The Workshop itself has been organized in the framework and with the financial support of Grant FP7-REGPOT-2007-1, no. 204697, of the European Community to the University of Patras within the Regional Potential part of its 7th Framework Program (2007–2013). The Grant, titled: "Advanced Centre of Excellence in Structural and Earthquake Engineering (ACES)" www.aces.upatras.gr, supports the Structures Group of the Civil Engineering Department at the University of Patras to develop further its material and human resources, enhance its international reputation and pave the way for its younger generation of faculty. A second Workshop, this time on "Innovative Materials and Techniques in Concrete Construction", is due to follow in late 2010. The International Scientific Committee of ACES, representing the networking partners of the University of Patras in the project, and composed of:

Michel Bouchon	(Un. J.Fourier, Grenoble)
Michel Geradin	(JRC, Ispra)
Haig Gulvanessian	(BRE)
Giuseppe Mancini	(Politecnico di Torino)
Urs Meier	(EMPA)
Artur Pinto	(JRC, Ispra)
Jean-Claude Quéval	(CEA, Saclay)
Joost Walraven	(TU Delft)

have greatly assisted the organizers of the Corfu Workshop and contributed to is success, and will do the same for the second one in 2010. Together with Peter Fajfar and Helmut Krawinkler and the editor of these Proceedings, they constituted the Scientific Committee of the 2009 ACES Workshop in Corfu. The University of Patras's thanks go also to all of them.

The co-ordinator of the ACES project on behalf of the University of Patras and editor of this volume gratefully acknowledges the full-hearted support of his colleague Prof. Stathis Bousias and his close co-worker Dr. Dionysis Biskinis. Without them the organization of the 2009 ACES Workshop and the preparation of its Proceedings would not have been made possible.

Patras, Greece
November 2009

Michael N. Fardis

Contents

Contributors

Bora Acun Department of Civil Engineering, Middle East Technical University, 06531 Ankara, Turkey, bacun@metu.edu.tr

Sinan Akkar Department of Civil Engineering & Earthquake Engineering Research Center, Middle East Technical University, Ankara 06531, Turkey, sakkar@metu.edu.tr

Stavros A. Anagnostopoulos Department of Civil Engineering, University of Patras, Patras GR26504, Greece, saa@upatras.gr

Anastasios Anastasiadis Department of Civil Engineering, Aristotle University, Thessaloniki 54124, Greece, anas@civil.auth.gr

Dimitra Angelopoulou Department of Civil Engineering, University of Patras, Patras GR 26500, Greece, dimitra.angelopoulou@gmail.com

Themistocles A. Antonopoulos Department of Civil Engineering, University of Patras, Patras GR26504, Greece, antonopoulos@upatras.gr

Rogerio Bairrao LNEC/DE/NESDE, Av. do Brasil 101, 1700-066, Lisboa, Portugal, bairrao@lnec.pt

Nikitas Bazeos Department of Civil Engineering, University of Patras, Patras GR26504, Greece, n.bazeos@upatras.gr

Dimitri E. Beskos Department of Civil Engineering, University of Patras, Patras GR26504, Greece, d.e.beskos@upatras.gr

Julian Bommer Department of Civil and Environmental Engineering, Imperial College London, South Kensington Campus, London SW7 2AZ, UK, j.bommer@imperial.ac.uk

Dionysis Bournas EUCENTRE, Via A. Ferrata, No. 1, Pavia, 27100, Italy, dionysis.bournas@eucentre.it

Oreste Salvatore Bursi DIMS, Università di Trento, via Mesiano, 77 38126 Trento, Italy, oreste.bursi@ing.unitn.it

Gian Michele Calvi Department of Structural Mechanics, European Centre for Training and Research in Earthquake Engineering (EUCENTRE), Università degli Studi di Pavia, Via A. Ferrata, No. 1, Pavia 27100, Italy, gm.calvi@eucentre.it

Carlo Cauzzi Department of Structural Engineering, Politecnico di Milano, p.zza L. Da Vinci, 32, 20133 Milano, Italy, cauzzi@stru.polimi.it

Rosario Ceravolo DISTR, Politecnico di Torino, c. Duca degli Abruzzi 24, 10124 Torino, Italy, rosario.ceravolo@polito.it

Charisis Chatzigogos Géodynamique et Structure, 157 rue des Blains, 92220 Bagneux, France, charisis.chatzigogos@geodynamique.com

Cheng Chen Department of Civil and Environmental Engineering, ATLSS Engineering Research Center, Lehigh University, Bethlehem, PA 18015, USA, chc4@lehigh.edu

Constantin Christopoulos Department of Civil Engineering, University of Toronto, 35 St. George Street, Toronto, M5SlA4, Canada, c.christopoulos@utoronto.ca

Mirko Corigliano European Centre for Training and Research in Earthquake Engineering (EUCENTRE), Via A. Ferrata, No. 1, Pavia, 27100, Italy, mirko.corigliano@eucentre.it

Edoardo Cosenza DIST – Dipartimento di Ingegneria Strutturale, Università degli Studi di Napoli Federico II, Via Claudio 21, 80125 Napoli, Italy, edoardo.cosenza@unina.it

Flavia De Luca DIST – Dipartimento di Ingegneria Strutturale, Università degli Studi di Napoli Federico II, Via Claudio 21, 80125 Napoli, Italy, flavia.deluca@unina.it

Gregory Deierlein Department of Civil and Environmental Engineering, Stanford University, Stanford, CA 94305, USA, ggd@stanford.edu

Matjaž Dolšek Faculty of Civil and Geodetic Engineering, University of Ljubljana, Jamova 2, Ljubljana, SI-1000, Slovenia, mdolsek@ikpir.fgg.uni-lj.si

Kenneth J. Elwood Department of Civil Engineering, University of British Columbia, Vancouver, BC, V6T 1Z4, Canada, elwood@civil.ubc.ca

Ezio Faccioli Department of Structural Engineering, Politecnico di Milano, p.zza L. Da Vinci, 32, 20133 Milano, Italy, faccioli@stru.polimi.it

Peter Fajfar Faculty of Civil and Geodetic Engineering, University of Ljubljana, Jamova 2, Ljubljana, SI-1000, Slovenia, pfajfar@ikpir.fgg.uni-lj.si

Paolo Franchin Department of Structural Engineering and Geotechnics, University of Rome La Sapienza, Via Gramsci 53, 00197, Rome, Italy, paolo.franchin@uniroma1.it

George Gazetas Soil Mechanics Laboratory, Department of Civil Engineering, National Technical University of Athens, Athens, Greece, gazetas@central.ntua.gr

Ahmed Ghobarah Department of Civil Engineering, McMaster University, Hamilton, ON L8S 4L7, Canada, ghobara@mcmaster.ca

Polat Gülkan Department of Civil Engineering & Earthquake Engineering Research Center, Middle East Technical University, Ankara 06531, Turkey, a03516@metu.edu.tr

Selim Günay Department of Civil and Environmental Engineering, University of California, Berkeley, CA, 94720-1710 USA, selimgunay@berkeley.edu

Ronald O. Hamburger Simpson Gumpertz & Heger Inc., The Landmark at One Market, Suite 600, San Francisco, CA 94105, USA, rohamburger@sgh.com

Terje Haukaas Department of Civil Engineering, University of British Columbia, Vancouver, BC, V6T 1Z4, Canada, terje@civil.ubc.ca

Po-Chien Hsiao Department of Civil Engineering, University of Washington, Seattle, WA, USA, pchsiao@u.washington.edu

Luis Ibarra Southwest Research Institute CNWRA, San Antonio, TX 78238, USA, luis.ibarraolivas@swri.org

Iunio Iervolino DIST – Dipartimento di Ingegneria Strutturale, Università degli Studi di Napoli Federico II, Via Claudio 21, 80125, Naples, Italy, iunio.iervolino@unina.it

Takahito Inoue National Research Institute for Earth Science and Disaster Prevention, Hyogo Earthquake Engineering Research Center, Miki, Hyogo, 673-0515, Japan, dinoue@bosai.go.jp

Toshikazu Kabeyasawa Department of Civil Engineering, Faculty of Engineering, University of Tokyo, 7-3-1 Hongo, Bunkyo-ku, Tokyo 113-8656, Japan, tosikazu@eri.u-tokyo.ac.jp

Toshimi Kabeyasawa Earthquake Research Institute, University of Tokyo, 1-1-1 Yayoi, Bunkyo-ku, Tokyo 113-0032, Japan, kabe@eri.u-tokyo.ac.jp

Kouichi Kajiwara National Research Institute for Earth Science and Disaster Prevention, Hyogo Earthquake Engineering Research Center, Miki, Hyogo, 673-0515, Japan, kaji@bosai.go.jp

Özkan Kale Department of Civil Engineering & Earthquake Engineering Research Center, Middle East Technical University, Ankara 06531, Turkey, ozkankale@gmail.com

Georgios Kampas Department of Civil Engineering, University of Patras, Patras GR 26500, Greece, geekampas@gmail.com

Andreas Kappos Department of Civil Engineering Aristotle, University of Thessaloniki, Thessaloniki GR54124, Greece, ajkap@civil.auth.gr

Theodore L. Karavasilis Department of Engineering Science, University of Oxford, Oxford, OX1 3PJ, UK, theodore.karavasilis@eng.ox.ac.uk

Helmut Krawinkler Department of Civil and Environmental Engineering, Stanford University, Stanford, CA 94305, USA, krawinkler@stanford.edu

Sashi Kunnath Department of Civil and Environmental Engineering, University of California at Davis, 1 Shields Avenue, 2001 Engineering III, Davis, CA 95616, USA, skkunnath@ucdavis.edu

Carlo G. Lai European Centre for Training and Research in Earthquake Engineering (EUCENTRE), Via A. Ferrata, No. 1, Pavia, 27100, Italy, carlo.lai@eucentre.it

Igor Lanese TREES Lab, European Centre for Training and Research in Earthquake Engineering (EUCENTRE), Via A. Ferrata, No. 1, Pavia, 27100, Italy, igor.lanese@eucentre.it

Alain Le Maoult CEA/DEN/SEMT/EMSI, 91191Gif sur Yvette, Paris, France, alain.lemaoult@cea.fr

Chao-Hsien Lee National Center for Research on Earthquake Engineering, 200, Sec.3, Xinhai Rd., Taipei, 10668, Taiwan

Abbie Liel Civil, Environmental and Architectural Engineering, University of Colorado, Boulder, CO, USA, abbie.liel@colorado.edu

Dimitrios Lignos Department of Civil and Environmental Engineering, Stanford University, Stanford, CA 94305, USA, dlignos@stanford.edu

Chih-Han Lin National Center for Research on Earthquake Engineering, 200, Sec.3, Xinhai Rd., Taipei, 1066/8, Taiwan, hanklin@ncree.org

Jui-Liang Lin National Center for Research on Earthquake Engineering, 200, Sec.3, Xinhai Rd., Taipei, 10668, Taiwan, jllin@ncree.org

Ker-Chun Lin National Center for Research on Earthquake Engineering, 200, Sec.3, Xinhai Rd., Taipei, 10668, Taiwan, kclin@ncree.org

Min-Lang Lin National Center for Research on Earthquake Engineering, 200, Sec.3, Xinhai Rd., Taipei, 10668, Taiwan, mllin@ncree.org

Ying-Cheng Lin Department of Civil and Environmental Engineering, ATLSS Engineering Research Center, Lehigh University, Bethlehem, PA 18015, USA, ycl206@lehigh.edu

Georges Magonette European Laboratory for Structural Assessment (ELSA), IPSC, JRC, EC, Via E. Fermi, 2749 21027 Ispra (VA), Italy, georges.magonette@jrc.ec.europa.eu

Nicos Makris Department of Civil Engineering, University of Patras, Patras GR 26500, Greece, nmakris@upatras.gr

Gaetano Manfredi DIST – Dipartimento di Ingegneria Strutturale, Università degli Studi di Napoli Federico II, Via Claudio 21, 80125 Napoli, Italy, gaetano.manfredi@unina.it

Eduardo Miranda Department of Civil and Environmental Engineering, Stanford University, Yang&Yamazaki Bldg. Rm 281, Stanford, CA 94305-4020, USA, emiranda@stanford.edu

Jack P. Moehle Department of Civil and Environmental Engineering, 775 Davis Hall, University of California, Berkeley, CA 94720-1710, USA, moehle@berkeley.edu

Francisco Javier Molina European Laboratory for Structural Assessment (ELSA), European Commission – JRC, Via E. Fermi, 2749 21027, Ispra (VA), Italy, francisco.molina@jrc.ec.europa.eu

Marco Molinari DIMS, Università di Trento, via Mesiano, 77 38126 Trento, Italy, marco.molinari@ing.unitn.it

Khalid Mosalam Department of Civil and Environmental Engineering, University of California, Berkeley, CA 94720-1710, USA, mosalam@ce.berkeley.edu

Takuya Nagae National Research Institute for Earth Science and Disaster Prevention, Hyogo Earthquake Engineering Research Center, Miki, Hyogo, 673-0515, Japan, nagae@bosai.go.jp

Masayoshi Nakashima National Research Institute for Earth Science and Disaster Prevention, Hyogo Earthquake Engineering Research Center, Miki, Hyogo, 673-0515, Japan, nakashima@archi.kyoto-u.ac.jp

Stefano Pampanin Department of Civil and Natural Resources Engineering, University of Canterbury, Christchurch, New Zealand, stefano.pampanin@canterbury.ac.nz

Peng Pan Department of Civil Engineering, Tsinghua University, Haidian District, Beijing 100084, China, panpeng@tsinghua.edu.cn

Marios Panagiotou Department of Civil and Environmental Engineering, University of California at Berkeley, 705 Davis Hall, Berkeley, CA, USA, panagiotou@berkeley.edu

Stavroula Pantazopoulou Department of Civil Engineering, Demokritus University of Thrace, Vas. Sofias Str., No. 12, Xanthi, 67100, Greece, pantaz@civil.duth.gr

Roberto Paolucci Department of Structural Engineering, Politecnico di Milano, Piazza Leonardo da Vinci, 32, Milano 20133, Italy, roberto.paolucci@polimi.it

George Papagiannopoulos Department of Civil Engineering, University of Patras, Patras GR26504, Greece, gpapagia@upatras.gr

Alberto Pavese Department of Structural Mechanics, Università degli Studi di Pavia, Via A. Ferrata, No. 1, Pavia 27100, Italy, alberto.pavese@eucentre.it

Alain Pecker Géodynamique et Structure, 157 rue des Blains, 92220 Bagneux, France, alain.pecker@geodynamique.com

Pierre Pegon European Laboratory for Structural Assessment (ELSA), IPSC, JRC, EC, Via E. Fermi, 2749 21027 Ispra (VA), Italy, pierre.pegon@jrc.ec.europa.eu

Paolo Emilio Pinto Department of Structural Engineering and Geotechnics, University of Rome La Sapienza, Via Gramsci 53, 00197, Rome, Italy, pinto@uniroma1.it

Dimitris Pitilakis Department of Civil Engineering, Aristotle University, Thessaloniki 54124, Greece, dpitilak@civil.auth.gr

Kyriazis Pitilakis Department of Civil Engineering, Aristotle University, Thessaloniki 54124, Greece, pitilakis@civil.auth,gr

Nigel Priestley ROSE School, IUSS Pavia, Pavia 27100, Italy, nigelpriestley@xtra.co.nz

Jean-Claude Queval CEA/DEN/SEMT/EMSI, 91191Gif sur Yvette, Paris, France, jcqueval@cea.fr

José Restrepo Department of Structural Engineering, University of California at San Diego, La Jolla, CA, USA, jrestrepo@ucsd.edu

James M. Ricles Department of Civil and Environmental Engineering, ATLSS Engineering Research Center, Lehigh University, Bethlehem, PA 18015, USA, jmr5@lehigh.edu

David Roke Department of Civil and Environmental Engineering, ATLSS Engineering Research Center, Lehigh University, Bethlehem, PA 18015, USA, dar305@lehigh.edu

Mahmoud Safar Department of Civil Engineering, McMaster University, Hamilton, ON L8S 4L7, Canada, safarmm@univmail.cis.mcmaster.ca

Jean Salençon Laboratoire de Mécanique des Solides, Ecole Polytechnique, 91128 Palaiseau, France, salencon@lms.polytechnique.fr

Heidy Sanchez L. Institute for Advanced Studies (IUSS), Centre for Post-Graduate Training and Research in Earthquake Engineering and Engineering Seismology, Via A. Ferrata, No. 1, Pavia, 27100, Italy, heidy.sanchez@eucentre.it

Richard Sause Department of Civil and Environmental Engineering, ATLSS Engineering Research Center, Lehigh University, Bethlehem, PA 18015, USA, rsause@lehigh.edu

Konstantinos Senetakis Department of Civil Engineering, Aristotle University, Thessaloniki 54124, Greece ksenetak@civil.auth.gr

Choung-Yeol Seo Department of Civil and Environmental Engineering, ATLSS Engineering Research Center, Lehigh University, Bethlehem, PA 18015, USA, cys4@lehigh.edu

Peter Stafford Department of Civil and Environmental Engineering, Imperial College London, South Kensington Campus, London SW7 2AZ, UK, p.stafford@imperial.ac.uk

Sotiria Stefanidou Laboratory of Concrete and Masonry Structures, Department of Civil Engineering, Aristotle University of Thessaloniki, Thessaloniki 54124, Greece, sot_stef@otenet.gr

Halûk Sucuoğlu Department of Civil Engineering, Middle East Technical University, 06531 Ankara, Turkey, sucuoglu@ce.metu.edu.tr

Timothy Sullivan Department of Structural Mechanics, Università degli Studi di Pavia, Via Ferrata 1, Pavia 27100, Italy, timothy.sullivan@unipv.it

Despina Syntzirma Department of Civil Engineering, Demokritus University of Thrace, Vas. Sofias Str., No. 12, Xanthi, 67100, Greece, dsyntz@civil.duth.gr

Shahrzad Talachian Department of Civil Engineering, University of British Columbia, Vancouver, BC, V6T 1Z4, Canada, talas@interchange.ubc.ca

Konstantinos Trevlopoulos Department of Civil Engineering, Aristotle University, Thessaloniki 54124, Greece, ktrevlop@civil.auth.gr

Thanasis Triantafillou Department of Civil Engineering, University of Patras, Patras, GR26504, Greece, ttriant@upatras.gr

Ching-Yi Tsai National Center for Research on Earthquake Engineering, 200, Sec.3, Xinhai Rd., Taipei, 10668, Taiwan, cytsai@ncree.org

Keh-Chyuan Tsai National Taiwan University, Taipei, Taiwan; National Center for Research on Earthquake Engineering, 200, Sec.3, Xinhai Rd., Taipei, 10668, Taiwan, kctsai@ncree.org

Panos Tsopelas Department of Civil Engineering, University of Thessaly, Pedion Areos, GR 38334, Volos, Greece, tsopelas@uth.gr

Manuela Vanini Department of Structural Engineering, Politecnico di Milano, p.zza L. Da Vinci, 32, 20133 Milano, Italy, vanini@stru.polimi.it

Michalis F. Vassiliou Department of Civil Engineering, University of Patras, Patras GR26504, Greece, mfvassiliou@gmail.com

Manuela Villani Rose School, IUSS, Pavia, Via A. Ferrata, No. 1, Pavia 27100, Italy, mvillani@roseschool.it

Yuan-Tao Weng National Center for Research on Earthquake Engineering, 200, Sec.3, Xinhai Rd., Taipei, 10668, Taiwan, ytweng@ncree.org

Farzin Zareian University of California, Irvine, CA 92697, USA, zareian@uci.edu

Aikaterini Ziotopoulou Soil Mechanics Laboratory, Department of Civil Engineering, National Technical University of Athens, Athens, Greece, kziotopoulou@ucdavis.edu

Part I
Ground Motions for Performance-Based Earthquake Engineering

Chapter 1
Mapping Seismic Hazard for the Needs of Displacement-Based Design: The Case of Italy

Ezio Faccioli, Manuela Villani, Manuela Vanini, and Carlo Cauzzi

1.1 Introduction

In structural design, the seismic action has been traditionally based on the elastic acceleration spectrum that, for the fundamental period of a single degree of freedom (SDOF) system, can be linked to the force acting on it through its mass, and to the displacement through an appropriate member stiffness. Selecting the stiffness represents one major problem in force-based design, because it is erroneously assumed to be independent of strength. Moreover the use of a unique force reduction factor, based on ductility, for a given structural type was shown to be invalid [20]. Since long time it is known that strength is less important than displacement for structures under seismic loading, because displacements and deformations of structural and non-structural members directly control damage. Thus, a description of seismic demand more suitable for performance based design becomes the displacement response spectrum (*DRS*).

Herein, we illustrate first an improved version of a recent long period *DRS* hazard map for Italy [15], followed by some significant additions. After recalling the input used for the Probabilistic Seismic Hazard Analysis (PSHA) and showing the revised hazard map for long period displacement, we discuss reduction factors for deriving spectra for damping ratios up to 30% from the 5% damped ones. A further section deals with basin amplification effects on the *DRS* and their quantification by a simple method in a recent observation case history.

1.2 Basic Input Data

The original sponsor of the spectral displacement hazard maps (Italian Civil Protection Department) indicated at the outset that the *DRS* maps should rely as much as possible on the same input data that had been used for developing

---

E. Faccioli (✉)
Department of Structural Engineering, Politecnico di Milano, p.zza L. Da Vinci, 32, 20133 Milano, Italy
e-mail: faccioli@stru.polimi.it

M.N. Fardis (ed.), *Advances in Performance-Based Earthquake Engineering*, Geotechnical, Geological, and Earthquake Engineering 13, DOI 10.1007/978-90-481-8746-1_1,

the current, conventional seismic hazard maps for Italy [22, 17]. These show 5% damped, horizontal acceleration response spectrum ordinates in the 0–2 s period range; Uniform Hazard (UH) spectra covering the whole country are also attached, together with the probabilistic map for a_g, to the current Italian building code [17] (hereinafter NT 2008).

Among the input data, the most important is the so-called ZS9 model of Seismic Source Zones (SSZs) [16], a revised assessment of the Italian seismo-tectonic setting, jointly with the updated earthquake catalogue CPTI04 [23]. The SSZs in ZS9 are depicted in Fig. 1.1, which also displays the maximum "observed" moment magnitude, $M_{W\max}$ (hereinafter for simplicity $M_{\max}$), associated to each zone.

As required by the computer code used for Seismic Hazard Analysis (SHA) (CRISIS2003, see [18]), the SSZs activity was quantified through the parameters a and b of the Gutenberg and Richter (G-R) relationship, using two criteria of catalogue completeness, as described in [15]. For both criteria, $M_{\min} = 4.76$ was taken in all SSZs, except for the Mt. Etna volcano zone ZS936 where $M_{\min} = 4.35$ was adopted. The upper magnitude bound, $M_{\max}$, assigned to each SSZ plays here a critical role, since *DRS* ordinates scale directly as 10^M at $T > 4$–5 s. Two values had been proposed by [22]: the largest observed magnitude ($M_{\max 1}$, shown as $M_{W\max}$ in

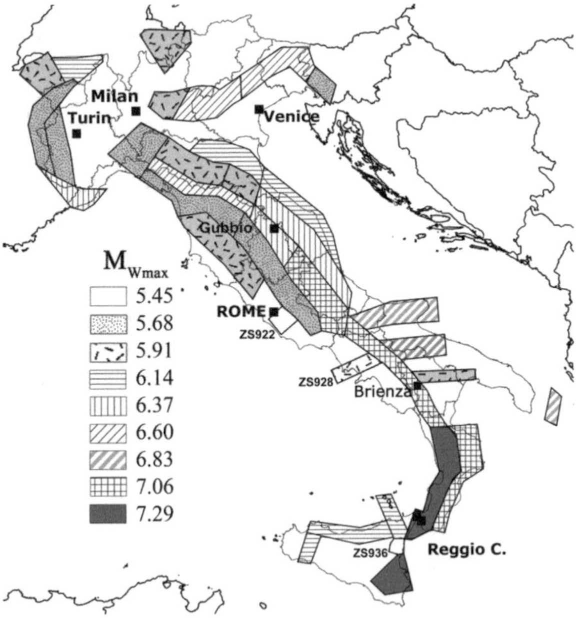

Fig. 1.1 Layout of seismic source zones (SSZs) in ZS9 model [16] for Italy used in this study, with their maximum observed moment magnitude ($M_{W\max}$=$M_{\max 1}$). For SSZs in *light gray shading* the expected maximum magnitude ($M_{\max 2}$) has been set to 6.14, while for all other SSZs $M_{\max 2}$=$M_{\max 1}$. Some SSZs discussed in the text are labeled by their identification number (e.g. ZS936). The sites shown in the map are used in the sequel

Fig. 1.1) and $M_{max2} \geq M_{max1}$ conservatively estimated at 6.14 in those SSZs where the observed value was lower, shown with a gray shading in Fig. 1.1 (except for the "volcanic" zones ZS922, ZS928 and ZS936). The influence of these two assumptions was assessed by different analyses [15] leading to the use of M_{max2} as the maximum magnitude.

The attenuation equations to be used for the PSHA were chosen on the basis of a simple analytical form that allows their application also to regions lacking a complete knowledge of active faults and the possibility of predicting the spectra up to at least 10 s. The attenuation model [8], hereinafter C&F08, has been the main tool used since it provides the *DRS* ordinates for 5%-, 10%-, 20%- and 30%-damping over a broad period range, based on digital high-quality data from 60 worldwide crustal earthquakes with $5.0 \leq M_W \leq 7.2$ and focal distance 15 km < R < 150 km. An analysis of variance showed scarce evidence for regional dependence of ground motions. Other recent attenuation equations suitable for application in this frame were evaluated. Among them, those in [4] (B&A08) for *PSA* spectra, belonging to the Next Generation Attenuation (NGA) family, were found to satisfy a number of requirements. Since, however, it was shown in [15] that they are not uniformly reliable at long periods, the use of the B&A08 equations was restricted to $T \leq 5$ s.

The hazard maps in [15] were created extending the use of the C&F08 equations also for $R < 15$ km because several SSZs in the ZS9 model are less than 15 km deep, some actually less than 10 km. Because the equations contain no distance saturation term, using them below the lower bound distance (~15 km) of their reference dataset led to hazard overestimation in some areas. To tackle this problem without recalculating all the coefficients of the attenuation equations as a result of changing the source-to-site distance measure, the original dataset of [8] was greatly enlarged, and the associated metadata improved by including both the fault distance, R_f, and R. Having thus assembled a dataset of 2,966 distance pairs for $R < 200$ km, a nonlinear regression of R on R_f and M_w was performed, yielding the correlation

$$R\ (\text{km}) = 2.122 + 0.991\, R_f + 0.0160 \exp(0.982\, M_w)\quad \sigma_R = 6.92\,\text{km}. \qquad (1.1)$$

Introducing Eq. (1.1) into the C&F08 equations [8] provides a model that predicts bounded ground motion amplitudes near the source and distance attenuation curves with a magnitude-dependent shape. We call this model "modified C&F08".

The epistemic uncertainties in present SHA analyses have been handled through a logic tree, as explained in [15].

1.3 Seismic Hazard Maps and UH *DRS*

Figure 1.2 illustrates the 475 years *DRS* map for T=10s, at the 50-percentile level (right), compared with the maximum ground acceleration (a_g) map, presently adopted in the NT2008 code (left). Over most of Northern and Central Italy the long period spectral displacement demand does not exceed 5 cm, and only in Calabria

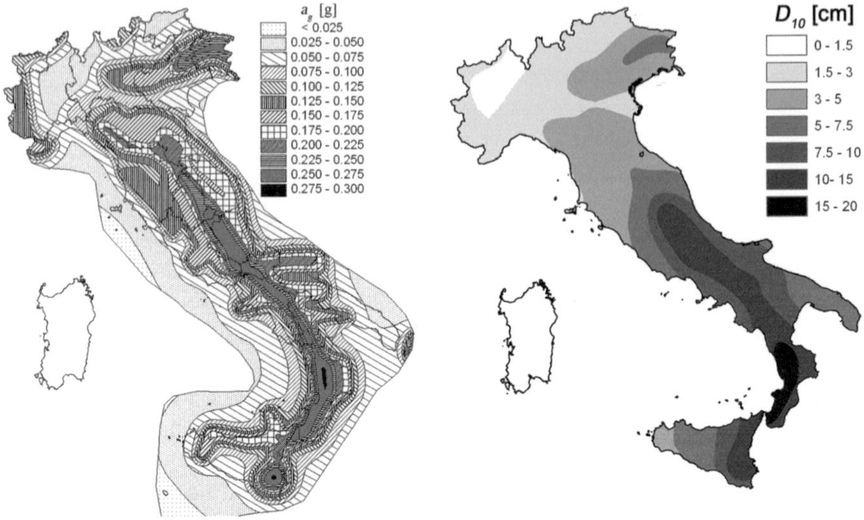

Fig. 1.2 Comparison of seismic hazard maps for Italy on ground type A: (*left*) at short period, i.e. the maximum ground acceleration (a_g) map in current Italian code NT2008 and (*right*) the long period spectral displacement (D_{10}) map (median value) for 10% exceedance probability in 50 years

(Southern Italy), where $M_{max} > 7$, does the spectral demand exceed 15 cm on hard ground. This region, with Eastern Sicily, and the Central Apennines (at the latitude of Rome) is the only one where D_{10} values are appreciably lower than those shown in [15], and believed to be more realistic. The long period values point to interesting consequences on structural design: for a return period of 475 years, reinforced concrete buildings properly designed for static loads were estimated to withstand a seismic displacement demand of 35 mm largely without damage, and masonry buildings one of 7.5 mm [19]. Figure 1.2 shows that 35 mm would not be reached anywhere in Northern Italy except Friuli.

Comparing short and long period hazard maps sheds light on the different hazard levels portrayed by different ground motion parameters chosen for design. Because $DRS \propto 10^{\alpha M}$, with α close to 1 for $T > 5$ s, the long period map displays prominently the influence of the SSZs with the largest M_{max} values. On the other hand, a_g scales as $10^{\alpha M}$ with α roughly around 0.3–0.4. Hence, in the a_g map the highest hazard zones in the NE (Friuli) and in the South (Calabrian Arc) both reach a value close to 0.3g. In contrast to this, in the long period *DRS* map the Calabrian Arc values are 3 times larger than in Friuli.

The results of SH analysis have also been represented as UH, 5%-damped *DRS* for all of the "comuni" (the smallest local administrations) in Italy, for the different return periods. Figure 1.3 illustrates average hard ground UH spectra (with $\pm$ 1σ bands) for different locations (symbols on map at centre of figure). Since one of the goals motivating this study is to define a simplified *DRS* model of seismic action, the UH spectra were carefully analyzed and from the shapes shown in Fig. 1.3 some

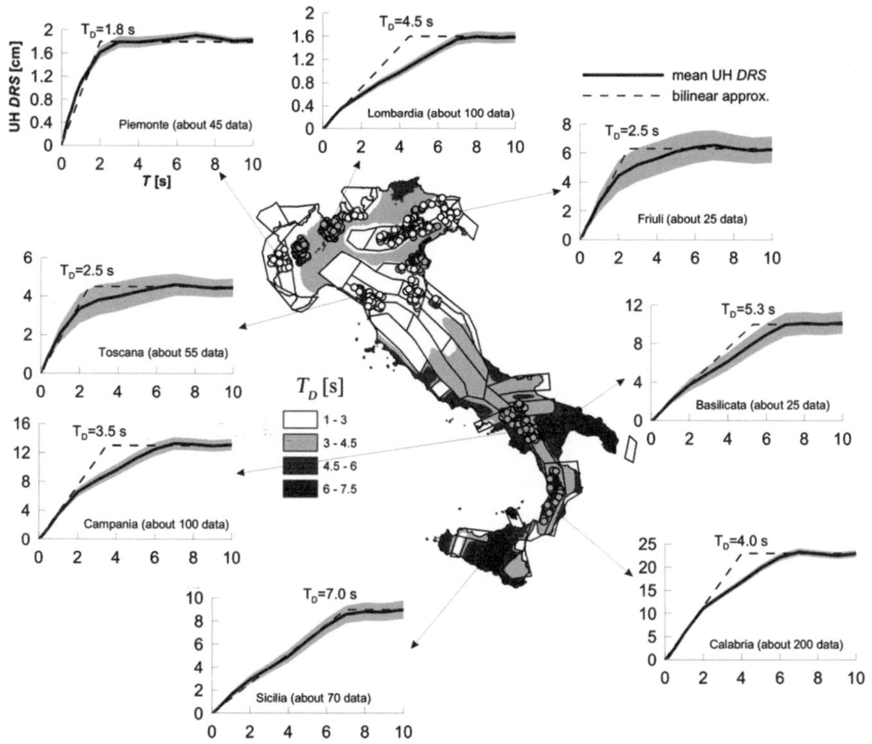

Fig. 1.3 At center, map of values of corner period T_D, derived from Eq. (1.2) for a return period of 475 years. In surrounding panels, plots of samples of 5% damped UH *DRS* at several groups of locations (indicated by bunches of *circular symbols* on the map) in Italy. *Dashed* bilinear curves in spectral plots were obtained as eye-fittings to the mean UH *DRS* of each group, shown by a *solid black curve*, with shaded stripe representing the $\pm\sigma$ dispersion band. T_D values determined from such fits (displayed on each panel) are shown by the color fillings of *circle symbols*, so that they can be directly compared with the underlying map values on same color scale

conclusions were drawn. A simple, bilinear approximation of the *DRS* spectra seems quite satisfactory for sites:

- inside SSZs with $6.0 \leq M_{max} \leq 6.5$ (e.g. Piedmont, Friuli and Tuscany), with the constant long period branch beginning between 1 and 3 s
- outside SSZs with $M_{max} > 6.5$ (e.g. as in Apulia and Sicily), but with the constant branch typically beginning at about 7 s

A trilinear approximation would do a better job e. g. in Northern Italy when the distance from SSZs is large (Lombardy), and in Southern Italy inside SSZs with large M_{max}, (as in Campania, E of Naples, and in the Calabrian Arc).

Thus, the simplest acceptable approximation for the *DRS* shape is bilinear, with a constant long period branch of amplitude D_{10} starting at a corner period, T_D

(Eq. (1.2) below), that separates such branch from that with constant velocity. At shorter periods, a linearly increasing branch is defined in Eq. (1.3) for $\zeta = 5\%$, that is:

$$T_D = \frac{2\pi D_{10}}{\max_T PSV} \tag{1.2}$$

$$DRS(T) = \frac{D_{10}}{T_\mathrm{D}} T \quad \text{for } 0 \leq T \leq T_\mathrm{D}. \tag{1.3}$$

The max *PSV* value in Eq. (1.2) can be calculated from the UH *DRS* at a site through the pseudo-spectral relation. The map of T_D values for ground type A and 475-year return period is displayed at center of Fig. 1.3 and its reliability visually checked by analyzing the UH spectra of about 700 sites (shown by circle symbols in Fig. 1.3), as described in more detail in [15]. The proposed bilinear approximation is strictly intended for the *DRS* and should not be used to back-calculate spectral acceleration at short periods [7]. One practical way to make the *DRS* approximation suitable also for the design of short-period structures, would be to impose that the total base shear should not exceed a limit, e. g. proportional to 2.5 × maximum ground acceleration a_g, if P-δ effects are neglected [7]. The dependence of the corner period on both magnitude and distance is at variance with the NEHRP [6] criteria used for mapping the long period spectral corner in the USA. In [6] T_D depends only on the modal magnitude governing hazard at the site, and just increases with magnitude. However, previous studies of response spectra have already pointed out the distance dependence of T_D [13].

1.4 Overdamped, Uniform Hazard Spectra

Overdamped spectra are generally approximated multiplying $DRS(\zeta=0.05)$ by a damping correction factor (herein R_ζ). Different proposals can be found in the literature for the correction of spectral ordinates for damping $\zeta \neq 0.05$ [3]. For example, in Eurocode 8 R_ζ therein denoted as η is constant for $T_\mathrm{B} < T < T_\mathrm{E}$, and linearly increasing for $T<T_\mathrm{B}$ and $T > T_\mathrm{E}$, reaching the value 1 at $T = 0$ and $T = T_\mathrm{F}$ (where T_B, T_C, ..., are spectrum control periods of the spectra [9]). Because attenuation equations for *DRS* ordinates were made available in [8] also for $\zeta = 0.10$, 0.20 and 0.30 (C&F08), the overdamped UH spectra at about 50 locations were calculated in the SH analysis, and the corresponding spectral ratios $DRS^{\mathrm{UH}}(T;\ \zeta)/DRS^{\mathrm{UH}}(T;\ \zeta=0.05)$ compared with the different R_ζ definitions. The comparison suggested that the Eurocode 8 [9] correction factors provide the best fit to the median ratios derived from the UH spectra for $T < 7$ s, while for $T > 7$ s R_ζ linearly increases reaching unity at roughly $T = 25$ s (see Fig. 1.4), implying that at such a period the oscillator response can be considered equal to the absolute value of the maximum ground displacement. Hence the recommendation, adopted also in [7], is that the 0.05 damped elastic *DRS* should be modified in accordance with the structure equivalent damping through multiplication of the displacement ordinates by the factor, R_ζ, defined as follows:

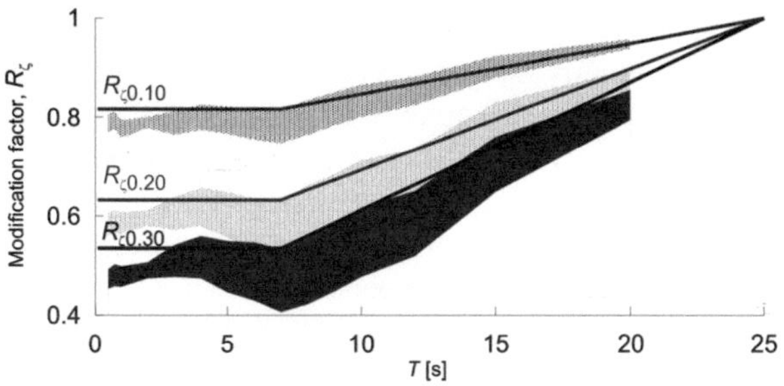

Fig. 1.4 Correction factors for overdamped *DRS* ordinates with respect to the 0.05 damped *DRS*: the bands represent the ratios derived from a representative sample of UH spectra, while *solid lines* are the ratios yielded by (1.4)

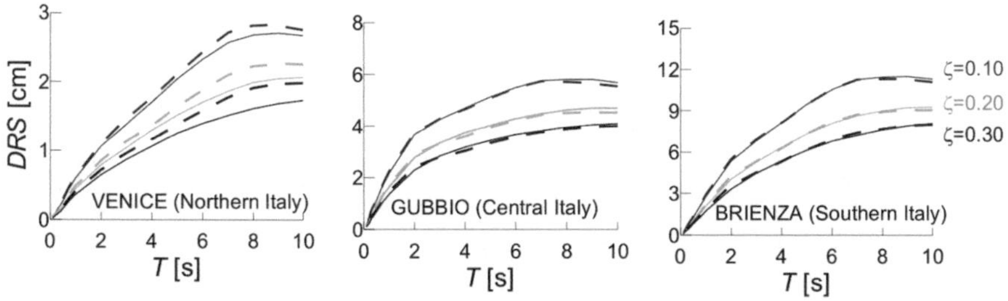

Fig. 1.5 Uniform hazard *DRS* on ground A for three selected locations (Fig. 1.1): *solid curves* denote spectra computed with attenuation relation for overdamped spectral ordinates, while *dashed curves* show the approximate spectra yielded by application of (1.4) to the 0.05 damped spectrum

$$R_\zeta = R_{\zeta 0} = \left(\frac{0.10}{0.05+\zeta}\right)^{0.5} \quad T < 7\text{ s} \tag{1.4a}$$

$$R_\zeta = \frac{1}{18}\left((1-R_{\zeta 0})T + 25R_{\zeta 0} - 7\right) \quad 7\text{ s} \le T \le 25\text{ s} \tag{1.4b}$$

Figure 1.5 compares the overdamped UH spectra calculated by directly using the attenuation equations for $\zeta \neq 0.05$ in the SH analysis with the spectra obtained by applying (1.4) to the *DRS*($\zeta = 0.05$) for three sites (different from those used in the calibration of R_ζ), displayed in Fig. 1.5. Even though some differences exist for sites with low seismicity, these can be neglected for most practical purposes. Hence, the use of Eq. (1.4) is recommended.

1.5 Basin Effects and UH Probabilistic Spectra: A Case Study

Several cases of strong motion recorded on alluvium filled basins show amplification effects that significantly exceed, at medium and long periods, those predicted by empirical relations or included in seismic codes. Extensive work [12, 10, 21] has been performed earlier aimed at quantifying these effects and providing some practical guidelines for modifying the *DRS* ordinates on sediment filled valleys. The recent Aquilano, Central Italy earthquake of April 6th 2009 (M_w 6.3) has provided an opportunity for testing such guidelines, as discussed below.

"Basin effects" mainly arise from the generation, and diffraction at the edges of the valley, of surface waves that travel horizontally in the upper sediments. While the influence of these effects on site response has been the subject of substantial theoretical studies, the resulting modification on the response spectra at the basin surface (especially for periods > 1– 2 s) aimed at practical application (see e. g. [10]) has received less attention.

In earlier work [21] parametric studies were carried out on alluvium filled valleys to understand how seismic motion is amplified as a function of valley geometry, of the fault mechanism and of different valley-fault configurations, using configurations representative of valleys in Italy, for focal distances at valley centre of about 15 and 20 km. A key parameter in all analyses is the fundamental 1D vibration period of the sedimentary column at valley centre, T_{01D}, acting as a theoretical upper bound to 2D amplification effects [10]. Both the wave field generated from the earthquake source and its propagation inside the basin sediments were computed by applying a Domain Reduction Method (DRM) [14]. The analyses were performed on the model shown in Fig. 1.6, borrowed from the European Sismovalp project (http://www-lgit.obs.ujf-grenoble.fr/sismovalp/), and chosen as representative of typical valleys of European Alps, both in shape and mechanical properties. The spectrum ordinates calculated at the valley surface receivers were mapped and averaged according to "centre" and "edge" zones, as shown in Fig. 1.6.

In Fig. 1.6, a 2D cross-section of the Aterno River Valley just S of the city of L'Aquila is also sketched, showing similar geometrical features (being about 3.5 km wide and 250 m deep [2, 11]). The upper soil for the model valley center zone was of type C, while the surface sediments at L'Aquila show more likely B type properties, with near-surface *Vs* values possibly of about 500 m/s.

In Fig. 1.7, an observed *DRS* obtained as geometric mean of the two horizontal components from the AQK (city centre) strong motion records of the Aquilano mainshock is compared with previously obtained [21] *DRS* envelopes of numerical simulations for M_w 6.3, in the valley central zone. Allowing for the difference in valley dimensions and upper soil properties, the agreement of observed and synthetic

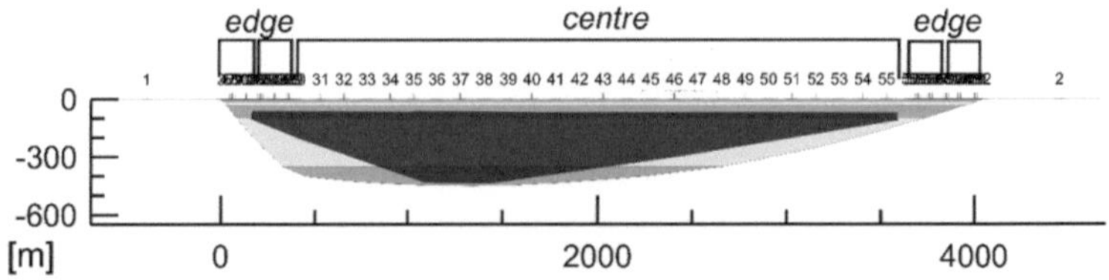

Fig. 1.6 Valley model used in parametric analyses with zone subdivision and surface receivers. A 2D sketch of the Aterno River Valley across the city of L'Aquila is superimposed in *darker shading* and *black outline* for comparison

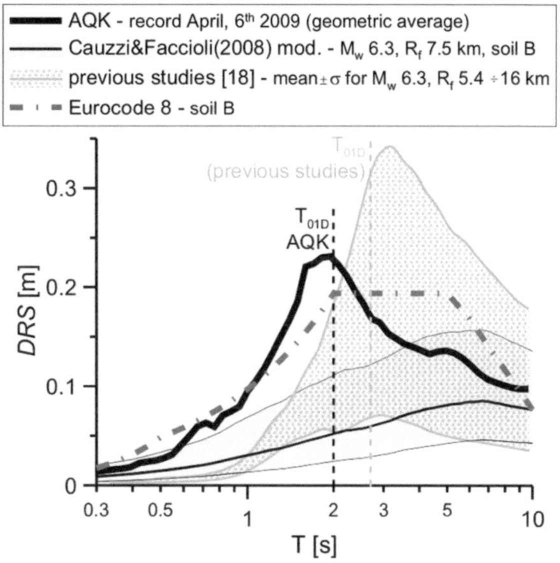

Fig. 1.7 Comparison between observed *DRS* at AQK site (geometric average of horizontal components) for L'Aquila main event, numerical simulations (mean $\pm\,\sigma$ of cases for M_{w} 6.3, R_{f} 5.4 $\div$16 km), mean $\pm\,\sigma$ prediction band from C&F08 attenuation relation and Eurocode 8 spectrum. *Vertical lines* are $T_{01\mathrm{D}}$ values for the Aterno River Valley and the reference valley model

data is fair; the spectral peak at around the corresponding $T_{01\mathrm{D}}$ values (shown as a vertical line) should be noted. Observed spectra as well as numerical simulations are decidedly outside the dispersion band of the C&F08 attenuation relation. The Eurocode 8 spectrum [9] accounts for most of the amplification observed, at periods close to $T_{01\mathrm{D}}$.

The earlier extensive numerical analyses [21] led to practical guidelines defining amplification curves for sites susceptible of basin amplification effects, using as a starting point the amplification bands predicted through the C&F08 equations.

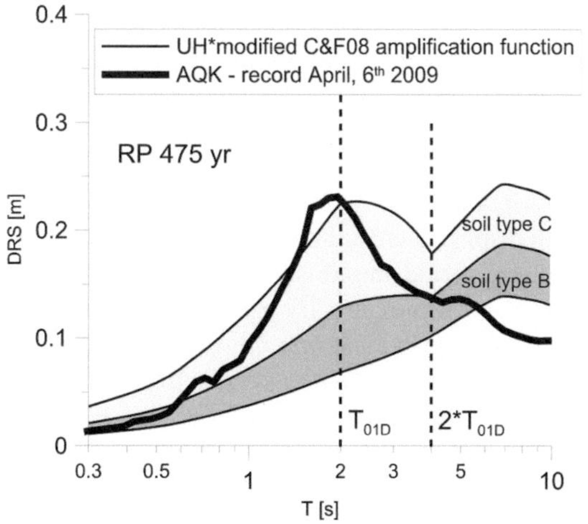

Fig. 1.8 Comparison between observed *DRS* at AQK site and UH spectra, corrected for basin effects with C&F08 amplification bands modified as explained in the text. The $T_{01\mathrm{D}}$ value shown is for the cross-section of the Aterno River Valley near AQK (city centre)

For valley centre zones, in particular, it was proposed that the period range with constant amplification should extend to T_{01D}, and should be followed by a decreasing amplification branch up to $2T_{01D}$. At sites with EC8 class B soil, the use of spectral amplification bands for the Class C (or D) soil had been suggested for higher magnitude events. Based on these recommendations, the UH *DRS* for rock (A) for L'Aquila city has been multiplied by the C&F08 amplification factor band, modified as explained, and compared with the observed spectrum at AQK (Fig. 1.8). The comparison of a probabilistic 475-year spectrum with an observed one, generated by a fault with unknown re-activation period (1000 years?) should be viewed with much caution. Nevertheless, the agreement of the predicted DRS band for soil C with the observed spectrum up to periods as high as 3-to-4 s is remarkable, especially at the peak amplification period, while an overestimate at the very long periods is apparent.

1.6 Conclusions

This article extends the recent study of [15] with significant modifications and additions. The new SH maps for Italy based on displacement response spectral ordinates over a very broad range of vibration periods (0.05–20 s), designed to meet the requirements of the Displacement Based Design, have been re-calculated and improved with respect to [15] using a more refined version of the empirical ground motion attenuation model, that avoids some shortcomings of [8] at near source distances. This should be seen also in the light of the increasing number of very tall buildings and great bridges being erected or designed in different regions of the world. As an example we recall the Burj Dubai Tower with a fundamental period of 11.3 s, [1], and the proposed, single-span suspension bridge across the Messina Strait, Italy, for which $T_1 > 30$ s, [5]. Moreover, a modification factor for the DRS with damping different than 0.05 has been suggested and already adopted in a recent model code for displacement based design.

It is hoped that the results of this work will have an impact on current building codes that anchor the whole design spectrum to a single intensity value (typically a_g) and, thus lead to wrong estimates of the long period branch. Since short and long period hazard pictures can be vastly different, both short and long period anchors should be furnished by a building code in order to have a (hopefully) accurate description of the design ground motion both in amplitude and in shape, useful for both displacement and force design applications.

Long period amplification on sediment filled basins and valleys was finally discussed, as it may modify significantly the *DRS* spectra. Relying on the results of previous studies, a practically oriented approach has been applied and tested to derive site specific *DRS* spectra, from UH spectra, accounting for basin amplification effects. The comparison with a key observation from the 2009 Aquilano earthquake indicates that the approach in question gives realistic results as to amplification levels and period range of high amplification.

References

1. Baker F, Stanton Korista D, Novak LC (2008) Engineering the world's tallest – Burj Dubai. Proceedings of CTBUH 8th world congress "Tall & green: typology for a sustainable urban future", Dubai, 3–5 March
2. Blumetti AM, Di Filippo M, Zaffiro P, Marsan P, Toro B (2002) Seismic hazard of the city of L'Aquila (Abruzzo — Central Italy): new data from geological, morphotectonic and gravity prospecting analysis. In: Dramis F, Farabollini P, Molin P (ed) Large-scale vertical movements and related gravitational processes: proceedings (of the) international workshop, Camerino – Rome, 21–26 June, 1999. Città di Castello, Edimond, 2003
3. Bommer JJ, Elnashai AS, Weir AG (2000) Compatible acceleration and displacement spectra for seismic design codes. Proceedings of 12th world conference on earthquake engineering, Auckland, New Zealand, Paper No. 207
4. Boore DM, Atkinson GM (2008) Ground-motion prediction equations for the average horizontal component of PGA, PGV, and 5%-damped PSA at spectral periods between 0.01 s and 10.0 s. Earthq Spectra 24(1):99–138
5. Brancaleoni F, Diana G (1993) The aerodynamic design of the Messina straits bridge. Int J Wind Eng Ind Aerodynamics 48:395–409
6. Building Seismic Safety Council (2003) The 2003 NEHRP recommended provisions for new buildings and other structures. Part 1: provisions (FEMA 450). http://www.nehrp.gov/pdf/fema450provisions.pdf
7. Calvi GM, Sullivan TJ (2009) A model code for the displacement-based seismic design of structure, DBD09. IUSS Press, Pavia
8. Cauzzi C, Faccioli E (2008) Broad band (0.05 s to 20 s) prediction equations for displacement response spectra calibrated on a worldwide digital database. J Seismol 12:453–475
9. CEN (Comité Européen de Normalisation) (2004) EN 1998-1:2004. Eurocode 8: design of structures for earthquake resistance – part 1: general rules, seismic actions and rules for buildings, Brussels
10. Chávez García F, Faccioli E (2000) Complex site effects and building codes: making the leap. J Seismol 4(1):23–40
11. De Luca G, Marcucci S, Milana GH, Sanò T (2005) Evidence of low-frequency amplification in the city of L'Aquila, Central Italy, through a multidisciplinary approach including strong-and weak-motion data, ambient noise, and numerical modeling. Bull Seismol Soc Am 95:1469–1481
12. Faccioli E, Cauzzi C, Paolucci R, Vanini M, Villani M, Finazzi D (2007) Long period strong ground motion and its use as input to displacement based design. In: Pitilakis K (ed) Earthquake geotechnical engineering. 4th international conference on earthquake geotechnical engineering – invited lectures, Chapter 2: 23–51, Springer, Netherlands
13. Faccioli E, Paolucci R, Rey J (2004) Displacement spectra for long periods. Earthq Spectra 20:347–376
14. Faccioli E, Vanini M, Paolucci R, Stupazzini M (2005) Comment on "Domain reduction method for three-dimensional earthquake modelling in localized regions, part I: theory" by J. Bielak, K. Loukakis, Y. Hisada, C. Yoshimura, and "Part II: verification and applications", by C. Yoshimura, J. Bielak, Y. Hisada, A. Fernández. Bull Seismol Soc Am 95:763–769
15. Faccioli E, Villani M (2009) Seismic hazard mapping for Italy in terms of broadband displacement response spectra. Earthq Spectra 25:515–539
16. Meletti C, Galadini F, Valensise G, Stucchi M, Basili R, Barba S, Vannucci G, Boschi E (2008) A seismic source model for the seismic hazard assessment of the Italian territory. Tectonophysics (online version) 450:85–108
17. Norme Tecniche per le Costruzioni (2008) DM 140108, Ministero delle Infrastrutture, Roma, Gazzetta Ufficiale n. 29 del 4.2.2008
18. Ordaz M, Jara JM, Singh SK (1991) Riesgo sísmico y espectros de diseño en el estado de Guerrero. Technical Report, Instituto de Ingenieria, UNAM, Mexico City

19. Pinho R, Crowley H, Faravelli M (2007) Preliminary study on the identification of thresholds to displacement spectra for the design of new buildings, Deliverable D7 of Project S5. http://progettos5.stru.polimi.it/Deliverables.html (password available on request)
20. Priestley MJN, Calvi GM, Kowalsky MJ (2007) Displacement-based seismic design of structures. IUSS Press, Pavia
21. Vanini M, Pessina V, Di Giulio G, Lenti L (2007) Influence of alluvium filled basins and edge effect on displacement response spectra. Deliverable D19 of Project S5. http://progettos5.stru.polimi.it/Deliverables.html (password available on request)
22. Working Group (2004) per la redazione della mappa di pericolosità sismica. INGV, Final Report. http://zonesismiche.mi.ingv.it/ with Appendices
23. Working Group CPTI (2004) Catalogo Parametrico dei Terremoti Italiani, versione 2004 (CPTI04) Istituto Nazionale di Geofisica e Vulcanologia (INGV), Bologna, Italy, http://emidius.mi.ingv.it/CPTI

Chapter 2
Some Examples of 1D, Fully Stochastic Site Response Analyses of Soil Deposits

Carlo G. Lai, Mirko Corigliano, and Heidy Sanchez L.

2.1 Introduction

The geologic and geotechnical characteristics of a site have a strong influence on the nature of the ground shaking experienced by a structure. In current engineering practice, site response analyses are performed with a deterministic approach at most parametrically. However, deterministic analyses of seismic amplification do not necessarily represent the response at a site where the uncertainties of the parameters used in the subsoil model are large. Moreover, the quality and completeness of the data used to construct the subsoil model strongly affect the reliability of the results. Every input parameter is affected by an uncertainty that propagates through the model, thereby influencing the results of site response analysis. It can be easily shown that the larger the uncertainty of the input parameters, the larger is the uncertainty of the results. The uncertainty of the input parameter can often be reduced by just performing other measurements or geotechnical tests. Some parameters influence the response more than others and reducing their uncertainty allows a reduction of the uncertainty of the final result. A correct estimate of the uncertainty, together with the assessment of reliability of results of ground response analysis may only be achieved through fully stochastic-based procedures. They allow assessing the sensitivity of results to both epistemic and aleatory uncertainty of model parameters as well as the variability of seismic input. The spatial variability of the physical properties and geotechnical parameters represent another source of uncertainty [1]; however by invoking the *ergodic hypothesis* this can be treated as a sort of equivalent uncertainty.

This work illustrates the results of 1D linear-equivalent fully stochastic site response analysis at two sites of great historical interest: the first in Kancheepuram, Southern India, the other in Vicoforte, Northern Italy. The stochastic analysis at these sites have been carried out using Monte Carlo simulations associated with the

C.G. Lai (✉)
European Centre for Training and Research in Earthquake Engineering (EUCENTRE), Via A. Ferrata, No. 1, Pavia, 27100, Italy
e-mail: carlo.lai@eucentre.it

M.N. Fardis (ed.), *Advances in Performance-Based Earthquake Engineering*, Geotechnical, Geological, and Earthquake Engineering 13, DOI 10.1007/978-90-481-8746-1_2,

Latin Hypercube sampling technique. Randomly generated geotechnical parameters varying within properly defined probability distributions were assumed to generate the seismic response of over 5,000 numerical simulations. The variability of seismic input was also taken into account by considering an appropriate set of seismo- and spectrum-compatible natural records. The seismic input to be used for each case study was defined by means of a standard Probabilistic Seismic Hazard Assessment (PSHA).

2.2 1D Stochastic Ground Response Analyses

The term "ground response analysis" is used here to refer to one-dimensional (1D) modeling of shear wave-propagating through a soil column. Such analyses are often carried out using equivalent-linear constitutive modeling. Alternatively, fully nonlinear modeling of material response is also considered, usually following a deterministic approach. Deterministic analyses do not allow estimation of uncertainty in the input parameters on the results. Therefore, after characterizing the site of study as one-dimensional, a stochastic methodology was implemented to carry out ground response analysis at two selected sites.

The adopted stochastic method [8] is based on Monte Carlo simulations, associated to the Latin Hypercube sampling technique. For each simulation the input parameters are defined from a set of randomly generated values sampled from predefined probabilistic distributions. A complete analysis may require as much as 5,000 deterministic evaluations of the model, a computational effort that can be easily achieved with today's computers computational capability.

The advantages of the Monte Carlo simulation applied to site response analysis over the deterministic approach are the following:

- It is possible to estimate the sensitivity of final results due to variation of model parameters used to carry out the simulations;
- It is possible to propagate and thereby assess the influence of the uncertainty of the model parameters and of their critical combinations onto the final results;
- It is possible to evaluate how the reduction of the uncertainty of a given input parameter reduces the uncertainty of the seismic response;
- It is possible to assess the influence of the seismic input, which is usually characterized by a large degree of variability [3].

The methodology however requires the definition of the probabilistic distribution of the input parameters and of their possible cross-correlation coefficients. Usually such distributions are difficult to define for geotechnical parameters whose spatial and aleatory uncertainty can rarely be determined from standard or ever refined ground investigation campaigns. Nevertheless, tools are available to overcome this problem or at least mitigate it. There is a rich and consolidated literature on this topic, not treated any further in this paper (e.g., Baecher and Christian [2]).

2.2.1 Procedure of Analysis

The procedure to carry out stochastic ground response analysis can be subdivided into the following steps:

(a) *Definition of the seismic input*: A set of 7 seismic- and spectrum-compatible real records were selected from the outcome of a standard probabilistic hazard study. The records are selected to be compatible with a uniform hazard spectrum computed at a specific outcropping rock site. Furthermore, selection of the records is based on the result of deaggregation of the seismic hazard. The set of 7 real accelerograms was taken to follow a uniform probabilistic distribution: at each run a record has the same probability of being selected than the others.
(b) *Geotechnical and lithostratigraphic characterization*: Definition of a subsoil model of the site from a thorough in-depth analysis of available geological, geotechnical and geophysical data. For 1D soil profile, the construction of the lithostratigraphic model requires the definition of the thickness and unit weight γ of the soil layers, the V_S profile and the damping- and shear modulus-degradation curves.
(c) *Statistical characterization*: Definition of the probabilistic distribution for the geotechnical parameters (e.g. layer thickness, V_S, γ of each layer) considering the statistics of these parameters, or at least an estimate of their range of variability.
(d) *Stochastic modelling*: Implementation of stochastic site response analysis via Monte Carlo simulation is accomplished via the Latin Hypercube sampling technique [9]. A minimum of 1,000 numerical simulations have to be carried out to stabilize the results. The analyses are carried out using SHAKE91 [11], a computer program that implements a equivalent linear constitutive model in a 1D framework.
(e) *Selection of spectrum compatible output records*: The final step concerns the selection of accelerograms that are compatible with the mean spectrum from the stochastic simulations. These records will be then used for dynamic analyses. The database used for the selection of the records consists of the acceleration time histories obtained after propagation of the signal through the soil profile.

2.3 Case Study 1: Archaeological Site at Kancheepuram, India

Kancheepuram is a municipality in the Indian state of Tamil Nadu, 72 km southwest of Chennai (see Fig. 2.1a). Kancheepuram has magnificent Indian temples of unique architectural beauty. One of the biggest and most important is Ekambareswara (see Fig. 2.1b). It is the tallest Gopuram (Indian Temple) in Kancheepuram and one of the tallest Indian temple towers in Southern India.

Probabilistic Seismic Hazard Analysis (PSHA) was performed using the standard Cornell-McGuire approach for PSHA. The zone-free approach by Woo [12] has also

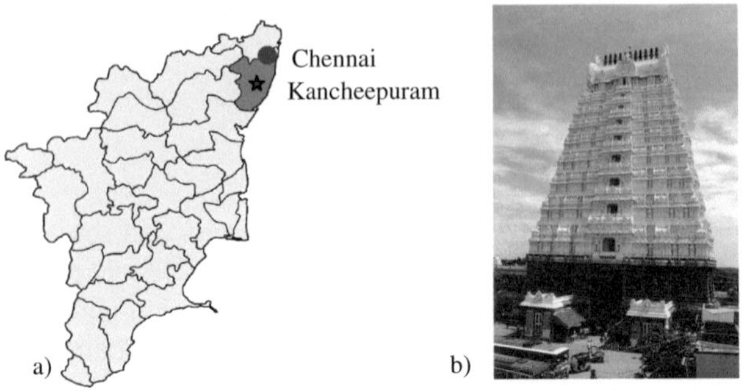

Fig. 2.1 (**a**) Position of Kancheepuram in the State of Tamil Nadu in Southern India; (**b**) Ekambareswara temple; the tallest Gopuram (57 m) built in 1509 AD at the Kancheepuram temple

been considered. Epistemic uncertainty has been addressed through the adoption of a logic tree. The outcome of the PSHA at the site consists of uniform hazard spectra on stiff soil and level ground surface. Deaggregation analysis of PSHA results was performed to identify the controlling earthquake for the site. A set of 7 spectrum-compatible accelerograms recorded on rocky sites were selected as input for site response analyses [7].

Geological, geotechnical and geophysical data obtained from specific site investigation were analyzed in order to assess lateral in-homogeneity of soil layers and to obtain geotechnical information and physical properties of soil and rock at the site. The investigation campaign included conventional geotechnical testing such as soil sampling, Standard Penetration Test (SPT), laboratory testing (Cyclic Triaxial Test) and geophysical investigations through the use of MASW (Multichannel Analysis Surface Waves), ReMi (Refraction Microtremor) and H/V Nakamura test. After integrating results of field and laboratory tests, the soil deposit was idealized as a 1D model. Figure 2.2a shows the 3D subsoil model generated using the borehole information from the geotechnical investigation, which shows how the layers can be approximated as a 1D model. Figure 2.2b shows the position in plan of the boreholes drilled at the temple site and the location of the cross sections that were traced to study the lateral variability at the site [10].

Owing to lack of detailed information, the uncertainty of the lithostratigraphic model adopted for the site was estimated by defining appropriate intervals within which the geotechnical parameters were assumed to vary. The lower and upper bounds of the intervals were established based on geotechnical data. A Gaussian distribution was assumed for each parameter. The uncertainty was computed as two times the standard deviation normalized by the mean value (e.g. coefficient of variation). Table 2.1 shows the estimated values of uncertainty for the soil profile used to model the site of Kancheepuram.

Randomly generated V_S and thickness profiles were considered using the probabilistic distribution of the input parameters described in Table 2.1. Figure 2.3 shows the mean profile and about 1,000 simulations for the V_S and thickness profiles. The

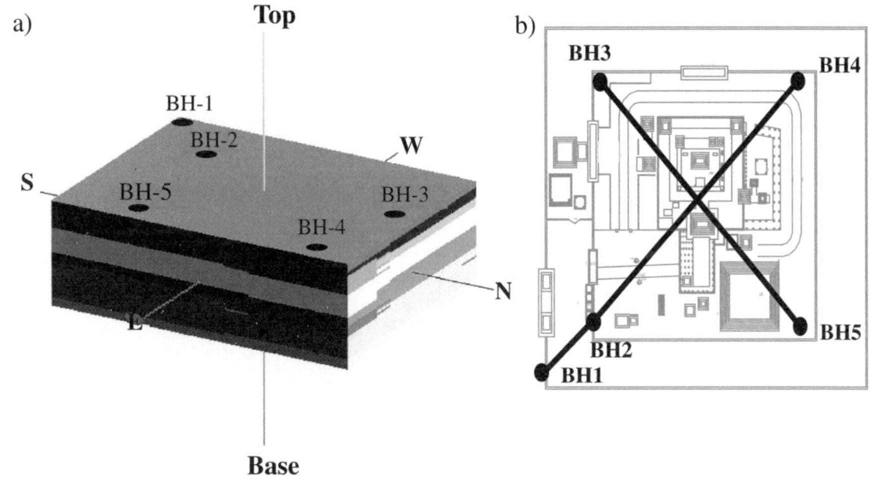

Fig. 2.2 (**a**) 3D block representation of the soil deposit at Kancheepuram; (**b**) Location at the temple site of the boreholes used to generate the cross sections adopted to assess the horizontal variability of the soil deposit [10]

Table 2.1 Mean values and coefficients of variation (CoV) used for the geotechnical parameters adopted in Monte Carlo simulations at Kancheepuram site

Layer	Thickness (m)	CoV (%)	V_s (m/s)	CoV (%)	γ (kN/m^3)	CoV (%)
1	1	15	150	35	15	3
2	2.5	25	200	40	16	5
3	9	50	250	22	18	5
4	25	60	480	38	19	5
5	Bedrock	–	750	20	20	5

position of the bedrock's roof has also been assumed variable, since also this datum was affected by uncertainty.

Degradation curves for shear modulus and damping ratio were taken deterministic and estimated from results of laboratory testing. The various geotechnical parameters were assumed to be uncorrelated, although the methodology could in principle account for possible cross-correlation among the variables.

Each profile has been automatically run through the program SHAKE91. The program computes the response in a horizontally layered soil-rock system subjected to transient and vertical shear waves and assumes that the soil cyclic behaviour can be simulated using an equivalent linear model. The mean acceleration response spectrum and the mean plus and minus one standard deviation, computed out of more than 1,000 numerical simulations is shown in Fig. 2.4 and compared with the UHS obtained from the hazard analysis. The maximum ground acceleration found in the PSHA for the 475 year return period was 0.08 g over hard rock, while the mean value resulting from site response is 0.14 g.

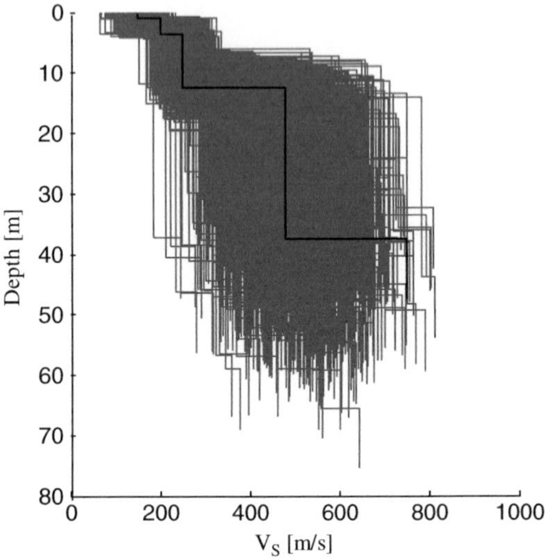

Fig. 2.3 Random profiles of shear wave velocity obtained using the Latin Hypercube sampling technique at Kancheepuram site [7]

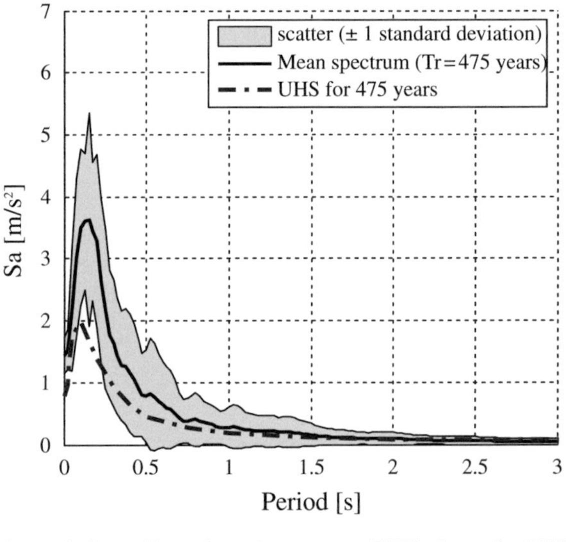

Fig. 2.4 Comparison of the uniform hazard spectrum (UHS) from the PSHA study at the Kancheepuram site for 475 years return period and the mean acceleration response spectrum from 1,000 numerical simulations (the *grey area* is the mean ± 1 standard deviation range) [7]

The mean spectrum computed at the free surface summarizes the results of the stochastic analysis. However, for dynamic analysis of the temple, the acceleration, velocity or displacement time-history may be needed. The definition of these time-histories may come from the records computed at the free surface, using them as a "database", from where a set of 7 accelerograms can be selected, imposing

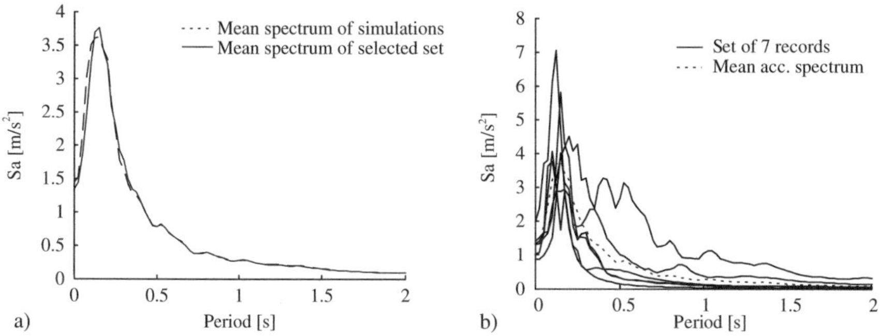

Fig. 2.5 (**a**) Comparison of the mean spectrum from the simulations and the mean spectrum of the selected set of acceleration time histories for dynamic analysis of the temple; (**b**) Acceleration response spectra of the 7 records selected to match the mean spectrum from the 1,000 numerical simulations for the 475-year return period at free surface [7]

compatibility with the mean spectrum obtained from the stochastic site response analysis. Figure 2.5 compares the mean spectrum from Monte Carlo simulations and the mean spectrum of the set of selected records.

For the dynamic analyses of the temple the second horizontal component of the accelerogram is required. Once the set of 7 records was selected for the first component, the corresponding component in the other direction of the input signal recorded on stiff soil was propagated through a subsoil model with the same shear wave velocity and thickness profile used for the first component. The input signals were linearly scaled according to the PGA predicted with the spectrum-compatible set using the corresponding scaling factors. However this approach does not guarantee the spectrum compatibility for the second component of the accelerograms.

2.4 Case Study 2: "Regina Montis Regalis" Basilica with the Largest Elliptical Dome at Vicoforte in Northern Italy

The "*Regina Montis Regalis*" Sanctuary of Vicoforte (Cuneo, Italy), is a Cathedral of great historical, architectural, and structural significance, owing its fame primarily to the great masonry elliptical dome (see Fig. 2.6), the largest in the world of this shape in terms of overall dimensions (internal axes 37.23 m by 24.89 m) [5]. Erected in 1732 on the late Renaissance structure of the Sanctuary, the dome-drum system has suffered from the start from significant structural problems, partly due to differential settlements arising from unfavourable geotechnical conditions and, to a large extent, from its very slender, bold structural configuration. Since 1973 the Sanctuary of Vicoforte has been the object of an extensive monitoring and strengthening program prompted by concern with the stability of its elliptical dome, badly damaged by cracking [4].

Fig. 2.6 "Regina Montis Regalis" Basilica located at Vicoforte, Northern Italy and detail of dome

Probabilistic Seismic Hazard Analysis (PSHA) was performed using the standard Cornell-McGuire approach for PSHA. Like for Kancheepuram, also for Vicoforte the seismic hazard study has been conducted using the logic tree approach and the zone-free method by Woo [12]. Also here the outcome of the PSHA study consisted of site-specific uniform hazard spectra on stiff soil and level ground surface. Deaggregation analysis of PSHA results was performed to identify the controlling earthquake for the site and a set of 7 spectrum-compatible accelerograms recorded on rocky sites were finally selected as inputs for site response analyses [6].

Throughout the years three main geotechnical investigation campaigns were carried out at the site. The first one, in 1976, consisted of borehole drilling, laboratory testing of the soil samples and survey of the foundation geometry. In 2004 a new site investigation campaign was carried out. The tests performed during this campaign included undisturbed soil sampling, laboratory testing and geophysical cross hole test. Finally in 2008 a geophysical investigation campaign was carried out, to complement the previous, mainly local-type, investigations. The tests performed included 2D seismic tomography (see Fig. 2.7a), 3D electric tomography, MASW, ReMi, Nakamura and seismic Cross Hole tests.

The soil below the Basilica consists of two main formations of different thickness, separated by a transition layer ("*cappellaccio*") consequence of the alteration of the bottom formation (marlstone). The latter is characterized by large values of stiffness and strength. As far as ground response analyses are concerned, the marlstone was assumed to be the bedrock formation. The shallow formation consists of silt-clayey layer and is characterized by poor mechanical properties. The 3D subsoil model in Fig. 2.7b was obtained by interpolating the lithological data from boreholes and geophysical data. Given the characteristics of the subsoil at the site, four different vertical cross sections were chosen to carry out 1D and 2D ground response analyses. In this paper only the results of 1D stochastic analysis are illustrated. The cross section chosen for the 1D analysis runs in the North-South direction, which is the least variable direction in terms of soil lateral inhomogeneity.

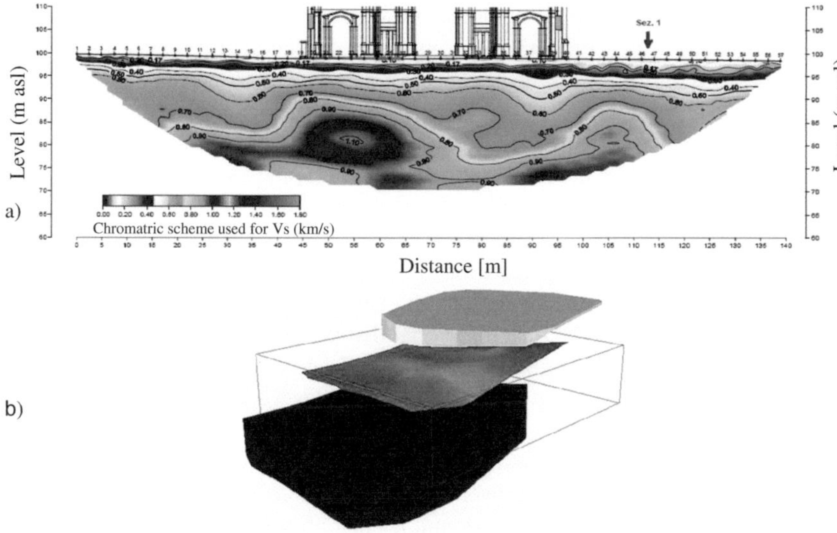

Fig. 2.7 (**a**) Shear wave velocity (V_S) 2D seismic tomography; (**b**) 3D subsoil model at Vicoforte [6]

The thickness of each layer was defined deterministically using information from borehole tests, which allowed to constrain the position of the interfaces among the layers for the vertical cross-section under study. Once the thickness of each layer was defined, the V_S properties of the 1D model have been described using the V_S profiles obtained from the cross hole test. Figure 2.8a shows the results obtained

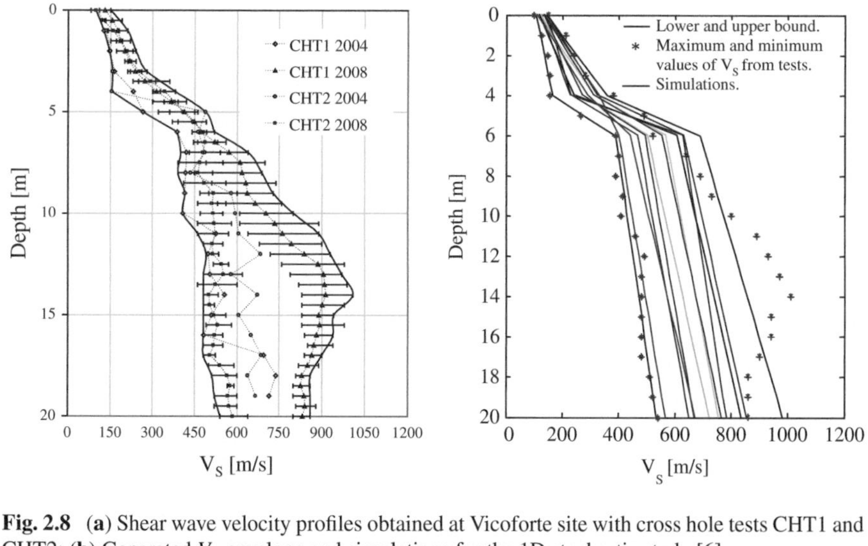

Fig. 2.8 (**a**) Shear wave velocity profiles obtained at Vicoforte site with cross hole tests CHT1 and CHT2; (**b**) Generated V_S envelope and simulations for the 1D stochastic study [6]

for both the 2004 and 2008 campaigns. From these values an envelope was built (Fig. 2.8b) inside which random V_S profiles were generated. Usually shear wave velocity is taken uniform over the thickness of a layer. Given that soils are pressure-sensitive materials, for this study, V_S changes not only from one material to the other, but within the same layer and is depth-dependent. Other parameters to be defined are the mass density (ρ) and the shear modulus and damping ratio degradation curves for each material. Degradation curves for shear modulus and damping ratio were assumed deterministic and estimated based on results of laboratory testing. The procedure that has been set-up however, would allow considering stochastic degradation curves [8]. As for Kancheepuram, the various geotechnical parameters were assumed to be uncorrelated. Once the probabilistic distribution of the geotechnical parameters is defined, the Latin hypercube sampling technique is used to perform Monte Carlo simulations. More than 1,000 V_S profiles were generated, all included in the envelope in Fig. 2.8b.

Figure 2.9 shows the comparison between the mean spectrum and the mean ± 1 standard deviation obtained from over 1,000 simulations and the UHS for the 475 years return period. The maximum ground acceleration found in the PSHA for the 475 year return period was 0.1 g over hard rock, while the mean value resulting from the site response is about 0.2 g.

As described for the case study of Kancheepuram, a set of 7 records can be selected by imposing compatibility with the mean spectrum from the stochastic site response analysis. Figure 2.10 shows the comparison for the Vicoforte site between the mean spectrum from Monte Carlo simulations and the mean spectrum of the set of selected records. As expected, agreement between the two spectra is excellent. The second horizontal component of the accelerograms has been chosen as described for the case study of Kancheepuram.

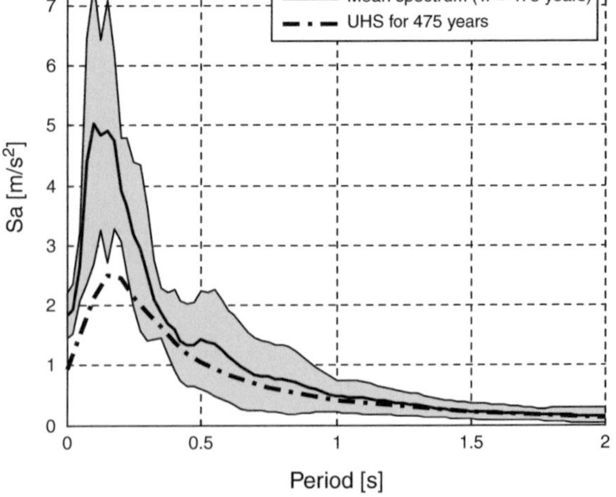

Fig. 2.9 Comparison of the uniform hazard spectrum (UHS) from the PSHA study at Vicoforte site for 475 years return period and the mean acceleration response spectrum out of 1,000 numerical simulations (the *grey area* is the mean ± 1 standard deviation range) [6]

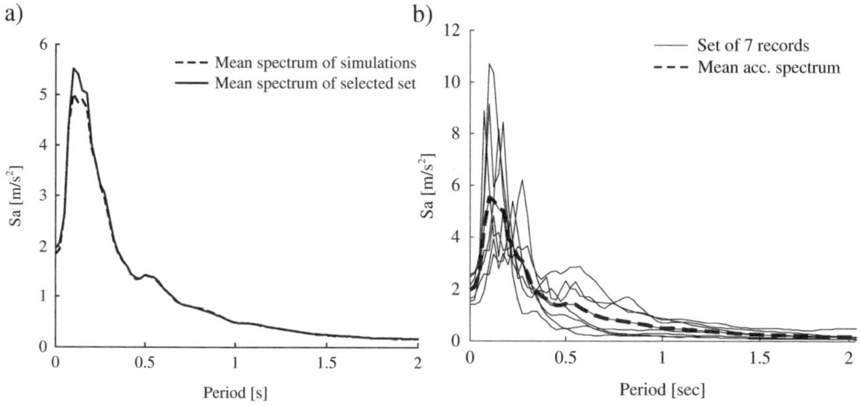

Fig. 2.10 (**a**) Comparison of the mean spectrum from the simulations and the mean spectrum of the selected set of acceleration time histories to be used for dynamic analysis of the Basilica; (**b**) Acceleration response spectra of the 7 records selected to match the mean spectrum from the 1,000 numerical simulations for the 475-year return period at free surface [6]

2.5 Concluding Remarks

Site response analyses at a given site is often characterized by large variability due to different seismic inputs and for a given input (predictions) due to uncertainty in subsoil modelling and geotechnical parameters. Single deterministic site response analyses are inadequate to handle this problem. Also a series of parametric analyses may be inadequate to fully address this issue, since some critical combinations of geotechnical parameters and seismic input may induce significant ground amplification that is completely overlooked by such type of analyses. Stochastic site response analysis allows assessing the sensitivity of results to the uncertainty of model parameters and of reference seismic input. This may also be used to optimize resources in geotechnical site investigation and characterization.

A methodology was set up to perform 1D, equivalent linear fully stochastic site response analyses, taking into account uncertainty of seismic input and model parameters. The procedure allows selection of spectrum-compatible records with reference to the mean spectrum obtained by the stochastic site response analysis.

Successful application of the procedure has been shown at two selected site in Southern India and Northern Italy.

References

1. Assimaki D, Pecker A, Popescu R, Prevost J (2003) Effect of spatial variability of soil properties on surface ground motion. J Earthq Eng 7(Special issue):1–44
2. Baecher GB, Christian JT (2003) Reliability and statistics in geotechnical engineering. Wiley, New York, NY
3. Boore DM (2004) Can site response be predicted. J Earthq Eng 8(special issue 1):1–41

4. Chiorino MA, Calderini C, Spadafora A, Spadafora R (2008) Structural assessment, testing, rehabilitation and monitoring strategies for the world's largest elliptical dome and sanctuary at Vicoforte. SCAoMaTiS, Como Lake, Italy
5. Chiorino MA, Spadafora A, Calderini C, Lagomarino S (2008) Modelling strategies for the world's largest elliptical dome at Vicoforte. Int J Arch Her 3(2):274–303
6. Lai CG, Corigliano M, Sanchez HL, Scandella L (2009) Definition of the seismic input at the "Regina Montis Regalis" Basilica of Vicoforte, Northern Italy. Research Report No. EUCENTRE–2009/03, IUSS Press, Pavia
7. Lai CG, Menon A, Corigliano M, Ornthammarath T, Sanchez HL, Dodagoudar GR (2009) Probabilistic seismic hazard assessment and stochastic site response analysis at the archaeological site of Kancheepuram in southern india. Research Report No. EUCENTRE–2009/01. IUSS Press, Pavia
8. Lai CG, Strobbia C, Dall'Ara A (2008) Terremoto di Progetto e Analisi di Risposta Sismica Stocastiche nei Territorio Toscani della Garfagnana e Lunigiana., IUSS Press, Pavia (in Italian)
9. McKay MD, Beckman RJ, Conover WJ (1979) A comparison of three methods for selecting values of input variables in the analysis of output from a computer code. Technometrics 21(2):239–245. doi:10.2307/1268522
10. Sanchez HL (2008) Stochastic 1D site response analysis at the archaeological site of Kancheepuram, Master thesis, European School for Advanced Studies in Reduction of Seismic Risk (ROSE School), University of Pavia, Italy
11. Schnabel PB, Lysmer J, Seed HB (1972) SHAKE: a computer program for earthquake response analysis of horizontally layered sites, Report No. UCB/EERC-72/12, Earthquake Engineering Research Center, University of California, Berkeley, December, pp 102
12. Woo G (1996) Kernel estimation methods for seismic hazard area source modeling. Bull Seismol Soc Am 86(2):353–362

Chapter 3
Evaluation of the Coherence of Strong Ground Motions Using Wavelet Analysis

Michalis F. Vassiliou and Nicos Makris

3.1 Introduction

During the last three decades an ever increasing database of recorded ground motions has shown that the kinematic characteristics of the ground motion near the fault of major earthquakes contain distinguishable velocity and displacement pulses [10, 1, 8, 11, among others]. In some events the pulse is also distinguishable in the acceleration history and in this case the ground motions are particularly destructive to most civil structures. In other cases acceleration records contain high frequency spikes and resemble random motions; however, their velocity and displacement histories uncover a coherent long-duration pulse that results from the nonzero mean of the acceleration fluctuations. These motions have a much smaller destructive potential to most civil structures ($T_s < 4$ s) even when they produce ground displacements as large as 3 m. A comprehensive comparison between the destructive potential of these two classes of near-source ground motions has been presented by Makris and Black [4]. The area under the acceleration pulse was coined by Bertero the ''incremental'' ground velocity in an effort to distinguish between the net increment of the ground velocity along a monotonic segment of its time history and the peak ground velocity that might be the result of a succession of high-frequency one-sided acceleration spikes. As an example, Fig. 3.1 (left) shows the fault parallel components of the acceleration, velocity, and displacement histories of the June 18, 1992 Landers, California earthquake (M_w=7.2) recorded at the Lucerne Valley station. The coherent 8.3 s long pulse, responsible for a 1.8 m forward displacement, can also be distinguished in the velocity history; whereas, the acceleration history is crowded with high-frequency spikes without exhibiting any distinguishable acceleration pulse. The acceleration pulse associated with the 8.3 s velocity pulse has so feeble inertia effects that is immaterial to most engineering structures.

M.F. Vassiliou (✉)
Department of Civil Engineering, University of Patras, Patras GR26504, Greece
e-mail: mfvassiliou@gmail.com

M.N. Fardis (ed.), *Advances in Performance-Based Earthquake Engineering*, Geotechnical, Geological, and Earthquake Engineering 13, DOI 10.1007/978-90-481-8746-1_3,

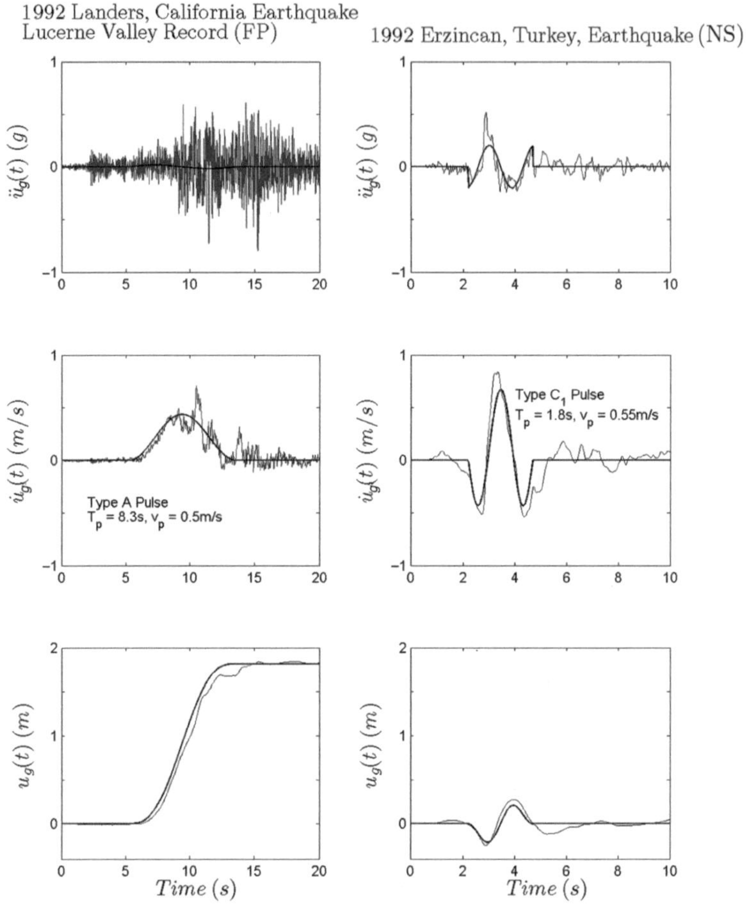

Fig. 3.1 *Left*: Fault-parallel component of the acceleration, velocity and displacement time histories recorded at Lucerne Valley during the June 18, 1992, Landers, California earthquake. Heavy line is a one-sine (Type A) acceleration pulse. *Right*: North-South component of the acceleration, velocity and displacement histories recorded during the 1992 Erzincan, Turkey earthquake. Heavy line is a Type-C_1 pulse (Makris and Chang [5] and Table 3.1 of this paper)

Figure 3.1 (right) shows the acceleration, velocity and displacement time histories recorded in the 1992 Erzinkan, Turkey earthquake with $M_w = 6.7$. What distinguishes this record is that the velocity pulse is a result of a distinguishable acceleration pulse – not the sum of many acceleration spikes with nonzero mean as is the case in the Lucerne Valley record in Fig. 3.1 (left). This paper focuses on matching and extracting coherent acceleration pulses (not velocity pulses) of strong ground motions.

Given that the maximum inelastic displacement of structures scales with T_p^2 (the period of the predominant acceleration pulse) the need for a mathematically formal, objective and easily reproducible procedure to extract pulse periods and pulse

Table 3.1 Elementary symmetric and antisymmetric wavelets used in this study

Name	Equation	Graph	Energy
One-cosine (Type B Cyclodial)	$\cos(2\pi t), \quad 0 \le t \le 1$	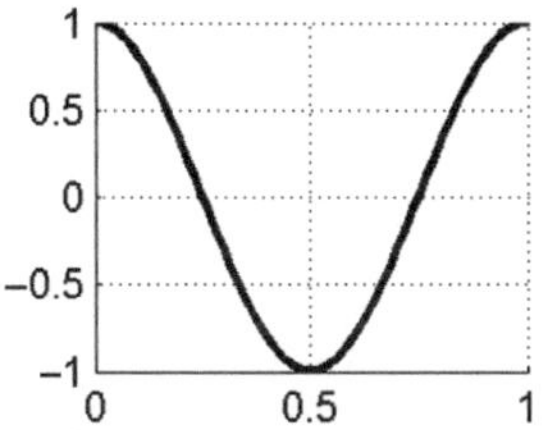	½
Symmetric Ricker (2nd derivative of Gaussian) Ricker [9]	$\left(1-t^2\right)e^{-\frac{t^2}{2}}$	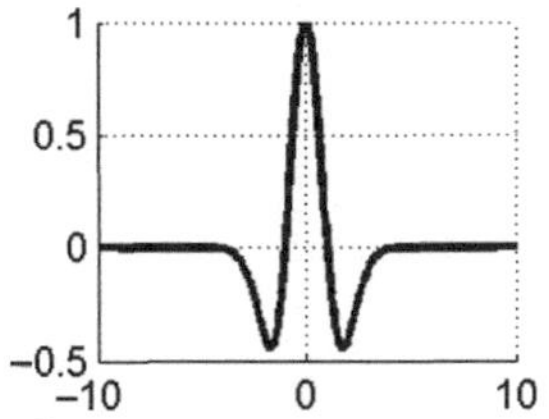	$\frac{3}{4}\sqrt{\pi}$
Antisymmetric Ricker (3rd derivative of Gaussian) Ricker [9]	$te^{-\frac{t^2}{2}}\left(t^2-3\right)$	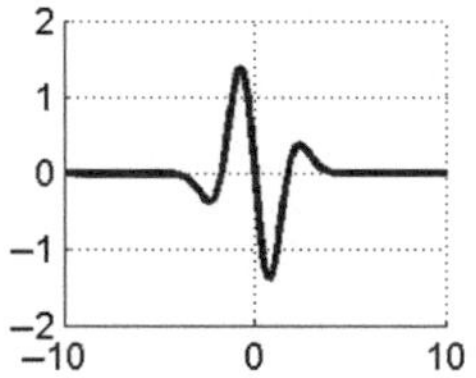	3.323
Type C1 cyclodial	$\cos(2\pi t + 0.0697\pi),$ $0 \le t \le 1.4303$	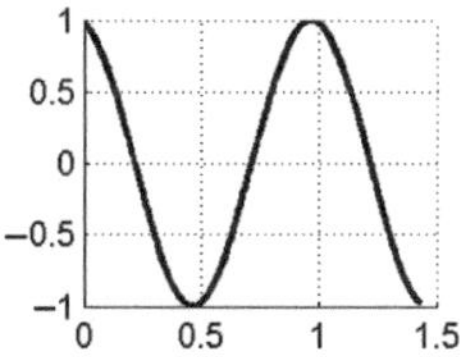	0.681
Type C2 cyclodial	$\cos(2\pi t + 0.0410\pi),$ $0 \le t \le 2.459$	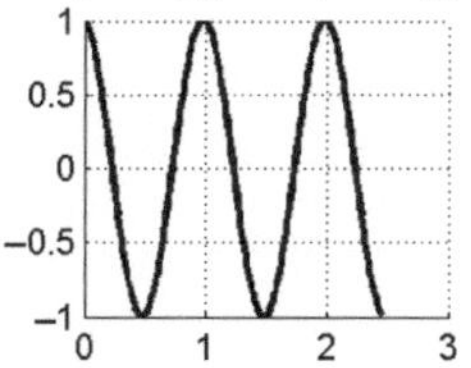	1.209

amplitudes from near source ground motions becomes apparent. This need is the main motivation for this work.

While the aforementioned studies focused on the mathematical representation of distinguishable acceleration and velocity pulses, the first systematic study for quantitatively identifying coherent velocity pulses in near-fault ground motions has been presented by Baker [2]. Baker's work uses wavelet analysis to extract automatically the largest velocity pulse (not acceleration pulse) in a given earthquake record; therefore, it offers some characteristic time and length scales of the ground motion.

This article essentially builds upon the work of Baker [2], Mavroeidis and Papageorgiou [7], Gabor [3] and Makris and Black [4] in an effort to extract in a mathematically rigorous way the most representative time scales and length scales in earthquake acceleration records. In this work we first examine with continuous wavelet transform 183 acceleration records using five (5) elementary mother wavelets and we conclude that not only the period (dilation-contraction of the wavelet) but also the phase and number of cycles (oscillatory character) need to be manipulated in order to achieve the best local matching of the prevailing acceleration pulse. Accordingly the acceleration records of strong ground motions are represented mathematically by the Gabor [3] wavelet (or its variation proposed by Mavroeidis and Papageorgiou [7]) and the concept of wavelet transform is extended so that the mother wavelet is not only translated and dilated but also subjected to an appropriate phase shift and enhanced with additional cycles.

3.2 Wavelet Analysis

Over the last two decades, wavelet transform analysis has emerged as a unique new time-frequency decomposition tool for signal processing and data analysis. Wavelets are simple wavelike functions localized on the time axis. Table 3.1 gives examples of wavelets. For a wavelike function to be classified as a *wavelet*, it must have (a) zero mean and (b) finite energy:

$$E = \int_{-\infty}^{\infty} |\psi(t)|^2 \, \mathrm{d}t < \infty; \tag{3.1}$$

In this work we are merely interested to achieve the best local matching of any acceleration record with a wavelet that will offer the best estimates of the period (T_{p} = time scale) and amplitude (a_{p}, since $a_{\mathrm{p}} T_{\mathrm{p}}^2$ = length scale) of the prevailing energetic pulse. Accordingly, we perform a series of convolutions of the ground acceleration signal, $\ddot{u}_{\mathrm{g}}(t)$, with the wavelet $\psi(t)$ by manipulating the wavelet through a process of translation (i.e. movement along the time axis) and a process of dilation-contraction (i.e. spreading out or squeezing the wavelet)

$$C(s,\xi) = w(s) \int_{-\infty}^{\infty} \ddot{u}_{\mathrm{g}}(t)\, \psi\left(\frac{t-\xi}{s}\right) \mathrm{d}t. \tag{3.2}$$

The values of $s = S$ and $\xi = \Xi$ for which $C(s,\xi) = C(S,\Xi)$ becomes maximum offer the scale and location of the wavelet $w(s)\psi\left(\frac{t-\xi}{s}\right)$ that locally match best the acceleration record, $\ddot{u}_{\mathrm{g}}(t)$. Equation (3.2) is the definition of the wavelet transform. The quantity $w(s)$ outside the integral in Eq. (3.2) is a weighting function. Typically $w(s)$ is set equal to $1/\sqrt{s}$, to ensure that all wavelets $\psi_{s,\xi}(t) = w(s)\,\psi\left(\frac{t-\xi}{s}\right)$ at every scale s have the same energy. The same energy requirement among all the daughter wavelets $\psi_{s,\xi}(t)$ is the default setting in the MATLAB wavelet toolbox [6] and what has been used by Baker [2]. However, the same energy requirement is, by all means, not a restriction. A weighting function $w(s) = 1/s^1 = 1/s$ suppresses the large scale wavelets; therefore it accentuates the presence of shorter period pulses; whereas a weighting function, $w(s) = 1/s^0 = 1$ accentuates the presence of longer period pulses. For instance, in order to capture the long velocity pulse of the Lucerne Valley record in Fig. 3.1, the weighting function needs to promote the long periods. Therefore a weighting function $w(s) = 1/s^0 = 1$ had to be used to capture the 8.3 s long velocity pulse of that record.

3.3 Selection of the Best Matching Wavelet

Figure 3.1 shows that a Type-A cycloidal pulse (one sine acceleration pulse) was selected to approximate the pulse of the 1992 Lucerne Valley record; whereas a Type-C_1 (one cycle displacement pulse) was selected to approximate the 1992 Erzikan record. The question that arises is which wavelet matches better the majority of records so that it can be used with confidence to extract invariably the pulse period and pulse amplitude of any pulse-like acceleration record. This paper examines the performance of the wavelets listed in Table 3.1. Note that the mathematical expressions appearing in Table 3.1 are for the mother wavelets. Each of these wavelets was used to match to the extend possible the 183 records listed in Vassiliou and Makris [12].

As an example, the top five graphs of Fig. 3.2 show the performance of each of the five candidate wavelets listed in Table 3.1 when matching the El Centro Array #5 record from the 1979 Imperial Valley Earthquake. The measure used to evaluate the capability of a wavelet to locally match the predominant acceleration pulse (matching index) is the inner product of the extracted mathematical pulse, $\lambda\,(S, \Xi)\,\psi_{S,\Xi}\,(t)$, with the acceleration record, normalized with the energy of the record. $\lambda\,(S, \Xi)$ is defined in (Vassiliou and Makris [12])

$$e = \frac{\int_{-\infty}^{\infty} \ddot{u}_g\,(t) \cdot \lambda\,(S, \Xi)\,\psi_{S,\Xi}\,(t)\,\mathrm{d}t}{\int_{-\infty}^{\infty} (\ddot{u}_g(t))^2 \mathrm{d}t} \tag{3.3}$$

The scores of the 5 wavelets appearing in Table 3.1 during the contest of best matching all 183 acceleration records are obtained as follows. When matching each record, the wavelet with the highest matching index, e, takes 4 points, the second best takes 3 points, the third takes 2 points, the one before last takes one point and the last one takes zero points. Figure 3.3 (left) portraits the scores of all 5 wavelets in the form of a histogram. Clearly there is no clear winner since the outcome of the contest depends on the set of ground motions which includes probably more records with antisymmetric coherent pulses than symmetric pulses. The above exercise indicates that in order to satisfy the request of matching satisfactorily the majority of records with a single mother wavelet; this wavelet should allow for a manipulation of its phase and number of cycles.

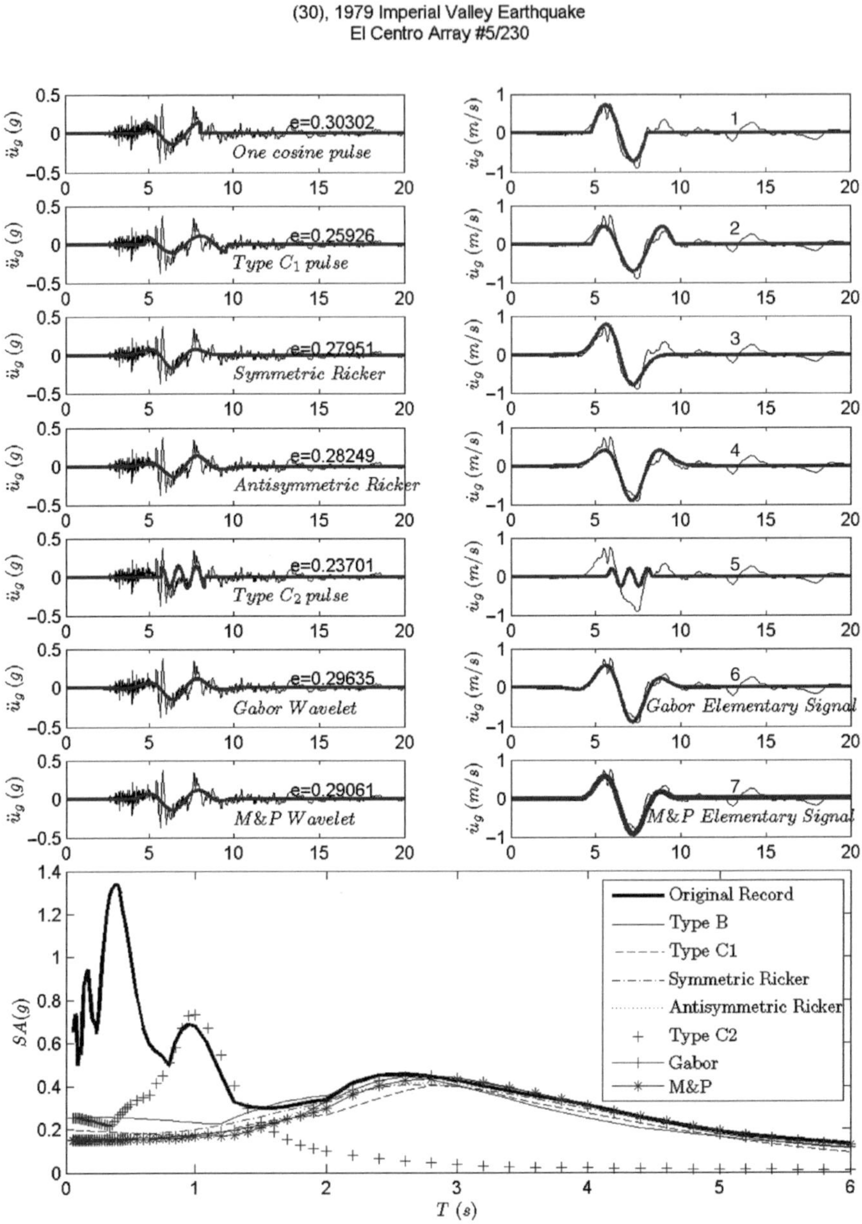

Fig. 3.2 Matching of the acceleration record with various elementary wavelets (*left*) and the resulting velocity signals (*right*). In the wavelet transform, all daughter wavelets have same energy. *Bottom*: Comparison of the elastic acceleration response spectra with 5% viscous damping

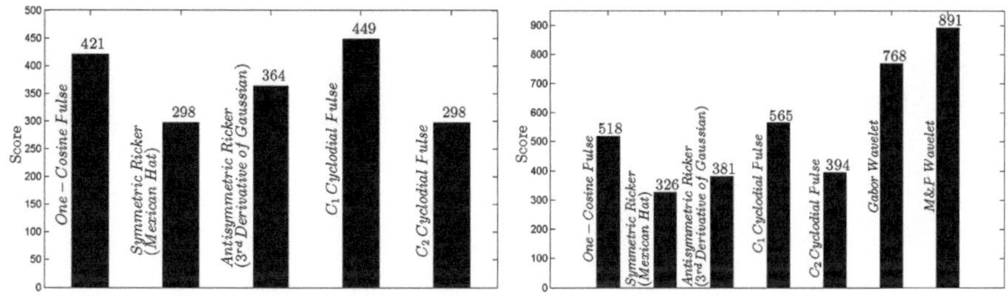

Fig. 3.3 (*Right*) Scores of the five wavelets in Table 3.1 in the contest of best matching the 183 records (*Left*) Scores of all seven wavelets of interest. The M&P wavelet ranks first ahead of the Gabor wavelet, while the other five wavelets in Table 3.1 which do not allow for phase modulation and manipulation of their oscillatory character are far behind

3.4 Extension of the Wavelet Transform by Also Modulating the Phase and the Oscillatory Character of the Elementary Signal

In the classical wavelet transform defined with Eq. (3.2) the mother wavelet is only subjected to a translation together with a dilation-contraction, $\psi\left(\frac{t-\xi}{s}\right)$ The dilation contraction is controlled with the scale parameter s; while, the movement of the wavelet along the time axis is controlled by the translation time, ξ. For instance, any daughter wavelet of the symmetric Ricker mother wavelet in Table 3.1 assumes the form

$$\psi\left(\frac{t-\xi}{s}\right)=\left[1-\left(\frac{t-\xi}{s}\right)^2\right]e^{-\frac{1}{2}\left(\frac{t-\xi}{s}\right)^2}. \tag{3.4}$$

The need to include four parameters in a mathematical expression of a simple wavelike function that is a good candidate to express the coherent component of a recorded ground motion has been presented and addressed by Mavroeidis and Papageorgiou [7]. They suggested that a sound analytical model for pulse like records should include four parameters – that is the pulse duration (or period), the pulse amplitude, the number of cycles and the phase of the pulse. They identified as the most appropriate analytical expression the Gabor [3] "elementary signal" which they slightly modified to facilitate derivations of closed-form expressions of the spectral characteristics of the signal and response spectra.

One of the earliest and most seminal publications in time frequency analysis has been presented by Gabor [3].

$$g(t)=e^{-\left(\frac{2\pi f_{\mathrm{p}}}{\gamma}\right)^2 t^2}\cos\left[2\pi f_{\mathrm{p}}t+\varphi\right] \tag{3.5}$$

which is merely the product of a harmonic oscillation with a Gaussian envelop. In Eq. (3.24), f_{p} is the frequency of the harmonic oscillation, ϕ is the phase angle and γ is a parameter that controls the oscillatory character of the signal. The Gabor wavelike signal given by Eq. (3.5) does not have a zero mean; therefore, it cannot be a wavelet within the context of the wavelet transform. Nevertheless, the derivative of the Gabor [3] elementary signal,

$$\frac{\mathrm{d}g(t)}{\mathrm{d}t} = -\frac{2\pi f_{\mathrm{p}}}{\gamma^2} e^{-\left(\frac{2\pi f_{\mathrm{p}}}{\gamma}\right)^2 t^2} \left(\gamma^2 \sin\left(2\pi f_{\mathrm{p}} t + \varphi\right) + 4\pi f_{\mathrm{p}} t \cos\left(2\pi f_{\mathrm{p}} t + \varphi\right)\right), \quad (3.6)$$

is by construction a zero-mean signal and is defined in this paper as the Gabor wavelet. According to the notation used in this paper for the wavelet functions the frequency, f_{p}, in Eq. (3.6) is replaced with the inverse of the scale parameter, $1/s$; while ξ denotes the translation time. Accordingly the Gabor wavelet is expressed as

$$\begin{aligned}\psi\left(\frac{t-\xi}{s},\gamma,\varphi\right) = -\frac{2\pi}{\gamma^2}\frac{1}{s} e^{-\left(\frac{2\pi}{\gamma}\right)^2\left(\frac{t-\xi}{s}\right)^2} &\left(\gamma^2 \sin\left(2\pi\left(\frac{t-\xi}{s}\right)+\varphi\right)\right. \\ &\left. + 4\pi\left(\frac{t-\xi}{s}\right)\cos\left(2\pi\left(\frac{t-\xi}{s}\right)+\varphi\right)\right)\end{aligned} \quad (3.7)$$

In the expression for the Gabor wavelet, Eq. (3.7), the dilation-contraction is controlled with the parameter s while the movement of the wavelet along the time axis is controlled with translation parameter ξ, the same way as is done in the Ricker wavelet given by Eq. (3.4). The novel attraction in the Gabor wavelet in Eq. (3.7) is that in addition to the dilation-contraction and translation $(t-\xi)/s$, the wavelet can be further manipulated by modulating the phase, φ, and the parameter γ, which controls the oscillatory character (number of half cycles). We can now define the four parameter wavelet transform as

$$C(s,\xi,\gamma,\varphi) = w(s,\gamma,\varphi) \int_{-\infty}^{\infty} \ddot{u}_{\mathrm{g}}(t)\, \psi\left(\frac{t-\xi}{s},\gamma,\varphi\right) \mathrm{d}t \quad (3.8)$$

The inner product given by Eq. (3.8) is performed repeatedly by scanning not only all times, ξ, and scales, s, but also various phases φ and various values of the oscillatory nature of the signal γ. When needed more values of φ and γ may be scanned. The quantity $w(s,\gamma,\varphi)$ outside the integral is a weighting function which is adjusted according to the application. Vassiliou and Makris [12] have demonstrated that the weighting function determines whether a long or a short duration pulse will be extracted.

The values of $s = S$, $\xi = \Xi$, $\gamma = \Gamma$, $\varphi = \Phi$ for which the coefficient $C(s,\xi,\gamma,\varphi) = C(S,\Xi,\Gamma,\Phi)$ becomes maximum offer the scale, location, phase and number of half cycles of the wavelet $\psi\left((t-\xi)/s,\gamma,\varphi\right)$ that locally matches best the acceleration record, $\ddot{u}_{\mathrm{g}}(t)$. Figure 3.4 plots the magnitude of the extended wavelet transform, $C(s,\xi,\gamma,\varphi)$, of the El Centro Array #5 acceleration record for various scales as a function of time and four different values of the phase $\varphi = \{0, \pi/4, \pi/2, 3\pi/4\}$

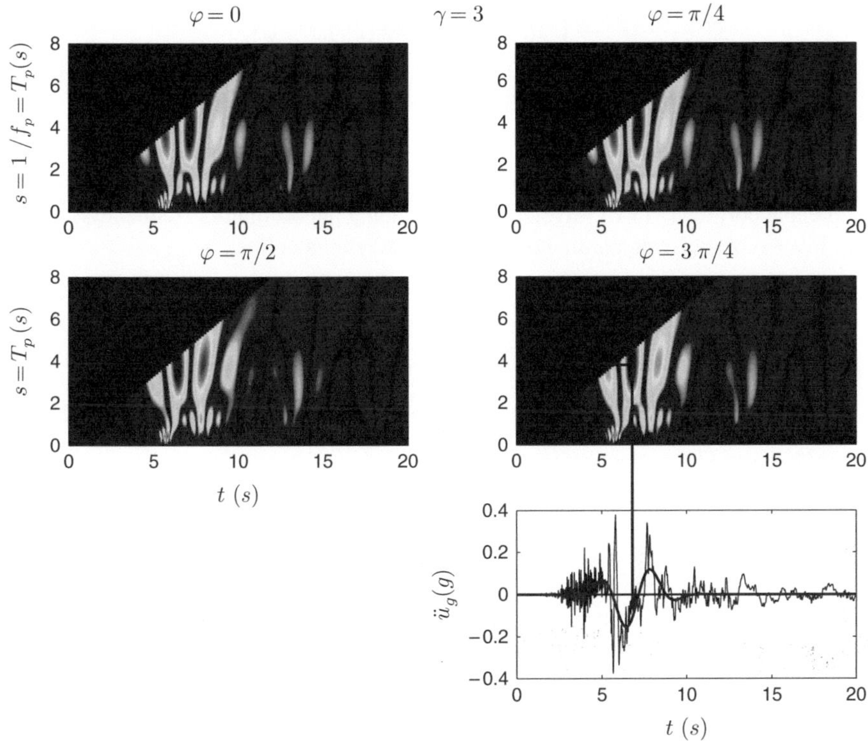

Fig. 3.4 Scalograms of the extended wavelet transform defined by Eq. (26) exercised on the El Centro Array #5 acceleration record from the 1979 Imperial Valley earthquake with the Gabor wavelet ($\gamma = 3$). The maximum value of the wavelet transform is located at $s = T_{\mathrm{p}} = 3.8$ s, $\gamma = 3$ and $\varphi = 3\pi/4$

when γ=3. The maximum value of the wavelet transform C(s,ξ,γ,φ) is located at $s = T_{\mathrm{p}} = 3.8$ s and $\varphi = 3\pi/4$.

Figure 3.2 (sixth plot from the top) plots the shape of the Gabor wavelet $\lambda\,(S, \Xi, \Gamma, \Phi) \cdot w\,(S, \Gamma, \Phi) \cdot \psi(\frac{t-\xi}{s}, \Gamma, \Phi)$ where the values (S, Ξ, Γ and φ) are those that maximize the extended wavelet transform given in Eq. (3.8) in which $\ddot{u}_{\mathrm{g}}(t)$ is the El Centro Array #5 acceleration record from the 1979 Imperial Valley earthquake.

The elementary signal proposed by Mavroeidis and Papageorgiou [7] to approximate velocity pulses is a modification of the Gabor signal given by (3.5)

$$v\,(t) = \frac{1}{2}\left(1 + \cos\left(\frac{2\pi f_{\mathrm{p}}}{\gamma}t\right)\right)\cos\left(2\pi f_{\mathrm{p}} t + \varphi\right) \qquad (3.9)$$

Clearly the wavelike signal given by Eq. (3.9) does not always have a zero mean; therefore it cannot be a wavelet within the context of wavelet transform. Nevertheless, the time derivative of the elementary velocity signal given by Eq. (3.10)

$$\frac{\mathrm{d}v(t)}{\mathrm{d}t} = \frac{\pi f_\mathrm{p}}{\gamma}\left(\sin\left(\frac{2\pi f_\mathrm{p}}{\gamma}t\right)\cos\left(2\pi f_\mathrm{p}t+\varphi\right) + \gamma\sin\left(2\pi f_E t+\varphi\right)\left(1+\cos\left(\frac{2\pi f_\mathrm{p}}{\gamma}t\right)\right)\right) \tag{3.10}$$

is by construction a zero-mean signal and is defined in this paper as the Mavroeidis and Papageorgiou (M&P) wavelet. After replacing the oscillatory frequency, f_p, with the inverse of the scale parameter the M&P wavelet is defined as

$$\psi\left(\frac{t-\xi}{s},\gamma,\varphi\right) = \left(\sin\left(\frac{2\pi}{s\gamma}(t-\xi)\right)\cos\left(\frac{2\pi}{s}(t-\xi)+\varphi\right) + \gamma\sin\left(\frac{2\pi}{s}(t-\xi)+\varphi\right)\right)\left(1+\cos\left(\frac{2\pi}{\gamma s}(t-\xi)\right)\right) \tag{3.11}$$

It is worth noting that the mathematical structure of both the Gabor and the M&P elementary signals forces us to conduct the extended wavelet transform on the acceleration records (not the velocity records) given that only the time derivatives of the elementary signals are wavelets within the context of the wavelet transform.

When selecting the appropriate weighting function the wavelet transform on the acceleration records proposed in this paper leads to the velocity pulses presented by Mavroeidis and Papageorgiou [7] and Baker [2]; yet, most importantly the coherence of the ground motion extracted with this work is directly related to the inertia effects on structures given that it originates from the acceleration records.

The last time-history plot on the left of Fig. 3.2 plots with a heavy line the shape of the M&P wavelet where the values of $s=S$, $\xi=\Xi$, $\gamma=\Gamma$ and $\varphi=\Phi$ are those that maximize the extended wavelet transform given by Eq. (3.8) in which $\ddot{u}_g(t)$ is respectively the 1979 El Centro, Array #5. The parameters (T_p=period=s=scale, γ=oscillatory character and φ=phase) of the Gabor and M&P wavelets that maximize the outcome of the extended wavelet transform given by Eq. (3.8) for three well known records are offered in Table 3.2. The extended version of Table 3.2 for 183 strong motions is presented in Vassiliou and Makris [12].

Table 3.2 Parameters of the Gabor and M&P wavelets that maximize the extended wavelet transform of Eq. (3.8)

			Gabor [3]				M&P [7]			
	Event	Station	a_p (g)	T_p (s)	γ	φ (rad)	a_p (g)	T_p (s)	γ	φ (rad)
1	Parkfield	CO2/065	0.47	0.7	3	1.57	0.41	0.6	3	1.57
2	S. Fernando	Pacoima Dam 164	0.42	1.5	3	0.79	0.39	1.4	2	0.79
3		Pacoima Dam 254	0.69	0.6	3	2.36	0.61	0.5	3	2.36

In order to address the question, which wavelet matches better the majority of records, the scores of the seven (7) wavelets under examination (5) wavelets appearing in Table 3.1 plus the Gabor and M&P wavelets defined by Eqs. (3.7) and (3.11) are obtained following the protocol defined in constructing Fig. 3.3. When matching each record, the wavelet with the highest matching index, e, takes 6 points the second best takes 5 points, all the way to the last matching index which takes zero points.

3.5 Conclusions

In this paper we develop and validate a mathematically formal procedure to extract the characteristic time scales and length scales of strong ground motions. The procedure uses wavelet analysis to identify and approximate energetic acceleration pulses (not velocity pulses). The study shows that the weighting function in the definition of the wavelet transform has a dominant role in extracting a specific pulse. For instance longer pulses which are often detected visually in the velocity records (and have attracted the attention of Mavroeidis and Papageorgiou [7] and Baker [2] among others) can be systematically captured with the wavelet transform of the acceleration records by implementing a less suppressive weighting function.

The capability of several popular symmetric and antisymmetric wavelets to locally match the energetic acceleration pulse is examined; and it is concluded that the exercise to identify the best matching wavelet shall incorporate, in addition to the standard translation and dilation-contraction of the traditional wavelet transform, a phase modulation together with a manipulation of the oscillatory character (addition of cycles) of the wavelet. This need leads to the extension of the wavelet transform to a more general wavelet transform during which the mother wavelet is subjected to the four abovementioned functions.

The paper examines the performance of two similar elementary signals – the seminal elementary signal proposed by Gabor [3] and its variation proposed by Mavroeidis and Papageorgiou [7] which in addition to a period (scale) parameter and an amplitude parameter, include a phase parameter, ϕ, and an oscillatory character parameter γ. The time derivatives of these abovementioned elementary signals satisfy the conditions for a wavelike function to be a wavelet and are defined as the Gabor and the Mavroeidis and Papageorgiou (M&P) wavelets.

The paper examines the capability of the Gabor anf M&P wavelets to locally match the energetic acceleration pulse of 183 strong ground motions and it shows that the performance of the proposed extended wavelet transform which convolves the acceleration record with the abovementioned four-parameter wavelets outshines the performance of the traditional wavelet transform which convolves the acceleration record with any two-parameter wavelet.

Acknowledgments Partial financial support for this study has been provided to the first author by the Alexander S. Onassis Public Benefit Foundation.

References

1. Abrahamson NA, Somerville P (1996) Effects of the hanging wall and foot wall on ground motions recorded during the Northridge earthquake. Bull Seism Soc Am 86:S93–S99
2. Baker WJ (2007) Quantitative classification of near fault ground motions using wavelet analysis. Bull Seism Soc Am 97:1486–1501
3. Gabor D (1946) Theory of communication. I. The analysis of information. J IEEE 93:429–441
4. Makris N, Black C (2004) Evaluation of peak ground velocity as a ''good'' intensity measure for near-source ground motions. J Eng Mech ASCE 130:1032–1044
5. Makris N, Chang S (2000) Effect of viscous, viscoplastic and friction damping on the response of seismic isolated structures. Earthquake Eng Struct Dyn 29:85–107
6. MATLAB (2002) High-performance language software for technical computation. The MathWorks, Natick, MA
7. Mavroeidis GP, Papageorgiou AS (2003) A mathematical representation of near-fault ground motions. Bull Seism Soc Am 93:1099–1131
8. Papageorgiou AS (1998) The character of near-source ground motions and related seismic design issues. Proceedings of the structural engineers world congress, San Francisco, CA, 18–23 July, 1998
9. Ricker N (1944) Wavelet functions and their polynomials. Geophysics 9:314–323
10. Somerville P, Graves R (1993) Conditions that give rise to unusually large long period ground motions. Struct Design Tall Buildings 2:211–232
11. Somerville P (1998) Development of an improved representation of near-fault ground motions, SMIP98-CDMG, Oakland, CA, pp 1–20
12. Vassiliou MF, Makris N (2009) Estimating time scales and length scales in earthquake acceleration records. Bull Seism Soc Am (Under second review)

Chapter 4
Real, Scaled, Adjusted and Artificial Records: A Displacement and Cyclic Response Assessment

Iunio Iervolino, Flavia De Luca, Edoardo Cosenza, and Gaetano Manfredi

4.1 Introduction

Seismic assessment of structures via non-linear dynamic analysis requires proper seismic input selection. Seismic codes suggest different procedures to select ground motion signals, most of those assuming spectral compatibility to the elastic design spectrum as the main criterion [12]. On the other hand, practitioners have several options to get input signals for their analysis; e.g., various types of synthetic, artificial, real or real-manipulated records [3]. Codes usually accept the use of different types of records and may provide additional criteria or limitations for each of those. In the Italian seismic code [6], for example, artificial records should have duration of at least 10 s in their pseudo-stationary part, and they cannot be used in the assessment of geotechnical structures. Synthetic generated by simulation of earthquake rupture and propagation process should refer to a characteristic scenario for the site in terms of magnitude, distance and other source seismological characteristics; finally, real records should reflect the event dominating the hazard at the site. However, practitioners cannot always accurately characterize the seismological threat to generate synthetic signals, or it is not possible to find a set of real records that fits properly code requirements in terms of a specific hazard scenario [5]. In fact, despite the increasing availability of databanks of real accelerograms in the last decades, i.e. of the most sound representation of ground motion, and the spread use of this type of records to characterize seismic input, it may be very difficult to successfully apply code provisions to natural record sets, especially those regarding spectral compatibility, if appropriate tools are not available [12]. This is why the relatively easy and fast generation of artificial records (i.e., via random vibration procedures) perfectly compatible with an assigned design spectrum, has become very popular for both practice and research purposes. More recently, algorithms to get the spectral compatibility of real records by wavelets adjustment were proposed [9]. This kind of manipulation is an extension of the more simple linear

I. Iervolino (✉)
DIST – Dipartimento di Ingegneria Strutturale, Università degli Studi di Napoli Federico II, Via Claudio 21, 80125 Napoli, Italy
e-mail: iunio.iervolino@unina.it

M.N. Fardis (ed.), *Advances in Performance-Based Earthquake Engineering*, Geotechnical, Geological, and Earthquake Engineering 13, DOI 10.1007/978-90-481-8746-1_4,

scaling of real records to modify (e.g., to amplify) the spectral shape to get a desired intensity level [10].

Although, several studies attempted to assess the reliability of each of these procedures (e.g., Schwab and Lestuzzi [16]), general conclusions seem hard to derive from literature. This work tries to address the spectral matching issue from the structural point of view in terms of non-linear peak and cyclic response, simply having as reference a code-based design spectrum. To this end, six categories of 28 accelerograms, each one consisting of four sets, were considered:

- un-scaled real records (URR);
- moderately scaled real records (SF5);
- largely scaled real records (SF12);
- wavelet-adjusted real records (RSPMatch);
- type 1 artificial records (Belfagor); and
- type 2 artificial records (SIMQKE).

The basis of this study is the elastic pseudo-acceleration design spectrum, that is, all sets are compatible with the same elastic code spectrum for a case study site in southern Italy (see following section). As structural response measures, or engineering demand parameters (EDPs), the peak inelastic displacement, the kinematic ductility and the equivalent number of cycles were considered to relate the structural response to both peak and cyclic content of the ground motion.

Analysis of a large number of single degree of freedom (SDOF) systems with an elastic-plastic with hardening behaviour aimed at assessing and comparing the bias, if any, associated to each typology of records for the three EDPs with respect to the un-scaled real ground motions which are considered as a benchmark.

4.2 Records

Six categories of records were selected assuming the same target spectrum built according to the new Italian seismic code for a case-study site (Avellino, southern Italy) having as geographical coordinates: lat. 40.914, long. 14.78. The 5%-damped elastic spectrum considered is that related to the life-safety limit state of an ordinary construction with a nominal life of 50 years on A-type (stiff) soil class (see CS.LL.PP. [6] for details).

For each category four spectrum compatible sets of seven records were selected (if real) or generated (if artificial). Assuming sets of seven records acknowledges the Italian and Eurocode 8 [4] prescriptions that allow using the mean structural response from non-linear dynamic analyses if at least seven records are employed. In the following the selection or generation of the sets are briefly reviewed.

URR, Un-Scaled Real Records – The sets of un-scaled real records were selected using REXEL, a software which allows to select combinations of multi-component real ground motion records contained in the European Strong-Motion Data Base (ESD – http://www.isesd.cv.ic.ac.uk/ESD) and in the Italian Accelerometric

Archive (ITACA – http://itaca.mi.ingv.it/ItacaNet), which on average match an arbitrary or code-based elastic spectrum [13]. By providing to the software the geographical coordinates of the site and the limit state of interest, it was possible to select four sets of records matching on average the target spectrum in the 0.15–2.0 s range. Moment magnitude (*M*) and source-to-site distance (R_e) range between 5.6–7.8 and 0–35 km, respectively. In Fig. 4.1a the four sets' means are

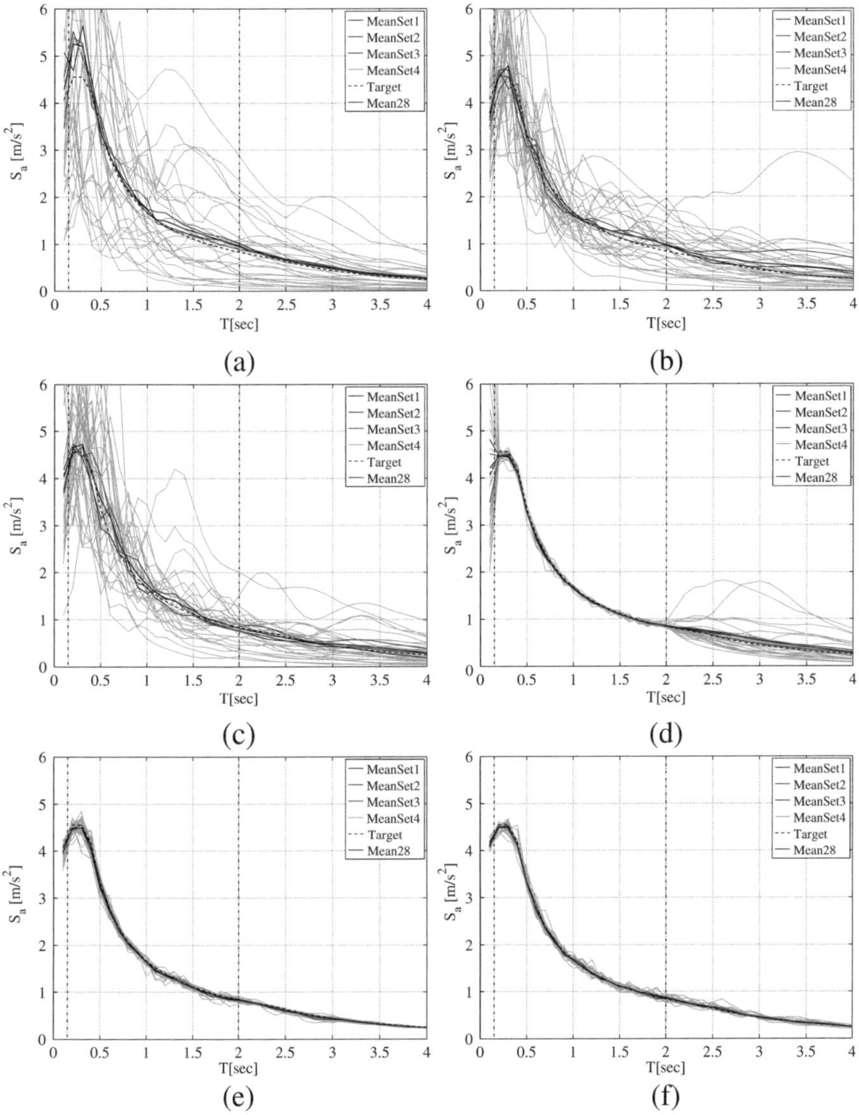

Fig. 4.1 Acceleration response spectra of URR (**a**), SF5 (**b**), SF12 (**c**), RSPMatch (**d**), Belfagor (**e**), SIMQKE (**f**) records

represented along with individual records and target spectrum. All the averages of the sets are within the [–10%, +30%] tolerance interval with respect to the target spectrum. In most of the compatibility interval they approximate very well the design spectral shape. The four URR sets have no registrations in common and come from 17 different earthquake events.

Scaled Real Records – Linearly (amplitude) scaled records were selected with REXEL. The range of periods considered is the same as per URR. The intent is to compare the responses to records moderately and significantly scaled. In particular two categories of four scaled records sets each, differing for the average scaling factor (SF), were selected: (1) SF = 5; (2) and SF = 12.

SF5 – In the same magnitude and distance ranges chosen for un-scaled records, four sets of seven compatible accelerograms each were selected through a specific option of REXEL, with a mean SF = 5 (Fig. 4.1b). The 28 different records (9 records in common with URR) come from 15 earthquake events (10 of them are in common with URR). Note that the variability of the scaled sets is smaller than those un-scaled as expected [12, 13].

SF12 – Using REXEL three sets of seven records, whose average SF was 12, were also defined. It was not possible to find another set with the desired characteristics. So, the fourth set of seven accelerograms was "manually" selected so that its average scaling factor was similar to the other three sets selected with the computer software. These four sets have no events in common with the URR sets and come from 17 different earthquakes (Fig. 4.1c).

RSPMatch, Wavelet Adjusted Records – RSPMatch2005 software [1, 9] was used to modify the URR sets. In this case the adjustment procedure was simply aimed at reducing the mismatch of individual records with respect to the target. The procedure was pursued only for 5% damping in the range of periods 0.15–2.0 s, in which records were already compatible on average and without the application of any linear scaling factor (Fig. 4.1d).

Artificial Records – Generally speaking, generation procedures for artificial accelerograms are based on the random vibration theory and the spectral matching is carried out via an iterative adjustment of the Fourier spectrum [15]. The two computer programs selected for this study generate different kind of signals: the first one, Belfagor [14] produces non stationary signals; the second one, SIMQKE [8] produces stationary signals that are subsequently enveloped in a trapezoidal shape.

Belfagor Records – Belfagor generates non stationary signals by using variable Fourier amplitudes empirically evaluated. In fact, the code asks for reference M, R_e values and soil type, even if the spectrum to match is a code spectrum. Using Belfagor, 28 accelerograms were generated. They all have the same duration, 21.48 s and a sampling time step of 0.005 s. Records were arranged in four sets of seven records (Fig. 4.1e).

SIMQKE Records – A second group of four sets of artificial records was generated by SIMQKE. This well-known software generates groups of stationary artificial records in a way they fit the target spectrum. In this case 28 records were generated together and then they were split in four groups of seven (Fig. 4.1f).

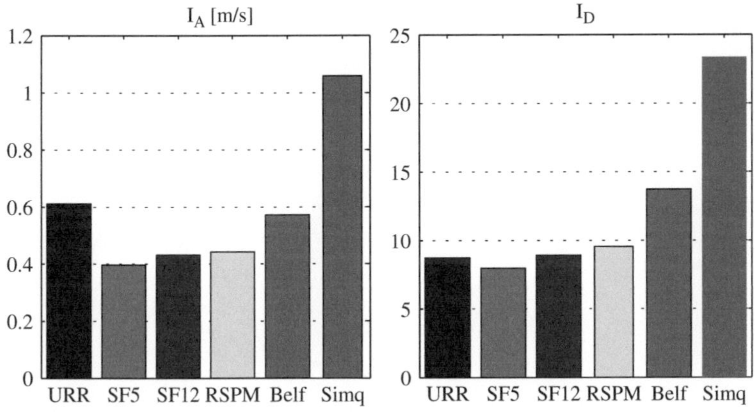

Fig. 4.2 Average values of I_A and I_D computed as mean value of 28 records

Each accelerogram of the six categories was also processed to evaluate characteristic (integral) parameters other than the spectral shape. Arias intensity (I_A), and the Cosenza and Manfredi index (I_D), Eq. (4.1), computed as the mean of the sample of 28 records for each category, are reported in Fig. 4.2. I_D is defined as a factor times I_A divided by the product of peak ground acceleration (*PGA*) times the peak ground velocity (*PGV*), Iervolino et al. [11].

$$I_D = \frac{2 \cdot g}{\pi} \cdot \frac{I_A}{PGA \cdot PGV} \tag{4.1}$$

Figure 4.2 shows that real records, both scaled and un-scaled, and RSPMatch records have similar mean values of I_D. Both categories of artificial records display higher values of I_D. The SIMQKE records show comparatively high values of I_A and I_D. Belfagor records compare better to real records at least in terms of I_A.

4.3 Analyses and Structural Response Measures

All records selected for each category were used as input for non-linear dynamic analyses of 40 inelastic SDOFs, whose periods (*T*) vary linearly from 0.1 to 2 s. Inelastic SDOFs have elastic-plastic backbone with linear hardening. The post-yield stiffness was taken as 0.03 times the initial stiffness (k_{el}) (see Fig. 4.3, where F_y and Δ_y are yield force and displacement respectively).

The peak elastic deformation experienced by an elastic structure is a ground-motion specific quantity. Therefore, one can achieve the same value of the strength reduction factor (*R*), either for each record in a dataset (constant *R* approach) or on an average sense for all the records, that is, relating the *R* factor to the target

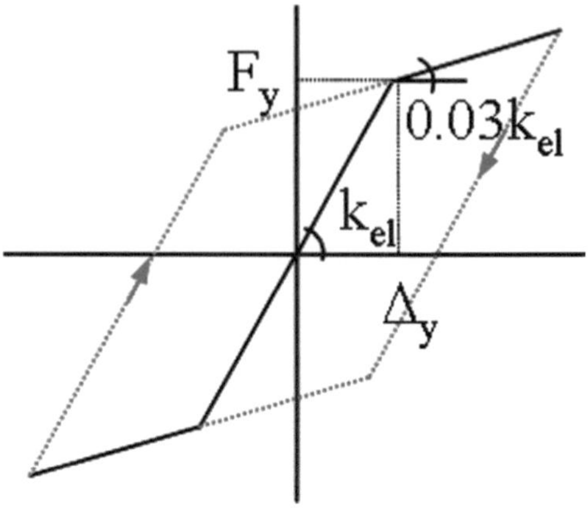

Fig. 4.3 SDOF backbone curve

spectrum matched (constant strength approach) as in Eq. (4.2) where $Sa_{e,t}$ is the acceleration ordinate in the code spectrum at the period of the SDOF and m its mass. The latter approach was considered herein, to simulate the effect of different sets of accelerograms on the same structure. In particular two R values were chosen, 4 and 10, to cover a wide range of non-linearity levels. However, it should be emphasized that the two different approaches may lead to different conclusions [2].

$$F_y = Sa_{e,t}(T) \cdot m/R \tag{4.2}$$

EDPs chosen were selected to investigate both peak and cyclic seismic response. Displacement-based parameters are the peak inelastic displacement ($Sd_{R=i}$) and the kinematic ductility (D_{kin}) evaluated as the ratio of $Sd_{R=i}$ to the yield displacement, Eq. (4.3). The equivalent number of cycles (N_e) was also considered. It includes the hysteretic energy (E_H) normalized with respect to the largest cycle, decoupling ductility demand (already considered above) and cyclic demand, Eq. (4.4).

$$D_{kin} = Sd_{R=i}/\Delta_y \tag{4.3}$$

$$N_e = E_H/\left[F_y \cdot \left(Sd_{R=i} - \Delta_y\right)\right] \tag{4.4}$$

4.4 Results and Discussion

Peak Response – The peak displacements of the SDOF systems are presented as mean value on 28 records pooled per typology. Figure 4.4a and b show inelastic result for the two R values of 4 and 10. Additional results relative to other R values and other engineering demand parameters (EDP) can be found in De Luca et al. [7]. Generally, the adjusted and artificial records seem to show a systematic underestimation of the displacement response compared to the URR for the higher

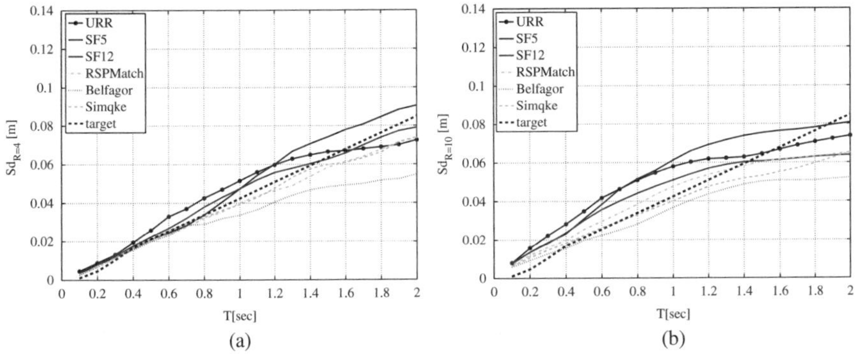

Fig. 4.4 Peak displacements for each category (28 categories) for $R = 4$ (**a**) and $R = 10$ (**b**)

non-linearity levels, and at least in the range of period of interest for the non-linear behavior of the most of common structures.

SF5 and SF12 do not show a systematic trend with the period. Belfagor records, in particular, lead systematically, and for both R values, to average inelastic displacements lower than the elastic target spectrum. However, hypothesis tests employed to assess quantitatively the significance of these results do not lead to the conclusions that any of these biases are statistically significant.

Ductility Demand – The kinematic ductility may be useful to assess the absolute displacement demand. Figure 4.5a and b show the same trend observed above, that is, artificial or adjusted records may lead to underestimation compared to URR only at high non-linearity levels. Ductility demands for each category are very close to each other for $R = 4$. Increasing the reduction factor leads to the same trend found for inelastic displacements. In this case URR ductility demand for $R = 10$ in the moderate periods range, is about two times ductility' of Belfagor records.

Equivalent Number of Cycles – More evident conclusions may be found when analyzing the trends of the equivalent number of cycles in Fig. 4.6.

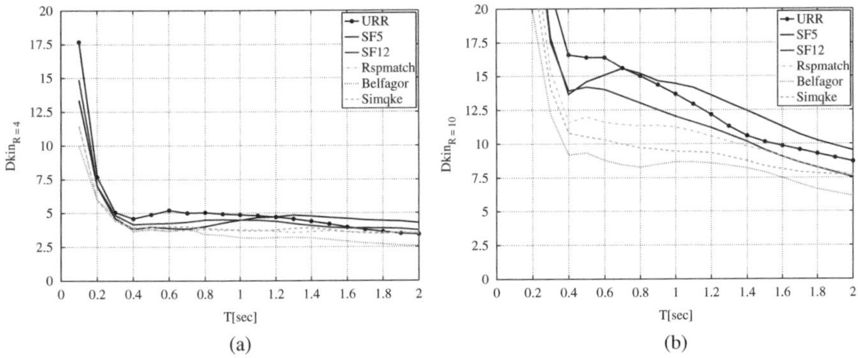

Fig. 4.5 Ductility demand for $R = 4$ (**a**) and $R = 10$ (**b**) (28 categories)

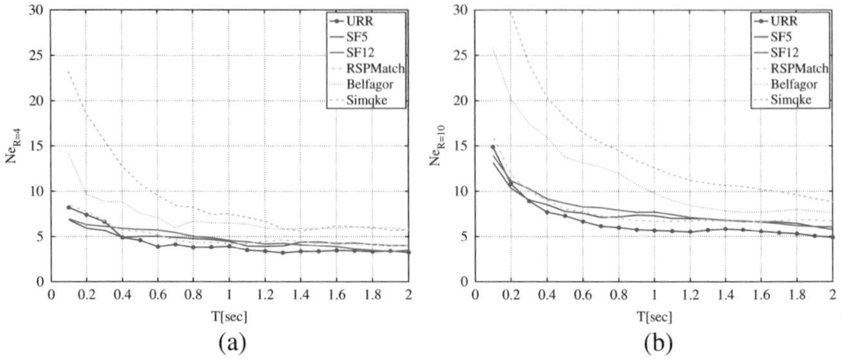

Fig. 4.6 Equivalent number of cycles for $R = 4$ (**a**) and $R = 10$ (**b**) (28 categories)

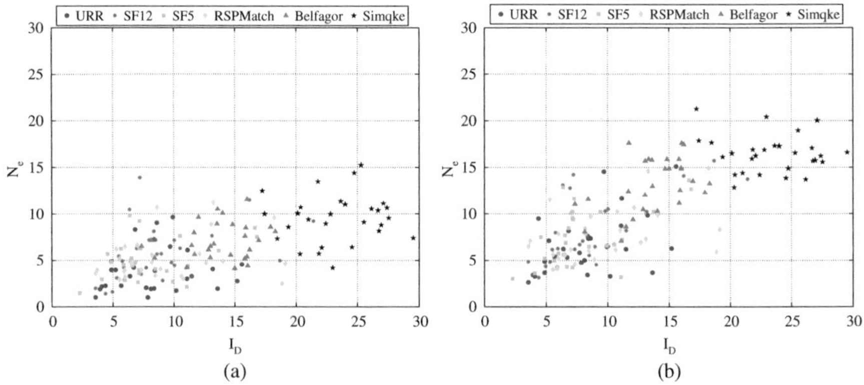

Fig. 4.7 N_e versus I_D for $R = 4$ (**a**) and $R = 10$ (**b**) and $T = 0.6$ s

At all non-linearity levels a significant overestimation in terms of cyclic response may be observed for both adjusted and artificial records. In this case SIMQKE records show the highest N_e values. Belfagor records give the same trend, although with a lower bias. Wavelet-adjusted records seem to not show any bias.

It is to note that trends found for N_e could have been predicted by the integral parameters discussed above; i.e., the I_D values of the sets. Figure 4.7 shows N_e versus I_D for the individual records for $T = 0.6$ s, for $R = 4$ (a) and 10 (b). The high values of I_D of the artificial records seem to agree with the high N_e values (more clearly for the SIMQKE records).

4.5 Conclusions

In this work different ways to achieve spectrum-compatible record sets were compared in terms of both peak and cyclic of inelastic seismic response of 40 non-degrading SDOFs.

Six typologies of records were considered: real un-scaled, real with limited average scaling factor, real with large average scaling factor, real adjusted with wavelets, and two different types of artificial records. The benchmarks were the design elastic spectrum for a case study site in southern Italy and the response to un-scaled records matching it on average.

Results seem to indicate that artificial and wavelet-adjusted records may underestimate peak displacement-related demand. However, this is evident only for high *R* values and it was not found to be statistically significant. On the other hand, when cyclic response is of concern, artificial records show a strong overestimation with respect to real records and wavelet-adjusted records.

All the trends for the linearly scaled records seem to be non-systematic, indicating that scaling does not bias the response if the spectral shape is a control factor.

References

1. Abrahamson NA (1992) Non-stationary spectral matching. Seismol Res Lett 63(1):30
2. Bazzurro P, Luco N (2004) Post-elastic response of structures to synthetic ground motions. Report for Pacific Earthquake Engineering Research (PEER) Center Lifelines Program Project 1G00 Addenda, CA, USA
3. Bommer JJ, Acevedo AB (2004) The use of real earthquake accelerograms as input to dynamic analysis. J Earthq Eng 8(S1):43–91
4. CEN (2004) EN 1998-1:2004 Eurocode 8, Design of Structures for earthquake resistance – part 1: general rules, seismic actions and rules for buildings. Comitè Europèen de Normalisation, Brussels
5. Convertito V, Iervolino I, Herrero A (2009) The importance of mapping the design earthquake: insights for southern Italy. Bull Seismol Soc Am 99(5):2979–2991
6. CS.LL.PP.; DM 14 (2008) Norme tecniche per le costruzioni. Gazzetta Ufficiale della Repubblica Italiana, 29 (in Italian)
7. De Luca F, Iervolino I, Cosenza E (2009) Un-scaled, scaled, adjusted and artificial spectral matching accelerograms: displacement- and energy-based assessment. Proceedings of XIII ANIDIS, "L'ingegneria Sismica in Italia", Bologna, Italy
8. Gasparini DA, Vanmarke EH (1976) Simulated earthquake motions compatible with prescribed response spectra. MIT civil engineering research report R76-4. Massachusetts Institute of Technology, Cambridge, MA
9. Hancock J, Watson-Lamprey J, Abrahamson NA, Bommer JJ, Markatis A, McCoy E, Mendis E (2006) An improved method of matching response spectra of recorded earthquake ground motion using wavelets. J Earthq Eng 10(S1):67–89
10. Iervolino I, Cornell CA (2005) Record selection for nonlinear seismic analysis of structures. Earthq Spect 21(3):685–713
11. Iervolino I, Manfredi G, Cosenza E (2006) Ground motion duration effects on nonlinear seismic response. Earthq Eng Struct Dyn 30:485–499
12. Iervolino I, Maddaloni G, Cosenza E (2008) Eurocode 8 compliant real record sets for seismic analysis of structures. J Earthq Eng 12(1):54–90
13. Iervolino I, Galasso C, Cosenza E (2009) REXEL: computer aided record selection for code-based seismic structural analysis. Bull Earthq Eng 8:339–362
14. Mucciarelli M, Spinelli A, Pacor F (2004) Un programma per la generazione di accelerogrammi sintetici "fisici" adeguati alla nuova normativa, Proceedings of XI Convegno ANIDIS, "L'ingegneria Sismica in Italia", Genova, Italy (in Italian)
15. Pinto PE, Giannini R, Franchin P (2004) Seismic reliability analysis of structures. IUSS Press, Pavia, Italy
16. Schwab P, Lestuzzi P (2007).Assessment of the seismic non-linear behaviour of ductile wall structures due to synthetic earthquakes. Bull Earthq Eng 5(1):67–84

Chapter 5
Theoretical Consistency of Common Record Selection Strategies in Performance-Based Earthquake Engineering

Peter J. Stafford and Julian J. Bommer

5.1 Introduction

Performance-based Earthquake Engineering (PBEE) is a very powerful conceptual framework within which one may conduct rigorous seismic risk analyses and make objective decisions. These decisions may be related to choosing between alternative designs, retrofitting options, or rehabilitation strategies, and the decision may be arrived at following economic, social or environmental considerations, or some combination of these. Frameworks for PBEE, such as that of the Pacific Earthquake Engineering Research (PEER) center [6, 8], allow this decision-making process to be followed in an objective and quantitative manner.

Nonlinear response time-history analyses are an integral component of PBEE and link the specification of seismic hazard to the response of structures, and to degrees of damage that they can be expected to sustain. The seismic hazard analysis is conducted by engineering seismologists, while the response, or seismic demand, analysis is undertaken by structural earthquake engineers. This partitioning of responsibilities is both a strength of PBEE and also a potential weakness. It is a strength because experts can focus upon those things that they know best, but it is a potential weakness as lack of communication can lead to situations where one party is making erroneous assumptions regarding what the other is doing. Any potential lack of communication is easily circumvented, in principle. All that is required is for engineering seismologists and structural earthquake engineers to work closely together to ensure that everyone understands the overall objective of the project. In addition, it is important that people who receive data (such as accelerograms) downstream understand the assumptions that have been made in generating this information. The approaches to record selection and modification that are most commonly implemented in practice suggest that the ideal situation is not always realized. This chapter aims to demonstrate this point by focusing upon the manner with which

P.J. Stafford (✉)
Department of Civil and Environmental Engineering, Imperial College London,
South Kensington Campus, London SW7 2AZ, UK
e-mail: p.stafford@imperial.ac.uk

M.N. Fardis (ed.), *Advances in Performance-Based Earthquake Engineering*,
Geotechnical, Geological, and Earthquake Engineering 13, DOI 10.1007/978-90-481-8746-1_5,

aleatory variability of earthquake ground-motion is handled within modern PBEE frameworks and assessing whether record selection strategies are consistent with the fundamental assumptions of these frameworks.

5.2 Treatment of Ground-Motion Variability in PBEE

In this chapter, we make use of the increasingly ubiquitous PBEE formulation of the PEER framework. We do so primarily because it will be familiar to most readers and for the sake of maintaining consistency with previous studies. However, it should be emphasized that the general concepts that we discuss are not limited to this particular framework. Irrespective of the adopted framework, the key issue of how aleatory variability of ground motion is addressed must be considered.

The full PEER PBEE framework can be described mathematically by Eq. (5.1) [6, 8]:

$$\lambda\left(\mathrm{dv}\right) = \int_{\mathrm{dm}}\int_{\mathrm{edp}}\int_{\mathrm{im}} G\left(\mathrm{dv}|\,\mathrm{dm}\right)\left|dG\left(\mathrm{dm}|\,\mathrm{edp}\right)\right|\left|dG\left(\mathrm{edp}|\,\mathrm{im}\right)\right|\left|d\lambda\left(\mathrm{im}\right)\right| \quad (5.1)$$

where $\lambda(x)$ denotes the mean annual rate at which the random variable X exceeds the value x; dv, dm, edp and im denote a decision variable, a damage measure, an engineering demand parameter, and an intensity measure respectively; and $G\left(y|\,x\right)$ represents the cumulative conditional distribution function of y given x. However, for the purposes of the current article, and more generally for record selection, only two components of the above expression are of interest. Following Cornell [7] we may state that the objective of PBEE is to estimate "the annual frequency, λ, that an earthquake induces in a particular structure some specified behavior state". Mathematically, this objective can be cast as in Eq. (5.2), with the behavior state being described fully by the edp.

$$\lambda\left(\mathrm{edp}\right) = \int_{\mathrm{im}} G\left(\mathrm{edp}|\,\mathrm{im}\right)\left|\frac{d\lambda\left(\mathrm{im}\right)}{dim}\right| dim \quad (5.2)$$

The two key expressions on the right-hand side of Eq. (5.2) mathematically represent the roles of structural earthquake engineers and engineering seismologists in PBEE. The term $G\left(\mathrm{edp}|\,\mathrm{im}\right)$ represents a fragility function that can be derived by conducting large numbers of time-history analyses and defines the probability of exceeding some level of a response measure, edp, given that a ground motion has a particular value of an intensity measure, im [1]. These functions are derived by structural earthquake engineers. The second term denotes the slope of a seismic hazard curve, which is calculated by engineering seismologists, and provides the mean annual rate at which the intensity measure (approximately) equals a value im.

Statements are often made that no account for ground-motion variability need be taken in the fragility analysis as it has already been accounted for in the probabilistic seismic hazard analysis (PSHA). However, this statement is *not* strictly true. The slope of the hazard curve can be expressed mathematically as in Eq. (5.3):

$$\left|\frac{d\lambda\,(\text{im})}{\text{dim}}\right| = \sum_{i=1}^{N} \nu_i \left\{ \iint f_{IM}(\text{im}|m, r) f_{M,R}(m, r) dmdr \right\}_i \tag{5.3}$$

In Eq. (5.3), N seismic sources are represented and are indexed by i; ν_i denotes the mean annual rate at which earthquakes having a magnitude, m, greater than some minimum value occur; and $f_{M,R}\,(m, r)$ represents the joint probability density function of combinations of magnitude and distance, r. All of the terms just defined are taken care of through the development of a seismic source model and through performing a seismicity analysis. The remaining term, $f_{IM}\,(\text{im}|\, m, r)$, is the probability density function (pdf) of an intensity measure given some earthquake scenario that is represented by magnitude and distance. This pdf is precisely what an empirical ground-motion model provides; it takes some seismological scenario and returns the pdf of a particular intensity measure for that scenario.

The above presentation has been labored, but it is essential to clearly understand what is “taken care of” in terms of aleatory variability of ground motions when a PSHA is carried out. With the notable exception of Vector-valued PSHA [4], which is very rarely encountered in practice, PSHA is only ever conducted for a single spectral ordinate at a time. Therefore, the aleatory variability of ground motion is only fully accounted for in PBEE if the variability of the single chosen intensity measure is able to describe all sources of variability in an earthquake record that could conceivably influence the structural response, as measured by a certain edp. It will hopefully be clear to the reader that it is extremely unlikely that a scalar intensity measure can be found for realistic applications that can fully describe the influence of ground-motion variability upon structural response. To reinforce this point, Fig. 5.1 shows empirical cumulative distribution functions with fitted cumulative normal distributions for six different intensity measures. These plots are generated for a particular scenario corresponding to $M_w = 6 \pm 0.1$, $R_{rup} = 50 \pm 10$ km and NEHRP site class C.

In Fig. 5.1 it can clearly be seen that there is significant variability of the intensity measures for this particular scenario. Not only is there a significant amount of variability in the intensity measures, these measures are generally not strongly correlated (see Table 5.1) indicating that knowledge of just one distribution is not sufficient to describe any of the other ground-motion characteristics. Of particular interest is the correlation between the two relatively closely-spaced spectral ordinates. While these ordinates are rather strongly correlated, with a correlation coefficient of 0.84 in this example, it is clear that knowing the nature of $Sa(1.0\text{ s})$ does not mean that the nature of $Sa(1.5\text{ s})$ is also known. More on this issue will be said in subsequent sections. For now it suffices to say that if one had a structure with an initial fundamental period of 1.0 s and nonlinear response was anticipated such that the spectral acceleration at a period of 1.5 s could influence the response, then

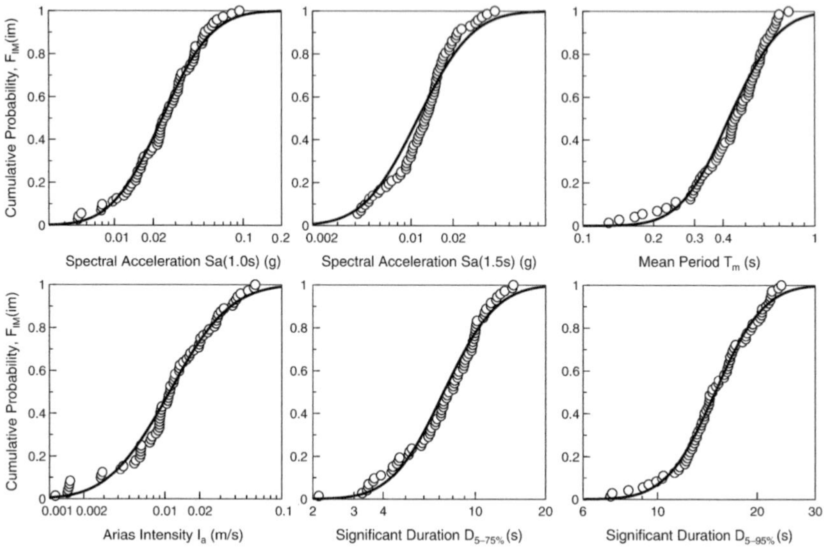

Fig. 5.1 Empirical cumulative distribution functions of intensity measures consistent with a seismological scenario of $M_w = 6 \pm 0.1$, $R_{rup} = 50 \pm 10$ km and NEHRP site class C. Data comprises 72 records from the NGA database

Table 5.1 Standard deviations of the logarithms of the intensity measures shown in Fig. 5.1 and the linear correlation coefficients among these intensity measures

Intensity measure, IM	Sigma, $\sigma_{\ln IM}$	$\rho_{IM,Sa(1.0\,s)}$	$\rho_{IM,Sa(1.5\,s)}$	ρ_{IM,T_m}	ρ_{IM,I_a}	$\rho_{IM,D_{5-75\%}}$	$\rho_{IM,D_{5-95\%}}$
$Sa(1.0\,s)$	0.71	1	0.84	0.58	0.51	0.35	0.13
$Sa(1.5\,s)$	0.70	0.84	1	0.64	0.44	0.34	0.26
T_m	0.39	0.58	0.64	1	–0.17	0.50	0.45
I_a	0.94	0.51	0.44	–0.17	1	–0.07	–0.13
$D_{5-75\%}$	0.41	0.35	0.34	0.50	–0.07	1	0.77
$D_{5-95\%}$	0.27	0.13	0.26	0.45	–0.13	0.77	1

conducting a PSHA for $Sa(1.0\text{ s})$ would *not* take care of all of the aleatory variability of ground motions. In this case, the influential variability of $Sa(1.5\text{ s})$ would not be fully accounted for.

5.3 Current Approaches to Record Selection

Having now defined the way in which the aleatory variability of ground motions is incorporated into PBEE, it is appropriate to turn to the current approaches to record selection that are most commonly encountered in practice. In considering these approaches we will be looking to see how consistent their consideration of ground-motion variability is with the theoretical framework just outlined.

5.3.1 Linear Scaling to Sa(T_1)

Shome et al. [13] made the recommendation that earthquake records should first be screened such that their metadata in terms of magnitude and distance broadly matched those of a design scenario (found through disaggregation) and should then be linearly scaled such that their spectral ordinate at the initial fundamental period of the structure (poorly known as this may be in reality), $Sa(T_1)$, matched a value defined from a PSHA.

This approach is perfectly consistent with the framework outlined previously. Some target value of $Sa(T_1)$ is defined from a PSHA conducted purely for $Sa(T_1)$ and hence the aleatory variability of this parameter is fully accounted for. All remaining aspects of the variability of the ground motions from the design scenario are implicitly accounted for by the nature of the records that are then selected.

If this approach was adopted for estimating the response of a linear elastic single degree-of-freedom (DOF) structure it would be perfect. However, if the structure is a multiple DOF system or if the structure is expected to undergo nonlinear response then other attributes of the ground motion will be important. For this reason analyses conducted using this approach can lead to biased estimates of the response by neglecting the variability of spectral ordinates away from the initial fundamental period [2, 11]. Figure 5.2 shows that records selected according to this approach lead to zero variability at the initial fundamental period (as desired, given that this variability is already accounted for) but that significant variability can exist at other periods. Note that this variability will result in more records being required in order to obtain a stable estimate of the median response [13, 10].

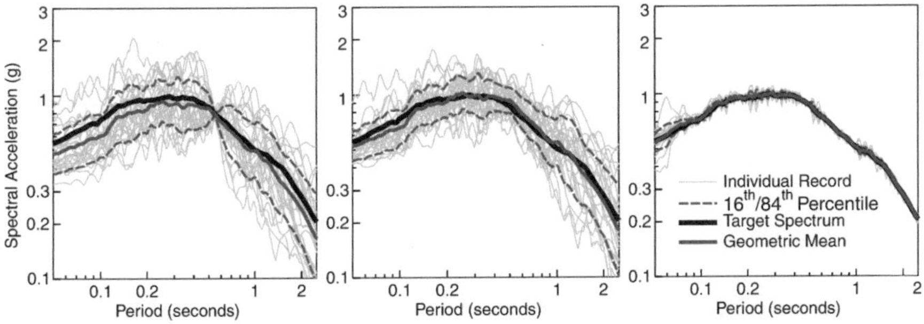

Fig. 5.2 Records chosen by: *left* linearly scaling spectra to a target value of spectral acceleration at the initial fundamental period of the structure; *middle* linearly scaling spectra to a target value of spectral acceleration over a range of periods, including the initial fundamental period; and *right* linearly scaling spectra to a target over a range of periods, and then improving this match through the addition of wavelets. Modified from Hancock et al. [10]

5.3.2 Linear Scaling to a Target Spectrum Over a Period Range

An approach that is often taken with a view to rectifying the shortcomings of the previous procedure is to select and scale records while constraining their spectral shape. In this approach, rather than scaling to a single ordinate the spectra are scaled such that individual spectra have a good fit to some target spectrum over a range of periods (see middle panel of Fig. 5.2).

In practice, the target spectrum is often taken to be a Uniform Hazard Spectrum (UHS) or a code spectrum. However, neither of these options are recommended and would be strongly advised against were they not required by most international design codes.

In terms of consistency with the treatment of aleatory variability in PBEE the approach is not as good as linear scaling to $Sa(T_1)$ as the variability at the initial fundamental period is not zero. However, some constraint on the variability at other periods is gained through consideration of the spectral shape. In addition, by accounting for spectral shape the possibility of introducing bias in the response through scaling is minimized [11, 10].

The most theoretically consistent approach to implementing this method is where the target spectrum is defined as a Conditional Mean Spectrum (CMS) [2]. The definition of the CMS implicitly accounts for the correlations that exist among spectral ordinates at different periods (see Table 5.1) but requires empirical models for these correlations [3]. By accounting for these correlations the approach becomes essentially the same as VPSHA, if the hazard is strongly dominated by a single scenario. However, even with the CMS as a target spectrum, the variability at $Sa(T_1)$ remains non-zero and is hence not perfectly consistent with the theoretical framework.

Despite this fact, the use of the CMS is currently regarded as being the optimal approach to record selection when the engineering demand parameter of interest is a peak response measure such as peak interstorey drift [12]. It must be noted, however, that the use of the CMS will not explicitly account for the correlations among spectral ordinates and other intensity measures such as duration. However, such relationships may be implicitly accounted for through initial screening of records on the basis of magnitude and distance. If these other intensity measures are influential for the response then this approach will suffer from similar shortcomings as the use of linear scaling to T_1 (but to a lesser extent).

5.3.3 Linear Scaling and Spectrum Matching

An extension of the previous approach is to select and scale records such that their spectra have a good approximation to the target over a range of periods, but to then modify these records using a spectrum matching approach to maximize the agreement between the records and the target (e.g. Hancock et al. [9]). This approach is more consistent with the theoretical framework than linearly scaling over a period range as the variability that remains at $Sa(T_1)$ is minimized (although it is generally still not zero). This approach also shares the advantages of not leading to biased estimates through taking account of the spectral shape [10]. However, this approach has been shown to introduce a small systematic bias when used for near-source earthquake scenarios that can result in pulse-like motions [5].

The major drawback of this approach is that the natural correlations among spectral ordinates are not retained (or at least, the correlations may still be correct, but the absolute magnitude of the fluctuations of spectral ordinates from period-to-period are lost). The result is that while the accelerograms may still look realistic in the time domain they clearly do not resemble real earthquake records when viewed as response spectra (see right panel of Fig. 5.2).

The main concern regarding spectrum matching arises when it comes to developing fragility curves rather than simply estimating the median response. The objective of fragility curves is to define the conditional variability of a response parameter given some intensity measure. Although it has not been formally proven, it is reasonable to assert that spectrum matching approaches suppress the natural peak-to-trough variability to the extent that the conditional variability in response is underestimated.

5.4 Implications for Record Selection

The preceding sections have highlighted the fact that the only record selection approach, of those commonly implemented in practice, which is entirely consistent with the theoretical framework of PBEE is that of selecting records that have properties consistent with a design scenario before linearly scaling them such that the variability of an intensity measure is zero. In the present article the example that has been shown is that most used in practice and concerns scaling to a target value of $Sa(T_1)$. However, studies such as those of Luco and Bazzurro [11], Hancock et al. [10] and the PEER Ground Motion Selection and Modification Working Group [12] have demonstrated that this approach is relatively inefficient and leads to large numbers of time-history analyses being performed. These studies also suggest that the optimal approaches in terms of reducing the numbers of required records are those that select records while accounting for spectral shape and then apply spectrum matching.

The implications in terms of estimating the median structural response are fairly clear in light of the recent studies just referred to. One should sacrifice a small degree of theoretical rigor in place of practical efficiency and should scale and match to a CMS when aiming to predict peak structural response measures. However, the implications for estimating the distribution of response (and hence developing accurate fragility curves) are less clear. PBEE frameworks rely heavily upon ensuring that aleatory variability is propagated from step to step throughout the overall process. From the preceding discussions it appears that the optimal approaches for estimating the median response will not remain optimal when attempting to estimate the conditional distribution of response. The issue of greatest importance appears to be associated with trying to retain an appropriate degree of peak-to-trough variability in the selected and modified records.

It seems clear that some degree of peak-to-trough variability must be retained in the records in order to ensure that the conditional distribution of response (fragility curves) are robustly defined. A comprehensive series of computationally intensive analyses must be undertaken in order to try to develop guidelines regarding how to best incorporate this peak-to-trough variability into time history analyses. However, for now it is still possible to speculate as to what sort of magnitude this variability should have and when or how it might be incorporated.

Figure 5.3 shows a measure of peak-to-trough variability, σ_{PT}, calculated by computing the standard deviation of the logarithm of spectral ordinates over a number of records and over a range of response periods defined by the "bandwidth". The range of periods is defined by first identifying some central period and then dividing and multiplying this period by the bandwidth in order to obtain the minimum and maximum periods of the range. This bandwidth reflects the degree of nonlinearity that is to be expected from the structural response as well as the contribution of higher mode effects. The greater the contribution of these effects, the larger the bandwidth. If a suitably large number of records are considered, the peak-to-trough variability tends to stabilize around a value of around 0.2 in natural logarithmic units as the bandwidth increases. For near-linear response this value is closer to zero (see left panel of Fig. 5.3).

Given the observations of Fig. 5.3 it may be that values of σ_{PT} can be defined in a systematic way. Once such values are defined there may be two options for accounting for this in PBEE. One approach may be to use a spectrum matching based approach and to later "add-in" the missing peak-to-trough variability, that is, to inflate the variance of the structural response using some correction factor. We do not propose a formal mechanism for doing this here. The other option could be to select suites of records that have an appropriate amount of peak-to-trough

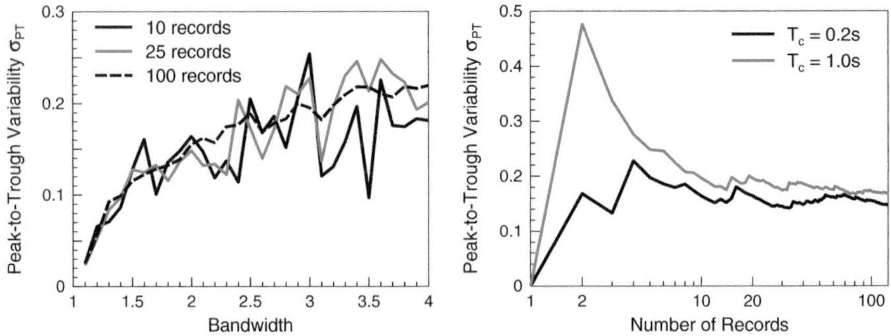

Fig. 5.3 Peak-to-trough variability, σ_{PT}, of records scaled to a target over a period range. The panel on the left shows σ_{PT} computed for three different numbers of records and against bandwidth where the bandwidth is a multiplicative factor applied to a central period, T_c, of 0.2 s. The panel on the right shows σ_{PT} against the number of records for two different central periods

variability. This variability might be achieved simply through careful selection, but could also be accounted for by using spectrum matching with a relatively high tolerance for the match. In this latter case one could obtain the correct general shape while also constraining the variability. These latter approaches are more straightforward, but are likely to be less efficient. However, before any firm recommendations can be made a comprehensive analysis must be undertaken that is well beyond of the scope of this brief contribution.

5.5 Conclusions

PBEE is strongly dependent upon the execution of response time-history analyses. However, the most popular current recommendations for selection and scaling of accelerograms do not strictly adhere to the theoretical framework of PBEE. This chapter has discussed this issue and suggests that, when attempting to estimate the median structural response, this inconsistency is not particularly important. Further work must be undertaken to understand whether approaches such as spectrum matching lead to under-estimates of the variance of the conditional distributions of response (fragility curves). The fact that these approaches suppress the natural peak-to-trough variability suggests that under-estimates will be obtained.

Acknowledgments The first author would like to acknowledge the support of the Willis Research Network.

References

1. Baker JW (2007) Probabilistic structural response assessment using vector-valued intensity measures. Earthq Eng Struct Dyn 36(13):1861–1883
2. Baker JW, Cornell CA (2006) Spectral shape, epsilon and record selection. Earthq Eng Struct Dyn 35(9):1077–1095
3. Baker JW, Jayaram N (2008) Correlation of spectral acceleration values from NGA ground motion models. Earthq Spectra 24(1):299–317
4. Bazzurro P, Cornell CA (2002) Vector-valued probabilistic seismic hazard analysis (VPSHA). Proceedings of the 7th U.S. national conference on earthquake engineering, Boston, MA
5. Bazzurro P, Luco N (2006) Do scaled and spectrum-matched near-source records produce biased nonlinear structural responses? Proceedings of the 8th U.S. national conference on earthquake engineering, San Francisco, CA, Paper No. 1029
6. Cornell CA, Krawinkler H (2000) Progress and challenges in seismic performance assessment. PEER Center News. http://peer.berkeley.edu/news/2000spring/index.html. Accessed 18 April 2009
7. Cornell CA (2005) On earthquake record selection for nonlinear dynamic analysis. Proceedings of the Esteva symposium, Mexico
8. Der Kiureghian A (2005) Non-ergodicity and PEER's framework formula. Earthq Eng Struct Dyn 34(13):1643–1652
9. Hancock J, Watson-Lamprey J, Abrahamson NA, Bommer JJ, Markatis A, McCoy E, Mendis R (2006) An improved method of matching response spectra of recorded earthquake ground motion using wavelets. J Earthq Eng 10:67–89

10. Hancock J, Bommer JJ, Stafford PJ (2008) Numbers of scaled and matched accelerograms required for inelastic dynamic analyses. Earthq Eng Struct Dyn 37(14):1585–1607
11. Luco N, Bazzurro P (2007) Does amplitude scaling of ground motion records result in biased nonlinear structural drift responses? Earthq Eng Struct Dyn 36(13):1813–1835
12. PEER Ground Motion Selection and Modification Working Group (2009) Evaluation of ground motion selection and modification methods: Predicting median interstorey drift response of buildings. PEER Report 2009/01, Berkeley, CA, p 288
13. Shome N, Cornell CA, Bazzurro P, Carballo JE (1998) Earthquakes, records, and nonlinear responses. Earthq Spectra 14(3):469–500

Chapter 6
Long-Period Earthquake Ground Motion: Recent Advances and Observations from the April 6 2009, M_w6.3 L'Aquila Earthquake, Italy

Roberto Paolucci

6.1 Introduction

The spread of performance-based design approaches, together with the fast increase of digitally recorded strong-motion (SM) accelerations worldwide, has promoted a growing interest in the determination of displacement spectra for design, the reliability of which has been traditionally limited to structural periods of maximum 3 to 4 s, especially due to long-period noise affecting analog instruments.

In the early years of this decade, to overcome the lack of a sufficiently extended set of reliable digital SM records, Faccioli et al. [8] (2004) proposed analytical displacement spectral shapes, simply based on the following ingredients:

(i) a theoretical prediction equation of peak ground displacement, d_{max}, based on a classical model of the seismic source, which, in its simplest expression, takes the form:

$$\log_{10} d_{max} = -4.3 + M_W - \log_{10} r \tag{6.1}$$

where d_{max} is in cm, M_w = moment magnitude and r = hypocentral distance (km);

(ii) an empirical relationship between M_w and the period τ of the largest velocity pulse of recorded ground motion, which, as proposed by Somerville [13], takes the form:

$$\log_{10} \tau = -3.17 + 0.5 M_W; \tag{6.2}$$

(iii) the analytical expression of the displacement response spectrum of simple acceleration pulses of period τ.

R. Paolucci (✉)
Department of Structural Engineering, Politecnico di Milano, Piazza Leonardo da Vinci, 32, Milano 20133, Italy
e-mail: roberto.paolucci@polimi.it

M.N. Fardis (ed.), *Advances in Performance-Based Earthquake Engineering*, Geotechnical, Geological, and Earthquake Engineering 13, DOI 10.1007/978-90-481-8746-1_6,

In the following years, such theoretical developments were supported by an ever increasing number of high quality digital records, especially coming from the Japanese K-Net (www.k-net.bosai.go.jp) and Kik-net (www.kik.bosai.go.jp) strong motion networks and from other world regions as well. After setting up a worldwide digital SM dataset consisting of about 1,200 three-components records, Cauzzi and Faccioli [5] calibrated a broadband, up to 20 s, empirical prediction equation of displacement spectral ordinates, which proved to yield reliable results upon validation against a number of recent earthquakes, including the one of L'Aquila, that will be dealt with in the sequel.

A common objection to calibration of empirical predicting tools for long period ground motion is that, even for high quality digital records, reliable displacement traces cannot be generally retrieved through simple double integration of the uncorrected accelerations. An answer to this objection has been given by a growing number of recent studies, starting from Boore [3], which have pointed out that the long-period spectral ordinates of digital accelerograms depend only weakly on the adopted baseline correction (BC) procedures, in contrast to the dramatic effect of the latter on the displacement waveforms. Even more pervasive for this purpose has been the evidence, pointed out by Wang et al. [15] and Paolucci et al. [12], that spectral ordinates calculated from co-located SM and broadband (BB) records coincide up to at least 10 s. Furthermore, differences are still practically negligible up to 20–30 s. One of the most significant examples of comparison of response spectra from available co-located records is shown in Fig. 6.1, referring to the 1997 Michoacán earthquake (M_W7.1, $r \sim$ 150 km) recorded in Zihuatanejo by the Mexican National Seismological Survey (www.ssn.unam.mx). Since the digital SM record was disturbed by a complicated baseline drift, it was not possible to find a suitable BC procedure to obtain a de-trended displacement trace without high-pass filtering. Nevertheless, the coincidence of the SM and BB spectra, up to at least $T = 10$ s, clearly points out the negligible influence of the trend of the displacement trace on the spectral ordinates of practical significance.

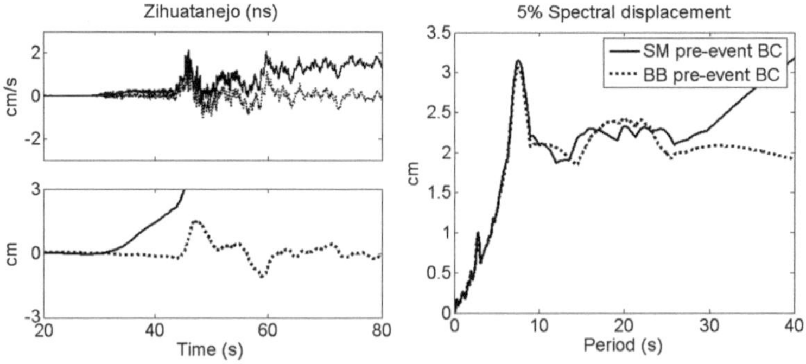

Fig. 6.1 Comparison of velocity and displacement time histories (*left*) and 5% damped displacement response spectra (*right*) obtained at Zihuatanejo, Mexico, during the Michoacán earthquake of January 11, 1997 (M_W7.1, $r \sim$150 km), by a strong motion (SM) accelerogram and by a broadband (BB) velocity meter. From Paolucci et al. [12]

These studies suggest that response spectra from digital records are reliable up to periods much longer than usually thought. Furthermore, the common technique of high-pass filtering the digital acceleration record from a cut-off period selected to avoid baseline drifts on the displacement traces appears to be in most cases too conservative and unduly depletes reliable information on long period spectral ordinates. However, the limited amount of data supporting this suggestion, and the difficulty in objectively determining the error with respect to unknown noise-free spectral ordinates, has restricted so far the potential impact of such observations.

To overcome these limitations, and to provide an objective and simple criterion for selecting reliable digital strong motion records, Paolucci et al. [12] used synthetic accelerograms contaminated by random long period noise to quantify the difference between the original accelerograms and the spurious ones in terms of response spectra. A noise index was introduced calculated from the uncorrected velocity trace of the record that was correlated with the probability of exceedance of a given level of error, in terms of response spectral ordinates at long periods. A similar approach was used by Akkar and Boore [1] as well, who reached the similar conclusion that, referring to digital SM records: "the elastic spectra from the most basic processing, in which only the pre-event mean is removed from the acceleration time series, do not diverge from the baseline-corrected spectra until periods of 10–20 s (. . .), implying that for many engineering purposes elastic spectra can be used from records with no baseline correction or filtering".

The previous advances to assess the reliability of long period response spectral ordinates and to calibrate up-to-date empirical prediction tools have recently led to the formulation of the new seismic hazard maps at long periods in Italy [9].

Further advances have been recently presented for quantification of site effects at long periods, e.g., by Manou et al. [11], Cauzzi and Faccioli [5] and Figini and Paolucci [10]. Trying to provide physical explanations of the observed long-period amplification factors for several Kik-net records, the latter work concludes that 1D soil models do not explain in general relevant site amplification effects at long periods, except for very soft soil conditions, and that the observed amplification levels at long periods are likely to be attributed to the effects related to more complex geological structures, such as deep and spatially extended alluvial basins. As will be shown later on, this observation is one of the keys to justify long period amplification of ground motion during the L'Aquila earthquake.

6.2 Near-Fault Strong Motion Records from the M_W 6.3 April 6 2009 L'Aquila Earthquake: Observations at Long Periods

During the night of April 6 2009, a M_W 6.3 earthquake struck the Abruzzi region and the whole Central Italy, causing about 300 deaths and vast destruction in the town and surroundings of L'Aquila, one of the largest urban centers in Central Italy, with about 80,000 inhabitants. As most of the largest earthquakes in the Italian Central and Southern Apennines, this was caused by a normal fault rupture, the epicenter of which was less than 5 km SW of the town center (Fig. 6.2).

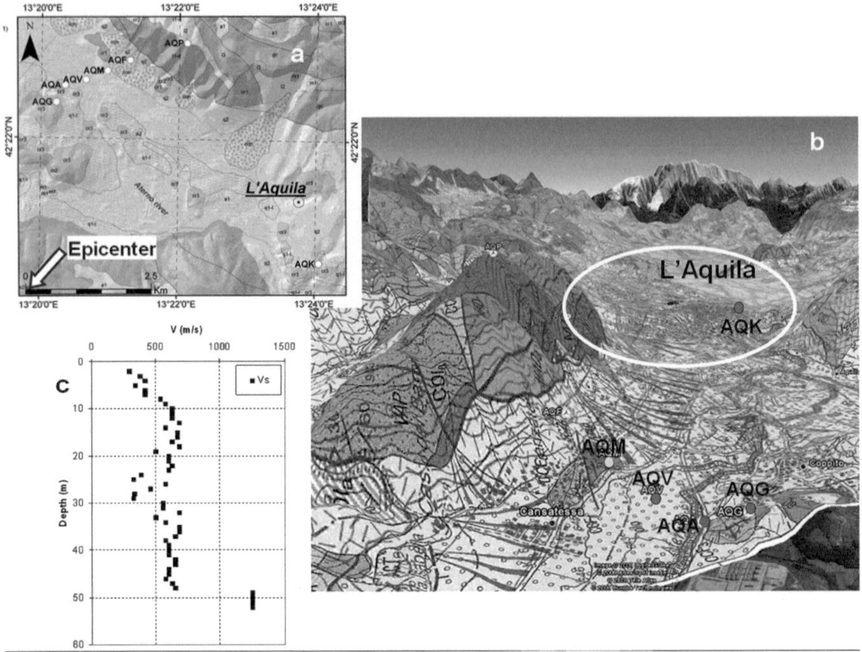

Fig. 6.2 (**a**) and (**b**) Geological sketch of L'Aquila, and location of the strong motion stations. The earthquake epicenter lies close to the *bottom-left* corner of figure (**a**). (**c**) Cross-hole Vs profile at AQV, available at the ITACA Italian strong motion database website

6.2.1 Geological Setting and Available Strong Motion Stations

L'Aquila lies on a fluvial terrace, some tens of m thick, consisting of calcareous breccias and conglomerates with limestone boulders and clasts in a marly matrix. The terrace lies on top of lacustrine sediments, mainly consisting of silty and sandy layers and minor gravel beds [7]. As shown in Fig. 6.2, the terrace is at the left bank of the Aterno river valley, which flows about 50 m below downtown L'Aquila.

A strong-motion array consisting of six stations was installed in 2001 by the National Department of Civil Protection (DPC) across the upper Aterno valley (Fig. 6.2). Recordings from this array, together with the AQK station located close to downtown L'Aquila (Fig. 6.2b), provide a near-fault strong-motion data set never recorded to date in Italy for events with $M > 5$, and one of the few ones worldwide. The data set has been integrated in the new Italian strong motion database ITACA (ITalian ACcelerometric Archive), available at http://itaca.mi.ingv.it.

For the single station AQV, located at the center of the valley, a cross-hole Vs profile is available (Fig. 6.2c). An alternation of gravels and sands, with average Vs $\sim$ 500 m/s, is present down to a depth of 47 m, where bedrock is found. In contrast, according to the available geological surveys, the lacustrine sediments reach their maximum thickness (around 250 m) below the center of L'Aquila [7], roughly corresponding to station AQK.

Owing to malfunctioning of the power supply, stations AQF and AQP of the network did not trigger during the main shock, while station AQM, set to 1 g full-scale, saturated although installed on outcropping rock. The reliability of the latter record is presently under investigation.

6.2.2 Observed Earthquake Ground Motion

Referring to Ameri et al. [2] for a more thorough presentation of the L'Aquila SM data, the attention is limited here to the long-period part of the observed ground motion. For this purpose, horizontal and vertical time series of velocity and displacement obtained at the array sites (AQG, AQA, AQV) and at AQK during the main shock are shown in Figs. 6.3 and 6.4, respectively. To avoid the onset of spurious arrivals in the displacement waveforms from acausal high-pass filtering and to possibly recover reliable permanent displacements via double integration of accelerations, records were processed by a baseline correction technique, consisting of least-squares fitting the velocity time-histories by three consecutive line segments, and removing them from the velocity itself [3]. As shown in Fig. 6.4, coherent displacement time series are obtained, especially along the Aterno river transept, showing a downwards permanent displacement in the SE direction, in agreement with the GPS-based findings by Cirella et al. [6].

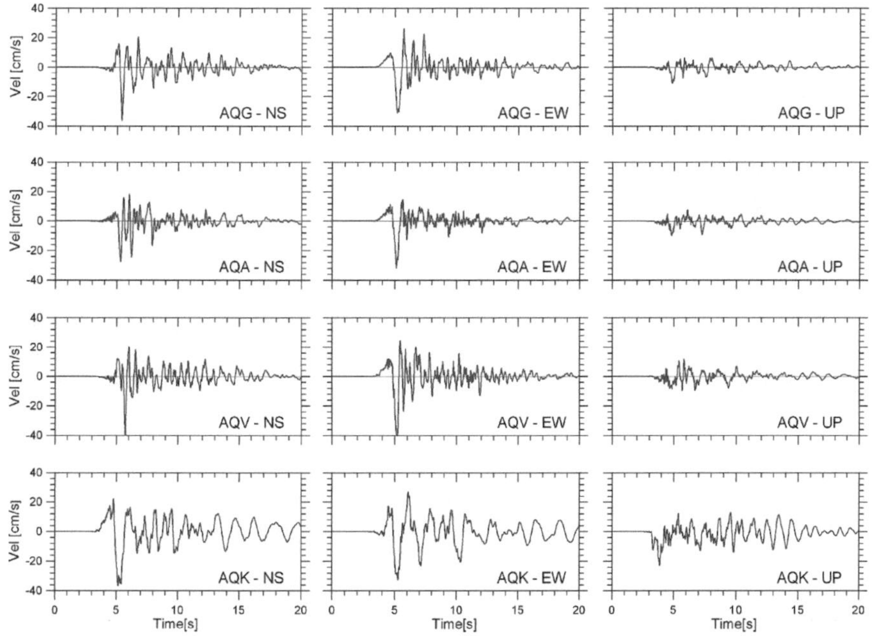

Fig. 6.3 Velocity time histories recorded at the array sites (AQG, AQA, AQV) and AQK: NS component (*left panel*), EW component (*centre*) and UP component (*right*)

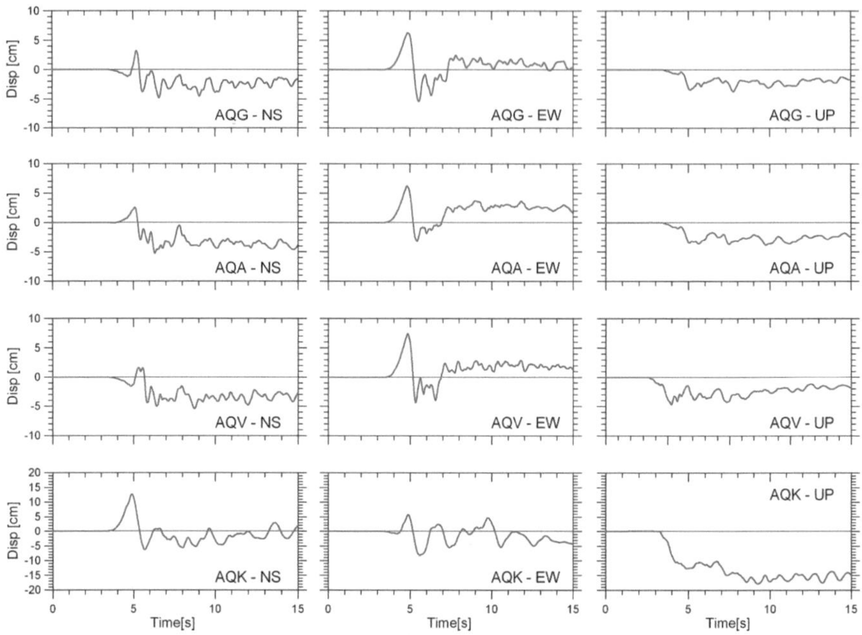

Fig. 6.4 Displacement time histories obtained at the array sites (AQG, AQA, AQV) and AQK: NS component (*left panel*), EW component (*centre*) and UP component (*right*)

Ground motion velocity pulses are present at all the array records, and, to a larger extent, at AQK, where long period ground motion is likely the combined effect of the seismic source radiation and the interaction with the deep lacustrine sediments of the Aterno valley, addressed to in the previous section.

To investigate in more detail the source-related effects on the L'Aquila records, the ground motion has been decomposed into its strike normal (SN) and strike parallel (SP) components, assuming a 147° fault strike angle, whence the predominant period T_{SN} and T_{SP} of the main velocity pulses, along with the corresponding peak ground velocities (PGV), were calculated. Results are summarized in Table 6.1, showing T_{SN} around 1.0 s for stations AQA, AQG and AQV, and increasing significantly up to about 1.5 s for station AQK. Note that Eq. (6.2) gives $T_{\mathrm{SN}} = 0.95$ s for $M_{\mathrm{w}} = 6.3$, in very good agreement with the array stations.

Table 6.1 Observed values of the period of the strike normal (SN) and strike parallel (SP) largest velocity pulse, along with the corresponding PGV

	AQG	AQV	AQA	AQK
T_{SN} (s)	1.10	0.90	1	1.55
T_{SP} (s)	0.78	0.80	0.7	1.30
$\mathrm{PGV_{SN}}$ (cm/s)	34.8	40.7	32.6	44.7
$\mathrm{PGV_{SP}}$ (cm/s)	28.2	31.6	21.0	20.5

The ratio T_{SP}/T_{SN} is about 0.75, in agreement with similar observations from other worldwide earthquakes recorded in near-fault. Observed PGVs from 30 to 40 cm/s in the SN and from 20 to 30 cm/s in the SP directions are also in reasonable agreement with available empirical prediction equations (e.g., Bray and Rodriguez-Marek [4]). Note that all records show a predominant velocity peak in the SN direction.

6.2.3 Displacement Response Spectra

Elastic 5%-damped displacement response spectra of the recorded ground motion were calculated for the SN and SP components and compared in Fig. 6.5 to the theoretical displacement response spectrum according to Faccioli et al. [8], using the values $d_{max} = 0.1$ m and $\tau = 0.95$ s obtained according to Eqs. (6.1) and (6.2) for M_w=6.3 and $r = 10$ km. For simplicity, the Aterno valley spectra are plotted in terms of average values for both components. There is a striking similarity of theoretical and observed spectral ordinates, both in terms of shape and values, showing that long period ground motion recorded in L'Aquila is close to the expectations based on simple physical considerations. Due to the location of the hypocenter and of the seismic fault with respect to the stations (see e.g. Ameri et al. [2]), the effect of directivity on recorded ground motion at stations considered in this paper is likely not relevant. The focal mechanism effect is more relevant, with a significant difference between SN and SP components.

It is not clear yet why such difference of SN and SP ground motion is so magnified at station AQK, as it is apparent from Fig. 6.5. Most likely, the AQK record is affected by a complex coupling of the source mechanism and the deep structure of the Aterno basin, where the lacustrine sediments reach around 250 m depth right underneath L'Aquila, as previously mentioned. Although numerical simulations of near-fault seismic wave propagation in such complex geological environments is still demanding (see e.g. Stupazzini et al., [14]), as it requires 3D numerical models

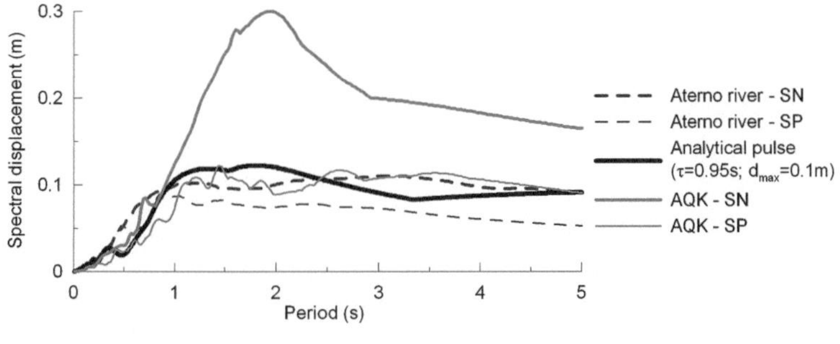

Fig. 6.5 Comparison of 5% damped displacement response spectra along the SN and SP components. *Gray continuous lines*: station AQK. *Dashed lines*: average spectra of the Aterno river array stations. *Black continuous line*: analytical pulse according to Faccioli et al. [8], for M_w=6.3 and $r = 10$ km

and computational tools, they are expected to shed light on this important effect on earthquake ground motion, that may play a major role in some of the most active seismic areas in Central and Southern Italy.

As a matter of fact, intra-mountain grabens filled by lacustrine sediments are a typical surface expression of the extensional tectonic regime that dominates the seismic activity in Central/Southern Italy along the Apennines chain. Besides the Aterno valley referred to in this paper, there are many examples of such basins, as shown in Fig. 6.6, that, due to their tectonic origin, lie in the vicinity of one or more active seismic faults. Before the L'Aquila earthquake, the best documented case is the one of Avezzano and the Fucino plain that was the greatest lake in Central Italy before being completely drained at the end of the nineteenth century. In 1915, a M7 earthquake struck the area, causing more than 30,000 deaths in Avezzano and surrounding villages, with consequences probably strongly magnified by the basin-induced ground motion amplification. In all basins illustrated in Fig. 6.6 strong motion records show long period ground motion amplification at around 2–3 s, similarly to what is clearly pointed out by the AQK spectrum in Fig. 6.5. The analysis of such records, together with 3D numerical simulations using the high-performance spectral element code GeoELSE (http://geoelse.stru.polimi.it), are presently under way. A proposal to account for complex site effects in practice, with special emphasis on long periods, can be found in Faccioli et al. (see Chapter 1, this volume).

Finally, Fig. 6.7 compares the displacement response spectra of horizontal records with the design spectra prescribed by the Italian Technical Norms for buildings (NTC08) and by Eurocode 8 (EC8), where Peak Ground Acceleration was

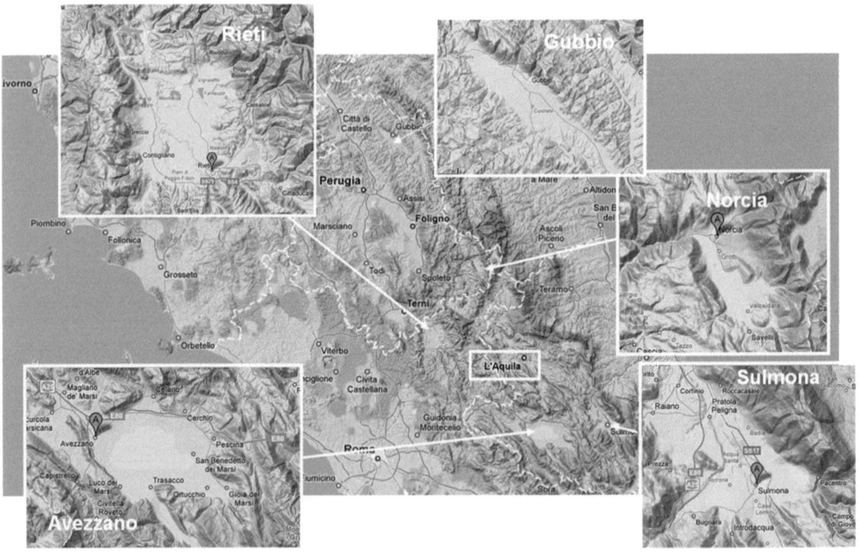

Fig. 6.6 Examples of closed-shape deep alluvial basins in Central Italy, related to an extensional tectonic regime

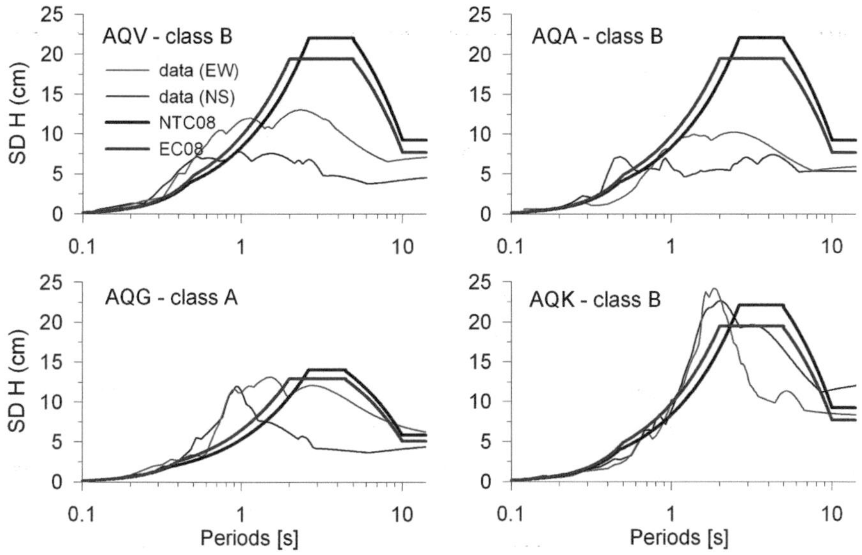

Fig. 6.7 Comparison of the 5% damped observed displacement response (H components) spectra with the design spectra prescribed by the European (EC8) and Italian (NTC08) seismic norms

anchored to the value assigned for L'Aquila by the Italian seismic hazard map for probability of exceedance of 10% in 50 years. Although observed spectral displacements are reasonably well delimited both by the NTC08 and EC8 design spectra, it is worth noting that displacement spectra for design deserve further improvements, both in terms of shape and of absolute values, based on the knowledge and findings of the last few years summarized in this paper.

Acknowledgments The author is indebted to his colleagues and co-workers at the Department of Structural Engineering, Politecnico di Milano for the cooperation on the ongoing research activities on long period ground motion, and to his colleagues at Istituto Nazionale di Geofisica e Vulcanologia, for the activities aiming at maintaining and improving the ITACA database within the DPC-INGV Project S4, 2007–2009.

References

1. Akkar S, Boore DM (2009) On baseline corrections and uncertainty in response spectra for baseline variations commonly encountered in digital accelerograph records. Bull Seismol Soc Am 99:1671–1690
2. Ameri G , Massa M, Bindi D, D'Alema E, Gorini A, Luzi L, Marzorati S, Pacor F, Paolucci R, Puglia R, Smerzini C (2009) The 6 April 2009, Mw 6.3, L'Aquila (Central Italy) earthquake: strong-motion observations. Seismol Res Lett, 80(6):951–966
3. Boore DM (2001) Effect of baseline corrections on displacements and response spectra for several recordings of the 1999 Chi-Chi, Taiwan, earthquake. Bull Seismol Soc Am 91:1199–1211
4. Bray JD, Rodriguez-Marek A (2004) Characterization of forward-directivity ground motions in the near-fault region. Soil Dyn Earthq Eng 24:815–828

5. Cauzzi C, Faccioli E (2008) Broadband (0.05 s to 20 s) prediction of displacement response spectra calibrated on worldwide digital records. J Seismol 12:453–475
6. Cirella A, Piatanesi A, Cocco M, Tinti E, Scognamiglio L, Michelini A, Lomax A, Boschi E (2009) Rupture history of the 2009 L'Aquila (Italy) earthquake from non-linear joint inversion of strong motion and GPS data. Geophys Res Lett 36:L19304, doi:10.1029/2009GL039795
7. De Luca G, Marcucci S, Milana G, Sanò T (2005) Evidence of low-frequency amplification in the city of L'Aquila, Central Italy, through a multidisciplinary approach. Bull Seismol Soc Am 95:1469–1481
8. Faccioli E, Paolucci R, Rey J (2004) Displacement spectra for long periods. Earthq Spectra 20:347–376
9. Faccioli E, Villani M (2009) Seismic hazard mapping for Italy in terms of broadband displacement response spectra. Earthq Spectra, 25:515–539.
10. Figini R, Paolucci R (2009) Site effects at long periods from digital strong motion records of the Kik-net, Japan. J Earthq Eng 13:567–584
11. Manou K, Anastasiadis A, Pitilakis K (2007) Elastic displacement response spectra. Proceedings of 4th International Conference on Earthquake Geotechnical Engineering, Thessaloniki, Greece, Paper No 1620.
12. Paolucci R, Rovelli A, Faccioli E, Cauzzi C, Finazzi D, Vanini M, Di Alessandro C, Calderoni G (2008) On the reliability of long-period response spectral ordinates from digital accelerograms. Earthq Eng Struct Dyn 37:697–710
13. Somerville PG (2003) Magnitude scaling of the near fault rupture directivity pulse. Phys Earth Planet In 137:201–212
14. Stupazzini M, Paolucci R, Igel H (2009) Near-fault earthquake ground motion simulation in the Grenoble Valley by a high-performance spectral element code. Bull Seismol Soc Am 99(1):286–301
15. Wang G-Q, Boore DM, Tang G, Zhou X (2007) Comparisons of ground motions from colocated and closely spaced one-sample-per-second global positioning system and accelerograph recordings of the 2003 M6.5 San Simeon, California, earthquake in the Parkfield region. Bull Seismol Soc Am 97(1B):76–90

Chapter 7
Uncertainty in Nonlinear SDoF Response Due to Long-Period Noise of Accelerograms

Sinan Akkar, Polat Gülkan, and Özkan Kale

7.1 Introduction

The nonlinear response of structures has been the subject of research for many years, because most structural systems are expected to behave in the post-elastic range under severe seismic action. With advances in displacement-based design and assessment procedures, significant amount of studies in this field have focused on calculating the expected peak nonlinear deformation demands on oscillators (inelastic spectral displacements, SD_{IE}). Recently, the estimation of SD_{IE} has been upgraded to more sophisticated prediction equations in the sense that they account for a more complete suite of seismological estimator parameters for their effects on nonlinear SDoF response (e.g., [9]). These equations have similar formats to those implemented in the conventional ground-motion prediction equations (GMPEs).

The continuous developments in the SD_{IE} predictive models bring forward the reliability of long-period content of strong motions because recent studies (e.g., [3]) have shown that the uncertainty in the long-period ground-motion components can be a serious limitation for nonlinear oscillator response. The major source of this uncertainty stems from the inherent long-period noise embedded in the strong-motion records. Figure 7.1 shows an example to demonstrate this effect. It presents the variations of elastic (SD_E) and inelastic spectral displacements (left and right panels, respectively) of a record that is subjected to a set of different data processing schemes to remove the existing long-period noise. The dispersion (scatter about the mean spectral curve defined by the solid black line) is more prominent in SD_{IE} and commences at relatively shorter vibration periods when compared to the corresponding deviations in SD_E. Thus, the uncertainty in the long-period ground-motion components is magnified more by nonlinear oscillators with respect to their linear counterparts.

S. Akkar (✉)
Department of Civil Engineering & Earthquake Engineering Research Center,
Middle East Technical University, Ankara 06531, Turkey
e-mail: sakkar@metu.edu.tr

M.N. Fardis (ed.), *Advances in Performance-Based Earthquake Engineering*, Geotechnical, Geological, and Earthquake Engineering 13, DOI 10.1007/978-90-481-8746-1_7,

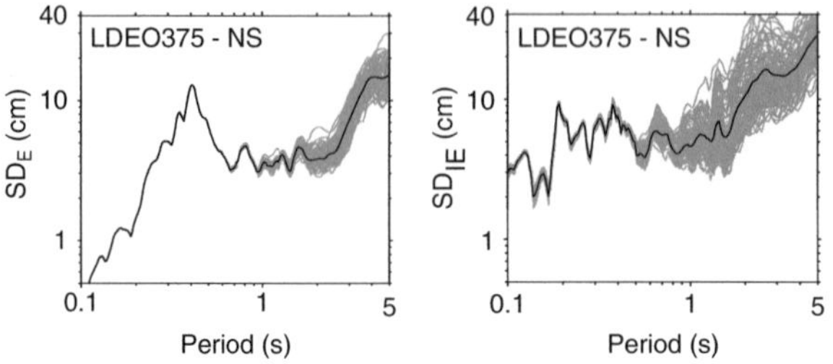

Fig. 7.1 Elastic (*left panel*) and inelastic (*right panel*) spectral displacements (*gray curves*) of a digital record subjected to different data processing schemes to remove the long-period noise. Inelastic spectra are calculated for elastoplastic hysteretic behavior with a normalized strength ratio of $R = 4$. Black solid lines show the mean of spectral displacements computed for each processing scheme (Modified from Akkar and Boore 2009 [3])

The influence of long-period noise and high-pass filtering (widely used data processing tool for removing the long-period noise) on reliable SD_E calculations has been discussed in various studies (e.g., [1, 7, 2]). These have revealed practical guidance for determining spectral bands where the peculiarities in elastic spectral response due to high-pass filtering can be disregarded. To the best of our knowledge, this discussion has not been extended systematically for variations in SD_{IE}. [Few studies showed how different filtering techniques might affect the nonlinear oscillator displacements (e.g., [6, 5]) but none of them has gone into the details of interaction between high-pass filter cut-offs and SD_{IE}]. This chapter contributes to this discussion by investigating the influence of high-pass filter cut-offs (T_c) on the nonlinear spectral and residual displacements. The latter parameter is used to validate our SD_{IE}-based observations for another nonlinear SDoF deformation quantity. Moreover, the residual displacement spectrum [8] has recently become an important deformation demand index for verifying the seismic performance of buildings. Random sets of T_c are generated in this paper for a suite of ground motions to represent the likely variation in the choice of this parameter while removing the long-period noise. SD_{IE} and SD_R are calculated then for two commonly used inelastic spectrum types: the constant-strength (R) and constant-ductility (μ) spectra to observe whether the level of uncertainty in SD_{IE} and SD_R changes due to the conceptual differences in the calculations of these two spectra. Bilinear hysteretic models are preferred in the spectral calculations to minimize the interference of secondary structural parameters on the probability calculations. Magnitude and level of inelasticity (for different R or μ values) that contribute to the dispersion in SD_{IE} and SD_R are also investigated within the context of the study for a complete picture of shortfalls invoked by high-pass filtering. In the final part of the article, usable spectral period ranges are derived, based on robust probabilistic methods where the risk of unreliable SD_{IE} due to high-pass filtering is below a certain level. The results and discussions of this paper are useful for the improvements of nonlinear spectral displacement GMPEs. The discussions may also be important to understand the limitations of high-pass filtered records for their implementation in the nonlinear response history analysis of long-period structural systems.

7.2 Strong Motion Data

The ground-motion data used comprises analog and digital records compiled from the Turkish strong-motion database.[1] A total of 528 records with source-to-site distances less than 200 km were subdivided into different magnitude bins (Fig. 7.2) to quantify the role of this seismological parameter on the uncertainty of SD_{IE} induced by the variations in high-pass filter cut-offs. The entire dataset is composed of NEHRP C and D type records with V_{S30} ranging between 360 m/s < V_{S30} < 760 m/s and 180 m/s < V_{S30} < 360 m/s, respectively. Thus, observations about the variations in SD_{IE} and SD_R stemming from the uncertainty imposed by T_c are constrained for site class contribution. This limitation could be important, since the soil profile can seriously modify the frequency content of ground motions. The records are mainly from strike-slip and normal faults. The effects of faulting style and source-to-site distance on the uncertainty of SD_{IE} and SD_R due to high-pass filtering are not considered.

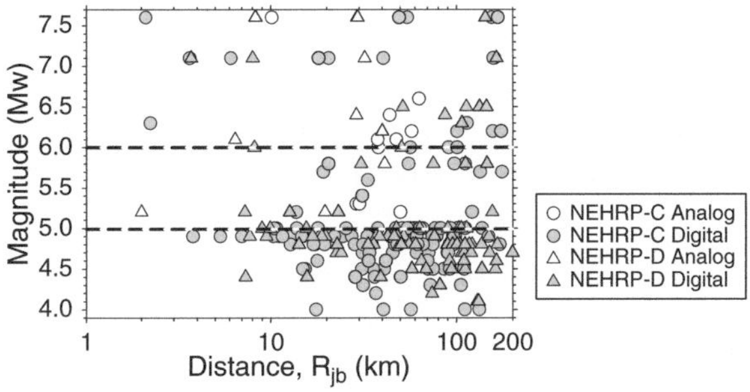

Fig. 7.2 Distribution of ground-motion data in terms of moment magnitude (M_w) and distance (R_{jb}, closest distance between the station and the horizontal projection of the ruptured fault surface). Different scatter symbols denote analog and digital accelerograms recorded at NEHRP C and D type site class. Magnitude bins are $4 \leq M_w \leq 5$ (digital), $5 < M_w \leq 6$ (analog and digital) and $M_w > 6$ (analog and digital). The scarce data at short distances and $M_w > 6$ resulted in uneven M_w bins in terms of analog and digital recordings

7.3 Uncertainty in Nonlinear SDoF Deformation Demands Caused by High-Pass Filtering

In this section, the methodology for generating the random high-pass filter cut-offs is first described. Next observations are presented about the contributions of magnitude, inelasticity level, recording quality and μ vs. R difference to the uncertainties

[1] The database compilation is carried out under the framework of the project entitled "Compilation of Turkish strong-motion network according to the international standards." This collaborative project is supported by the Scientific and Technical Research Council of Turkey.

in SD_{IE} and SD_R that are due to the randomness of T_c. The post-yield stiffness ratio, α, is set as 3% in all nonlinear spectral calculations.

The variations in high-pass filtering (to echo the views of different researchers) are accounted for by the random generation of T_c values. An optimum high-pass filter cut-off ($f_{c,opt}$) was determined for every record in the dataset through a methodology that considers magnitude-dependent frequency domain features of ground motions [2]. In brief, the procedure requires the high-pass filtered records to decay proportional to f^2 gradient at the low frequencies (long periods), because this is justifiable in terms of theoretical source spectrum behavior and it advocates minimum interference of high-pass filtering to long-period ground-motion components. The random generation of high-pass filters was achieved by considering $f_{c,opt}$ and the magnitude dependent theoretical corner frequency (f_a) of Atkinson and Silva [4] source spectrum that controls the size of the finite fault during the rupture process. For each magnitude cluster, (a) the minimum and maximum $f_{c,opt}/f_a$ ratios are determined, and (b) the corresponding minimum and maximum high-pass filter cut-offs ($f_{c,min}$ and $f_{c,max}$) are computed by multiplying the minimum and maximum $f_{c,opt}/f_a$ with the f_a belonging to each record in the considered bin.[2] High-pass filter cut-offs are termed as "severe" and "relaxed" ($f_{c,svr}$ and $f_{c,rlx}$, respectively) when the randomly generated filter values fall within the f_a–$f_{c,max}$ and f_a–$f_{c,min}$ intervals, respectively. Severe filtering represents the conservative approach in data processing for long-period noise. Relaxed filtering simulates more tolerable analyses in

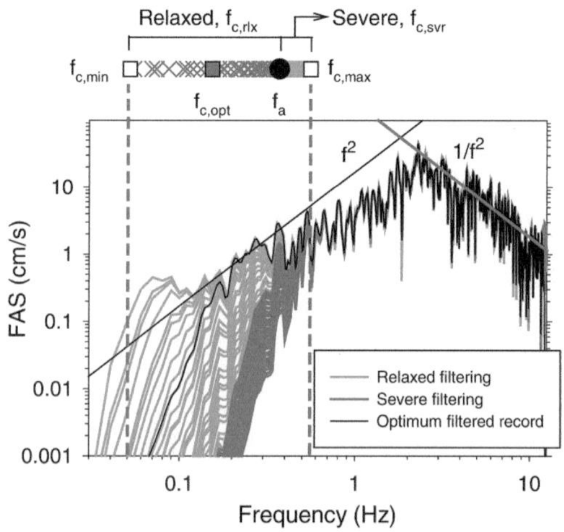

Fig. 7.3 Determination of random "relaxed" and "severe" high-pass filter cut-offs. Relaxed filtering values tend towards the lower-end of frequency domain, thus removing relatively lesser frequency components with respect to severe filtering. The long-period (low-frequency) components of almost all of the filtered acceleration time series decay faster than the f^2 gradient, suggesting the consistency of the present random generation procedure with the theoretical behavior

[2]Frequency (f) and period (T) terms are used in an interchangeable manner in this narrative. For example, the reciprocal of T_c is f_c and vice versa.

high-pass filter cut-off choices. The random generation of filter cut-offs follows uniform distribution with Latin Hypercube sampling method. The number of T_c values generated for the relaxed and severe filtering is 30 and 40, respectively. The unequal number of filter-cut offs is due to the uneven bandwidths of f_a–$f_{c,max}$ and f_a–$f_{c,min}$ intervals: the bandwidth of severe filtering interval is larger with respect to the relaxed filtering and it requires more filter cut-off values to cover the entire f_a–$f_{c,max}$ range. An example for the generation of high-pass filter cut-offs for an arbitrary record in our database is given in Fig. 7.3. The procedure is illustrated using Fourier acceleration spectrum (FAS) for a better visualization of the discussion above.

Figure 7.4 shows the scatter plots of some selected cases to emphasize the significance of magnitude, inelasticity level, μ and R difference as well as analog vs.

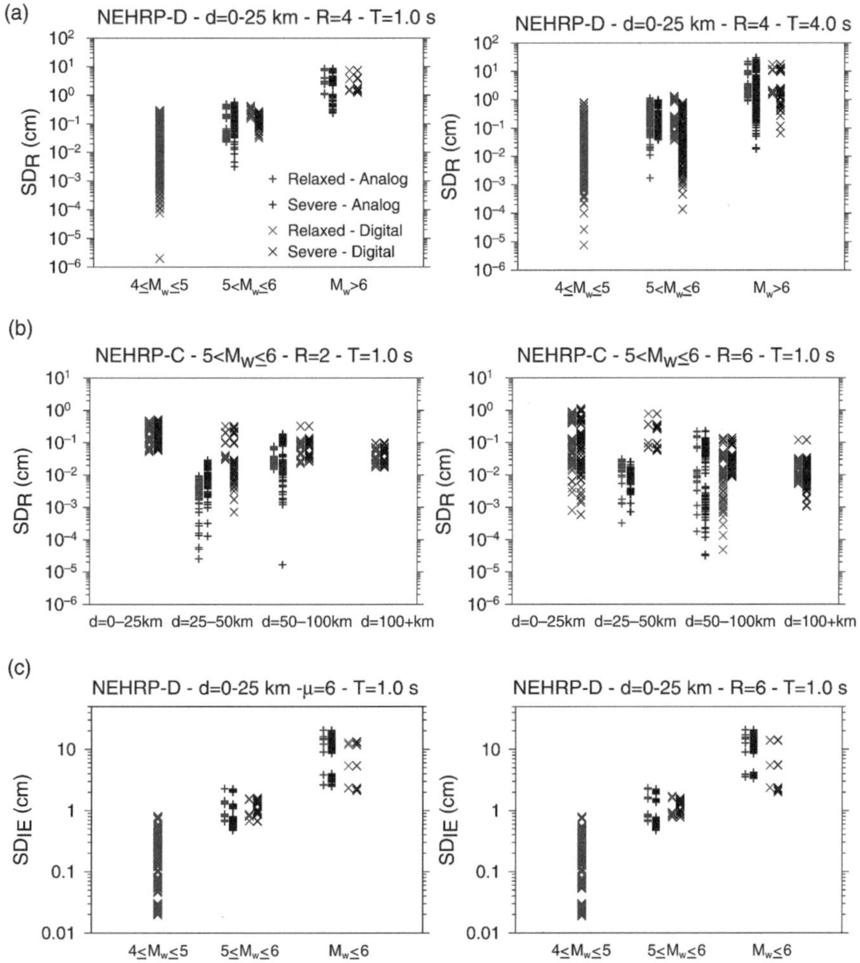

Fig. 7.4 Selected cases from different site class and distance intervals to emphasize the influence of (**a**) magnitude, (**b**) inelasticity level and (**c**) μ vs. R difference on the dispersive behavior of nonlinear oscillator demands due to high-pass filter cut-offs

digital record quality on the reliability of nonlinear oscillator deformations that is affected by the variations in high-pass filter values. The scatter in Fig. 7.4a indicates that the dispersion in nonlinear deformation demands is larger for small magnitude events. The increase in vibration period (left panel vs. right panel) also increases the sensitivity of nonlinear deformation demands to the variations in T_c. The significance of inelasticity level on nonlinear deformation uncertainty is emphasized in Fig. 7.4b. As the level of inelasticity increases from $R = 2$ to $R = 6$, the dispersion in SD_R increases for the source-to-site distance range considered here. In the light of nonlinear response of oscillators, this observation is expected as higher inelasticity level would result in longer period shifts causing the prominence of high-pass filter cut-offs. (Note that the data in Fig. 7.4b are subdivided into different distance bins for a better illustration of the observations). As depicted in Fig. 7.4c constant ductility (left panel) and constant strength (right panel) spectra would be equally influenced from variations in high-pass filter cut-offs. Note that the recording quality (analog vs. digital) is important for reducing the uncertainty in nonlinear deformations due to high-pass filtering. The dispersion pertaining to analog records is higher with respect to those of digital accelerograms for all cases presented in Fig. 7.4.

7.4 Spectral Period Ranges for the Minimum Influence of High-Pass Filtering

In order to quantify the level of reliability in SD_{IE}, period-dependent probability distributions are used for spectral points resulting from the pre-determined T_c values of each record in the database. Given a constant ductility or normalized strength, let $\mu_{SD_{IE}}$ be the average spectrum of all nonlinear spectral curves of an accelerogram due to the implementation of pre-determined high-pass filter cut-offs (Fig. 7.5). For a given vibration period, the probability of spectral points falling into an interval $\mu_{SD_{IE}} \mp \chi\mu_{SD_{IE}}$ is calculated, with the fractional factor χ taking values less than 1. Note that when the calculated probability attains a high value, it is an indication of almost all spectral points falling within the interval defined by $\mu_{SD_{IE}} \mp \chi\mu_{SD_{IE}}$. This concept is presented on the left panel in Fig. 7.6. When this methodology is repeated for a large range of vibration periods, one can obtain the probability curves for each $\mu_{SD_{IE}} \mp \chi\mu_{SD_{IE}}$ interval. A typical set of such probability curves is displayed on the right panel of Fig. 7.6 for the case presented in Fig. 7.5. The fraction χ is chosen as 0.05, 0.1, 0.15 and 0.2 for all vibration periods and these constants are also kept the same for the entire probability calculations in this chapter.

When the dispersion at $T = 0.6$ s and $T = 3.0$ s (Fig. 7.5) is compared with the probability curves in Fig. 7.6, one realizes that higher dispersion in $T = 3.0$ s is associated with the decaying portion of probabilities at any χ level. Such decay in the probability curves would suggest low levels of reliability for SD_{IE}. Thus for this particular case one may speculate that high-pass filtering interferes with the nonlinear oscillator response for $T > 2.5$ s (where the decay in the probability curves is steep), and use of SD_{IE} at vibration periods longer than $T = 2.5$ s might result

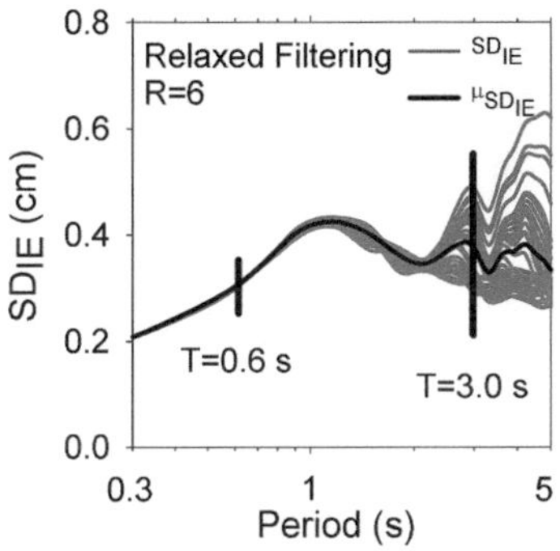

Fig. 7.5 Nonlinear constant strength ($R = 6$) displacement spectra of the record whose high-pass filter cut-offs are determined through the methodology in Fig. 7.3. The SD_{IE} curves are due to the implementation of relaxed filter cut-offs to the concerned accelerogram. The *black solid line* is the mean variation of all SD_{IE} curves ($\mu_{SD_{IE}}$). As discussed in Fig. 7.4a, the dispersion in SD_{IE} (described by the scatter about $\mu_{SD_{IE}}$) increases with increasing vibration period. (Compare the divergence of spectral curves about $\mu_{SD_{IE}}$ at $T = 0.6$ s and $T = 3.0$ s). This behavior suggests a decrease in the reliability of SD_{IE} after a certain vibration period. The reason for the decreased reliability (increased uncertainty) in SD_{IE} towards longer vibration periods is the pronounced interaction between the filter cut-offs and nonlinear oscillator response that is magnified further with different high-pass filter cut-offs. Though not shown here for space limitations, the dispersive behavior of SD_{IE} is more prominent in the severe high-pass filtering case since those filter cut-offs would remove relatively larger amounts of long-period components to emphasize the filter cut-off influence more on the nonlinear oscillator response

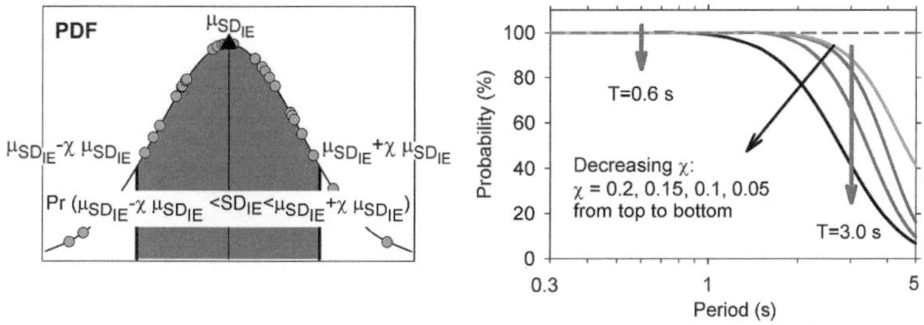

Fig. 7.6 *Left panel*: Illustration of the probability distribution of SD_{IE} values about $\mu_{SD_{IE}}$ for a given vibration period. The *gray shaded region* is the probability of spectral points falling into an interval $\mu_{SD_{IE}} \mp \chi\mu_{SD_{IE}}$. *Right panel*: Probability curves of different χ for the case presented in Fig. 7.5. They are computed by applying the presented probability concept (*left panel*) for a large range of periods

in erroneous conclusions on the nonlinear deformation demands. Use of larger χ results in a more gradual decay in the probability curves (compare $\chi = 0.2$ and $\chi = 0.05$ curves). However, their gradients are approximately the same revealing a similar assessment about the range of spectral periods where high-pass filtering starts dominating the nonlinear oscillator response.

Table 7.1 Proposed vibration periods for constant ductility and strength SD_{IE} for analog and digital records when relaxed and severe filtering criteria are applied

	Relaxed filtering					Severe filtering			
	Analog		Digital			Analog		Digital	
	$5<M_w\leq6$	$M_w>6$	$4\leq M_w\leq5$	$5<M_w\leq6$	$M_w>6$	$5<M_w\leq6$	$M_w>6$	$5<M_w\leq6$	$M_w>6$
Elastic	2.80	15.61	1.36	5.52	21.06	1.65	3.63	2.55	6.02
$\mu=2$	2.07	12.56	1.14	5.45	16.89	1.41	2.83	2.31	4.14
$\mu=4$	1.88	12.11	0.92	4.65	15.06	1.16	2.01	1.88	2.90
$\mu=6$	1.45	10.67	0.76	3.96	13.32	1.13	1.51	1.92	2.63
$R=2$	2.08	13.29	1.17	4.98	17.76	1.33	2.55	2.27	4.13
$R=4$	2.13	12.21	0.97	4.35	14.70	1.31	1.69	1.81	2.54
$R=6$	1.90	10.75	0.84	3.98	13.58	1.14	1.42	1.62	2.16

Table 7.1 lists the spectral periods determined from the mean probability curves that are computed by the application of above concept to each magnitude cluster considered in this study. The periods are determined for analog and digital records per discussions in Fig. 7.4. Confined to the rationale in our methodology the reliability of SD_{IE} will not be affected by the chosen high-pass filter cut-off within the spectral bands bounded by these periods. The spectral period ranges are based on a probability of $\Pr(\mu_{SD_{IE}} - \chi\mu_{SD_{IE}} < SD_{IE} < \mu_{SD_{IE}} + \chi\mu_{SD_{IE}}) = 80\%$. The choice of this probability level was an arbitrary decision. However, for most cases, the decay in the probability curves becomes steep in the vicinity of this probability level giving an indication for the significant interference between high-pass filtering and nonlinear oscillator response. Consequence of this argument was to choose this probability level in the present methodology. A usable spectral period range for small magnitude ($4 \leq M_w \leq 5$) analog records cannot be recommended, due to the insufficient data to derive meaningful statistics.

To verify the consistency of the proposed methodology, usable period ranges for elastic oscillator response are also derived and compared with the results of similar studies in the literature. The recommended period values for elastic response are given in Table 7.1. The corresponding comparisons are presented in Fig. 7.7. The scatter in this figure suggests a fairly good comparison between the present recommendations and the ones suggested by previous studies in particular for relaxed filtering criterion. This observation may be an indication of the stability of our proposed methodology. Note the discrepancy between the usable spectral periods of elastic and nonlinear response in Table 7.1 that emphasizes the importance of high-pass filter cut-off influence on nonlinear SDoF deformations.

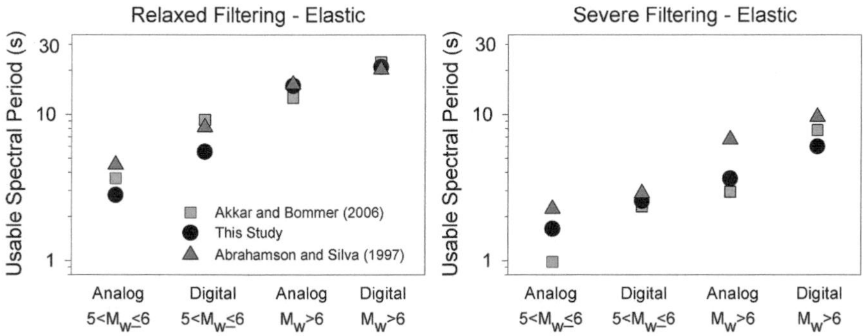

Fig. 7.7 Usable spectral periods of this study for elastic response and their comparisons with other recommendations in the literature

7.5 Summary and Conclusions

The uncertainty in the nonlinear oscillator deformations due to existing long-period noise in the accelerograms has been explored. The present analyses reveal that the variations in high-pass filter cut-offs that are used for removing the long-period noise introduce significant dispersion to the nonlinear deformation demands. The level of uncertainty depends primarily on the variations in magnitude, level of inelasticity and recording quality (i.e. analog vs. digital records). Small magnitude events and higher levels of inelasticity (large μ and R values) cause adverse effects of high-pass filtering on nonlinear deformations. The concerned dispersion in analog records supersedes the corresponding scatter of digital records regardless of magnitude and level of inelasticity. These observations strongly suggest the necessity of usable period range definitions for reliable nonlinear spectral calculations that are presented in Table 7.1 in terms of different μ and R levels as well as for different magnitude intervals. We believe that the use of these usable ranges may result in more accurate ground-motion models for the estimation of nonlinear response.

Acknowledgments The study has been supported by the project entitled "Compilation of Turkish strong-motion network according to the international standards" with Grant No. 105G016. The authors express their sincere gratitude for this support provided by the Scientific and Technological Research Council of Turkey (TÜBİTAK). The collaborating national agency was the Earthquake Department of the Disaster and Emergency Management State Agency.

References

1. Abrahamson NA, Silva WJ (1997) Empirical response spectral attenuation relations for shallow crustal earthquakes. Seismol Res Lett 68:94–127
2. Akkar S, Bommer JJ (2006) Influence of long-period filter cut-off on elastic spectral displacements. Earthq Eng Struct Dyn 35:1145–1165
3. Akkar S, Boore DM (2009) On baseline corrections and uncertainty in response spectra for baseline variations commonly encountered in digital accelerograph records. Bull Seismol Soc Am 99:1671–1690
4. Atkinson GM, Silva W (2000) Stochastic modeling of California ground motions. Bull Seismol Soc Am 90:255–274
5. Bazzurro P, Sjoberg B, Luco N, Silva W, Darragh R et al (2004) Effects of strong motion processing procedures on time histories, elastic and inelastic spectra. Invited Workshop on Strong Motion Record Processing, Consortium of Organizations for Strong-Motion Observation Systems (COSMOS). http://www.cosmos-eq.org/Projects/Bazzurro_Sjoberg_Paper.pdf. Accessed 15 July 2009
6. Boore DM, Akkar S (2003) Effect of causal and acausal filters on elastic and inelastic response spectra. Earthq Eng Struct Dyn 32:1729–1748
7. Boore DM, Bommer JJ (2005) Processing of strong-motion accelerograms: needs, options and consequences. Soil Dyn Earthq Eng 25:93–115
8. Kawashima K, MacRae GA, Hoshikuma J, Nagaya K et al (1998) Residual displacement response spectrum. J Struct Eng ASCE 124:523–530
9. Tothong P, Cornell CA (2006) An empirical ground-motion attenuation relation for inelastic spectral displacement. Bull Seismol Soc Am. doi:10.1785/0120060018

Chapter 8
Are Current Design Spectra Sufficient for Soil-Structure Systems on Soft Soils?

Aikaterini Ziotopoulou and George Gazetas

8.1 The Problem: Code Spectra Versus Reality

It is well known how important soil effects are on the intensity and frequency content of ground motions [1, 17–19, 22, 26, 28, 33, 34]. In civil engineering practice these effects are often computed theoretically (wave propagation analysis assuming equivalent-linear or nonlinear soil behavior). Yet, seismic codes have universally faced the problem of soil amplification in a purely empirical and (unavoidably) oversimplified way:

- The soil deposits are classified in a few broad categories, each of which encompasses a wide range of soil layer stiffness and thickness down to bedrock.
- The response spectra $S_a(T)$ from numerous world wide accelerograms recorded on top of soils belonging to each category, were statistically processed. The shape of the design spectrum for the particular soil category was based on the average of the normalized spectrum, $S_a(T)/A$, for each period T, after some "conservative smoothing" [27].

The design spectra that have thus resulted share a crucial characteristic: the more "flexible" a soil deposit (i.e. the smaller its stiffness and/or the larger its thickness), the flatter the design spectrum. If this were the reality, ignoring SSI for a structure on soft ground would have led to conservative results: SSI effect would always be helpful [2, 16].

Yet, reality has repeatedly shown the opposite trend. Numerous records on "soft" soils have produced response spectra of a sharp rather than flat shape, with well defined peaks around the site fundamental period. Figure 8.1 highlights the discrepancy between seismic codes and reality. The consequences of such a disparity, especially on SSI systems may be significantly detrimental.

A. Ziotopoulou (✉)
Soil Mechanics Laboratory, Department of Civil Engineering, National Technical University of Athens, Athens, Greece; University of California at Davis, Davis, CA, USA
e-mail: kziotopoulou@ucdavis.edu

M.N. Fardis (ed.), *Advances in Performance-Based Earthquake Engineering*, Geotechnical, Geological, and Earthquake Engineering 13, DOI 10.1007/978-90-481-8746-1_8,

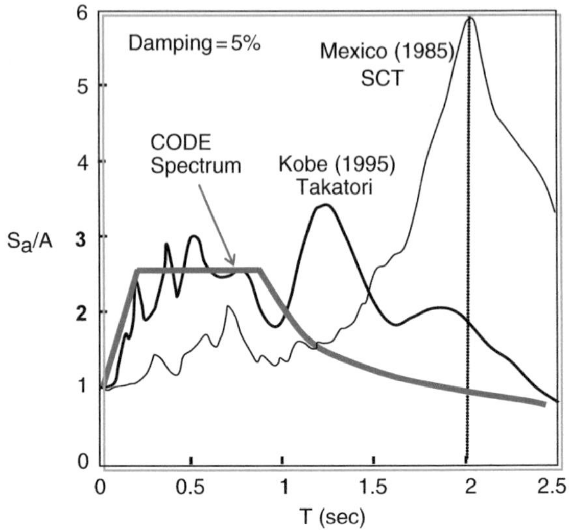

Fig. 8.1 The discrepancy between a code design spectrum typical for soft soils and the response spectra of two actual soil amplified motions

8.2 Why This Discrepancy?

As illustrated in the sketch of Fig. 8.2, the culprit behind the discrepancy is the averaging of dissimilar response spectra; its accomplish: the very broad range of stiffness and thickness of each soil category. A range of natural periods in the ratio of 1–4 is quite possible within one single category, say category D (according to NEHRP) [21, 29]. The actual seismic motions in a number of (soft) soil profiles belonging to category D but with so vastly different fundamental periods are likely to have response spectra with sharp peaks at well-separated periods [4, 14]. Thus, at the period for which one spectrum has a peak the spectra on sites with different periods are likely to have very small values. Hence, by averaging all these different

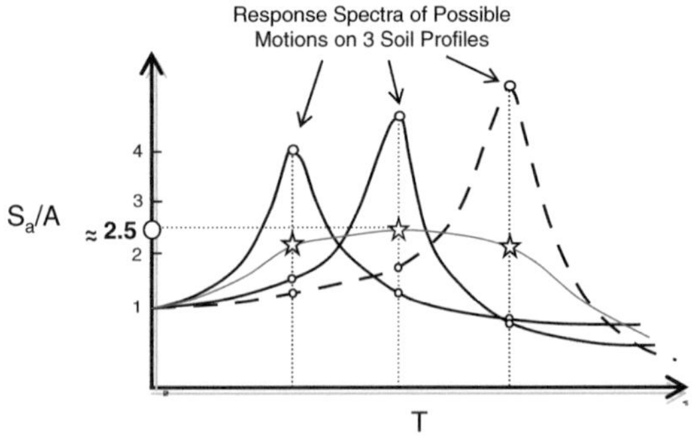

Fig. 8.2 Sketch illustrating the derivation of code spectra from the average (for each specific period) of the S_a/A values of all recorded spectra. The three individual idealized spectra are from possible motions in three soft soil profiles, all belonging to the same soil class (category), and all bearing the effects of resonance but at different periods. The resulting spectrum spuriously suppresses the soil-excitation resonance

values we simply "annihilate" the real sharp peaks. In other words, spuriously and against safety, we disregard (or rather depress) the resonance between soil deposit and excitation [30–32].

The topic has already been brought to light by Mylonakis and Gazetas [20] and Gazetas [10], in an attempt to reevaluate the importance of soil-structure interaction (SSI). They showed that the effects of SSI have been incorrectly predicted on the basis of the Code Spectra as being always beneficial ; and recalled many failures in Mexico (1985) and Kobe (2005) that have persuasively been shown to be to a large extent the (detrimental) effect of SSI [11]. More recently, Xu and Xie [35] along similar lines developed a unique average bi-normalized spectrum for 206 strong-motion records of the Chi-Chi (1999) earthquake. Each and every individual acceleration response spectrum was doubly normalized : the ordinate, S_a, with respect to the peak ground acceleration, A; the abscissa, T, with respect to the predominant period T_p of the spectrum. The average of the individual "$S_a/A : T/T_p$" spectra exhibited indeed a sharp peak, at $T/T_p = 1$, with a maximum value of the order of 4, rather than the 2.5 of the code spectra. The practical indirect conclusion from the above studies was that the increase of the period of a structure-soil system with decreasing soil stiffness would not necessarily lead to reduced intensity of shaking, as presently implied by the code spectra.

8.3 Summary of the Analytical ("Remedial") Study

In contrast with the purely empirical method with which the Code Spectra have been developed, an analytical methodology is followed, comprising the following steps:

- For a particular soil category (for example C according to EC8, or D according to NEHRP [3, 5, 6, 8, 15, 23, 24]) we "construct" a number of idealized generic soil profiles having the following characteristic parameters:
  - velocity : $V_{S,30} = 180$ m/s, 260 m/s, 360 m/s. $V_{S,30}$: average shear wave velocity from the ground surface down to a depth of 30 m
  - distribution of V_s with depth : uniform, trapezoidal, with-crust (see Fig. 8.3)
  - depth to "rock" : H = 30 m and 60 m.
  - "rock" to soil wave velocity ratio : $V_{S,\,ROCK} \,/\, V_{S,30} = 1.5$ and 5
- Seven accelerograms recorded on "rock" are utilized as ("rock-outcrop") excitation after being scaled (up or down) to achieve peak ground acceleration : $A = 0.20g, 0.40g, 0.60g$. These records are (Fig. 8.4):

  Stone Canyon Reservoir, Northridge 1994
  Aegion-Rock, Aegion 1995
  Sakarya, Izmit 1999
  Dayhook, Tabas 1978
  Gilroy-1, Loma Prieta 1989
  Lucerne, Landers 1992
  Superstition Mountain, Imperial Valley 1979.

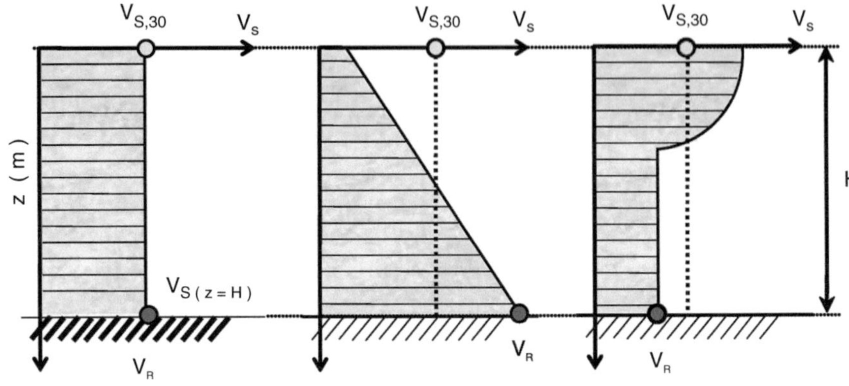

Fig. 8.3 The three types of generic soil profiles used in our parametric investigation

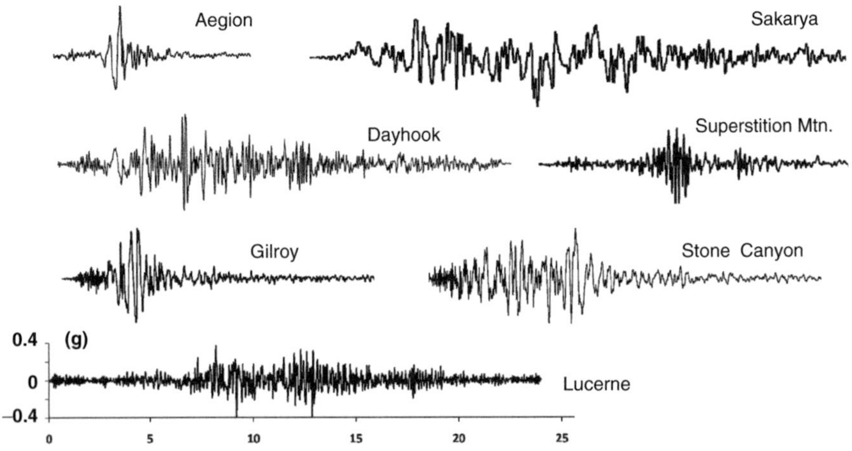

Fig. 8.4 "Rock" accelerograms used as excitation (scaled to 0.40*g*)

- By exciting all the aforesaid soil profiles with each record in all possible combinations we obtain results in 1009 cases. The analysis is first done, with the well-known equivalent-linear method of Schnabel et al. [25] (SHAKE) and, second, with the inelastic method introduced by Gerolymos and Gazetas [13] (NL-DYAS).

The response spectra of the ground surface motions resulting from each of the 2×1009 analyses are utilized in two different ways:

(a) Only the spectral accelerations is normalized, by dividing with the corresponding peak ground acceleration, S_a/A – the established conventional normalization used for deriving the current design spectra $(S_a/A : T)$.

(b) Both the spectral acceleration, S_a/A, and the period, T, are normalized, the latter by the predominant period T_p of the ground surface motion. The plot $S_a/A : T/T_p$ is called Bi-Normalized Spectrum (BNS).

- The average for each period T of the 1009 simply normalized spectral values (type (a)) give a mean response spectrum $(S_a/A : T)$ which is expected to be quite similar with the current code spectrum for this soil category.
- The average for each period ratio T/T_p of the 1009 doubly normalized spectra (type (b)) give a mean response spectrum $(S_a/A : T/T_p)$ which is expected to differ both in shape and in amplitude from the conventional spectrum.

8.4 Results: Towards a More Rational Spectrum

All the 1009 response spectra obtained with the equivalent-linear soil response analyses and simply or doubly normalized as afore-explained, are portrayed in Fig. 8.5a

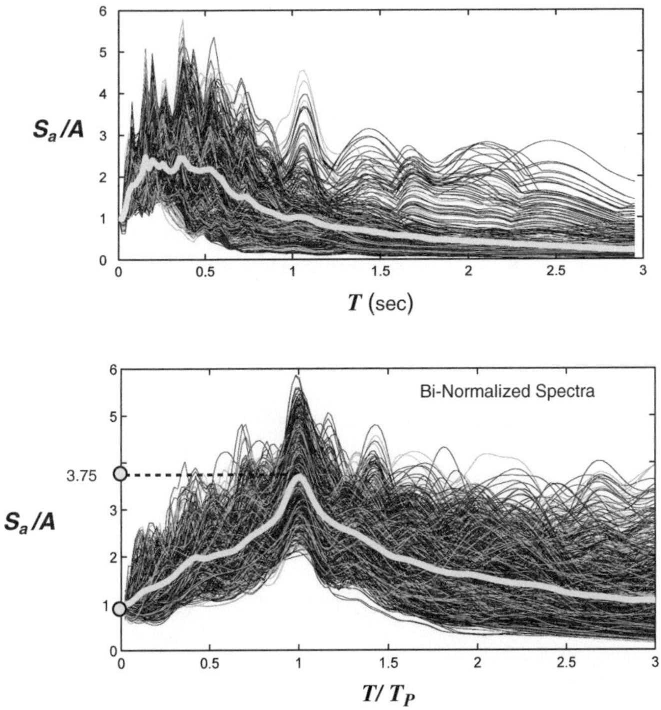

Fig. 8.5 Compilation of response spectra of ground surface motions from all the equivalent-linear analyses. (**a**) Conventionally normalized spectra; (**b**) bi-normalized spectra. The *thick curves* are the mean response spectra

and b respectively. Their average response spectra, after some "conservative smoothing" could serve as the design spectra. The following conclusions emerge from the two figures:

(a) Regarding the conventionally derived spectrum – as anticipated, its shape is indeed quite similar with the smooth shape of the code spectrum for this soil category : a nearly constant ordinate, approaching (from below) $S_a/A \approx 2.5$, for the range of periods from 0.15 to 0.60 s, approximately. (If more excitations had been employed, and additional and more realistic soil profiles had been considered, the period range of nearly constant S_a would have likely increased, and the spectrum would have been even smoother.)
(b) Regarding the Bi-Normalized Spectrum – its shape is vastly different from the conventional spectrum : a sharp peak at $T/T_p \approx 1$ dominates. Its maximum value, max (S_a/A), reaches 3.75, i.e. it is 50% greater than the peak value of the conventional spectrum.

Evidently, the (true or pseudo) resonance between soil and excitation is well preserved only in the bi-normalized spectrum. The conventional Spectrum does not reflect the physics of the problem, while being unsafe for many structures (with $T \approx T_p$) and leading to erroneous conclusions on the possible effects of soil-structure interaction.

8.5 The Uniqueness of the Bi-Normalized Spectrum

Several interesting attributes of the Bi-Normalized Spectrum (BNS) have been demonstrated analytically by Ziotopoulou and Gazetas [36]. Specifically:

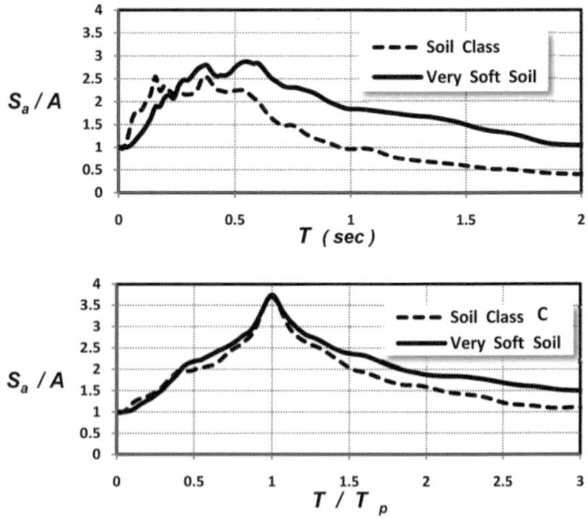

Fig. 8.6 Comparison of the mean spectra of Fig. 8.5 (Soil Class: $V_{S,30} = 180$–360 m/s) with the mean spectra for a much softer soil ($V_{S,30} = 100$ m/s). The differences of the conventionally normalized spectrum (*top*) almost disappear in the normalized spectra (*bottom*)

- The BNS is hardly influenced by soil category, i.e., it is practically the same for all soil categories! The same conclusion was drawn by Xu and Xie [35] for the strong records of the Chi-Chi (1999) earthquake. (Of course, T_p may change significantly from soil to soil, decreasing with soil stiffness ; and moreover, it is often affected by the nature of seismic excitation. Its estimation is a totally different ball game.)
- The BNS is only marginally influenced by the nature of the performed wave propagation analysis: equivalent-linear and truly nonlinear analyses differ appreciably only in the low-period range ($T/T_p < 0.5$), not in the basic shape of the spectrum.
- The BNS is only marginally influenced by the nature of seismic excitation. (Of course, again, the above argument does not extend to T_p which is affected by the dominant excitation periods.)

Indicative of the uniqueness of this BNS is Fig. 8.6, which reveals the practical independence of BNS from the soil category – contrary to the behavior of the conventional spectrum.

8.6 Conclusion, Limitations

One unique Bi-Normalized Spectrum (BNS), for all soil categories and most likely seismic excitations, emerged from the comprehensive set of wave-propagation analyses reported in this article. This unique spectrum is sketched in Fig. 8.7 and is approximated with the following expressions:

$$S_a/A = \exp\left(1.35\left[T/T_p\right]\right) \quad \text{for } T/T_p < 1 \qquad (8.1a)$$

$$S_a/A = 3.75\left(T/T_p\right)^{-1.2} \quad \text{for } T/T_p \geq 1 \qquad (8.1b)$$

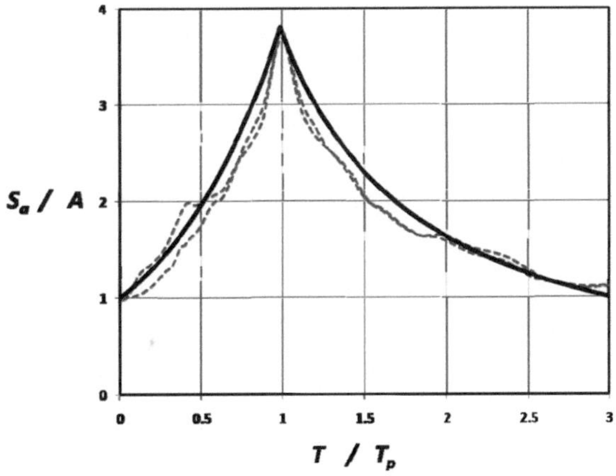

Fig. 8.7 Mean bi-normalized spectra (BNS) from the equivalent linear wave propagation (SHAKE) and from the inelastic wave propagation (NL-DYAS) studies, and idealized smooth spectrum proposed for design, Eq. (8.1)

The potential benefits from adopting this simple spectrum have been highlighted in the chapter. However, the imprecise definition of T_p and the profound difficulty in predicting T_p in reality remain serious obstacles in adopting it at present. And of course, empirical support from recorded motions must be (statistically) significant, to arrive at a robust such design spectrum.

Acknowledgments This work forms part of an EU 7th Framework research project funded through the European Research Council (ERC) Programme "Ideas", Support for Frontier Research – Advanced Grant, under Contract number ERC-2008-AdG 228254-DARE.

References

1. Afra H, Pecker A (2001) Calculation of free field response spectrum of a non-homogeneous soil deposit from bed rock response spectrum. Soil Dyn Earthq Eng 22:157–165
2. Anastasopoulos I (2005) Fault rupture soil–foundation–structure–interaction (FR-SFSI), Doctoral Thesis, National Technical University of Athens, Greece
3. Bachman R, Bonneville D (2000) The seismic provisions of the 1997 Uniform Building Code. Earthq Spectra 16(1):85–100
4. Borcherdt RD (1994) Estimates of site-dependent response spectra for design (methodology and justification). Earthq Spectra 10(4):617–653
5. Building Seismic Safety Council (BSSC) The 2003 NEHRP recommended provisions for new buildings and other structures. Part 1: provisions, Federal Emergency Agency (FEMA 450), Washington DC
6. Dobry R, Borcherdt RD, Crouse CB, Idriss IM, Joyner WB, Martin GR, Power MS, Rinne EE, Seed RB (2000) New site coefficients and site classification system used in recent building seismic code provisions. Earthq Spectra 16(1): 41–67
7. Drosos V, Gerolymos N, Gazetas G (2007) Calibration and verification of nonlinear wave propagation method. In: Pitilakis K (ed) Proceedings of the 4th International Conference on Earthquake Geotechnical Engineering, Paper No. 1594, Thessaloniki, Greece
8. EC-8 (2001), Part 1 – general Rules, seismic actions and rules for buildings, Eurocode 8, Draft 4 (prEN 1998), European Committee for Standardization
9. Gazetas G (1982) Vibrational characteristics of soil deposits with variable wave velocity. Num Anal Meth Geomechanics 6(1):1–20
10. Gazetas G (2006) Seismic design of foundations and soil-structure interaction, 1st European Conference on Earthquake Engineering and Seismology, Geneva, Switzerland, 3–8 September.
11. Gazetas G, Anastasopoulos I, Gerolymos N, Mylonakis G, Syngros L (2004) The collapse of the Hansin Expressway Fukae Bridge, Kobe 1995: soil-foundation-structure interaction, reconstruction, seismic isolation, Entwicklungen in der Bodenmechanik, Bodendynamic und Geotechnik, Festschrift zum 60. Geburtstag von Univ.-Prof. Dr.-Ing. Habil. Stavros Savidis, Berlin, pp. 93–120
12. Gazetas G, Bianchini G (1979) Field evaluation of body and surface-wave soil-amplification theories, Proceedings of the 2nd US National Conference on Earthquake Engineering, EERI, pp. 603–612, Stanford University
13. Gerolymos N, Gazetas G (2005) Constitutive model for 1-D cyclic soil behavior applied to seismic analysis of layered deposits. Soils Found 45:147–159
14. Hartzell S, Bonilla LF, Williams RA (2004), Prediction of nonlinear soil effects. Bull Seism Soc Am 94:1609–1629
15. Holmes WT (2000) The 1997 NEHRP Provisions for seismic regulations for new buildings and other structures. Earthq Spectra 16(1):101–114
16. Housner GW (1959) Behavior of structures during earthquakes. Journal of Applied Mechanics. ASCE, Proceedings Paper 2220, 85, EM4, pp. 109–129, October, 1959

17. Idriss IM (1990), Response of soft soils during earthquakes. In: Michael Duncan J (ed) Proceedings of H. Bolton seed memorial symposium, Vol. 2, Bitech, Vancouver
18. Kramer SL (1996) Geotechnical earthquake engineering. Prentice–Hall, Upper Saddle River, NJ
19. Li S, Xie L (2007) Effect of hanging wall and forward directivity in 1999 Chi-Chi earthquake on inelastic displacement response of structures. Earthq Eng Eng Vib 6(1):77–84
20. Mylonakis G, Gazetas G (2000) Seismic soil-structure interaction: beneficial or detrimental? J Earthq Eng, 4 (3):277–301
21. NEHRP (2000) Recommended provisions for seismic regulations for new buildings and other structures: Part 1, provisions, FEMA 368, Building Seismic Safety Council (BSSC), Washington, DC
22. Pitilakis K (2004) Site effects. In: Ansal A (ed) Recent advances in earthquake geotechnical engineering and microzonation. Kluwer Academic Publishers, Dordrecht, pp 139–193
23. Pitilakis K, Gazepis C, Anastasiadis A (2006) Design response spectra and soil classification for seismic code provisions. In: Bouckovalas G (ed) Geotechnical Evaluation and Application of EC8, pp 37–52, Athens, National Technical University of Athens Publications.
24. Rodriguez-Marek A, Bray JD, Abrahamson NA (1999) Task 3: characterization of site response general site categories. Pacific Earthquake Engineering Research Center (PEER) Report 1999/2003
25. Schnabel PB, Lysmer J, Seed HB (1972) SHAKE: a computer program for earthquake response analysis of horizontally layered sites. Rep. EERC 72-12, University of California, Berkeley, CA
26. Schnabel PB, Seed HB, Lysmer J (1972) Modification of seismological records for effects of local soil conditions. Bull Seismol Soc Am 62:1649–1664
27. Seed HB, Ugas C, Lysmer J (1976) Site-dependent spectra for earthquake-resistant design. Bull Seismol Soc Am 66:221–243
28. Somerville P (1998) Emerging art: earthquake ground motion, geotechnical special publication no. 75. In: Dakoulas P, Yegian M, Holtz RD (eds) Geotechnical Engineering & Soil Dynamics III, ASCE, Reston, VA, USA
29. Stewart JP, Liu AH, Choi Y, Baturai MB (2001) Amplification factors for spectral acceleration in active regions. Pacific Earthquake Engineering Research Center (PEER) Report 2001/2010
30. Tazoh T, Sato M, Shimizu K, Koyama K (1988) Nonlinear seismic response analysis of horizontally layered soil deposits. Proceedings of 9th world conference on earthquake engineering, Tokyo-Kyoto, II, pp 495–501
31. Tazoh T, Shimizu K, Sato M, Hirosse T, Kouama K (1988) Identification of time-varying predominant period and damping constant in nonlinear seismic response of soft soil deposit. Proceedings of 9th world conference on earthquake engineering, Tokyo-Kyoto, II, pp 483–488
32. Towhata I (1996) Seismic wave propagation in elastic soil with continuous variation of shear modulus in the vertical direction. Soils Found JGS 36(1):61–72
33. Travasarou Th, Gazetas G (2004) On the linear seismic response of soils with modulus varying as a power of depth – the Maliakos marine clay. Soils Found JGS 44(5):85–93
34. Vucetic M, Dobry R (1991) Effect of soil plasticity on cyclic response. J Geotech Eng ASCE 117:89–107
35. Xu L, Xie L (2004) Bi-normalized response spectral characteristics of the 1999 Chi-Chi Earthquake. Earthq Eng Eng Vib 3(2):47–155
36. Ziotopoulou A, Gazetas G (2009) Non-Linear Seismic response analysis of soil deposits and piles and proposed unique bi-normalized spectrum. Research Report LSM. NTUA-09-01

Chapter 9
Elastic Demand Spectra

Kyriazis Pitilakis, Anastasios Anastasiadis, Dimitris Pitilakis, Konstantinos Trevlopoulos, and Konstantinos Senetakis

9.1 Introduction

Performance-based seismic design requires estimation and selection of elastic acceleration and displacement spectra for different damping ratios in a wide range of periods. In this study, demand spectra from a database of strong ground motion records from Japan, Greece and the United States over well defined soil conditions, are computed. The results are presented for different soil categories and two different seismic excitation levels, as it is suggested in Eurocode 8 (EC8). The discussion focuses on the differences between the design elastic demand spectra of EC8 and the computed demand spectra from a large data set of real records.

Traditionaly, the demand spectra are calculated at the foundation level for the free-field soil response, assuming linear soil behavior. However this does not correspond to reality. The soil behavior is not linear elastic and the the soil-foundation-structure interaction may modify considerably the design input motion and hence the elastic demand spectra. In this paper, an equivalent linear approximation is implemented in the substructure approach [4] of soil-foundation-structure interaction (SFSI). The combined effects of the SFSI and the nonlinear soil behavior on the spectral demand of the system are highlighted through parametric analyses in the context of performance based design.

In many real cases where important structures are founded on soft soils of low strength, it is necessary to improve soil conditions using different techniques. One of the most popular techniques is the construction of stone columns. A recently developed alternative technique replaces at a given depth the initial soil with a compacted mixture of rubber and sand material (RSM) [1]. The elastic demand spectra under these conditions may differ considerably from the demand spectra of the original soil. The effect of the RSM reinforcement of soft soil on the demand spectra of a Single-Degree of Freedom (SDOF) system on a soft cohesive soil corresponding to Soil Type C according to Eurocode 8, is depicted for the seismic input of the Athens,

K. Pitilakis (✉)
Department of Civil Engineering, Aristotle University, Thessaloniki 54124, Greece
e-mail: pitilakis@civil.auth.gr

M.N. Fardis (ed.), *Advances in Performance-Based Earthquake Engineering*, Geotechnical, Geological, and Earthquake Engineering 13, DOI 10.1007/978-90-481-8746-1_9,

1999 earthquake. Similar analysis where performed for the traditional stone column improvement.

The aim of these three parts of the work is to validate the reliability of EC8 demand spectra with a large sample of high quality worldwide records and highlight some important aspects of the selection of the design elastic demand spectra in view of performance-based design. Real structures on good or poor soil conditions are rarely subjected to the elastic demand spectra proposed by seismic codes, which correspond to an ideal free-field ground motion. SFSI, soil non-linearity and soil improvement may modify considerably the design elastic demand spectra.

9.2 Comparison of EC8 Elastic Demand Spectra with Worldwide Strong Ground Motion Records

9.2.1 Selection of Strong Ground Motion Records

A large set of high quality, mainly digital, records worldwide has been selected and processed in order to calculate separately the horizontal elastic acceleration and displacement spectra directly from the respective time histories after proper filtering. The results are presented in terms of elastic demand spectra.

A large number (about 1,000) well-documented strong ground motion recordings with different PGA values, Mw magnitudes and epicentral distances were selected for the soil classes studied in this paper (Fig. 9.1a). Selection is based on very good knowledge of the dynamic soil properties (mainly V_S values) and the of the depth to bedrock. In this paper we consider a selected sample of about 300 records. They come from Japan (164 KiK-Net records), Greece, (58 records) and from California (86 strong ground records from the COSMOS Virtual Data Center and the National Strong Motion Project Datasets). The seismic records are classified in two main categories depending on the level of seismic intensity, $M_W > 5.5$, i.e. approximately Peak Ground Acceleration (PGA) $> 0.2g$) and $M_W < 5.5$, or PGA about less than

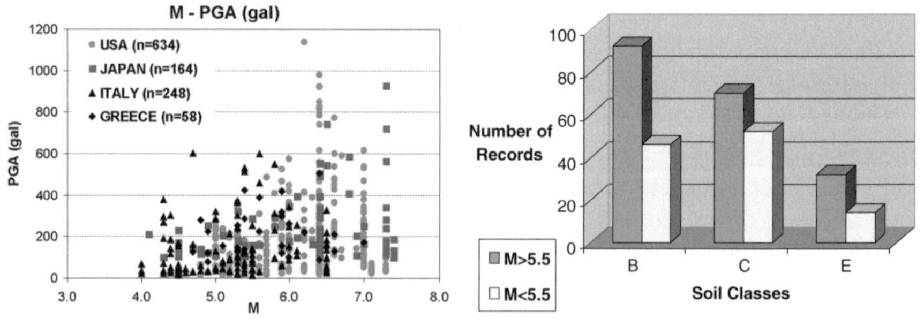

(a) (b)

Fig. 9.1 (**a**) Mw-PGA pairs for the selected strong motion records and (**b**) number of records used for soil type B, C and E according to the EC8 classification, for different seismic magnitudes M_W

0.2g. Actually this is not always true, as in some cases even for low magnitude it is possible to have higher accelerations. In Fig. 9.1b the total number of seismic records used for each Soil Type B, C and E according to EC8 are presented, classified in two main categories depending on the level of seismicity. In total, 138 records were selected for Soil Type B, 122 for Soil Type C and 48 for Soil Type E.

9.2.2 Processing of Strong Ground Motion Records

Acceleration and displacement response spectra of the records from Japan and Greece were computed separately for every seismic record [3], in order to estimate the corresponding demand spectra. The response spectra were computed for a period range up to 4 s for $M_W > 5.5$ and $M_W < 5.5$ for Soil Type B and C per EC8. The digital and analogue records were baseline corrected and filtered with a band-pass Butterworth filter.

In order to compare the demand spectra derived from the strong ground motion database with the design elastic spectra of the current seismic codes, acceleration and displacement spectral values were normalized. The normalized demand spectra were then processed in order to estimate the mean and mean plus one standard deviation (Mean + σ) curve.

9.2.3 Soil Type B

Based on the available detailed geotechnical soil profiles 138 records were classified in Soil Type B. 92 records are classified in the high seismicity ($M > 5.5$) and 46 in the low seismicity range. Among the 92 records there are 56 records from Japan, 18 from the Greece and 18 from California. The PGA values vary between 62.7 and 924.5 gal, a range that includes certain very strong earthquake recordings. The absolute S_d values vary from 10 to 200 mm, while absolute peak PSA values vary from 2 to 32 m/s^2 (Fig. 9.2a). The normalized demand spectra for Soil Type B are presented

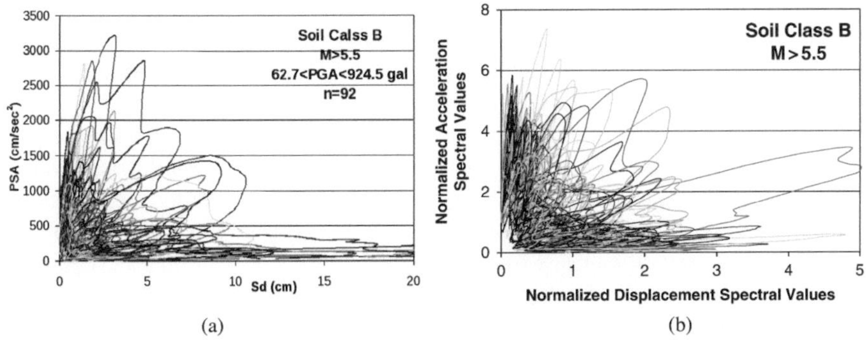

Fig. 9.2 Soil type B: (**a**) Absolute demand spectra ($M_W > 5.5$, $n = 92$ records); and (**b**) normalized demand spectra

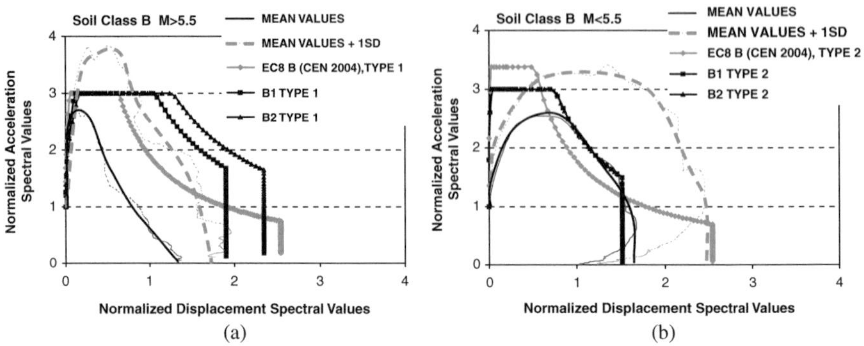

Fig. 9.3 Soil type B: Mean + σ values of normalized elastic spectra for (**a**) $M > 5.5$ ($n = 92$ records); and (**b**) $M < 5.5$, compared with the design elastic spectra of EC8 and Pitilakis et al. [5]

in Fig. 9.2b. Figure 9.3 presents the mean and mean plus one standard deviation demand spectra (smoothing lines as well) for the two intensity levels, together with the corresponding demand spectra according to EC8 and Pitilakis et al. [5].

Given the scatter of the values, for the set of strong earthquakes ($M > 5.5$) the design elastic demand spectra of EC8 and the proposed demand spectra by Pitilakis et al. [5] are better correlated with the mean plus one standard deviation spectra (Fig. 9.3a). For earthquakes with $M < 5.5$ the mean values of the recorded demand spectra are in good agreement with both design curves (Fig. 9.3b), while the mean plus one standard deviation curve gives much higher values.

9.2.4 Soil Type C

The total number of records in Soil Type C is 122. For $M > 5.5$ there are 70 records: 28 records from Japan, 14 from the Greece and 28 from California. The range of PGA values varies between 59 and 741 gal. For $M < 5.5$ there are 22 records from Japan, 12 from Greece and 12 from USA; 52 records in total. The PGA values range between 10 and 320 gal, meaning that this sample includes both weak and strong recordings, mainly due to variable near field conditions.

Figure 9.4 presents the computed absolute values of demand spectra for $M > 5.5$. The peak displacement spectral values are between 20 and 200 mm, while the range of the maximum acceleration spectral values is between 5 and 15 m/s^2.

Equally large scatter of spectral values is also observed at the normalized curves of the demand spectra (Fig. 9.5a). From Fig. 9.4 it is observed that the mean values of the computed demand spectra for M > 5.5 compare well with the corresponding design elastic spectra of EC8 and the proposed demand spectra by Pitilakis et al. [5], especially for spectral values which correspond to the constant plateau. For small earthquakes with $M < 5.5$ (Fig. 9.5b), the mean values of the computed demand spectra appear significantly lower than the design elastic spectral values, while mean plus one standard deviation values are practically identical with the corresponding design elastic spectra of EC8.

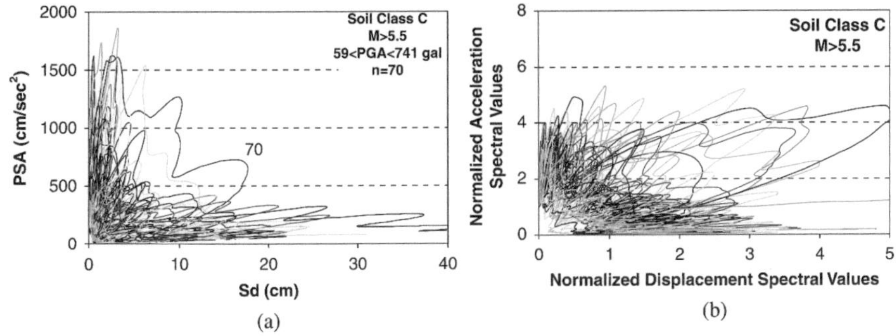

Fig. 9.4 Soil type C ($M > 5.5$, $n = 70$ records): (**a**) absolute and (**b**) normalized elastic spectra

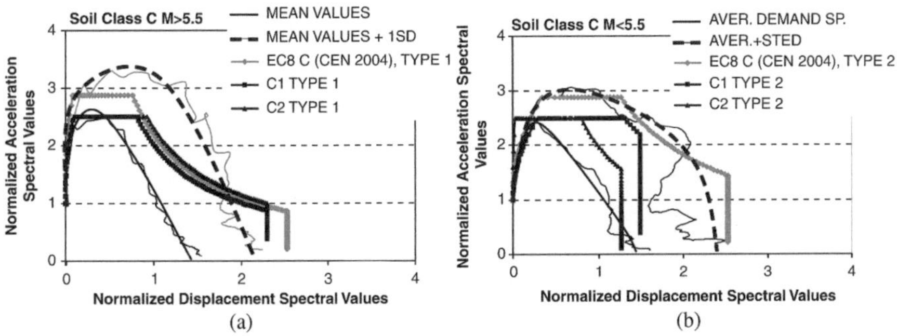

Fig. 9.5 Soil type C: Mean + σ values of normalized elastic spectra for (**a**) $M > 5.5$ and (**b**) $M < 5.5$, compared with the design elastic spectra of EC8 and Pitilakis et al. [5] (2004)

9.2.5 Effect of SFSI and Nonlinear Soil Behaviour on Seismic Demand

A simple soil-foundation-structure model is used in this study to elucidate the combined effects of the nonlinear soil behaviour and the soil-foundation-structure interaction on the system response, in the context of performance-based design. The simplified structure is a lumped mass of 100 t fixed at a height of 3.8 m over a rigid surface foundation with plan dimensions 4 × 4 m. The fixed base natural frequency of the elastic structure is fixed at 3.19 Hz and structural damping at 5%. The soil profile consists of 30 m of clayey sand to sandy clay, separated into four layers with different properties, overlaying a rigid bedrock with significant interface impedance with the soil. The properties of the soil and the structure are shown in Fig. 9.6a. The soil profile is classified as type C Ground according to the EC8 with $V_{s30} = 209$ m/s. For linear elastic behaviour, the natural frequency of the soft shallow profile is 2.1 Hz. The shear modulus reduction and damping curves which are assigned to this benchmark soil profile are shown in Fig. 9.6b.

In the nonlinear domain, the input motion and its dynamic characteristics play significant role in the soil response. In order to promote the nonlinear soil response,

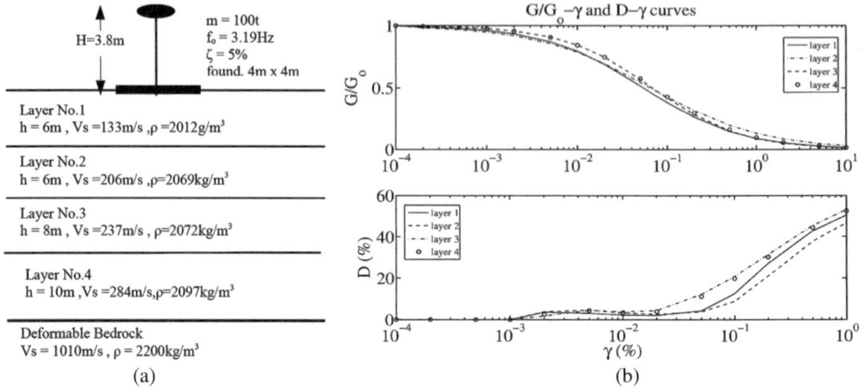

Fig. 9.6 (**a**) Simplified soil-foundation-structure system used in the analyses and (**b**) shear modulus reduction and damping curves of the soil profile

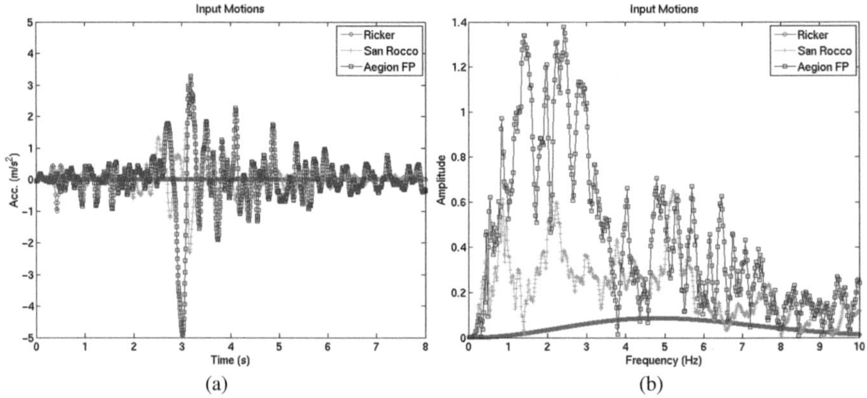

Fig. 9.7 Ricker wavelet with PGA 1 m/s^2 and period 0.2 s, San Rocco, 1976 Friuli earthquake record with PGA 2.3 m/s^2 and Aegion, 1995 earthquake record with PGA 4.9 m/s^2, in the time (**a**) and the frequency domain (**b**)

the system is subjected to a sequence of three ground motions. Initially a 2nd order Ricker wavelet is assigned as input ground motion. It has PGA amplitude of 1 m/s^2 and period 0.2 s, centered at 0.43 s. Then the system is subjected to two real earthquake ground motion records. The first one was recorded in San Rocco, during the 1976 Friuli, Italy earthquake and the second was recorded in Aegion, during the 1995 Aegion, Greece earthquake. The three ground motions are shown in Fig. 9.7, in the time and frequency domains.

9.2.6 Comparative Results

The response of the soil-foundation-structure system was calculated adopting the equivalent linear approach in the substructure approximation of the SFSI [4]. The

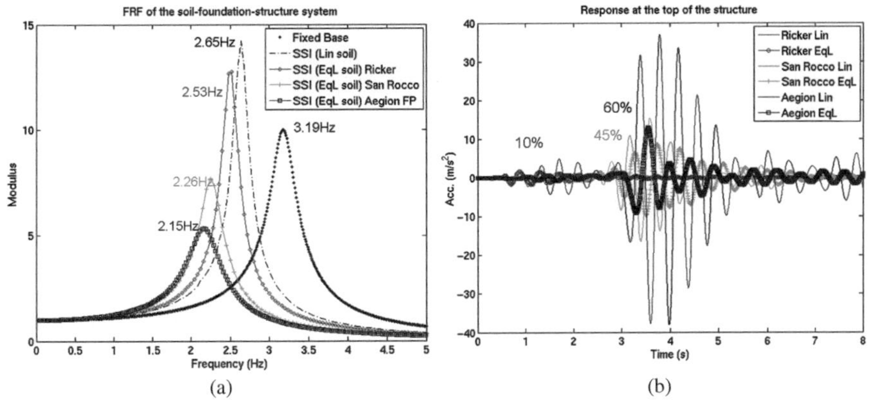

Fig. 9.8 (**a**) Frequency response function of the system for the fixed-base case, the linear SFSI case and the SFSI equivalent linear case. (**b**) Acceleration response at the top of the structure for the three ground motions. Note the reduction in the acceleration amplitude due to the equivalent linear soil behaviour

response in time and frequency domain when subjected to the three ground motions is shown in Fig. 9.8.

The fundamental frequency of the fixed base system decreases from 3.19 to 2.65 Hz due to SFSI. When the equivalent linear soil-foundation-structure system of Fig. 9.6 is subjected to the Ricker wavelet, the natural frequency of the system shifts further down to 2.5 Hz. This decrease in the fundamental frequency is caused by the foundation compliance, which is taken into consideration in the case of interaction with the soil.

The stronger earthquake record of San Rocco causes the soil-foundation stiffness to further decrease, leading to a fundamental frequency of the system at 2.25 Hz, i.e. 30% lower from the fixed base case and 15% from the linear viscoelastic case. According to the Frequency Response Function (FRF) of the system, the response at the top of the structure decreases with the implementation of the equivalent linear soil model (Fig. 9.8b). Finally the strong Aegion earthquake record mobilizes the nonlinear soil behaviour. The soil shear modulus reduces by more than 85% at a depth of 4.5 m. This eventually results in a decrease of the fundamental frequency by more than 19% from the linear SFSI case and by 33% from the fixed base case. The soil softening and the increased energy dissipation in the soil, through hysteretic material action and radiation damping, cause the response at the top of the structure to decrease, as evidenced in Fig. 9.8b, where the acceleration amplitude is reduced by at least 60%. These differences affect the elastic demand spectra.

Inspection of the demand spectra presented in Fig. 9.9 for the two actual recordings shows that the seismic demand generally decreases with the SFSI and with the nonlinear soil behaviour. That is, when the interaction of the structure with the soil is taken into account together with non-linear soil behaviour, the seismic demand may be reduced considerably in certain period of major engineering importance ($T < 0.7$ s).

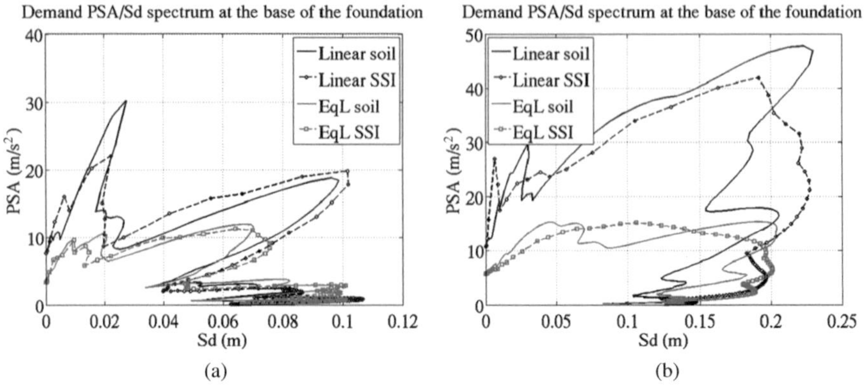

Fig. 9.9 Elastic demand spectra at the foundation level for (**a**) the San Rocco, and (**b**) the Aegion earthquake records. The free-field soil response is denoted with *solid line*, the SFSI response with *dashed marked line*, while the linear soil behaviour is denoted with *black colour* and the equivalent linear soil case with *gray*

The SFSI is more important in the linear elastic soil case, as not only it decreases the spectral acceleration, but it changes the shape of the demand spectrum as well, amplifying the displacement. This is apparent mostly for short-to-intermediate structural periods. In the high period range, the SFSI might increase the displacement spectral values.

On the other hand, the nonlinear soil behaviour reduces mainly the pseudo-acceleration spectral values. In the case of the San Rocco earthquake record (Fig. 9.9a) more significant is the effect of the nonlinear soil behaviour. The shape of the demand spectrum does not change significantly when assuming interaction. The spectral acceleration values, however, diminish due to the increased energy dissipation in the nonlinear soil, especially in the small displacement range.

The effects of the equivalent linear soil behaviour and the SFSI are more pronounced when the system is subjected to the strong earthquake record of Aegion (Fig. 9.9b). Incorporation of the SFSI increases the spectral displacement in the intermediate-to-long period range, especially in the linear soil case. On the other hand, the assumption of a nonlinear behaviour of the soil itself changes significantly the spectral values. The de-amplification of the spectral acceleration values from the linear case is significant, as it may be of the order of 400%.

9.3 Effect of Soil Improvement and SFSI on Seismic Demand

9.3.1 Replacement with Rubber – Soil Mixtures

The behaviour of a simple SFS system founded on a soft cohesive soil corresponding to Soil Type C according to Eurocode 8 replaced with rubber-sand compacted layer, is examined next. Figure 9.10a shows the effect of soil replacement with Rubber – Soil Mixtures (RSM) on the elastic demand spectra, at the foundation level of a

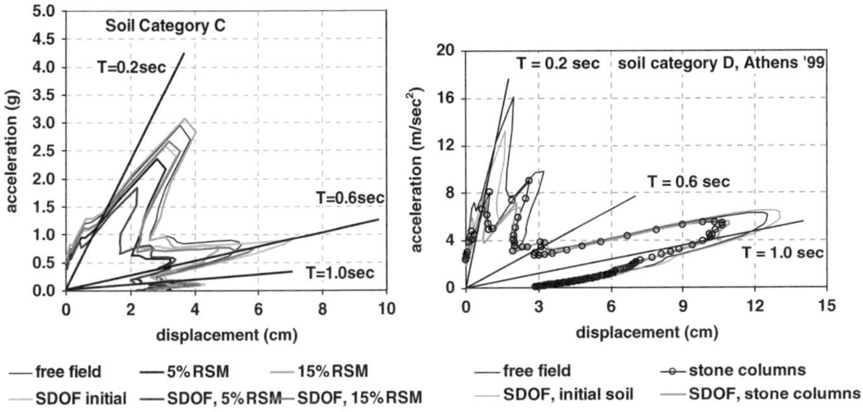

(a) (b)

Fig. 9.10 (**a**) Effect of foundation soil replacement with RSM on the demand spectra calculated at the foundation level of a SDOF system [7] and (**b**) effect of a stone column improvement on the demand spectra calculated at the foundation level of a SDOF system

SDOF structure founded on a Class C soil, for the seismic input of the Athens, Greece 1999 earthquake. They were computed using a coupled soil-structure FEM model under plane strain conditions, in the frequency domain. The shear moduli of the soil layers were reduced from their initial values according the last iteration of equivalent linear 1D analysis. The damping ratio was set accordingly for each layer. The dynamic soil properties of the RMS mixtures were deduced from specific resonant column tests [1,7].

Four characteristic dynamic analyses are considered: two with the initial or replaced soil without the SDOF structure and two with the structure. In particular: (i) a EC8 C type soil in its original or free field state (noted as "free field"), (ii) the same soil replaced in a depth of 5 m by RSM (noted as 5% and "15% RSM" according to the weight ratio of the mixture), (iii) the SDOF on the original soil ("SDOF initial") and (iv) the SDOF on the replaced soil ("SDOF, RSM").

In the case of Soil Type C, the appropriate use of a well-selected RSM replacement (RSM5 instead of RSM15) may produce significant beneficial effects in the coupled system seismic demand, over almost all frequencies of interest. In the case of Soil Type B, which is not presented here, the RSM replacement leads to a considerable increase of the seismic demand, especially in high frequencies. The softer the replacement material (RSM15) the greater the seismic demand. However the increase was lower compared to the initial soil when SSI effects are considered.

9.3.2 Stone Columns

The same numerical approach has been also used to calculate the demand spectra of a similar SDOF structure founded in Soil Type D according to EC8 (very loose soil), which is deemed necessary to be improved with stone columns. As previously the seismic event of Athens, Greece 1999 (Fig. 9.10b) was used.

Again, four characteristic cases are considered: (i) an EC8 Type D Soil in its original or free field state (noted as "free field"), (ii) the same soil improved with stone columns without any structure whatsoever (noted "stone columns"), (iii) the SDOF on the original soil ("SDOF, initial soil") and (iv) the SDOF on the replaced soil ("SDOF, stone columns").

The demand in terms of acceleration is lower in the presence of stone columns at low periods and it is even lower when soil-foundation-structure interaction is taken into account. At long periods (around 1.0 s in this particular case) the stone columns lead to displacement reduction, while at short periods they seem to have no effect. Further analyses have shown that displacement demand reduction, due to soil improvement with stone columns, appears to be more prominent for heavier structures. This may be attributed to the restraint of foundation rocking due to the stiffening of the foundation soil, as the "rocking component may increase the structural demands especially for soil-structure systems with deep embedded foundations" [2].

It is worth noting that interventions that modify foundation soil stiffness do not only affect seismic demand but structural capacity as well. It has been shown that as superstructure drifts increase due to foundation flexibility, displacement ductility capacity is reduced [6]. So, soil stiffening interventions, such as stone columns are expected to have favourable effects on structural capacity.

9.4 Conclusions

Demand spectra from a database of strong ground motion records from Japan, Greece and United States, are computed. The discussion focuses on the differences between the design elastic spectra of EC8 and the computed demand spectra from a large data set of real records. It is shown that the EC8 demand spectra need further improvement, in order to be better correlated with actual recordings. In this direction, further research is required in order to enrich the available data with digital records from well-documented sites, focusing on the contribution of near field conditions, as well as source and path effects, to the design displacement and the respective demand spectra.

The combined effects of the SFSI and the nonlinear soil behaviour on the spectral demand of the system are highlighted in the context of performance based design. An equivalent linear approximation is implemented in the substructure approach of SFSI. It is shown that SFSI may modify the shape of the demand spectra, mainly in the intermediate to long period range, while the nonlinear soil behaviour decreases the acceleration spectral values for short periods.

Finally, a study case presented herein demonstrates that the influence of the SFSI with the improvement of the the subsoil conditions using stone columns, may modify considerably the elastic demand spectra at the foundation level, in particular in the short period range. Moreover the effect of soil improvement on the demand spectra is also demonstrated for the case of ground replacement with compacted rubber and sand mixtures. It is shown that the appropriate use of a well selected RSM replacement may result in significant beneficial effects in the coupled system seismic demand in almost all frequencies of interest.

Acknowledgments The fourth co-author acknowledges the support provided by the Greek State Scholarships Foundation (iky.gr).

References

1. Anastasiadis A, Pitilakis K, Senetakis K (2009) Dynamic shear modulus and damping ratio curves of sand/rubber mixtures. Proceedings of the XVII ICSMGE satellite conference on earthquake geotechnical engineering, Alexandria, Egypt
2. Mahsuli M, and Ghannad MA (2009) The effect of foundation embedment on inelastic response of structures. Earthq Eng Struct Dyn 38:423–437
3. Manou D, Anastasiadis A, Pitilakis K (2007) Elastic displacement response spectra. Proceedings of 4th international conference on earthquake geotechnical engineering., Thessaloniki, 25–28 June 2007, Paper ID 1620
4. Pitilakis D, Clouteau, D (2009) Equivalent linear substructure approximation of soil-foundation-structure interaction: model presentation and validation. Bull Earthq Eng. doi:10.1007/s10518-009-9128-3
5. Pitilakis K, Gazepis C, Anastasiadis A (2004) Design response spectra and soil classification for seismic code provisions. Proceedings of 13th world conference on earthquake engineering, Vancouver, 1–6 August 2004, Paper No. 2904
6. Priestley MJN, Calvi GM, Kowalski MJ (2007) Displacement-based seismic design of structures. IUSS Press, Pavia
7. Senetakis K, Anastasiadis A, Trevlopoulos K, Pitilakis K (2009) Dynamic response of SDOF systems on soil replaced with sand/rubber mixture. In: Papadrakakis M, Lagaros ND, Fragiadakis M (eds), COMPDYN 2009 ECCOMAS thematic conference on computational methods in structural dynamics and earthquake engineering, 22–24 June, Rhodes, Greece

Part II
Performance-Based Seismic Design and Retrofitting – Methodologies

Chapter 10
A Dynamic Macro-Element for Performance-Based Design of Foundations

Alain Pecker, Charisis T. Chatzigogos, and Jean Salençon

10.1 General Context – Motivations

Since Newmark's fundamental remark [8] that action exceeding resistance during transient loading does not necessarily mean failure, displacement and performance-based design have been established as major trends in modern earthquake engineering. Most of the developments made to date, have mainly focused on structural aspects without consideration of soil-foundation-structure interaction. Nonetheless, the latter has proved to influence the response of the overall structure significantly, both in terms of safety and in terms of serviceability. Non-linear phenomena arising at the foundation level will usually function as desirable isolation mechanisms for the superstructure unless they lead to excessive displacements and rotations. In certain cases, taking this positive effect into account may prove necessary for achieving a viable design. On the other hand, uncontrolled displacement/rotation accumulation at the foundation level may constitute a severely detrimental agent for the structure. Prescribing acceptable limits and possessing tools for an efficient prediction of displacements and rotations lies in the heart of performance-based design applied to soil structures and, in particular, to foundations.

Current engineering practice offers two possibilities for displacement/rotation evaluation at the foundation level: either a detailed FEM model (encompassing the superstructure, the foundation and at least part of the foundation soil) or a simplified Newmark type of analysis. The first option stands as the most rigorous from a mechanical point of view, but it is also the most difficult: it implies more or less a FEM model with consideration of phenomena such as soil plasticity, interface sliding, loss of soil-footing contact (geometric non-linearity), absorbing boundaries etc. Such features increase significantly the complexity and the uncertainties of modelling, especially for 3D foundation configurations, and require significant modelling skills and computational resources. The second option is evidently much simpler; it is bound however to a number of significant limitations: it presupposes

A. Pecker (✉)
Géodynamique et Structure, 157 rue des Blains, 92220 Bagneux, France
e-mail: alain.pecker@geodynamique.com

M.N. Fardis (ed.), *Advances in Performance-Based Earthquake Engineering*, Geotechnical, Geological, and Earthquake Engineering 13, DOI 10.1007/978-90-481-8746-1_10,

a heuristically defined failure mechanism and a known force history applied on the foundation. Since this force history is usually computed from a linear soil structure interaction analysis, it does not account for foundation yielding that actually modifies the force history at the foundation level. Moreover, it can only provide post-yield displacements/rotations of the foundation and neglects pre-yield displacements that may be significant for frequent (small) excitations.

10.2 Dynamic Macro-Element

The dynamic macro-element is an alternative and a compromise to the two aforementioned methods. It can be thought of as a link element attached at the base of the structure. Its scope is to reproduce the non-linear effects that arise along the soil-footing interface during dynamic soil-structure interaction. It preserves the simplicity of the structural model by encapsulating all the non-linear mechanisms within a unique constitutive law affected to the introduced link element. Force evaluation at the foundation level is performed incrementally and it is fully coupled with the superstructure response. This approach presents the advantage of translating complexity onto developing a sufficiently sophisticated constitutive law for the foundation macro-element which is written once for all. This element can then be used as a typical link/spring element in structural FEM modelling.

Modelling principle. Figure 10.1 schematically presents the main modelling principle for the foundation macro-element. It consists in dividing the soil domain in two distinct virtual fields: the near field and the far field. The near field is the part of the soil domain in the vicinity of the foundation. It is considered that all non-linearity in the system is concentrated there, whereas the far field remains linear. The decomposition allows introducing the notion of foundation dynamic impedance describing far field behaviour. Foundation impedance accounts for wave propagation, both from the incoming and scattered wave field, and provides the linear visco-elastic component of the macroelement constitutive law.

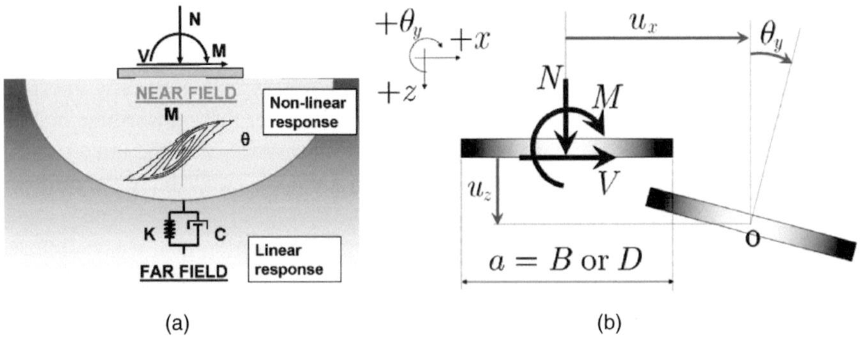

Fig. 10.1 (**a**) Modelling principle for macroelement; (**b**) forces and displacements for planar loading

Focus is hitherto put on describing nonlinear near field response for shallow perfectly rigid footings. Chatzigogos et al. [3] overviewed the literature on the topic and introduced a macroelement model for shallow circular footings on cohesive soils. In Chatzigogos et al. [4] the proposed macroelement is further generalized to encompass the most usual soil and footing-soil interface conditions: both cohesive and frictional soils, two-dimensional or three-dimensional foundation geometries and interface conditions allowing for foundation uplift or not.

The macroelement model development consists in a two-step procedure: initial description of each nonlinear mechanism that contributes to the overall response independently from one another; then introduction of the surface of ultimate loads of the foundation (calculated separately in the Yield Design theory context) as a means to calibrate coupling between the non-linear mechanisms. In its present state of development, the model comprises three nonlinear mechanisms: (a) sliding along the soil-footing interface, (b) yielding in the vicinity of the footing due to soil irreversible behaviour and (c) uplift as the footing may get detached from the soil surface. For simplicity, planar loading and cohesive soil conditions are considered in the following. For frictional soil conditions we refer to Chatzigogos et al. [4]. The model is formulated in terms of dimensionless parameters with the dimensionless forces assembled into a vector $\underline{Q}$ and dimensionless displacements into a vector $\underline{q}$:

$$\underline{Q} = \left[Q_N \; Q_V \; Q_M \right] = \frac{1}{aN_{\max}} \left[aN \; aV \; M \right] \tag{10.1}$$

$$\underline{q}^T = \left[q_N \; q_V \; q_M \right] = \frac{1}{a} \left[u_z \; u_x \; a\theta_y \right] \tag{10.2}$$

In Eqs. (10.1) and (10.2), a is a characteristic dimension of the footing (width B for strip footings, or diameter D for circular footings) and the dimensional forces and displacements are as in Fig. 10.1b. The quantity $N_{\max}$ represents the maximum vertical centered force that can be supported by the foundation.

Uplift mechanism. The uplift mechanism is described with a nonlinear elastic model that respects its fully reversible and non-dissipative character. In fact, footing detachment introduces a non-linearity of geometric nature: as the footing is uplifted, the soil-footing contact area is diminished, which leads to a reduction of the apparent stiffness of the foundation. This reduction is reproduced by means of an appropriately calibrated tangent elastic stiffness matrix $\underline{\underline{K}}$, function of the level of elastic displacements in the system:

$$\dot{\underline{Q}} = \underline{\underline{K}} \left(\underline{q}^{\mathrm{el}} \right) \dot{\underline{q}} \tag{10.3}$$

The tangent elastic stiffness matrix is determined through finite element analyses for strip footings conducted by Crémer et al. [5] and for circular footings conducted by Wolf and Song [12]. Under certain simplifying assumptions and using the above results, Chatzigogos et al. [2] have shown that the tangent elastic stiffness matrix describing uplift on an elastic soil may be written as follows:

$$\begin{pmatrix} \dot{Q}_N \\ \dot{Q}_V \\ \dot{Q}_M \end{pmatrix} = \begin{bmatrix} K_{NN} & 0 & K_{NM} \\ 0 & K_{VV} & 0 \\ K_{MN} & 0 & K_{MM} \end{bmatrix} \begin{pmatrix} \dot{q}_N \\ \dot{q}_V \\ \dot{q}_M \end{pmatrix} \tag{10.4}$$

In Eq. (10.4), elements of $\underline{\underline{K}}$ are defined though the following relationships:

$$K_{NN} = K_{NN}^0 \tag{10.5}$$

$$K_{VV} = K_{VV}^0 \tag{10.6}$$

$$K_{NM} = K_{MN} = \begin{cases} \varepsilon K_{NN}\left(1 - \dfrac{q_{M,0}^{el}}{q_M^{el}}\right) & si\ \left|q_M^{el}\right| > \left|q_{M,0}^{el}\right| \\ 0 & si\ \left|q_M^{el}\right| \leq \left|q_{M,0}^{E}\right| \end{cases} \tag{10.7}$$

$$K_{MM} = \begin{cases} \gamma\delta K_{MM}\left(\dfrac{q_{M,0}^{el}}{q_M^{el}}\right)^{\delta+1} + \varepsilon^2 K_{NN}\left(1 - \dfrac{q_{M,0}^{el}}{q_M^{el}}\right)^2 & \text{if } \left|q_M^{el}\right| > \left|q_{M,0}^{el}\right| \\ K_{MM}^0 & \text{if } \left|q_M^{el}\right| \leq \left|q_{M,0}^{el}\right| \end{cases} \tag{10.8}$$

$$q_{M,0}^{el} = \pm\frac{Q_N}{\alpha K_{MM}^0} \tag{10.9}$$

In the above expressions, $K_{NN}^0, K_{VV}^0, K_{MM}^0$ represent the real part of the foundation impedance for quasistatic loading. The quantity $q_{M,0}^{el}$ represents the rotation angle of the foundation at the instant of uplift initiation. For an elastic soil, this quantity is linear with respect to the applied vertical force on the foundation. The parameters α, β, γ, δ and ε are numerical parameters that depend on footing geometry. Their values for strip and circular footings are given in Table 10.1.

Soil plasticity mechanism. The soil plasticity mechanism is described through a bounding surface hypoplastic formulation following Dafalias and Hermann [6]. The yield surface of classical plasticity is replaced by a bounding surface f_{BS}: in the interior of this surface a continuous plastic response is obtained as a function of the distance between the actual force state represented by the loading point $\underline{Q}$ and an image point **P** on the bounding surface, defined through an appropriately chosen

Table 10.1 Numerical values for the parameters in Eqs. (10.5), (10.6), (10.7), (10.8), and (10.9)

Geometry	α	β	γ	δ	ε
Strip	4.0	2.0	1.0	1.0	0.50
Circular	6.0	3.0	2.0	0.5	0.75

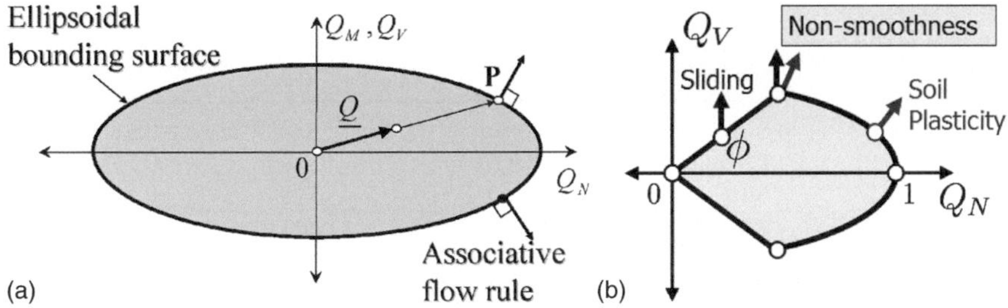

Fig. 10.2 (**a**) Ellispoidal bounding surface in the space of force parameters and mapping rule and (**b**) coupling of interface sliding and soil plasticity in the plane $Q_N - Q_V$

mapping rule (cf. Fig. 10.2a). As the loading point moves towards the bounding surface, the plastic response becomes more and more pronounced until a plastic flow is eventually produced when the loading point reaches the bounding surface: this situation corresponds to bearing capacity failure of the foundation. We can thus identify the bounding surface f_{BS} with the ultimate surface of a footing resting on a cohesive soil with a perfectly bonded interface (no uplift or sliding allowed). Gouvernec [7] has presented numerical results offering a detailed determination of this surface for various footing shapes. An extremely simple approximation that proves sufficient for practical applications is obtained by considering that this ultimate surface is an ellipsoid centered at the origin:

$$f_{\mathrm{BS}} = Q_N^2 + \left(\frac{Q_M}{Q_{M,\max}}\right)^2 + \left(\frac{Q_V}{Q_{V,\max}}\right)^2 - 1 = 0 \tag{10.10}$$

The functional form in Eq. (10.10) remains approximately independent of footing geometry and soil heterogeneity as shown by Bransby and Randolph [1]. The only parameters that change are $Q_{V,\max}$ and $Q_{M,\max}$, which define the maximum horizontal force and moment respectively: they occur for a zero vertical force. The quantity $N_{\max}$ (necessary for the definition of Q_i) is retrieved from solutions presented by Salençon and Matar [11]. $Q_{V,\max}$ is obtained by the condition of sliding along the interface. Finally, $Q_{M,\max}$ is obtained for strip and circular footings from solutions presented by Bransby and Randolph [1] and Gouvernec [7] respectively. Table 10.2 provides values of these parameters for three values of soil heterogeneity expressed through the dimensionless parameter $k = a\nabla c/c_0$, where a is the characteristic footing dimension, ∇c the cohesion gradient with respect to depth and c_0 the cohesion at the soil surface.

Sliding mechanism. For a frictional interface obeying the Mohr-Coulomb strength criterion and characterized by a friction angle ϕ, sliding of the footing has to be addressed as well. Consideration of frictional interface will induce Mohr-Coulomb branches in the Q_N–Q_V space, as it is shown in Fig. 10.2b. The global domain of admissible force states will thus be obtained by the intersection of the bounding surface and the Mohr-Coulomb branches. This domain is convex

Table 10.2 Parameters for the definition of the bounding surface (*cf.* Eq. (10.10))

	Geometry	$N_{\max}$	$Q_{V,\max}$	$Q_{M,\max}$
$k=0$	Strip	$5.14c_0a$	0.195	0.111
	Circular	$6.05c_0(a^2\pi/4)$	0.165	0.111
$k=2$	Strip	$8.01c_0a$	0.125	0.119
	Circular	$7.63c_0(a^2\pi/4)$	0.131	0.115
$k=6$	Strip	$10.29c_0a$	0.097	0.131
	Circular	$9.68c_0(a^2\pi/4)$	0.103	0.129

but non-smooth. Non-smoothness is treated within the multi-mechanism plasticity framework. For the examined case, two plastic mechanisms are introduced: the associated bounding surface hypoplastic model presented above (related to the soil response) and a non-associated perfectly plastic model related to the interface response. For the numerical implementation of multi-mechanism plasticity, the algorithm developed by Prévost and Keane [9] is used.

Coupling between mechanisms and ultimate surface of foundation. The three non-linear mechanisms are combined together in a fully coupled way. Sliding and soil plasticity interact through the multi-mechanism plasticity formulation. Plasticity and uplift are coupled through dependence of the uplift initiation parameter on the vertical force, which in turn derives from the elastoplastic response of the system. Moreover, Crémer et al. [5] have shown that in the presence of plastic soil behavior, uplift initiation is no longer linear with respect to Q_N, as is shown in Fig. 10.3b. To address this additional uplift-plasticity coupling effect, Eq. (10.9) is replaced by a nonlinear relationship of the form:

$$q_{M,0}^{\mathrm{el}} = \pm \frac{Q_N}{\alpha K_{MM}} e^{-\zeta Q_N} \tag{10.11}$$

In Eq. (10.11), ζ is a numerical constant calibrated with respect to the foundation ultimate surface.

The above nonlinear mechanisms have been formulated independently from one another. However, these formulations are implicitly based on the notion of foundation ultimate surface. The foundation ultimate surface defines the domain of all

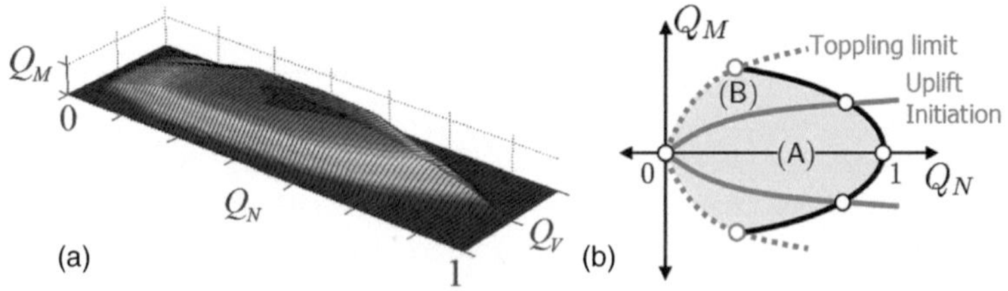

Fig. 10.3 (**a**) Foundation ultimate surface (cf. Chatzigogos et al. [2]) and (**b**) uplift-plasticity coupling in the plane $Q_N - Q_M$

possible combinations of loads that can be supported by the foundation. It can be determined on the basis of geometry and strength criteria of the system and independently of any constitutive law for the constituents of the system. This is achieved within the context of the Yield Design theory as for example in Chatzigogos et al. [2] who presented the ultimate surface of a circular footing on a cohesive soil under general loading with possible uplift. A representation of this surface is shown in Fig. 10.3a. Knowledge of the foundation ultimate surface is essential in the sense that it is identified with the hypoplastic bounding surface in the absence of uplift and also because it guides parameter calibration (parameter ζ and parameters of plasticity model) for uplift-plasticity coupling. The condition it supplies is that the toppling limit shown in Fig. 10.3b has to be identified with the boundary of the foundation ultimate surface (and certainly should not exceed it).

10.3 Numerical Application

Results pertaining to the validation of the presented model with respect to experimental and numerical results have been presented in Chatzigogos et al. [3, 4]. A simple numerical application of the model is presented herein in order to demonstrate its versatility for non-linear dynamic soil-structure interaction analyses. The macroelement is used to model a typical bridge pylon subjected to horizontal earthquake excitation as is shown in Fig. 10.4. The pylon model exhibits 4 degrees of freedom (DOFs): One horizontal translation of the superstructure and three DOFs for the translations and rotation of the foundation, which are reproduced by the macroelement. Three analyses are conducted with different base conditions considered at the foundation level: (a) elastic analysis, (b) elastic analysis with activation of uplift mechanism only and (c) full elastoplastic analysis with uplift. The results of the three analyses are given in Fig. 10.5, which provides the obtained displacement time histories at the footing centre and the force-displacement diagrams at the foundation level.

Fig. 10.4 (**a**) Typical bridge pylon and (**b**) simplified model for nonlinear dynamic analyses with macroelement

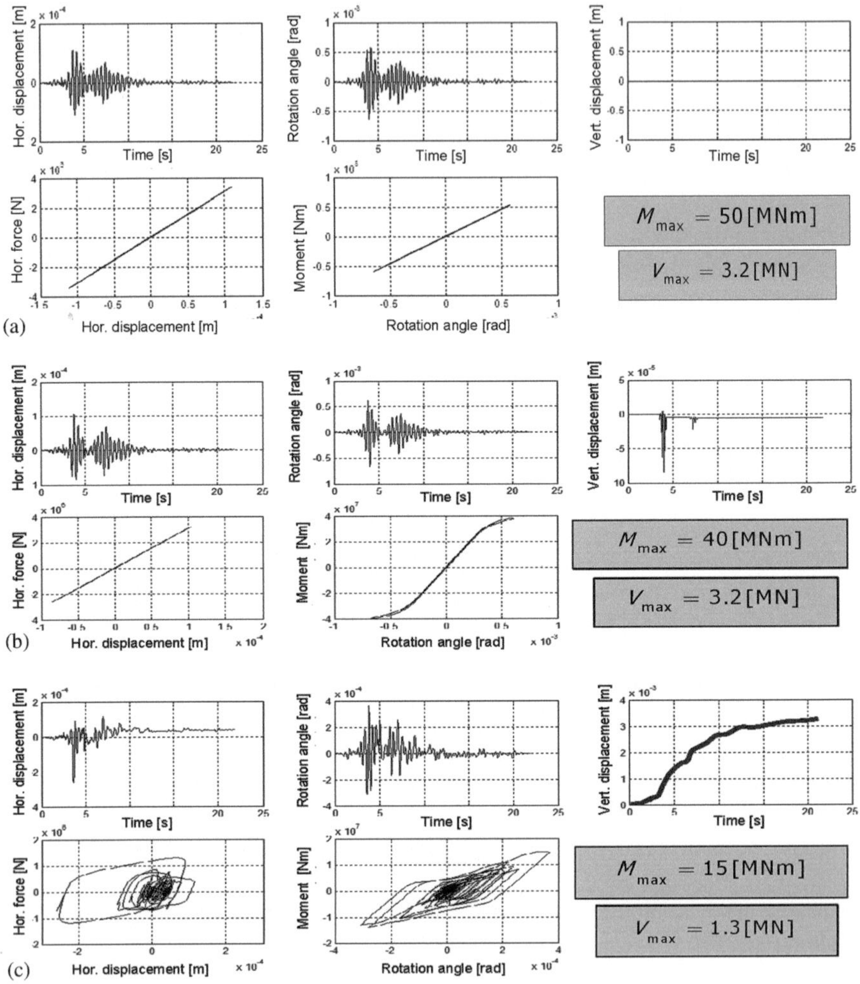

Fig. 10.5 Macroelement used for modelling a typical bridge pylon subjected to horizontal seismic excitation: (**a**) linear elastic base conditions, (**b**) elastic response with uplift and (**c**) full elastoplastic response with uplift

Consideration of uplift on an elastic soil (Fig. 10.5b) reveals the reversible and non-dissipative character of this mechanism and leads to an initial reduction of the maximum moment applied on the footing. One can also notice the coupling between the rotation and the vertical displacement (heave) as the footing is uplifted. The maximum horizontal force is not affected since this force parameter is not coupled with uplift unless plastic behaviour is considered.

In the case of full elastoplastic response with uplift (Fig. 10.5c), a significant reduction is computed both for the maximum moment and the maximum horizontal force on the footing. This emphasizes the "isolation" effect for the foundation offered by the consideration of non-linearities, which are anyway present in the real structure. The price to pay is larger maximum and permanent displacements. As a matter of fact, in the context of displacement based approaches, forces are no longer of interest because the model gives direct access to the (peak and residual) displacements. In the examined case, it is interesting to notice the accumulation of vertical settlement as the system is excited horizontally: this is an immediate consequence of the strong coupling between all degrees of freedom within the adopted plasticity formulation. It is an important feature of real behaviour that cannot be captured by other types of simplified foundation models such as Winkler spring models.

Additionally, it is interesting to notice that accumulation of permanent displacement is taking place continuously and not only when ultimate resistance is exceeded as in a Newmark type model. The macroelement is thus capable of efficiently reproducing both pre- and post-yield displacements of the footing. Finally, the model simulates the cycles of energy dissipation at the foundation level (notice the difference in cycle shape between moment and horizontal force due to presence of uplift affecting rocking response) allowing for a combined consideration of ductility demand in both the superstructure and the foundation.

10.4 Concluding Remarks

The macroelement model is intended to serve as a practical tool for performance-based design of shallow foundations. It can facilitate permanent displacement evaluation at the foundation level within standard FEM structural modelling and offers the flexibility of activating/deactivating, at will, independent non-linear mechanisms that may take place at the foundation level during dynamic soil-structure interaction. It exhibits sufficient generality to be applicable to most usual shallow foundation configurations and in parallel preserves simplicity that facilitates numerical implementation. Finally, it may prove useful in tackling new research questions such as assessing direct-displacement based design methodologies with consideration of SSI as in the spirit of Priestley et al. [10].

References

1. Bransby MF, Randolph MF (1998) Combined loading of skirted foundations. Géotechnique 48(5):637–655
2. Chatzigogos CT, Pecker A, Salençon J (2007) Seismic bearing capacity of a circular footing on a heterogeneous cohesive soil. Soils Found 47(4):783–798
3. Chatzigogos CT, Pecker A, Salençon J (2009) Macroelement modeling of shallow foundations. Soil Dyn Earthq Eng 29(5):765–781
4. Chatzigogos CT, Figini R, Pecker A, Salençon J (2010) A macroelement formulation for shallow foundations on cohesive and frictional soils. Int J Num Anal Meth Geomechanics (accepted for publication)

5. Crémer C, Pecker A, Davenne L (2001) Cyclic macro-element for soil-structure interaction: material and geometrical non linearities. Int J Num Anal Meth Geomechanics 25:1257–1284
6. Dafalias YF, Hermann LR (1982) Bounding surface formulation of soil plasticity. In: Pande GN, Zienkiewicz OC (eds) Soil mechanics-transient and cyclic loading. Wiley, New York, NY, pp 173–218
7. Gouvernec S (2007) Failure envelopes for offshore shallow foundations under general loading. Géotechnique 57(9):715–728
8. Newmark N (1965) Effects of earthquakes on dams and embankments. Géotechnique 15(2): 139–160
9. Prévost JH, Keane CM (1990) Multi-mechanism elasto-plastic model for soils. ASCE J Eng Mech 116(8):1255–1263
10. Priestley MJN, Calvi GM, Kowalsky MJ (2007) Displacement-based seismic design of structures. IUSS Press, Pavia
11. Salençon J, Matar M, (1982) Capacité portante des fondations circulaires. Journal de Mécanique théorique et appliquée 1(2):237–267
12. Wolf JP, Song C (2002) Some cornerstones of dynamic soil-structure interaction. Eng Struct 24:13–28

Chapter 11
New Concept on Fail-Safe Design of Foundation Structure Systems Insensitive to Extreme Motions

Toshimi Kabeyasawa and Toshikazu Kabeyasawa

11.1 Introduction

The seismic performance objective and the design level of earthquake loading on building structures have been selected and revised in the conventional code of practice, to reflect damage and ground motions observed in past earthquakes. However, strong earthquake motions far exceeding the current design standard level have been recorded during recent earthquakes in Japan and worldwide. Recent simulations of strong motion from source models also show possible maximum ground motion exceeding the design level.

In the recent design philosophy for RC building structures, such as the second phase of Japanese BSL, seismic design of members is conducted to provide enough deformation capacity so that the building structure can survive expected or unexpected levels of extreme motions without collapse, even though large inelastic deformations might possibly be induced. Therefore, it has been an important design objective to ensure enough ultimate deformation capacity of RC members with rational calculation, proper requirements and detailing. A rigorous method of evaluating the members' deformation capacity is still under investigation, while RC structures have been designed with enough safety margins.

On the other hand, the observed damage of existing RC building structures was in general clearly less than the damage calculated for free field ground motions records, e.g., using detailed inelastic time-history analyses. The discrepancy between the damage observation and analyses has not yet been interpreted through rational research with reliable data. The reasons have been considered to be a combination of the following: (1) an actual strength of the existing building generally higher than the calculated one used in the analysis, (2) underestimation of the maximum response from the residual damage in post-earthquake inspection, and (3) earthquake input motion to the buildings smaller than what is recorded in the

T. Kabeyasawa (✉)
Earthquake Research Institute, University of Tokyo, 1-1-1 Yayoi, Bunkyo-ku, Tokyo 113-0032, Japan
e-mail: kabe@eri.u-tokyo.ac.jp

M.N. Fardis (ed.), *Advances in Performance-Based Earthquake Engineering*, Geotechnical, Geological, and Earthquake Engineering 13, DOI 10.1007/978-90-481-8746-1_11,

free field, due to soil-structure interaction or input loss at the foundation. However, laboratory tests or field observations with reliable results to verify the reasons have been very limited [7, 8].

Regarding the third reason, notably the input loss of the ground motion at the base of the building, a shake table test was conducted at E-Defense, the world largest three-dimensional earthquake simulator, to verify the effect of the base slip behaviour at the base level. The second phase tests of full-scale RC concrete building structures were conducted on two three-storey school buildings with flexible foundation from September to November 2006 [11, 9, 12, 10]. In the shake table tests outlined in this paper, a clear reduction of damage to the building structure was observed owing to the slip behaviour at the base. Test results show the feasibility of the development and practical application of a "hyper-earthquake resistant system" positively incorporating the slip behaviour under extreme motions.

11.2 Recent Earthquake Motions

Strong earthquake motions exceeding the current design standard level have been recorded during recent severe earthquakes in Japan and US, notably in Northridge 1994, Hyogoken-Nanbu 1995, Tottori-Seibu 2000, Niigata-Chuetsu 2004, Notohanto-Oki 2007 and Niigata-Chuetsu-Oki 2007 [13]. However, the observed damage of existing RC building structures were generally smaller than estimated from the hazard analyses, or from the inelastic time-history analyses using as input motion at the base the accelerograms recorded at the free-field, especially for low-rise buildings with slight or minor damage. For example, from an inventory survey on 3911 reinforced concrete buildings in the most high intensity region after the 1995 Hyogoken-Nanbu earthquake, the percentage of all buildings with damage less than minor was 88.6%, or 93% for new regular type buildings, constructed after 1981 and conforming to the current design code in Japan. The estimated intensity of the ground motion in the region was VII in JMA standard, with maximum acceleration of 0.5–0.8g and maximum velocity of 80–100 kine, which is 1.5–2 times higher than the BSL standard level. Therefore, it has been concluded that new RC buildings in Japan have adequate strength to avoid collapse even under extreme motions. So, our concern has moved to prevent damage control under the design level [2].

Although the capability for realistic collapse analysis is still under development, it may be concluded that most of the RC buildings in Japan conforming to the current code would probably survive motions well exceeding the design motion specified as very rare and have ample margin of deformation capacity up to collapse. This is partially due to the assumed spectral shape of the design motion, which is reduced very much in the long period region. However, it has not been very rare in recent earthquakes, such as 1995 Kobe, 2007 Noto and Niigata-Chuetsu-oki, to record motions with strong long period components. It should be noted that the frequency characteristics of the design motion is based only on old and fewer strong motion observations. As shown in Fig. 11.1, many accelerograms exceeding the

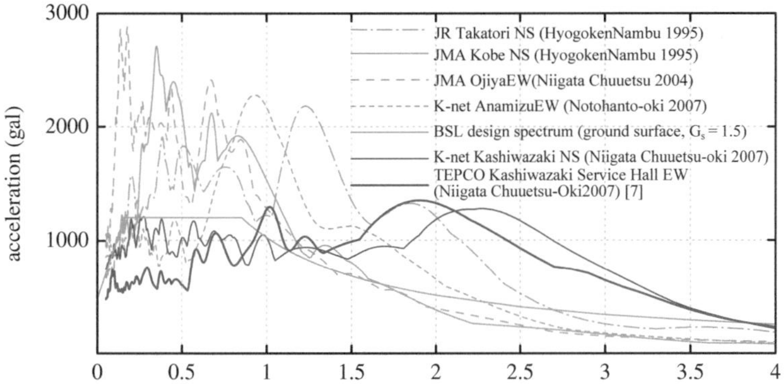

Fig. 11.1 Recent earthquake records and BSL design spectrum for very rare motion (Level 2)

design level have been recorded, even in earthquakes with Magnitude around 7. So, extreme motions are conceivable not only in future earthquakes, but also to have occurred in the largest past earthquakes, such as the Noubi earthquake in 1891 of Magnitude 8. Intensive research should continue, including observations in buildings and in the near field, development of rational and rigorous analytical tools up to collapse, experiments on soil-structure systems, and also improvement of methods for the evaluation of the capacity of members and structures, to ensure the ultimate safety of structures against potential extreme motions in future.

11.3 Full-Scale Dynamic Test of 3-Storey Buildings at E-Defense

11.3.1 Plan of the Test Specimens

Two three-storey specimens were designed to be tested with the following specific objectives: (1) simulation of progressive collapse of existing school buildings, (2) verification of strengthening by the added steel frames, and (3) soil-structure interaction with flexible foundation. Birds-eye views of the specimens are shown in Fig. 11.2. The first specimen in Fig. 11.2 was a bare RC school building designed following the 1970 Building code of Japan. Its failure mode under an extreme motion is expected to be by shear and axial collapse of columns in the first storey, starting with failure of the short columns, then inducing progressive structural failure with redistribution of column axial loads. The second was a specimen originally designed and constructed as identical as the first bare RC specimen, strengthened with attached steel frames, simulating seismic retrofit of existing buildings. The effectiveness of conventional details and new details for strengthening were to be verified and compared.

Both specimens have been constructed on the pool-shaped container with backfill soil without fixing at the base. The concrete is placed on the base concrete surface with the construction joint simulating the load-bearing foundation, so that the

Fig. 11.2 Overview of bare RC specimen in the container with backfill soil on the shake table

shear transfer at the joint would be by friction and cohesion at the concrete surface. The flexible boundary condition for the shaking table test would be the world first attempt to simulate the rocking and sway behaviour with the adjacent soil. This could be realistically done only in a full-scale test.

The structural floor plans (1st and 2nd floor plans, foundation level with the base container) of the bare RC specimen are shown in Fig. 11.3. The structure has three 4 m spans in the longitudinal (Y) direction and two 6 m spans in the orthogonal (X) direction. The specimen models an end part of typical Japanese schools in the

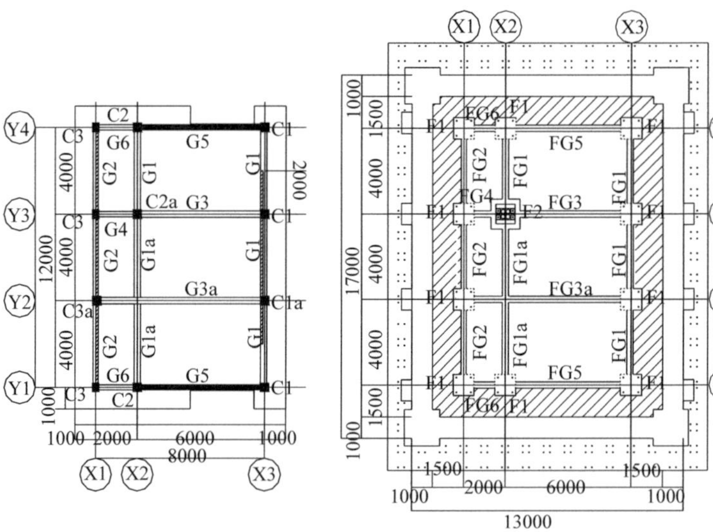

Fig. 11.3 Floor plan and container at the base

longitudinal direction, where an irregular layout of columns is often adopted for a special classroom with longer span. The total number of columns is 11 and a column at X2 and Y2 is missing from the regular pattern. 6 m long structural walls located in the outer two frames in the X-direction, are also a typical feature.

The inter-storey height is 2.5 m in each of the 1st to the 3rd storey, which is not corresponding to the full-scale but to 1:1.2 scale. The scale of the specimen was selected considering the limitation of the crane capacity of 800 ton and the area of the shake table 20 m × 15 m. Steel weights of 370 kN were mounted on the roof after setup on the table, to adjust the scaling effect. Both outer frames in the longitudinal direction have spandrel walls or standing walls, which are also typical in Japanese school buildings. The spandrel wall heights in these frames are different: 1.2 m in X1-frame, as is typical on the north side, and 0.8 m in X3-frame as on the south side.

The depth of foundation beams and of the footings is 0.8 m. The backfill soil is infilled in the surrounding area at the base of the specimens, up to a distance of 1.0 m from the footing or 1.4 m from the foundation beam to the side face of the container beam. No reinforcement is placed across the construction joint at the bottom of the footings. Insert for bolts were embedded at the base of the container to fix the footings to the container in case of the test with the fixed foundation.

11.3.2 Design Details and Seismic Evaluation of the Specimen

Design was based on the allowable stress method in accordance with the AIJ standard, 1975 edition [1] and the Building Standard Law and the Corresponding Enforcement Order of 1970s. A heightwise uniform lateral load pattern used, for a seismic base shear coefficient of 0.2. The calculated weights of the roof, 3rd, 2nd and base level including steel weight on the roof and other instruments are 1,103 kN, 789 kN, 789 kN and 855 kN, respectively assuming 24 kN/m^3 for the concrete.

Member sectional dimensions and reinforcement details are described in [10]. The slab thickness is 150 mm at the roof, 120 mm at the 2nd and 3rd floors and 100 mm at the 1st floor. The standard column section is 400 mm × 400 mm with eight 19 mm bars ($\rho_L = 1.4\%$, $\rho_w = 0.356\%$) and that of the girders is 300 mm × 500 mm with three 19 mm bars at the top ($\rho_{top} = 0.573\%$) and two 19 mm ones at the bottom ($\rho_{bottom} = 0.382\%$). Section sizes are smaller than in the old or current design practice, because of the 1:1.2 scale. Therefore, the lateral load-carrying capacities of real school buildings constructed in the 1970s are mostly not much different from that of the test specimen.

As a part of preliminary analyses, such as with pushover and dynamic analysis, seismic evaluation of the buildings was conducted based on the Japanese Standard [3], during the planning and the design of the test specimens. The evaluation is based on the calculated strengths and deformation capacity of the columns, assuming a storey mechanism. The cumulative seismic strength coefficient C_T is the sum of the ultimate strengths of the first storey columns (largest of the shear and flexural strengths), expressed in terms of the storey shear coefficient. The calculated seismic index of the designed specimen is $I_s = 0.51(F = 0.8,\ C_T S_D = 0.63)$ owing to the short columns in the X1-frame on the corridor side, which is less than the standard objective performance levels: $I_{so} = 0.6$ for ordinary buildings and $I_{so} = 0.7$ or 0.75 for school buildings. According to Japanese practice, the bare RC building should then be retrofitted. The calculated maximum base shear coefficient, equal to 0.63 at failure of short columns, was made a little less than or equal to the estimated friction coefficient at the base.

11.3.3 Static Pushover Test on Base Slip Behaviour

Static tests were carried out to determine the friction coefficient between the foundation base and the surface of the container slab, including the passive resistance of the surrounding infill soils. This was also to recenter the foundation after the dynamic test with sliding at the base. Therefore the static test was conducted on the table after that dynamic test. Then the foundation was fixed to the container for the dynamic test with fixed base. Oil jacks were placed on one side of the container at the upper level of the footings and the foundation beams to push the building, with reaction from the container beam. From the static load versus displacement relation under constant vertical load shown in Fig. 11.4, a value of 0.7–0.8 was obtained for the friction coefficient between the foundation base and the surface of the container slab, including the resistance by the backfill soil, estimated to contribute with a value around 0.1.

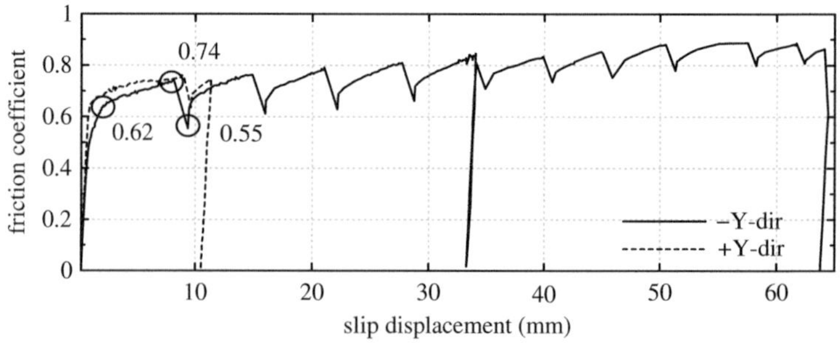

Fig. 11.4 Friction coefficient-slip relation from the static test

11.4 Dynamic Test Results of the Bare Specimen

The dynamic tests for the two specimens were conducted at E-Defense from September to November 2006. The test procedure and the observed structural damage of the bare RC specimen in the first dynamic test are outlined in this Section. The first test on the bare specimen was carried out six times, from September 22 to

Table 11.1 List of the shake test runs: input motions, base conditions, damage and responses

Run	Date	Motion (% of original)	Base	Damage (based on [4])	Peak 1st storey shear coefficient in *Y*	1st storey drift in *Y*
1	September 25	Kobe 10	Free	No	0.13	1/5,000
2	September 25	Kobe 25	Free	Slight	0.28	1/2,000
3	September 27	Kobe 50	Free	Minor	0.60	1/700
4	September 29	Kobe 100	Free	Minor	0.96	1/250
5	October 2	Kobe[a] 75	Bolted	Moderate	1.08	1/180
6	October 2	Kobe[b] 100	Bolted and propped	Collapse	1.30	1/20

[a]The target amplitude was 100% while the input in the test was 75% due to mistake
[b]The target amplitude was 130% and the effective input was equivalent to 100%

October 2. The target earthquake record of all the test runs was the same, JMA_Kobe 1995 [5]. The scale factor to the original level was varied as 0.1, 0.25, 0.5, 1.0, 0.75, and 1.0 for Runs 1−6. The base was not fixed through Runs 1–4, but the construction joint to the container allowed sway and rocking. Although the bolts were used to fix the footings before Run 5, the sway mechanism occurred because the stiffness and the pretension of the bolts were not sufficient. In Run 6, therefore, the steel plates were placed in the backfill soils between the footing and the container to prop the sway movement.

The damage level was evaluated on the basis of the standard [4] after each run, as listed in Table 11.1. Representative measured maximum response values are also listed in the table, such as the 1st storey drift ratio and shear coefficient. Minor damage was observed in the bare specimen after Run 4 under 100% of JMA Kobe, as shown in Fig. 11.5. A storey collapse occurred due to shear failure of short columns in Run 6, as shown in Fig. 11.6 under almost the same level of input motion as

Fig. 11.5 After Run 4 (Sway Foundation)

Fig. 11.6 After Run 6 (Fixed foundation)

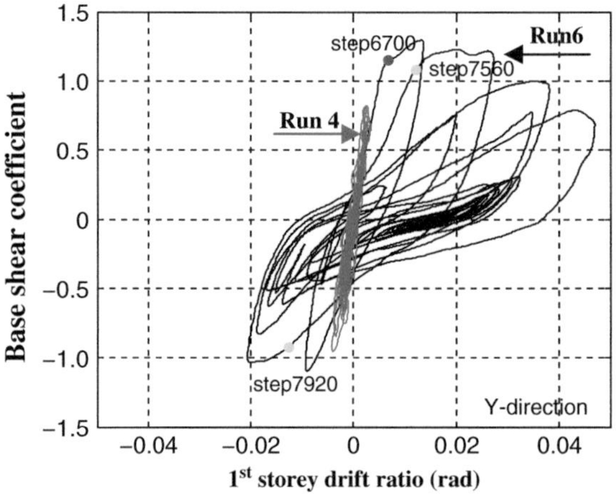

Fig. 11.7 Hysteretic relations of RC specimen

in Run 4. The hysteretic relations between the inter-storey shear force and the displacement are shown in Fig. 11.7 for Runs 4 and 6. The response of the building structure remained below yielding in Run 4, while the response attained inelastic displacement near collapse during Run 6.

Figure 11.8 shows the hysteretic relations between the slip deformations and the shear forces in Y-direction at the base on the container slab during Run 4. The shear forces are expressed in terms of friction coefficient, which are calculated as the ratio of the total shear to the total normal force on the slab, which were derived from the horizontal and vertical accelerations measured at the floors and bases. Although the slip deformation was small, the obvious reduction of the acceleration input at the first floor level from that of the shake table was observed as shown in Fig. 11.9. The acceleration response spectra in Run 4 at the 1st story shown in the dotted lines are apparently reduced from those at the shake table in Run 4, especially in the frequency region lower than the elastic periods of the structure, which were 4.2 and 6.0 in Y and X directions. The reduction of the responses in the building in

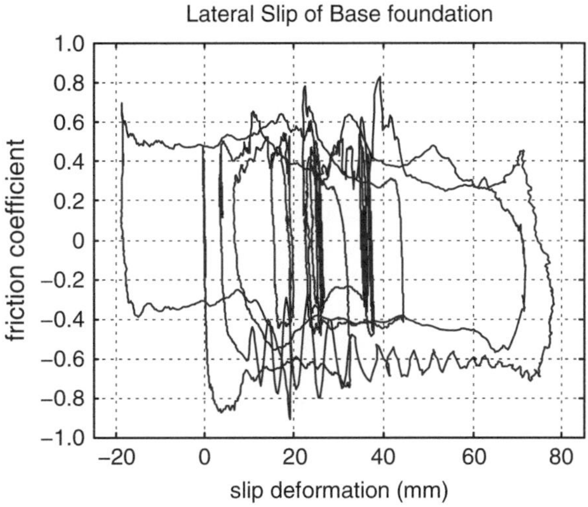

Fig. 11.8 Hysteretic relations of slip deformation at base

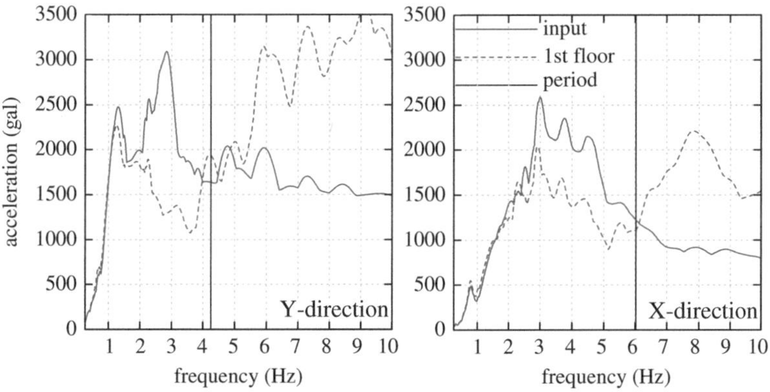

Fig. 11.9 Response spectra of input motion and at 1st floor

Run 4 compared with Run 6 is owing to the slip and the input loss at the bases. The higher amplitudes in the higher frequency region were estimated to be due to the scratching at each footing base, which might be less effective and negligible to the overall responses of the building.

11.5 Fail-Safe Design against Extreme Motions

It might still be controversial whether the clear slip behaviour observed at the base of the specimen would occur or not at the foundation base, especially in existing building structures on site with actual foundations. However, the base detail with a construction joint at the concrete surface can easily be designed and constructed similar to the specimen, in case of spread or and pile foundation. If we take into account in addition the elastic stiffness of the underlying soil, the effect of the input loss would be much more, even under the lower level of acceleration. Therefore, the behaviour could be positively taken into account as a fail-safe design against extreme earthquake motions exceeding the design level, especially for low-rise or medium-rise typical RC concrete structures with relatively high strength and limited ductility. As verified above by the full-scale shake table tests, the sway or the slip behaviour at the construction joint of the concrete base would obviously reduce the damage to the superstructure under extreme motion, compared to the case with fixed foundation. The reduction was not effective up to the design level (Level 2: 0.4*g*) or lower, but was obvious under the higher level, such as twice the design level with maximum acceleration of 0.8*g*.

Let us call here this simple system as "Hyper-earthquake-resistant system", for fail-safe design against extreme motions exceeding the design level. The conceptual elevation of the system is shown in Fig. 11.10. As shown there, a flat concrete slab with standing wall and a pool-shaped container, is to be constructed at the bottom of the base foundation supported by piles or foundation soil. The building may be designed normally but preferably with relatively higher lateral strength, or strength-dominant concept using walls and columns with wing walls. The slab surface should be leveled and smooth to control the friction coefficient. The coefficient would be around 0.6 statically, which would be reduced during the dynamic response down to 0.4 or less, as observed in the shake table test. The slip deformation would be 50–100 mm even in case of a motion like JMA Kobe, while the damage to the building would be minor owing to the slip behaviour at the base. The slip deformation and the clearance may be 100 mm or less, i.e. much less than required for a base isolation system. The backfill soil may be replaced with alternative and economic material, such as styrofoam etc. Damping effect could be expected by the

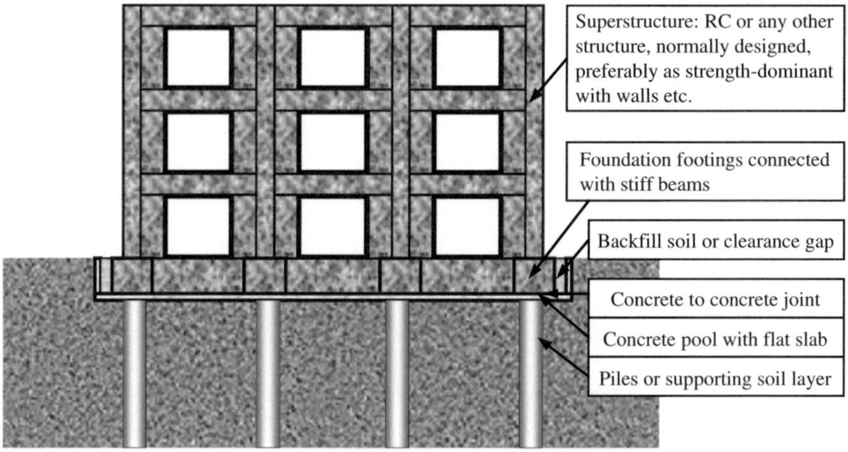

Fig. 11.10 "Hyper Earthquake Resistant System" against extreme motion

surrounding soil, which should be investigated in detail in future research. Flexible joint should be used for the piping system of the building, to ensure the functional use even after an extreme motion. It would be much easier and economic to ensure the fail-safe performance against extreme motions than to design and construct with special devices, such as with base isolation, dampers and so on.

If a constant friction coefficient is assumed, the required strength can be calculated for elastic response under extreme motions. It has been derived theoretically that the upper bound of the required strength could be constant, regardless of the fundamental period of the structure, derived from the relevant parameters, such as the friction coefficient and mass ratios of the building and the foundation [6]. Note that the theoretical required strength of the building in terms of shear coefficient is higher than the friction coefficient at the base. Because the theoretical upper bound is basically controlled by the friction coefficient, the designer does not need to care about the level and the frequency characteristics of the potential extreme motion in the future.

As for the friction at the base, there has been ample experimental research and data on the static friction coefficient between the concrete joint surfaces, depending on the surface condition. Past research was conducted mainly on joints of precast concrete structures to ensure the shear transfer at the joint, in other words to achieve a higher friction coefficient via a rough surface. In this case, the friction coefficient should be controlled to be as low as possible. From past research, even the static friction coefficient could be reduced down to around 0.4 through perfect surface smoothing. There is still more to be investigated even for static coefficients. As for the reduction with dynamic loading, the experimental data are sparse. Microscopic theoretical background or rigorous constitutive laws on friction including rate effects have not yet been established, although it has been a long time research theme in physics.

The "Hyper-Earthquake-Resistant System" is a system with a kind of safety fuse against potential very rare and yet unseen earthquake motions. The basic concept may be classified as an application of "Capacity Design Philosophy, " proposed by Tom Paulay, because the system is to ensure the hierarchy of explicit failure modes. The research on the system would be worthwhile in the future with various themes to be verified by experiment, observation and analysis.

11.6 Conclusions

Full-scale three-dimensional earthquake simulation tests on three-storey RC school building structures were conducted at E-Defense. The plan, specimens and results of the test are outlined in this paper. The damage observed in the bare RC specimen was clearly minor under the extreme motion, owing to the input loss due to the slip behaviour at the base. The observed behaviour could be used for the fail-safe seismic design of RC buildings with higher strength and limited ductility. The "Hyper-Earthquake-Resistant System" is proposed, which positively incorporates the slip behaviour as a safety fuse to control the response of the structure in a manner insensitive to potential extreme earthquake motions.

Acknowledgements The full-scale tests were carried out in September to November 2006 at E-Defense, NIED, as a part of "National research project on mitigation of major disaster in major city" Theme II Improvement of seismic performance of structures using E-defense/ Reinforced concrete structures (DaiDaiToku/RC project) of MEXT, Ministry of Education, Sports, Science and Technology, Japanese Government. Valuable discussions and collaborative efforts for the test planning by the members of the DaiDaiToku research committee on RC building structures are gratefully acknowledged.

References

1. AIJ (1975) AIJ standard for structural calculation on reinforced concrete structures (1975 and Japanese version). AIJ (Architectural Institute of Japan), Tokyo
2. AIJ (2004) Guidelines for performance evaluation of earthquake resistant reinforced concrete buildings (draft in Japanese version). AIJ (Architectural Institute of Japan), Tokyo
3. JBDPA (2001) Standard for seismic evaluation of existing reinforced concrete buildings (2001 Japanese version and 2004 English version). JBDPA (Japan Building Disaster Prevention Association), Tokyo
4. JBDPA (2001) Standard for damage level evaluation and guidelines for recovery technologies of buildings damaged by earthquake. JBDPA (Japan Building Disaster Prevention Association), Tokyo
5. JMA (1995) An earthquake record in the south part of Hyogo-prefecture Jan 17. JMA (Japan Meteorological observatory Agency), Tokyo
6. Kabeyasawa T, Kabeyasawa T (2008) Nonlinear soil-structure interaction theory for low-rise reinforced concrete buildings based on the full-scale shake table test at E-Defense. Paper submitted to proceedings of 14th world conference on earthquake engineering, Beijing, China
7. Kabeyasawa T, Kabeyasawa T, et al (2005) Observation of aftershocks and damage in buildings after 2004 Niigata-Chuuetsu Earthquake (in Japanese). Proceedings of AIJ annual convention, Osaka, Vol. IV, No. 2C, pp 719–720
8. Kabeyasawa T, Kabeyasawa T, et al (2006) Identification of input loss in base motion based on observation of aftershocks and damage to buildings after Niigata-Chuuetsu Earthquake (in Japanese). Collected Papers of Struct Eng 52B, 305–312
9. Kabeyasawa T, Kabeyasawa T, Matsumori T, Kabeyasawa T, YouSok K (2007) 3-D collapse test and analyses of the three-story reinforced concrete buildings on flexible foundation. 2007 Structures congress, SEI, ASCE, 14 pp
10. Kabeyasawa T, Kabeyasawa T, Matsumori T, Kabeyasawa T, Yousok K (2007) Shake table tests on the three-story reinforced concrete buildings with flexible foundations. Proceedings of the 2nd NEES/e-defense workshop on collapse simulation of reinforced concrete building structure, Miki, October 2006, pp 243–254
11. Kabeyasawa T, Kabeyasawa T, Matsumori T, YouSok K, Kabeyasawa T (2007) Plan of 3-D dynamic collapse tests on three-story reinforced concrete buildings on flexible foundation. 2007 Structures congress, SEI, ASCE, 16 pp
12. Kabeyasawa T, Matsumori T, Kabeyasawa T, Kabeyasawa T, Yousok K (2007) Design of the three-story reinforced concrete buildings with flexible foundations for testing at e-defense. Proceedings of the 2nd NEES/e-defense workshop on collapse simulation of reinforced concrete building structure, March 2007, Miki, October 2006, pp 225–242
13. TEPCO (2007) Time-history data recoded at Kashiwazaki-Kariwa Nuclear Power Plant during 2007 Niigata-Chuetsu-Oki Earthquake, TD-034, Japan Association for Earthquake Disaster Prevention.

Chapter 12
Performance-Based Seismic Design of Tall Buildings in the Western United States

Ronald O. Hamburger and Jack P. Moehle

12.1 Introduction

The initial development of performance-based seismic design procedures in the United States occurred in the 1980s and 1990s in response to societal reactions to the nearly annual occurrence of damaging earthquakes in the Western United States during this period. In a 15-year period, California experienced the 1979 Imperial Valley, 1982 Morgan Hill, 1983 Coalinga, 1986 North Palm Springs, 1987 Whittier-Narrows, 1989 Loma Prieta, 1992 Landers, Big Bear, and Petrolia, and 1994 Northridge earthquakes. These earthquakes provided many illustrations of both the strengths and weaknesses of US building code requirements for seismic resistance and spurred substantial evolution and improvement of the code provisions. These earthquakes did not cause collapse of many modern structures, and in general demonstrated an ability of contemporary code procedures to protect life safety, the primary goal of the building code, at least for moderate magnitude events. However, these earthquakes also amply demonstrated that the building code provisions permitted too much damage and economic loss and could readily impair the functionality of important facilities. These earthquakes also provided frequent reminders that the inventory of buildings in the United States included many older structures that were susceptible to life-threatening damage and which posed unacceptable seismic risks.

Some building owners and tenants, notably corporations and institutions that conducted high-value operations in their facilities, were interested in voluntarily upgrading their existing buildings. Engineers working on behalf of these owners and tenants however, quickly found that before committing to upgrade a facility these decision-makers wanted to know how their buildings would perform if they did not undertake retrofit. Further, these decision-makers often desired to tailor the level of retrofit to optimize their costs and benefits. These same owners and tenants

R.O. Hamburger (✉)
Simpson Gumpertz & Heger Inc., The Landmark at One Market, Suite 600, San Francisco, CA 94105, USA
e-mail: rohamburger@sgh.com

M.N. Fardis (ed.), *Advances in Performance-Based Earthquake Engineering*, Geotechnical, Geological, and Earthquake Engineering 13, DOI 10.1007/978-90-481-8746-1_12,

quickly became interested in bringing these same concepts forward in the design of new facilities, to assure that their important facilities would adequately protect their business and operational needs and that they did not develop new buildings that would cause unacceptable future economic losses.

Many owners and tenants, of course, were not interested in seismic upgrades of buildings, prompting state and local governments to adopt mandatory upgrade programs. These governments quickly found that as part of the political process of adopting such programs, it was again necessary to demonstrate the likely performance of buildings if no action were taken and the expected improved performance if the proposed retrofit programs were undertaken. Thus, performance-based seismic engineering was developed as an answer to the need of engineers and decision-makers to define the performance of buildings before and after retrofit, as a means of encouraging mitigation of existing building earthquake risks, and as a means to enable the development of buildings that would perform better than those designed to the minimum criteria contained in the building code.

The primary development of performance-based seismic design procedures occurred with the support of the Federal Emergency Management Agency (FEMA), which was interested in reducing the costs of future earthquakes and other disasters. FEMA funded the Applied Technology Council (ATC) to develop a series of performance-based engineering design criteria and guidelines. Working with the professional and research communities ATC developed a series of documents including ATC-40 [4] and FEMA-273/274 [5, 6] that defined the basis for the present state of performance-based seismic engineering practice throughout much of the world. In the US, these methodologies subsequently were converted by the American Society of Civil Engineers into ASCE-31 [1] and ASCE-41 [3] standards that could be adopted by building codes.

The technologies and procedures contained in ASCE-41 and its predecessor documents experienced widespread application in their intended use, the evaluation and upgrade of existing buildings. Although these procedures were never intended for use in design of new buildings, by 2003 engineers began to adopt and adapt these procedures for exactly that purpose. However, rather than using performance-based design as a means of designing better performing buildings, these engineers adopted these procedures to design new buildings that could provide performance equivalent to that intended by the building code, but at lower cost or with other advantages attractive to building developers.

This adoption of performance-based design procedures became particularly popular for very tall buildings, contributing to the development of many of these structures in the period 2003–2008 in Los Angeles, Seattle, San Francisco and other western cities with significant seismic hazard. The most common application of these procedures was to tall residential buildings, having post-tensioned concrete flat slab floor systems supported by a ring of perimeter reinforced concrete columns and reinforced concrete bearing walls surrounding the central elevator/utility core. The prescriptive provisions of US building codes prohibit such construction for buildings in excess of 50 m tall, without a dual special moment-resisting frame capable of resisting at least 25% of the code-specified seismic design forces. Many of these new structures extend to nearly 200 m tall. By using performance-based procedures, engineers were able to eliminate the moment-resisting frame, saving cost, and more importantly, permitting exterior designs that accommodated floor to ceiling windows and reduced story heights.

12.2 Building Regulation Issues

Western US cities generally adopt and rigorously enforce building regulations based on the International Building Code [11]. The International Building Code adopts prescriptive provisions for seismic design through reference to the *ASCE 7* [2] standard. The seismic requirements of *ASCE-7* are themselves based on the *NEHRP Recommended Provisions for Buildings and Other Structures* [9]. These requirements include force-based procedures that require demonstration of sufficient strength to resist specified earthquake forces in combination with other loads without exceeding story drift limits. The required earthquake forces are determined using elastic acceleration response spectra reduced by response modification (R) coefficients intended to represent the capacity of the selected structural system to withstand inelastic response without collapse or endangerment of life safety. To qualify for use of a given value of R, the structure must be detailed to conform to criteria specified in referenced materials industry standards published by the American Concrete Institute, American Institute of Steel Construction and other organizations. *ASCE-7* requires tall structures in zones of high seismic risk to include a special moment-resisting frame of either steel or reinforced concrete construction, capable of resisting at least 25% of the specified seismic design forces. The code design procedure evaluates earthquake loading for only a single design-level earthquake. This design-level earthquake is represented by an acceleration response spectrum having 2/3 the intensity of the Maximum Considered Earthquake (MCE) shaking upon which the code is based. For most regions of the US this MCE shaking is defined as having a 2%-50 year exceedance probability, though at sites close to major active faults, the intensity is capped by a deterministic estimate of the shaking that would result from a characteristic earthquake on those faults.

In addition to the prescriptive procedures described above, the *International Building Code* includes permissive language that enables the use of alternative procedures demonstrated to provide performance equivalent to the prescriptive provisions. The burden of proof of equivalence is on the design professional and acceptance of the proof as sufficient is at the discretion of the building official. The use of performance-based approaches to design tall buildings in the Western US has generally been permitted through these permissive "alternative procedure" provisions of the code.

The building code does not limit the procedures that can be used to demonstrate equivalence, nor in general does the building code state, other than in generic and qualitative terms, what performance is deemed acceptable. Most engineers and building officials seeking to apply performance-based approaches adopt target performance contained in commentary to t*he NEHRP Recommended Provisions.*

This commentary states that for ordinary structures, the objective of the seismic design provisions is to provide a low conditional probability of collapse, given the occurrence of Maximum Considered Earthquake shaking, and to preclude, to the extent practicable, economic losses associated with more frequent and moderate events. The recently published *FEMA P695* [7] and *FEMA P750* [10] reports clarify that the target conditional collapse probability is 10% and specify rigorous statistical methods for quantification of collapse probability. However, these statistical methods have not yet been adopted in practice and instead, engineers designing tall buildings have adopted procedures based on ASCE-41 as described further below.

The performance-based design procedures adopted by engineers have generally involved the use of nonlinear response history analysis with acceptance criteria based on available laboratory testing. Most US building officials do not have the technical expertise either to review complex analyses or interpret laboratory test reports. As a result, building officials have generally required third party, expert review of alternative designs as a condition of acceptance. Though procedures vary from city to city, the third party reviews typically include multiple persons regarded as experts in earthquake-resistant design. Usually, the third party review team includes a practicing engineer with expertise in tall building design and seismic technology, a researcher with particular knowledge of the types of structural systems to be employed (e.g., reinforced concrete walls, steel frames, etc.) and a geotechnical engineer. The reviews are rigorous and include consideration of the design criteria, ground motion selection and scaling, analytical modeling and results, and structural detailing. The review process has considerably lengthened the design period for many structures and has often resulted in substantive change to the design.

12.3 First-Generation Procedures

Engineers initially developed ad hoc procedures for design of tall buildings using performance-based approaches. Later, these approaches were formalized in documents produced by engineers in Northern California [14] and Southern California [12]. Generally, tall buildings designed using the alternative procedures conformed to the prescriptive code provisions with limited exceptions. These exceptions typically included exceedance of code-specified height limits, violation of code requirements with regard to redundancy and occasional use of materials, e.g., high strength steel and detailing procedures that were not specifically permitted in the code. Given the general similarity of these buildings to code-prescriptive designs, the procedures that developed typically included the following steps:

1. Development and approval of formal design criteria
2. Preliminary design
3. Code-level analysis
4. Verification of adequacy for Maximum Considered Earthquake shaking

Development and approval of the design criteria is an important first step in the process. The design criteria are stated in a formal document that includes a description of the overall structure and its intended load-resisting mechanisms; identification of any exceptions that will be taken to the building code requirements and the justification for these exceptions; and identification of analytical procedures, load combinations, design ground motions, material properties and detailing. The intent is that all substantive discussion of procedures and acceptance criteria be obtained before the designer has expended large effort in actually performing the design. In theory, if all procedures and assumptions are agreed to at the inception of the project, approval of the finished design should be straightforward and attainable without controversy. In practice, however, it is rarely possible to foresee all issues that will arise during the design development, and many substantive criteria issues are resolved through cooperative efforts of the designers and reviewers throughout the project.

The preliminary design is used as a basis for the succeeding steps. Capacity-based design procedures are typically central to the design of these buildings, with a preferred yield mechanism identified and other elements of the structure proportioned to remain elastic or essentially so, following yielding in the intended mechanism. For tall building design, the initial sizing of elements is often controlled by considerations of dead, live, and wind loads. In many structures, lateral design for wind forces controls even the final sizing of many elements.

The code-level design is used to confirm the adequacy of preliminary sizing and also to provide building officials with confidence, at a primary level, that the structure's system is comparable to one that would be derived using the code procedures. In this step, the engineer typically performs the code-prescribed analysis, and all relevant code strength, deformation, and detailing evaluations, except those which were specifically exempted in the formal design criteria. Since the building systems used in these structures are not strictly code-compliant, *R*-coefficients and other factors required in the code procedures are typically selected jointly by the designers and reviewers based on judgment.

Verification analyses are performed using three-dimensional nonlinear response history analyses. Typically, suites of seven horizontal ground motion pairs are used to perform the analyses. Ductile modes of behavior, including wall, slab, and beam flexure are typically evaluated using the mean of the maxima for relevant demand parameters (flexural strain, plastic rotation, etc.). Brittle modes of behavior, including wall shear, column axial force, slab punching shear, etc. are typically evaluated using either maximum demands obtained from the suites of analyses or mean demands that have been amplified by an estimated value of the standard deviation with the intent to provide a low probability of failure. Following procedures contained in *ASCE-41*, models and acceptance criteria for ductile modes of behavior are typically constructed using expected (mean) values of material properties, considering potential variability and strain hardening effects. Acceptance criteria for brittle modes of behavior are typically developed using lower bound material properties and sometimes using resistance factors to account for potential dimensional variability and construction quality issues.

12.4 PEER Tall Buildings Initiative

The PEER Tall Buildings Initiative (TBI) is a cooperative program of research and development undertaken by researchers at the Pacific Earthquake Engineering Research Center and practicing professional engineers experienced in tall building design. Spurred by the rapid growth in the use of performance-based seismic design methodologies for the design of tall buildings, the goal of this initiative is to provide a sound and reliable basis for these procedures and to help assure appropriate seismic performance of the resulting new generation of tall buildings.

The program encompasses a range of tasks intended to investigate: the dynamic characteristics of tall buildings; the performance capability of buildings designed using alternative procedures; societal preferences for tall building performance; alternative means of developing ground motions for design; soil-foundation-structure interaction effects, modeling and analysis procedures; and development of design guidelines. The program, initiated in 2006, is funded by a variety of government and private agencies and includes participation by the National Science Foundation, Federal Emergency Management Agency, United States Geologic Survey, California Geologic Survey, California Seismic Safety Commission, the City of Los Angeles, City of San Francisco, Applied Technology Council, Los Angeles Tall Buildings Council, Structural Engineers Association of California, and Southern California Earthquake Consortium, among others. The Charles Pankow Foundation provided funding for the development of design guidelines. An important companion report on analysis, modeling and acceptance criteria for tall buildings [8] is available from the Applied Technology Council. Reports on other task activities can be obtained at http://peer.berkeley.edu/tbi/index.html.

12.5 PEER Tall Building Design Guidelines

The PEER Guidelines for Seismic Design of Tall buildings represent an evolutionary step in the practice of performance-based seismic design of tall buildings. The guidelines embrace the same analytical technologies adopted by engineers following the San Francisco AB-083 and Los Angeles Tall Buildings Council criteria, but provide more guidance on structural modeling, acceptance criteria, and ground motion selection and scaling. An important departure from prior procedures is that the guidelines do not require a code-level analysis in that it is anticipated that the guidelines may be applied to structural systems for which the code response modification coefficients will not be defined, leaving the code analysis with questionable value. In so doing, the guidelines also shift focus to the performance objectives most commonly adopted by leading earthquake professionals today as the intent of the building code, that is, serviceability with minimal repair for frequent earthquake shaking levels and safety for rare earthquake shaking levels. With the exception of exterior cladding systems, the guidelines address structural performance only. The guidelines presume that nonstructural components and systems will be designed to conform to the building code criteria, but do caution that if a building's response characteristics are substantially different from that of typical code-conforming buildings, additional precautions may be required. The guidelines are written in a "recommendation" and commentary format. Recommendations are written in mandatory language, while commentary explains the basis for the recommendations and warns of significant design issues that may not be adequately covered by the recommendations.

As with the AB-083 and Los Angeles Tall Buildings Council criteria, designers must prepare a formal, project-specific design criteria document. The guidelines recommend third party review of the criteria, the analyses, and the design. The guidelines employ two levels of analysis: a Service level and a Maximum Considered level. The purpose of the Service-level check is to assure that the buildings will not experience significant damage from frequent earthquakes. Much controversy surrounded the selection of a Service-level shaking intensity. The 2008 edition of the Los Angeles Tall Buildings Council guidelines [13] specified service-level shaking with a 50% exceedance probability in 30 years (43-year mean recurrence interval), but permitted Service-level analyses to use 5% viscous damping. Studies conducted by the Applied Technology Council as part of the TBI effort and summarized in the ATC-72 report, suggest that 5% viscous damping is excessive for tall buildings. Instead, 2.5% equivalent viscous damping is more justifiable and, in keeping with this, some participants argued for use of a Service-level event with a 25-year mean recurrence, arguing that the response spectrum for such an event, when used with 2.5% damping, would be comparable to the 50%-30 year spectrum. Other participants believed that a 25-year recurrence for onset of damage to these buildings was not an appropriate design objective. Eventually consensus support was achieved for the use of a 2.5%-damped, 50%-30 year spectrum as the Service-level loading.

The stated performance goal for the Service-level event is to avoid onset of damage that would reduce the building's ability to withstand Maximum Considered-level shaking or which would require repair that would necessitate removing the building from service. It is expected that some repair of structural elements may be necessary to restore cosmetic appearance, and fire and weather resistance. Nonstructural damage is anticipated to be minor, but is not specifically evaluated.

The Service-level event does not provide an effective floor for a structure's base shear strength. Although in some highly active seismic regions, such as Los Angeles and San Francisco, the 2.5% damped 50%-30 year spectrum will result in strength demand comparable to that obtained following the prescriptive code criteria, in regions of lower seismicity, such as Portland, Oregon, and Salt Lake City, Utah, the Service-level spectrum will result in substantially less strength than would be required for a code-conforming building. Commentary warns designers in these regions that additional strength may be required to provide adequate margin against collapse at the Maximum Considered level.

The guidelines require an elastic, three-dimensional response spectrum analysis for the Service-level because the desired behavior is intended to be essentially elastic and also because it is desired to assure that an elastic analysis is available to benchmark and evaluate nonlinear models used in the Maximum Considered-level evaluation. Analytical models must extend to the structure's true base, which for most tall buildings is located several levels below grade. Soil-foundation-structure interaction effects need not be explicitly modeled, though it is permitted to do so. Based on analytical studies of typical buildings conducted under the Tall Buildings Initiative, when soil-foundation-structure interaction effects are not modeled explicitly, the mass of subgrade levels is permitted to be neglected.

Acceptance criteria include both strength and deformation. Strength is evaluated by comparing computed strength demands against expected strength. Expected strength is computed using the strength formula contained in the design specifications referenced by the building code; however, mean estimates of material properties are permitted to be used instead of minimum specified values. For example, the design engineer can propose using concrete and reinforcing steel strengths equal to 1.25 times the specified values, and structural steel strengths ranging from 1.1 to 1.3 times the specified values depending on the grade of steel used. Resistance factors are not used in determining element strength for this evaluation. Computed demand to capacity ratios may be as large as 1.5 for ductile elements and must be less than 1.0 for other elements. Story drift at any level is not permitted to exceed 0.5% of the story height.

If some computed demand to capacity ratios exceed a value of 1.5, designers are permitted to use three-dimensional nonlinear response history analysis to demonstrate acceptable Service-level performance. When such analyses are performed, a suite of not less than three horizontal ground motion pairs must be selected and modified to be compatible with the Service-level spectrum previously discussed. Either amplitude scaling or spectral matching may be used to achieve spectrum compatibility following procedures presented in the guidelines. Acceptance criteria must be developed based on suitable laboratory test data. Mean values of response parameters obtained from the suite of analyses cannot exceed demand levels at which the test data suggest the onset of strength degradation or damage, the appearance or repair of which would result in occupancy loss.

MCE-level evaluations are performed for the same level of shaking specified by the building code for this hazard level. The intent of the MCE evaluation is to demonstrate that the structure is capable of surviving this level of shaking with low probability of collapse. However, since the procedure does not include explicit collapse analyses, building adequacy is implied through limiting nonlinear response to levels at which significant margin would seem to remain. MCE evaluations are performed using nonlinear response history analysis and at least seven pairs of motions that are modified to be compatible with the MCE spectrum.

The guidelines provide extensive discussion of structural modeling techniques and assumptions. The subject of strength degradation, in particular, receives extensive discussion. Where strength degradation is explicitly modeled in a manner that reasonably predicts the hysteretic behavior obtained from testing using varied loading protocols, permissible levels of nonlinear response are relaxed relative to analyses conducted with models that have less explicit incorporation of cyclic strength degradation. As with Service-level evaluations, models must extend to the true base level. Modeling of soil-foundation-structure interaction is not required, but can be performed. If soil-foundation-structure interaction is not explicitly modeled, ground motions must be imparted at the structure's true base. Models are based on mean material properties.

As with Service-level evaluation, acceptance criteria include both strength and deformation considerations. Demands for ductile modes of behavior are evaluated using the equations:

$$Q \leq Q_{\mathrm{CP}} \tag{12.1}$$

$$Q = D + L_{\exp} + \overline{E} \tag{12.2}$$

where D is the effect due to dead load, $L_{\exp}$ is the effect due to expected live load, which may be taken as 25% of the code-specified live load, and $\bar{E}$ is the mean earthquake effect obtained from the suite of analyses. Q_{CP} is the permissible level of action, which is limited to the deformation at which strength would degrade to 80% of its peak value if explicit modeling of strength degradation is not performed. If strength degradation is modeled explicitly, mean demand up to the ultimate deformation is acceptable, as long as loss of element strength beyond the ultimate deformation is modeled, its effect on other elements is accounted for, gravity load carrying capacity is not lost, and story strength in any story does not degrade below 80% of the undamaged strength.

Modes of behavior that are not ductile are evaluated using demand obtained from the equation:

$$Q = D + L_{\exp} + F_E \tag{12.3}$$

where D and $L_{\exp}$ are as previously defined and F_E is taken either as 1.5 $\bar{E}$ or, for actions with strength demand limited by yielding of other elements, F_E may be taken as:

$$F_E = \overline{E} + 1.3\sigma \geq 1.2\overline{E} \tag{12.4}$$

where $\overline{E}$ is as previously defined and σ is the standard deviation of the response parameter as obtained from the suite of analyses. It is widely recognized that the true dispersion of responses cannot be adequately gauged using only seven earthquake ground motion pairs. The factor 1.5 applied to the mean response is intended to produce a low probability (around 10%) of exceeding the reliable strength in any one earthquake ground motion at the MCE level. It would be applicable, for example, to wall shear strength. The alternative equation is applicable, for example, to shear in an outrigger beam designed by capacity design methods to be limited by flexural strength. Strength capacities are computed using expected material properties and a resistance factor. The resistance factor may be taken as unity where failure of the element would not result in endangerment of life safety and must be taken in accordance with the building code otherwise.

The mean story drift from the suite of analyses in any story is not permitted to exceed 3% and the story drift for any single analysis is not permitted to exceed 4.5%. These limits were selected somewhat arbitrarily, based on the guideline writers' comfort with the ability of present analytical methods to predict response at very large deformation. In addition to limits on maximum transient story drift, the guidelines also limit maximum residual drift. The mean value of residual drift from the suite of analyses cannot exceed 1% of story height in any story and the maximum residual drift in any story from any analysis cannot exceed 1.5% of story height.

12.6 Summary

The PEER Tall Buildings Initiative has been a successful collaboration of earthquake engineering researchers, practicing structural and geotechnical engineers, and building code officials to address the need for appropriate consensus criteria for performance-based design of tall buildings in the Western US. Though evolutionary rather than revolutionary in nature, the PEER Tall Building Design Guidelines introduce significant improvements to practice in the design of these buildings. Of particular note is the provision of modeling guidelines that more realistically account for the nonlinear behavior of buildings than approaches previously used by the profession, together with incorporation of more rational acceptance criteria. The authors believe that the new guidelines will permit the development of tall buildings that are more likely to meet the intended performance objectives embedded in the building code, either than buildings designed to the prescriptive code provisions, or buildings that have been recently designed using performance-based approaches. Future developments in this area should include further guidance on selection and scaling of ground motions, direct consideration of nonstructural behavior and incorporation of explicit collapse margin investigations.

References

1. ASCE (2002) Seismic evaluation of existing buildings, ASCE-31. American Society of Civil Engineers, Reston, VA
2. ASCE (2005) Minimum design loads for buildings and other structures, ASCE-7. American Society of Civil Engineers, Reston, VA
3. ASCE (2006) Seismic rehabilitation of existing buildings, ASCE-41. American Society of Civil Engineers, Reston, VA
4. ATC (1996) Seismic evaluation and retrofit of buildings, report no. ATC-40. Applied Technology Council, Redwood City, CA
5. ATC (1997) NEHRP guidelines for the seismic rehabilitation of buildings, FEMA-273. Applied Technology Council for the Federal Emergency Management Agency, Washington, DC
6. ATC (1997) Commentary on the guidelines for seismic rehabilitation of buildings, FEMA-274. Applied Technology Council for the Federal Emergency Management Agency, Washington, DC

7. ATC (2009) Qualification of building seismic performance factors, FEMA P695. Applied Technology Council for the Federal Emergency Management Agency, Washington, DC
8. ATC (2009) Interim guidelines on modeling and acceptance criteria for seismic design and analysis of tall buildings. Applied Technology Council, Redwood City, CA
9. BSSC (2003) NEHRP recommended provisions for seismic regulation of buildings and other structure, FEMA 450. Building Seismic Safety Council for the Federal Emergency Management Agency, Washington, DC
10. BSSC (2009) NEHRP recommended provisions for seismic regulation of buildings and other structure, FEMA P750. Building Seismic Safety Council for the Federal Emergency Management Agency, Washington, DC
11. IBC (2006) International building code. International Code Council, Whittier, CA
12. LATBC (2006) An alternative procedure for seismic analysis and design of tall buildings located in the Los Angeles region. Los Angeles Tall Buildings Council, Los Angeles, CA
13. LATBC (2008) An alternative procedure for seismic analysis and design of tall buildings located in the Los Angeles region. Los Angeles Tall Buildings Council, Los Angeles, CA
14. SEAONC (2007) Administrative bulletin – requirements and guidelines for the seismic design of new tall buildings using non-prescriptive seismic-design provisions AB-083. Structural Engineers Association of Northern California, City of San Francisco, San Francisco, CA

Chapter 13
Introduction to a Model Code for Displacement-Based Seismic Design

Timothy Sullivan, Nigel Priestley, and Gian Michele Calvi

13.1 Introduction

Despite increasing awareness that structural and non-structural damage under seismic attack can be directly related to material strain levels, or drift respectively, and hence related to displacement, the design approach in current codes is still largely based on force (and hence acceleration) rather than displacement. Improvements in design codes have remained within the framework of a force-based design environment. However, as demonstrated by Priestley [27, 28] there are several conceptual drawbacks associated with the use of force-based methods in seismic design, even when concepts of ductility capacity are included. The direct displacement-based design (DDBD) method developed by Priestley et al. [30] and their co-researchers, appears to be a very promising alternative to force-based design. In recent years the Direct Displacement-Based Design (DDBD) method has undergone considerable development with extensive research leading to the publication of a book on DDBD [30]. More recently, a research project in Italy (RELUIS – linea IV) has led to the publication of a Draft Model Code [3] outlining the design approach for a range of structural typologies. In this paper, the background and motives for the new draft model code are reviewed, the main guidelines are introduced, and important areas for future research are identified.

13.2 Fundamentals of Direct Displacement-Based Design

Before describing the draft code in any detail, it is appropriate to review the fundamentals of DDBD. This review is carried out with reference to Fig. 13.1.

T. Sullivan (✉)
Department of Structural Mechanics, Università degli Studi di Pavia, Via Ferrata 1, Pavia 27100, Italy
e-mail: timothy.sullivan@unipv.it

M.N. Fardis (ed.), *Advances in Performance-Based Earthquake Engineering*, Geotechnical, Geological, and Earthquake Engineering 13, DOI 10.1007/978-90-481-8746-1_13,

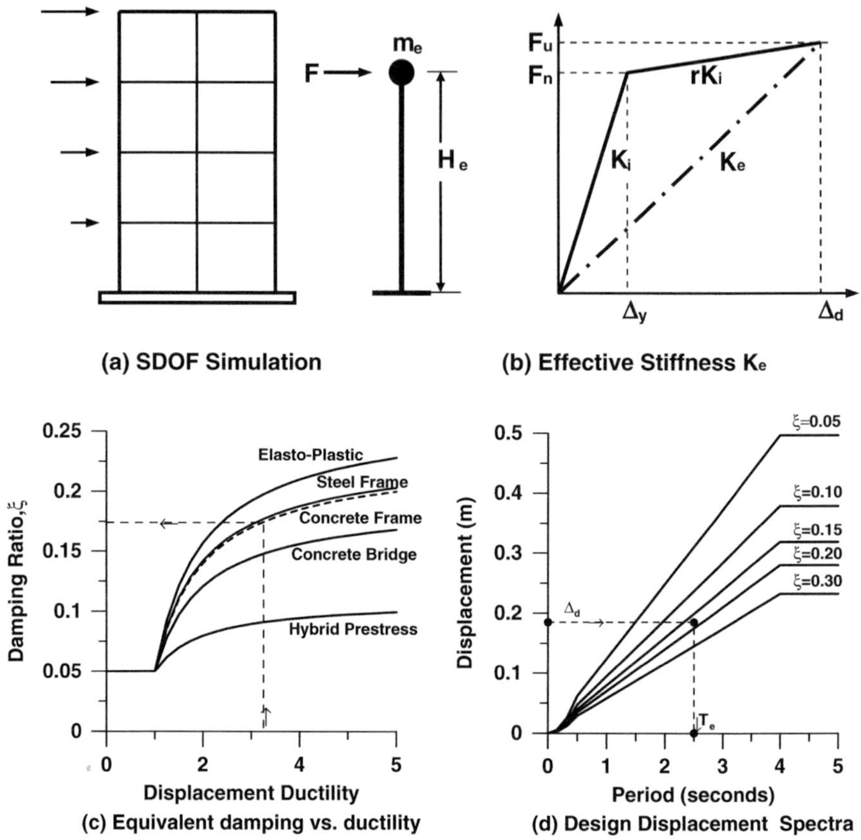

Fig. 13.1 Fundamentals of direct displacement-based design [30]

There are four components to the procedure:

- Representation of the MDOF structure (shown in Fig. 13.1a as a frame building, though the procedure is identical for all structures) as an equivalent SDOF structure, in terms of equivalent mass and characteristic displacement.
- Representation of the force-displacement response of the equivalent SDOF structure by the secant stiffness to maximum design displacement response rather than the pre-yield elastic stiffness, as shown in Fig. 13.1b.
- Adoption of relationships between displacement ductility demand and equivalent viscous damping, based on results of non-linear time-history analyses (NTHA), shown in Fig. 13.1c instead of nominal elastic damping of (typically) 5% critical.
- Use of design displacement spectra for different levels of equivalent viscous damping (Fig. 13.1d).

The design procedure requires the design displaced shape corresponding to the limit state, based on extensive inelastic analyses of typical structural forms, scaled to the displacement of a critical point of the structure, corresponding to limit-state material strain or drift limits. The following equations define the base shear demand of the equivalent SDOF structure:

Characteristic displacement: $$\Delta_{\mathrm{d}} = \sum_{i=1}^{n}\left(m_i\Delta_i^2\right)/\sum_{i=1}^{n}(m_i\Delta_i) \quad (13.1)$$

Equivalent mass: $$m_{\mathrm{e}} = \sum_{i=1}^{n}(m_i\Delta_i)/\Delta_{\mathrm{d}} \quad (13.2)$$

Effective height (buildings): $$H_{\mathrm{e}} = \sum_{i=1}^{n}(m_i\Delta_i H_i)/\sum_{i=1}^{n}(m_i\Delta_i) \quad (13.3)$$

Design displacement ductility: $$\mu = \frac{\Delta_{\mathrm{d}}}{\Delta_{\mathrm{y}}} \quad (13.4)$$

Equivalent viscous damping: $$\xi_{\mathrm{eq}} = 0.05 + C\cdot\left(\frac{\mu-1}{\mu\pi}\right) \quad (13.5)$$

Effective stiffness: $$K_{\mathrm{e}} = 4\pi^2 m_{\mathrm{e}}/T_{\mathrm{e}}^2 \quad (13.6)$$

Design base shear: $$F = V_{\mathrm{B}} = K_{\mathrm{e}}\Delta_{\mathrm{d}} \quad (13.7)$$

In Eqs. (13.1), (13.2), (13.3), (13.4, (13.5), (13.6), and (13.7), m_i, Δ_i, and H_i are the mass, design displacement and height (for buildings) of the subscript i mass locations of the structure, the yield displacement Δ_{y} is found from knowledge of the yield curvature or critical sections or yield drift of critical stories, which are both independent of strength, and dependent only on geometry and yield strain of flexural reinforcement, and C is a constant, found from extensive NTHA dependent on hysteretic characteristics of the structure. With reference again to Fig. 13.1, the effective period is found from Fig. 13.1d, entering with the characteristic displacement and selecting the appropriate level of damping given by Eq. (13.7). The procedure thus generally does not require iteration to achieve a valid solution; hence the procedure is termed “Direct” Displacement-Based Design. Full details are available in Priestley et al. [30].

Most seismic design codes have provisions for design based on NTHA. As a consequence there is already a mechanism available for DDBD provided a NTHA is used as design verification to establish conformity to code performance issues (typically drift limits and capacity design requirements). However, there is a need for codified design procedures which can be used for structures for which NTHA would represent excessive design effort.

13.3 Overview of the New Model Code

The draft model code codifies the direct displacement-based design methodology in a form that is in general compatible with key clauses of Eurocode 8 [10]. The text is presented in a traditional “Code + Commentary” format, on a split two-column page,

A Model Code for the Displacement-Based Seismic Design of Structures 1

CODE

1. DESIGN SEISMICITY

1.1 General: Structures shall be designed such that performance criteria defined in Section 2 are met for the levels of seismic intensity specified in this section for the designated seismic zone.

1.2 Seismic Zone Considerations: For structures in Zone A (moderate to high seismicity) performance criteria for Level 1 and 2 intensity shall be met. For structures in Zone B (low seismicity) performance criteria for Level 3 intensity shall be met.

1.3 Design Intensity Levels: The defined probability of exceedence for a given intensity level depends on the structural occupancy usage and damage consequences, as defined in Table 1.1.

COMMENTARY

C1 Two seismic zones are specified. Zone A is representative of reasonably high seismicity, where serviceability (Level 1 EQ) and damage control (Level 2 EQ) considerations govern the design.
Clause 1.2 assumes that the satisfaction of damage control criteria will govern over collapse prevention criteria, despite the fact that the level 3 seismic hazard is of greater intensity than level 2. If future research indicates that the collapse prevention limit state may govern, then the requirements of Clause 1.2 should be updated.
In Zone B earthquakes are rare. However, at very long return periods, significant seismic intensity may occur. For these structures it is inappropriate to design for serviceability or damage criteria, and the single requirement is that the building must not collapse, and there should be no loss of life under the Level 3 earthquake.

Fig. 13.2 Extract from the draft model code illustrating the two-column page "Code + Commentary" format

as can be seen from the extract presented in Fig. 13.2. The commentary aims to generally clarify the code requirements and indicate references where the background and further discussion on the requirements can be found. Uncertainties associated with the model code recommendations are also identified, thereby emphasising required areas for future research.

The bulk of the text and requirements in the Model Code have been developed from the recommendations provided in the book on DBD by Priestley et al. [30]. Chapter 14 of this book contains a draft code for a limited number of building systems which provided the starting point of the model code discussed here. Additional scope and critical review was provided by the various research units of the *RELUIS* project, reported by Calvi and Sullivan [4].

A general overview of the format, scope and current limitations of the draft model code follows:

13.3.1 Design Seismicity

Three levels of design seismicity are defined: Level 1 (serviceability), Level 2 (damage control) and Level 3 (collapse prevention). Structures in regions of moderate to high seismicity are designed to Levels 1 and 2; structures in regions of low seismicity are designed only for Level 3. Design probabilities of exceedence for the EC8 importance classification (classes I to IV) are defined in Table 13.1. Note that the approach taken is not to define the acceptable level of risk for a given seismicity

Table 13.1 Specified probability of exceedence for different structural categories and performance levels

	Earthquake design intensity		
Importance class	Level 1	Level 2	Level 3
I	Not required	50% in 50 years	10% in 50 years
II	50% in 50 years	10% in 50 years	2% in 50 years
III	20% in 50 years	4% in 50 years	1% in 50 years
IV	10% in 50 years	2% in 50 years	1% in 50 years

level as a variable with importance class, but to have uniform performance criteria for all structures at a specified intensity level. This is felt to be more compatible with probabilistic considerations than the use of an importance factor that increases the design base shear. It has been shown [30] that there is little correlation between strength and damage potential.

The form of the design displacement spectrum has been set, through consultation with members of the Research Unit 6 of the Project S5 [8], to be that shown in Fig. 13.3.

The shape of the displacement spectrum indicated in Fig. 13.3 is very simple in that it only requires definition of a spectral displacement corner period, T_D, together with the magnitude of the displacement, $\Delta_{D,05}$, at the corner period (note that suitable values for such parameters were one of the objectives of Project S5; refer to Cauzzi et al. [8], Cauzzi and Faccioli [7]). One initial concern with such a spectrum was that corresponding acceleration demands in the short period range are considerably higher than normal acceleration spectra would specify. This is not normally a problem for displacement-based design since it is the period corresponding to the effective stiffness rather than the elastic stiffness that is used. However, to overcome any problems for very stiff structures, the model code sets a limit on the design base shear, as indicated in Eq. (13.8), that corresponds to the plateau of the equivalent acceleration spectrum.

$$V_b = K_e \Delta_d + C\frac{P\Delta_d}{H_e} \leq 2.5 R_\xi.PGA.m_e + C\frac{P\Delta_d}{H_e} \qquad (13.8)$$

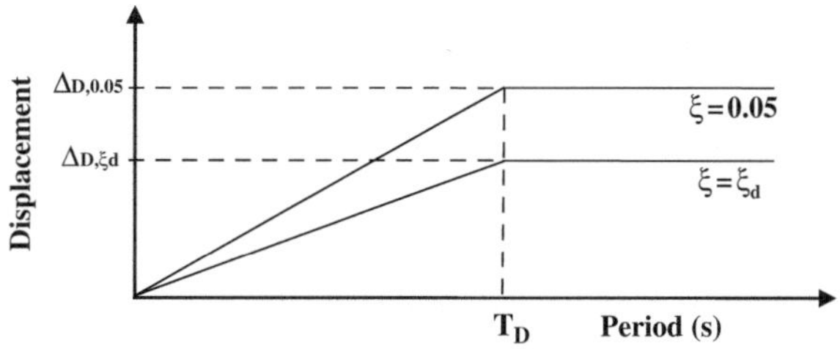

Fig. 13.3 Form of the design displacement spectrum specified in the model code

Where V_b is the design base shear, K_e is the effective stiffness, Δ_d is the design displacement, C is a constant used to control P-Δ effects, P is the total expected load and H_e is the effective height of the equivalent SDOF system. The limit to the base shear is set using R_ξ, the spectral reduction factor associated with the expected energy dissipation, the peak ground acceleration (PGA) expected at the site and m_e, the effective mass of the equivalent SDOF system. As such, the approach assumes that the elastic spectral acceleration plateau can be taken as 2.5 times the PGA. While such an approach may need refinement in the future to account for different soil types, this new proposal is clearly a simple but effective way of obtaining reasonable design base shear values for short period structures whilst maintaining the simplified form of the design displacement spectrum.

13.3.2 Performance Criteria

In order to achieve the desired performance, deformation limits have been set for each limit state. The draft model code has taken the initiative to set a number of strain limits for the first time, as detailed guidance on strain limits for certain material types (e.g. soils and structural steel) could not be found in the literature. This fact perhaps emphasises the lack of importance that has traditionally been placed on the role of deformations for design. While effort has been made to ensure that the limits indicated in the draft model code will provide the intended structural performance, it is expected that as the relationships between damage and deformation become better quantified (particularly for non-structural elements) the model code will be updated and revised.

In addition to maximum drift and strain limits, the model code has also proposed residual drift limits. A considerable amount of research into the control of residual deformations has been undertaken by MacRae [17], Kawashima et al. [15], Pampanin et al. [20], Christopoulos et al. [11], Ruiz-Garcia and Miranda [34], Uma et al. [37], and Pettinga et al. [24] amongst others. However, except for the case of retaining structures (for which the residual drift is considered similar to a peak displacement) the residual drift limits are not obligatory in the code owing to the current difficulty that exists in proving compliance for normal structural solutions and uncertainty as to appropriate residual drift limits.

13.3.3 Structural Typologies Considered in the Draft Code

Design recommendations for a wide range of structural typologies are presented in the draft Model Code. In many cases these are based on detailed and extensive research results, but in some cases the research base is less extensive. Table 13.2 defines the typologies considered, and makes comments about the status of supporting research background, highlighting areas where additional research is needed.

Table 13.2 Status of guidelines for structural typologies included in draft model code

Structural typology	Status of direct DBD guidelines in model code
RC frames	Comprehensive guidelines developed and verified (refer [31, 25, 30]).
RC walls	Comprehensive guidelines developed and verified (refer [31, 30, 2]).
Coupled RC walls	Comprehensive guidelines developed and verified (refer [22, 30]).
RC frame-walls	Comprehensive guidelines developed and verified (refer [36, 30]).
Composite structures	Few guidelines for DBD of composite systems and therefore additional research required.
Steel MRFs	Guidelines provided for MRFs possessing connections that provide the system with Ramberg-Osgood hysteretic characteristics [30]. Recommendations required for different joint types.
Steel CBFs	Guidelines provided for inverted V brace systems, verified on a limited number of case studies by Della Corte and Mazzolani [13]. Calibrated expressions for the equivalent viscous damping specific to CBF systems still under development.
Steel EBFs	Little work done on DBD of EBF systems and therefore additional research required.
Steel BRB frames	Equivalent viscous damping expressions currently based on bi-linear hysteresis and could be refined. Displacement profiles of Della Corte [12].
Pre-cast concrete structures	Guidelines from work of Belleri and Riva [1] on precast column base connections. Guidelines to be extended as part of future developments.
Hybrid RC Walls and Frames	Expressions for equivalent viscous damping from Priestley et al. [30] and Pennucci et al. [23]. References also to work by Priestley et al. [33]. Further refinements expected in future.
Timber portal structures	Guidelines developed for annular bolted joints by Zonta et al. [39]. Recommendations required for other connection types.
Timber framed wall structures	Guidelines provided based on general findings and recommendations [14].
Unreinforced masonry	Further work required to verify deformation limits. Current limits stem from work of Magenes and Calvi [18].
Isolated buildings	Recommendations taken from initial work of Priestley et al. [30] and developed by Cardone et al. [5].
Structures with added damping	Limited guidelines provided in model code and therefore additional research required.
RC bridges	Recommendations taken from initial work of Kowalsky et al. [16], and Priestley and Calvi [29] with refinements in line with findings of Ortiz Restrepo [19] and Zapata Montoya [38].
Isolated Bridges	Recommendations taken from initial work of Priestley et al. [32] and developed by Pietra et al. [26] and Cardone et al. [6].
Shallow Foundations	General DDBD approach formulated by Priestley et al. [30] to consider SSI and developed and tested for bridge piers by Paolucci et al. [21]. Further development required.
Deep Foundations	Little work done on DBD of deep foundation systems and therefore additional research required. Some guidelines provided for certain bridge pier-pile systems based on work of Suarez and Kowalsky [35].
Retaining Structures	Preliminary guidelines provided by Cecconi et al. [9] for diaphragm cantilever walls. Further work in this area is required to validate expressions and provide guidelines for other retaining systems.

13.3.4 Design Displacement Profiles

Direct displacement-based seismic design relies on determining the fundamental inelastic mode shape, and accounting for the influence of higher modes on displacement amplification, where significant, by the incorporation of design-displacement (or drift) reduction factors. Design profiles are thus defined in the Model Code for regular structures, based on the results of NTHA or derived from first principles. For example, for regular frame structures, the design displacement profile is

$$\Delta_i = \omega_\theta \theta_c h_i . \frac{(4H_n - h_i)}{(4H_n - h_1)} \tag{13.9}$$

Where ω_θ (= 1.15−0.0034 H_n < 1.0) is a reduction factor for higher mode amplification of drift, H_n is the total building height, h_i and h_1 are the heights of level i and 1 respectively, and θ_c is the code drift limit for the limit state considered. Note that it is assumed in this equation that material strain limits will not govern design, which is expected to be the case for regular frame structures. In cases of spans with very deep beams and short bay lengths this may not be the case, and drifts corresponding to strain limits may need to be substituted for θ_c in Eq. (13.9).

For some structures, particularly when the structural form is irregular (as, for example, is commonly the case with bridges) the Code-specified shape is expected to be approximate only, and an iterative approach to design is needed to finalize the required base shear strength.

13.3.5 Equivalent Viscous Damping

In general, the code uses relationships between equivalent viscous damping and displacement ductility demand in the format defined by Eq. (13.5), with the coefficient C defined for different structural systems based on the results of NTHA. Exceptions occur for timber structures, unreinforced masonry structures, and certain seismic isolation systems where the format of Eq. (13.5) is inappropriate. For example, the equivalent viscous damping of walls formed from timber framing with plywood sheathing is defined, based on work of Filiatraut and Folz [14], as

$$10\% \leq \xi_{eq} = 6\theta_c \leq 18\% \tag{13.10}$$

where θ_c is again the design drift. Code provisions also define the system damping when a number of different structural elements, with different equivalent viscous damping values, contribute to the seismic resistance.

13.3.6 Capacity Design

Common capacity design requirements for determining the minimum safe strength of members and actions to be protected against inelastic response, such as those included in EC8, have been critically examined in Priestley et al [30]. In particular it is noted that using elastic modal analysis results divided by a behaviour factor is non conservative for RC wall structures as higher mode contributions to actions are not significantly reduced by ductility in the fundamental mode of response. The recommendations in Priestley et al. [30] have largely been adopted in the model Code, and consist of three alternative approaches:

1. Capacity-protected actions (i.e. moments and shears) may be determined by NTHA of the structure, where conservatively high estimates of material strength in plastic hinge regions must be used.
2. A modified modal superposition approach in which higher mode forces obtained from modal analyses of the structure possessing effective stiffness characteristics are combined with overstrength 1st mode forces.
3. Capacity-protected actions are estimated using empirical dynamic amplification factors, in a similar fashion to approaches in current codes.

13.3.7 Additional Aspects of the Draft Model Code

P-Δ Effects: In the seismic design of a structure there are many complex phenomena that should be considered. However, in order to ensure a practical set of guidelines, efforts have been made to maintain simplified procedures whenever possible. For example, in order to control *P*-Δ effects, special analyses are not specified in the code and instead the design base shear includes a single *P*-Δ component which can be seen on the right side of Eq. (13.8). This term is in line with the recommendations of Priestley et al. [30] and accounts for different hysteretic types through a *P*-Δ constant *C*. Values of *C* of 1.0 and 0.5 are specified for steel and concrete structures respectively, thereby enabling designers to quickly account for *P*-Δ effects in calculating the required design strength.

Torsion: Another example of the desire to maintain simplicity is the simple procedure that has been provided to account for the twist that can develop for structures that are irregular in plan. Twist of a structure tends to increase the deformations on one side of a structure, and if not accounted for peak drifts would exceed the design drifts in this region. To account for such behaviour in a simplified manner, the design displacement of a system is modified according to the expected torsional rotation, θ_N, as shown in Eq. (13.11).

$$\Delta_{\mathrm{d}} = \sum_{i=1}^{n} \left(m_i \Delta_i^2\right) / \sum_{i=1}^{n} (m_i \Delta_i) - \theta_N . x_{\mathrm{CP-CM}} \tag{13.11}$$

Where $x_{\mathrm{CP-CM}}$ is the distance between the critical point (CP) on plan and the centre of mass (CM). The critical torsional rotation of the floor, θ_N, should be determined considering the deformations that occur due to torsion at the different points of the structure relative to the centre, and are positive when the CP displaces further than the CM, or negative when the CP displaces less than the CM.

Research by Beyer et al. [2] indicated that the displacements of the perimeter of a plan-irregular building are typically not greater than 10% the centre of mass displacements if the strength eccentricity is zero. Effectively, the findings of Beyer et al. [2] indicate that the inelastic twist of a structure can be controlled by designing for zero strength eccentricity. As such, the model code states that the expected floor rotation can be obtained from elastic analyses if the structure is provided with zero strength eccentricity. Alternatively, the model code suggests (in the Commentary) that torsion be allowed for through advanced analyses or following the recommendations provided by Priestley et al. [30].

13.4 Conclusions

Developments in the field of displacement-based design have lead to the recent publication of a draft model code for DBD that provides a useful tool for performance-based design. This paper has reviewed the background and motives for the new code and has provided a general preview of the key code requirements. The model code aims to provide simple means of controlling complex phenomena and this has been highlighted by reviewing the code guidelines for P-Δ effects and torsion. It has been shown that a large range of structural typologies are covered by the code but that recommendations for certain structural types still require considerable development and research.

References

1. Belleri A, Riva P (2008) Seismic behaviour of grouted sleeve precast column to foundation connections: results applied to the direct displacement based design. Proceedings of 14th world conference on earthquake engineering October 12–17, Beijing, China
2. Beyer K, Dazio A, Priestley MJN (2008) Seismic design of torsionally eccentric structures with U-shaped RC walls. MResearch Report ROSE-2008/03, IUSS Press, Pavia
3. Calvi GM, Sullivan TJ (eds) (2009a) A model code for the displacement-based seismic design of structures, DBD09, DRAFT Subject to public enquiry, IUSS Press, Pavia.
4. Calvi GM, Sullivan TJ (2009b) Development of a model code for direct displacement based seismic design, Atti di Linea IV, Convegno Finale del progetto RELUIS, 1–3 Aprile, Napoli, Italia
5. Cardone D, Dolce M, Palermo G, (2009) Direct displacement-based design of seismically isolated bridges. Bull Earthq Eng 7(2):391–410
6. Cardone D, Palermo G, Dolce M (2010) Direct displacement-based design of buildings with different seismic isolation systems. J Earthq Eng 14(2):163–191.
7. Cauzzi C, Faccioli E (2008) Broadband (0.05 to 20 s) prediction of displacement response spectra based on worldwide digital records, J Seismol 12:453–475. doi:10.1007/s10950-008-9098-y
8. Cauzzi C, Faccioli E, Paolucci R, (2008) Deliverable D2 – a reference model for prediction of long-period response spectral ordinates, Project S5 – Seismic input in terms of expected spectral displacements, Convenzione INGV – DPC 2004 – 2006, Politecnico di Milano, Italy
9. Cecconi M, Vecchietti S, Pane V (2007) The DDBD method in the design of cantilever diaphragm walls, Ottawa 2007, 60th Canadian Geotechnical Conference and 8th

Joint CGS/IAH-CNC Groundwater Conference – The Diamond Jubilee, Ottawa, Canada, pp 912–919
10. CEN (2004) Eurocode 8 – design provisions for earthquake resistant structures, EN-1998-1:2004: E, Comite Europeen de Normalization, Brussels, Belgium
11. Christopoulos C, Pampanin S, Priestley MJN (2003) Performance-based seismic response of frame structures including residual deformations. Part I: single-degree-of-freedom systems. J Earthq Eng 7(1):97–118
12. Della Corte G (2006) Vibration mode vs. collapse mechanism control for steel frames. Proceedings of the 4th international specialty conference on behaviour of steel structures in seismic areas (STESSA 2006), Yokohama, Japan, pp 423–428
13. Della Corte G, Mazzolani FM (2009) Direct displacement-based design of steel chevron bracing. Proceedings of the XIII Italian congress on seismic engineering (ANIDIS), Bologna, Italy
14. Filiatrault A, Folz B (2002) Performance-based seismic design of wood framed buildings. ASCE J Struct Eng, 128(1):39–47
15. Kawashima K, MacRae GA, Hoshikuma J, Nagaya K (1998) Residual displacement response spectrum. J Struct Eng, May, 523–530
16. Kowalsky MJ, Priestley MJN, MacRae GA (1994) Displacement-based design of RC bridge columns. Proceedings of 2nd international workshop on seismic design of bridges, Queenstown, New Zealand, pp 138–163
17. MacRae GA (1998) Residual displacements of reinforced concrete bridge columns subject to seismic loading. Proceedings of the 6th US National Conference on Earthquake Engineering, Seattle, WA, USA, Paper 155
18. Magenes G, Calvi GM (1997) In-plane seismic response of brick masonry walls. Earthq Eng Struct Dyn 26:1091–1112
19. Ortiz Restrepo JC (2007) Displacement-based design of continuous concrete bridges under transverse seismic excitation, Master Thesis, ROSE School, European School for Advanced Studies in Reduction of Seismic Risk, IUSS, Pavia, Italy
20. Pampanin S, Christopoulos C, Priestley MJN (2003) Performance-based seismic response of frame structures including residual deformations, part II: multi-degree-of-freedom systems. J Earthq Eng 7(1):119–147
21. Paolucci R., Di Prisco C, Figini R, Petrini L, Vecchiotti M (2008) Interazione dinamica nonlineare terreno-struttura nell′ambito della progettazione sismica agli spostamenti. Progettazione Sismica 1(2), Italy
22. Paulay T (2002) The displacement capacity of reinforced concrete coupled walls. Eng Struct 24:1165–1175
23. Pennucci D, Calvi GM, Sullivan TJ (2009) Displacement-based design of pre-cast walls with additional dampers. J Earthq Eng 13(S1):40–65
24. Pettinga JD, Pampanin S, Christopoulos C, Priestley MJN (2007) The role of inelastic torsion in the determination of residual deformations. J Earthq Eng 11:133–157
25. Pettinga JD, Priestley MJN (2005) Dynamic behaviour of reinforced concrete frames designed with direct displacement-based design. Research Report ROSE – 2005/02, IUSS Press, Pavia
26. Pietra D, Calvi GM, Pinho R (2008) Direct displacement-based seismic design of isolated bridges. Research Report ROSE-2008/01, IUSS Press, Pavia
27. Priestley MJN (1993) Myths and fallacies in earthquake engineering – conflicts between design and reality. Bull NZ Nat Soc Earthq Eng 26(3):328–341
28. Priestley MJN (2003) Myths and fallacies in earthquake engineering, Revisited, The ninth mallet Milne lecture, IUSS Press, Pavia, Italy
29. Priestley MJN, Calvi GM (2003) Direct displacement-based seismic design of bridges. Proceedings of ACI Special Seminar on Seismic Design of Bridges, San Diego, CA, USA
30. Priestley MJN, Calvi GM, Kowalsky MJ (2007) Displacement based seismic design of structures. IUSS Press, Pavia, Italy, 721 pp

31. Priestley MJN, Kowalsky MJ (2000) Direct displacement-based seismic design of concrete buildings. Bull NZSEE 33(4):421–444
32. Priestley MJN, Seible F, Calvi GM (1996) Seismic Design and Retrofit of Bridges. John Wiley & Sons, New York, US, 686 pp
33. Priestley MJN, Sritharan S, Conley JR, Pampanin S (1999) Preliminary results and conclusions from the PRESSS five-story precast concrete test building. PCI J, 44(6):42–67.
34. Ruiz-Garcia J, Miranda E (2006) Residual displacement ratios for assessment of existing structures. Earthq Eng Struct Dyn 35(3):315–336
35. Suarez V, Kowalsky MJ (2007) Direct displacement-based seismic design of drilled shaft bents with soil-structure interaction. J Earthq Eng 11(6): 1010–1030
36. Sullivan TJ, Priestley MJN, Calvi GM (2006) Seismic design of frame-wall structures. Research Report ROSE-2006/02, IUSS Press, Pavia
37. Uma SR, Pampanin S, Christopoulos C (2006) A probabilistic framework for performance-based seismic assessment of structures considering residual deformations. Proceedings of the 1st ECEES, Geneva, Switzerland, paper 731
38. Zapata Montoya RA (2008) Direct displacement-based design on bridges with foundation flexibility. Master Dissertation, ROSE School, European School For Advanced Studies in Reduction of Seismic Risk, Pavia, Italy
39. Zonta D, Piazza M, Zanon P, Loss C, Sartori C (2008) Direct displacement-based design of glulam timber frame buildings. Proceedings of 14th world conference on earthquake engineering (14WCEE), Beijing, China

Chapter 14
A Performance-Based Seismic Design Procedure for 3D R/C Buildings, Explicitly Accounting for Deformation Control

Andreas J. Kappos and Sotiria Stefanidou

14.1 Introduction

Performance-based seismic design (PBSD) is a conceptual design framework wherein design criteria are expressed in terms of performance objectives (such as serviceability, damage limitation, and life safety) for a structure subjected to distinct levels of seismic action [17]. A comprehensive presentation of PBSD and displacement-based design procedures is given in an *fib* document [9], where an interesting comparative study of eight methods applied to five different idealised building types is included, also reported in [19]. It is worth noting that "direct displacement-based design" is also included as an alternative design method in the provisions of the SEAOC Blue Book [18].

The method addressed herein, initially proposed in [10, 12], requires, as a first step, the use of conventional elastic analysis to obtain a basic strength level, while a detailed partially inelastic model is then developed, wherein members are permitted to exhibit inelastic behaviour only at predetermined locations. The procedure relates the detailing of critical sections to the local deformation demand (rotations or curvatures); the latter are determined using inelastic response-history analysis (inelastic static analysis can also be used for regular buildings, as suggested in [12]).

In the new version of the method, presented herein, a number of improvements are introduced, the key one being that the design of structural members is not only carried out for different performance levels, but it also ensures that their inelastic deformations fall within the range of values selected for keeping the damage level for the specific performance level adequately low. Further improvements include a revised safety format and the scaling procedure used in response-history analysis. The new method is applied to two alternative configurations of an irregular 10-storey R/C frame building, which is also designed according to the provisions of Eurocodes 2 and 8 [4, 5] for two different ductility classes. The seismic design of the building

A.J. Kappos (✉)
Laboratory of Concrete and Masonry Structures, Department of Civil Engineering,
Aristotle University of Thessaloniki, Thessaloniki 54124, Greece
e-mail: ajkap@civil.auth.gr

M.N. Fardis (ed.), *Advances in Performance-Based Earthquake Engineering*, 149
Geotechnical, Geological, and Earthquake Engineering 13, DOI 10.1007/978-90-481-8746-1_14,

according to the Code provisions is compared with that to the proposed method; then the seismic performance of all designs is assessed using a fully inelastic model and additional ground motions not used in design. Pertinent conclusions are drawn, regarding both economy and structural performance issues.

14.2 Description of the Proposed Method

In earlier versions [10, 12, 13] of the proposed method inelastic deformations were included as a design verification, not as a design parameter. To overcome this weakness, the writers sought a direct deformation-based design method, maintaining the key features of the aforementioned performance-based procedure. Moreover, while the application of earlier versions of the design method was restricted to regular buildings, the method is applied here to multistorey irregular buildings with setbacks, noting that response-history analysis based procedures appear better suited to irregular structural systems [19]. The steps involved in the proposed "direct deformation-based design method" are described in the following.

Step 1: Flexural Design of Plastic Hinge Zones Based on Serviceability Criteria

The purpose of this step is the establishment of a basic level of strength in the structure that would ensure that the structure remains serviceable ("immediate occupancy" requirement in FEMA 273 [8] and ASCE Standard 41-07 [2]) after an earthquake having a high probability of exceedance (usually taken as 50%/50 years). The verifications include specific limits for member ductility factors and plastic hinge rotations of critical members (see Step 4) and the corresponding demands are estimated from inelastic analysis of a reduced inelastic model of the structure (described in Step 3). Hence, an initial analysis is required, which would provide the strength of the members (energy dissipation zones) that will respond inelastically during the serviceability verification; this analysis constitutes Step 1 and is a vital part of the proposed procedure.

The design of selected dissipation zones like the beam ends and the bases of ground storey columns, is carried out using conventional elastic analysis. The strength of these zones is estimated taking into consideration the range within which the inelastic deformations should fall, which corresponds to the degree of damage allowed for the selected performance level. The procedure proposed in the following leads to reaching the permissible values of inelastic deformations (expressed through rotational ductility factors), since the latter are directly related to the reduction of element forces corresponding to elastic behaviour. This is a critical feature, not included in previous versions of the method that simply included a serviceability check, the result of which typically was that most members either remained elastic or were well below the allowable deformation limits [13].

To reach the aforementioned goal, element forces and rotations are first obtained from the results of an elastic analysis. Design for flexure is carried out in terms of design values, using commonly available design aids. On the other hand, serviceability checks are based on the results of inelastic analysis, for which mean values are commonly adopted; furthermore, several members are expected to posses some overstrength with respect to the design moments used in their dimensioning, due to detailing requirements, i.e. rounding (upwards) of required reinforcement areas and use of minimum reinforcement specified by codes. For these two reasons, the initial elastic analysis should be carried out for an appropriate fraction ν_o of the earthquake level associated with the serviceability performance level (50%/50 years); the suggested value is ν_o=2/3.

Subsequently, elastic rotations (θ_{el}) are related to the corresponding inelastic ones (θ_{inel}), using an empirical procedure (like that proposed in [15]). Referring to Fig. 14.1, having defined the target rotational ductility factor (μ_θ) and the maximum inelastic rotation, θ_{inel} (this is the total chord rotation, not the plastic one), from the θ_{el} found in the elastic analysis, the yield rotation (θ_y) is calculated for every structural member. For simplicity of the procedure one could assume first that *M*-θ response is elastic-perfectly plastic (as in Fig. 14.1) and second that the slope of the elastic and the elastoplastic *M*-θ diagram is the same. Then the corresponding yield moment (M_y) can be easily computed, as the intersection of the elastic part of the diagram and the vertical line drawn at θ_y, as shown in Fig. 14.1; this is the moment to be used for (flexural) design. A more accurate, and somewhat more involved, procedure, accounting for the fact that an increase in deformation does not come with a proportional decrease in design force, i.e. the slope of the first branch in the elastic and the elastoplastic diagram is different, is described in [14]. The differences in the yield moments resulting from the accurate procedure from those from the simplified one are not large (less than 10% on the average, but in some instances they are higher, especially for the positive M_y).

According to the aforementioned procedure, the reduced design forces are computed for every beam element, and they are directly related to the target rotational ductility selected for the serviceability performance level. The longitudinal reinforcement demand for the beams is calculated using standard flexural design procedures [16] and compared to the minimum requirements according to code

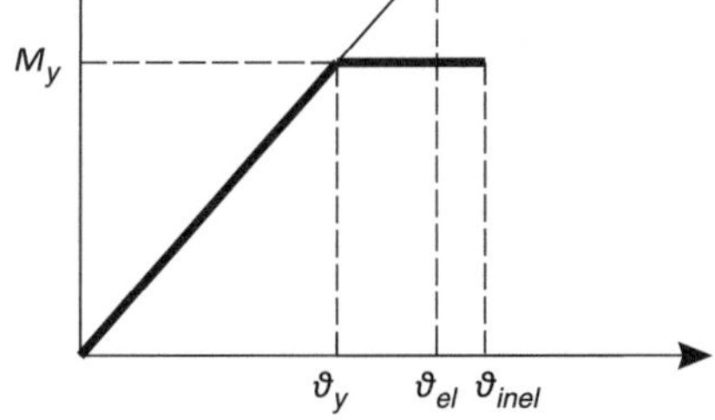

Fig. 14.1 Elastic and elastoplastic *M*-θ diagram for beams

provisions. In case the longitudinal reinforcement demands are found to be less than the minimum requirements, reduction of cross-sections is in order (reduction of stiffness), otherwise deformations for the considered performance level will be less than the allowable ones; clearly, this stage involves striking a balance between economy and performance.

Step 2: Selection of Seismic Actions

The response-history analyses necessary for seismic design according to the proposed method require the definition of appropriately selected input seismic motions. The accelerogram set used for the analysis should include a pair of components for every seismic motion and it is recommended that it be selected based on the results of a seismic hazard analysis ("deaggregation" phase, wherein M and R for the site in consideration are determined). Hence the selected input seismic motions should conform to certain criteria concerning magnitude (e.g. $M_S = 6.0\sim6.5$), and epicentral distance (e.g. $R = 10\sim25$ km), and also peak ground acceleration (PGA>$\sim$0.1g).

The earthquake motions used for design, should be properly scaled in order to correspond to the level associated with the limit state examined ("serviceability" limit state for the design of energy dissipation zones, and "life safety" for the other members). Several scaling procedures have been explored [11] and the one adopted by EC8-Part 2 [6] is used here, duly tailored to the needs of the performance-based design method.

Step 3: Set-Up of the Partially Inelastic Model

During this step a partially inelastic model (PIM) of the structure is set up, where the beams and the base of ground storey columns (and walls, if present) are modelled as yielding elements, with their strength based on the reinforcement calculated for reduced element forces according to the inelastic deformations allowed for the serviceability limit state (step 1). In the same model, the remaining columns (and walls) are modelled as elastic members.

Step 4: Serviceability Verifications

The usage of inelastic dynamic response-history analysis in the PIM, involves a set of recorded motions scaled to the intensity corresponding to the serviceability level. The verifications include specific limits for maximum drifts and plastic hinge rotations of critical members; recommended interstorey drift values range from 0.2 to 0.5% the storey height, while permissible plastic hinge rotations vary between 0.001 and 0.005 rad for columns and about 0.005 rad for beams. The purpose of this step, apart from checking the inelastic performance of the structural system, is the verification that the required rotational ductility factor (μ_θ) of beams and bases of ground storey columns is consistent with the values considered during the design. Hence, this step is basically an assessment (or verification) of the seismic response of the structure for the "serviceability" level; in principle, it can be skipped if adequate calibration of the method is carried out in the future.

Since inelastic dynamic analysis is used in order to check the seismic response of the structure for the aforementioned performance level, mean values of material strength are considered (f_{cm} and f_{ym} for concrete and steel respectively).

Step 5: Design of Longitudinal Reinforcement in Columns (and walls) for the "Life Safety" Limit State

The design of members (such as columns at locations other than the base of the structure) considered elastic in setting up the PIM, is based on the results of inelastic response-history analyses of the aforementioned model for each of the selected sets of input motions properly scaled to the intensity of the earthquake associated with the "life safety" requirement (probability of exceedance 10%/50 years). Simultaneous values of M_1, M_2, N are considered (biaxial bending and axial force), while the design is based on the most critical combinations. Consideration of mean values of material strength during the design leads to an overestimation of the longitudinal reinforcement of columns [11]. Since the input to the columns directly depends on the strength of the adjoining beams (designed to form plastic hinges) and the latter's yield moments are based on the mean value of steel strength (f_{ym}), then design column moments are over-estimated by the ratio f_{ym}/f_{yd} (equal to 1.26), which is deemed as over-conservative. The specific performance objective to be satisfied is that for the considered seismic action (10%/50 years) columns should not yield (except at the base), and mean values of column yield moments are used for this verification; hence the 1.26 factor is redundant. Since design for biaxial bending was carried out using commonly available design aids (based on f_{cd}, f_{yd}) [16] it was more convenient to use design values of material strength in the dynamic analysis of the PIM as well as in the design of the columns.

Step 6: Design for Shear

To account for the less ductile nature of this mode of failure, shear forces should correspond to seismic actions corresponding to the 2%/50 years earthquake (associated with the "collapse prevention" performance level). However, to simplify the design procedure, design and detailing for shear can be carried out using shear forces calculated from inelastic response-history analysis for the seismic action associated with the "life safety" performance level, and implicitly relate them to those corresponding to the 2%/50 years earthquake through appropriately selected magnification factors (γ_v); recommended γ_v factors [13] for beams and columns are equal to 1.20 and 1.15 respectively.

Step 7: Detailing for Confinement, Anchorages and Lap Splices

Detailing of all members for confinement, anchorages and lap splices, is carried out with due consideration of the level of inelasticity expected in each member. Structural members where the development of extended inelastic performance is anticipated (bases of ground storey columns or walls), are detailed according to the provisions of EC8 [5] concerning ductility class “Medium” (“DCM”), while others where inelastic performance is expected to be restricted (columns of upper storeys) are detailed according to the provisions for ductility class “Low” (“DCL”), essentially those of EC2 [4].

14.3 Application to Ten-Storey Buildings with Setbacks

The geometry of the ten-storey reinforced concrete building with setbacks at the two upper storeys, having a 3D frame structural system is shown in Fig. 14.2. The building was first designed according to the provisions of EC8 [5] for ductility classes “M” and “H”, and then redesigned to the performance/deformation-based procedure described in Section 14.2.

14.3.1 Discussion of Different Design Aspects

The design ground acceleration was taken equal to 0.24g, while ground conditions were assumed to be type “B” according to EC8 classification. The materials used for design were concrete class C25/30 and steel S500 [4]. The structure is classified as irregular in both directions according to the provisions of EC8, which has repercussions on the behaviour factor q and the type of analysis to be used for design. The q-factors for the DCH and DCM structures, were found equal to 4.14 and 2.76, respectively. The method of analysis used was the response spectrum method,

Fig. 14.2 3D view (*left*) and geometry of typical frames of 10-storey building

since the equivalent static method is not allowed in the case of irregular buildings. The rigidity of structural members was taken equal to $0.5EI_g$ for all members, as prescribed in EC8.

In applying the direct deformation-based method, both elastic and inelastic analyses of the structure were carried out using the software package *Ruaumoko* 3D [3]; modelling of members' inelastic performance was done by means of a spread plasticity model and bilinear elastoplastic hysteresis rule. The effective rigidity was taken equal to 50% the gross section rigidity (EI_g) for T-beams and for columns (same as in EC8). For the dynamic response-history analyses, a set of six pairs of actually recorded motions was selected from the European Database [1] and a synthetic record was added to form the final set of 7 records. All input motions were scaled to the intensity of the design spectrum (the same used for EC8 design), and pairs of horizontal components were applied simultaneously in each horizontal direction of the structure. The resulting longitudinal reinforcement demands were found to be generally less than the minimum Eurocode requirements. This hinted to the need for re-dimensioning the cross sections initially selected for the structural members (especially beams). Therefore, the proposed design method was additionally applied to a second structure ("Building 2") having the same geometry as Building 1 depicted in Fig. 14.2 and properly reduced cross sections (details are given in [14]).

Design according to the provisions of EC8 was applied mainly with a view to comparing the required reinforcement to the one resulting from the proposed design procedure, and providing a basis for evaluating the performance of complex structures designed to different methods.

14.3.2 Evaluation of Different Designs

The quantity of steel required in each member type is shown in Fig. 14.3 for the three different designs; it is clear that the application of the PBD method led to lower total reinforcement demands, the more important difference being in the transverse reinforcement in columns, which also implies easier detailing on site.

The seismic performance of the alternative designs was assessed by carrying out inelastic response-history analysis of fully inelastic models of the 3D R/C buildings (as opposed to the partially inelastic model used in design). A total of eight pairs of ground motion records were used (an extra pair was added to those used for design, and scaling factors were all adjusted accordingly in the new set). Verifications regarding interstorey drifts and plastic rotations were carried out for different levels of seismic action (50%/50 years, 10%/50 years and 2%/50 years), related to serviceability, life safety and collapse prevention objectives. Additional to the set of analyses based on stiffness assumptions corresponding to moderate levels of inelasticity ($EI_{ef} = 0.5EI_g$), extra analyses were carried out, where the secant stiffness of the cracked section at yield, $EI_{ef} = M_y/\varphi_y$, was used for all R/C members.

From the drifts at the serviceability-related earthquake shown in Fig. 14.4, it is clear that the seismic performance of both the EC8 designs and the building

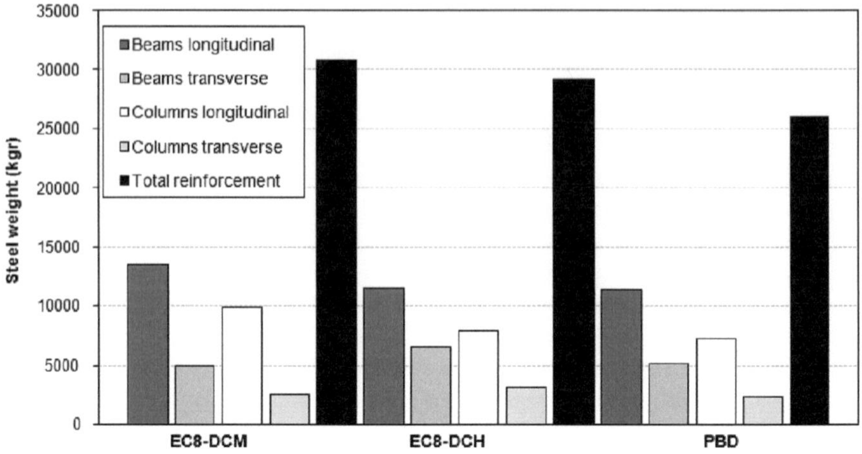

Fig. 14.3 Required amount of steel in beams and columns for code design (EC8) and PBD

designed for target deformations having the same cross-sections was very satisfactory. Moreover, the maximum value (average of 7 motions) of interstorey drift ratio, was equal to 0.32% for the PBD Building 1 (recorded at the 9th storey, i.e. at the setback), and increased to only 0.35% when a number of cross-sections were reduced ("Building 2"). As far as the development of plastic hinge rotations is concerned, the values obtained from the results of inelastic response-history analysis are significantly lower than the adopted serviceability limits (maximum value equal to about 0.002 and 0.003 for buildings 1 and 2, respectively).

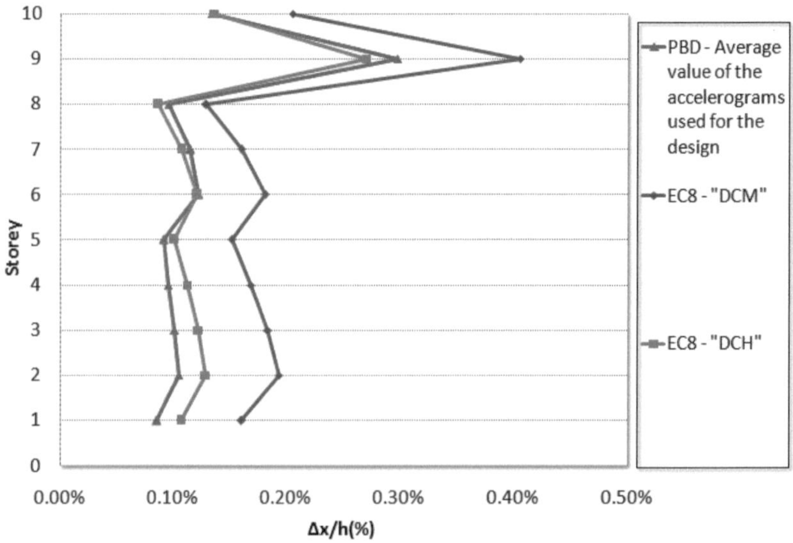

Fig. 14.4 Serviceability verification: (**a**) Building 1 – interstorey drifts in *x*-direction

From the several results of the performance assessment of the alternative designs of the irregular 10-storey building for the various levels of earthquake intensity, reported in detail in [14], which showed that both the EC8-designed buildings and those designed to the PBD satisfied the "life safety" criteria for the 10%/50 years event and the "collapse prevention" criteria for the 2%/50 years event, a potentially critical situation is shown in Fig. 14.5. It refers to the case that the 8 pairs of records were scaled to the intensity of the 2%/50 years earthquake and all R/C members were modelled with the reduced stiffness ($EI_{ef} = M_y/\varphi_y$), i.e. lower than those used for design; furthermore, the results are for Building 2 (reduced cross-sections), hence this is expected to be a critical case.

It is noted in Fig. 14.5 that even in this extreme case the maximum drift value (average of 8 records) is equal to 1.4% for Building 2 (and 1.3% for Building 1, not shown in Fig. 14.5), values that fall well below the allowable limits for R/C frame structures [7]. It is noted that analysis results should be interpreted on the basis of the average of the calculated values of each response-history analysis set, since the scaling procedure was based on the consideration of a mean spectrum. As depicted in Fig. 14.5, some analysis results (typically the ones concerning the synthetic ground motion in the set) can lead to an overestimation of interstorey drift values.

Finally, regarding the plastic hinges developed, the corresponding rotations were quite low in all cases, while the values of column plastic hinge rotations are very low compared to those in beams [14]; hence a ductile failure mechanism is ensured for all limit states considered.

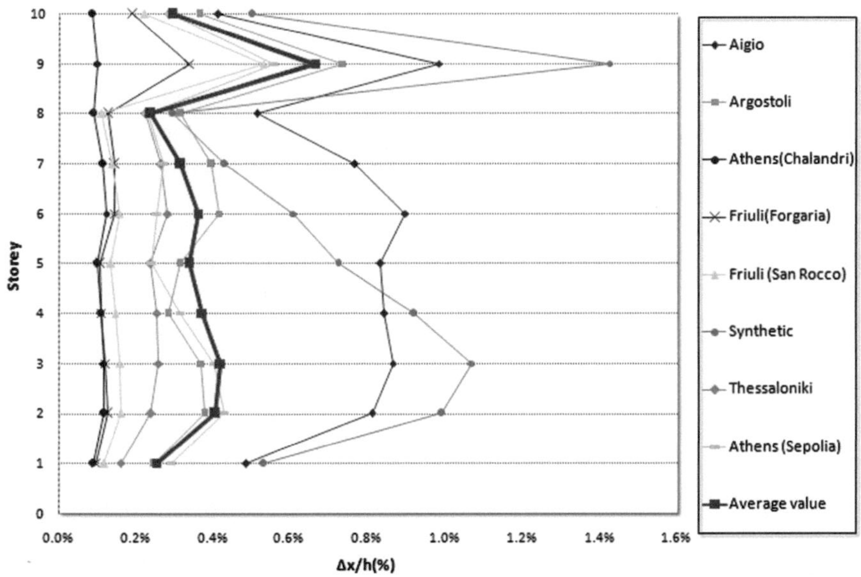

Fig. 14.5 Assessment of building 2: interstorey drifts in x-direction for the "life safety" performance level, $EI_{eff} = 0.5EI_g$

14.4 Concluding Remarks

The main goal of the presented study was the development of a seismic design procedure for 3D R/C structures that leads to adequate seismic reliability, concerning the inelastic deformations as a parameter of design and not as an a-posteriori control parameter for the verification of seismic performance. A key point of the proposed method is that the degree of damage allowed for every distinct performance level can be related to the allowable values of inelastic deformations and eventually to reduced design forces of structural elements. Hence, the seismic design, in addition to ensuring the development of a ductile plastic mechanism, also predefines a desirable level of inelastic deformation. Extension of the method to multi-storey buildings with setbacks was deemed necessary in order to explore its applicability. Furthermore, the structures considered were also designed according to the provisions of EC8 for two different ductility classes (M and H), to compare, in economic and performance terms, the results of the two design methods.

Assessment of the multi-storey buildings with setbacks designed according to the proposed deformation-based design method, was found to lead to a very satisfactory seismic performance under earthquake levels associated with life safety and collapse prevention. Furthermore, a worst-case scenario assuming secant values at yield for member rigidities and the "rare" earthquake level (related to the collapse-prevention objective) has shown that the PBD-designed building still performs satisfactorily, since the estimated drifts are well within the allowable values for R/C frames.

Notwithstanding its greater complexity compared to the current code procedures, the results of applying the proposed method to the design of irregular structures were encouraging. Since the deformation-based method accounts for the design according to the inelastic deformations anticipated for every performance level, basically the ductility of each member, the cross-sections required for the specific performance can be defined. Eventually, by designing according to the proposed deformation-based design method, economy is obtained (in comparison to Code design), concerning not only the cross-sections used but also the reinforcement requirements (especially the transverse reinforcement of columns).

Finally, it should be noted that, so far, the validity of the procedure has been verified for the case of regular and irregular 3D R/C buildings having a frame structural system. Clearly the relevance and feasibility of the proposed approach will emerge by applying it also to dual (wall + frame) systems.

References

1. Ambraseys N, Smit P, Berardi R, Rinaldis D, Cotton F, Berge C (2000) Dissemination of European strong-motion data, CD-ROM collection, European Commission, DGXII, Science, Research and Development, Bruxelles
2. ASCE/SEI (2007) Seismic rehabilitation of existing buildings – ASCE Standard 41-07. American Society of Civil Engineers, Reston, VA
3. Carr A (2004) RUAUMOKO, Manuals, "Vol. 1 Theory and User Guide to Associated programs, Vol. 3 User Manual for the 3-Dimensional Version" , University of Canterbury, New Zealand

4. CEN (Comité Européen de Normalisation) (2004a) Eurocode 2: design of concrete structures – part 1: general rules and rules for buildings (EN 1992-1-1: 2004). CEN, Brussels
5. CEN (2004b) Eurocode 8: design of structures for earthquake resistance – part 1: general rules, seismic actions and rules for buildings (EN 1998-1: 2004). CEN, Brussels
6. CEN (2005) Eurocode 8: design provisions of structures for earthquake resistance – part 2: bridges (EN1998-2:2005). CEN, Brussels
7. Dymiotis C, Kappos AJ, Chryssanthopoulos MC (1999) Seismic reliability of R/C frames with uncertain drift and member capacity. J Struct Eng ASCE, 125(9):1038–1047
8. FEMA (1997) NEHRP guidelines for the seismic rehabilitation of buildings, FEMA 273,Washington, DC, October 1997
9. *fib* (2003) Displacement-based seismic design of reinforced concrete buildings, *fib* Bull 25 Lausanne
10. Kappos AJ (1997) Partial inelastic analysis procedure for optimum capacity design of buildings. Proceedings of international workshop on seismic design methodologies for the next generation of codes (Bled, Slovenia, June 1997), Balkema, pp 229–240
11. Kappos AJ, Goutzika E, Stefanidou S (2007) An improved performance-based design method for 3D R/C buildings using inelastic analysis. ECCOMAS thematic conference on computational methods in structural dynamics and earthquake engineering, Rhodes, Greece
12. Kappos AJ, Manafpour A (2001) Seismic design of R/C buildings with the aid of advanced analytical techniques. Eng Struct 23(4):319–332
13. Kappos AJ, Panagopoulos G (2004) Performance-based seismic design of 3D R/C buildings using inelastic static and dynamic analysis procedures. ISET J Earthq Technol 41(1):141–158
14. Kappos AJ, Stefanidou S (2009) A deformation-based seismic design method for 3D R/C irregular buildings using inelastic dynamic analysis. Bull Earthq Eng (Springer), published online 29-12-09, DOI: 10.1007/s10518-009-9170-1
15. Panagiotakos TB, Fardis MN (2001) A displacement-based seismic design procedure for R/C buildings and comparison with EC8. Earthq Eng Struct Dyn 30:1439–1462
16. Penelis GG, Kappos AJ (1997) Earthquake-resistant concrete structures. E&FN SPON, London
17. Priestley MJN (2000) Performance based seismic design. Bull New Zealand Soc Earthq Eng. 33(3):325–346
18. SEAOC Seismology Committee (1999) Recommended lateral force requirements and Commentary (Blue Book). Sacramento, CA
19. Sullivan TJ, Calvi GM, Priestley MJN, Kowalski MJ (2003) The limitations and performances of different displacement based design methods. J Earthq Eng 7(Special Issue 1):201–244

Chapter 15
A New Seismic Design Method for Steel Structures

Theodore L. Karavasilis, Nikitas Bazeos, and Dimitri E. Beskos

15.1 Introduction

The current procedure for seismic design of new building structures is termed force-based design (FBD) since it uses seismic forces as the main design parameters. This design method [2] demands the design of the building against structural failures which might endanger human life on the basis of recommended constant values of the behavior (or strength reduction) factor, q, and checks deformations beyond which service requirements are no longer met after the detailing of the structure. The tentative guidelines for performance-based seismic design (PBSD) according to SEAOC [4] present two alternative forms of displacement-based design of new structures, namely, the direct displacement-based design (DDBD) method and the equal-displacement-based design (EDB) one. Contrary to FBD, those displacement-based procedures employ the maximum interstorey drift ratio (*IDR*) for describing performance levels and also treat user-defined values of the *IDR* as input variables for the initiation of the design process. However, these methods are limited in that they are applicable only to regular frames, adopt an equivalent linear (DDBD) or nonlinear (EBD) SDOF representation of the building, do not recognise basic differences in the response due to different lateral load resisting systems and try to control both structural and non-structural damage by imposing limits only on drift demands.

This paper proposes a performance-based seismic design methodology for steel building frames which combines the advantages of the well-known force-based and displacement-based seismic design methods in a hybrid force/displacement (HFD) design scheme. The method has been evolved from previous preliminary works of the authors [5, 6] and its present (latest) version is firmly supported by extensive parametric studies of the authors [9–11, 7] on the inelastic seismic response of planar steel frames. The main characteristic of the proposed method are: (1) it treats both drift and ductility demands as input variables for the initiation of the design

T.L. Karavasilis (✉)
Department of Engineering Science, University of Oxford, Oxford OX1 3PJ, UK
e-mail: theodore.karavasilis@eng.ox.ac.uk

M.N. Fardis (ed.), *Advances in Performance-Based Earthquake Engineering*, 161
Geotechnical, Geological, and Earthquake Engineering 13, DOI 10.1007/978-90-481-8746-1_15,

process; (2) it does not use a substitute single degree of freedom system; (3) it makes use of current seismic code approaches as much as possible (e.g., conventional elastic response spectrum analysis and design); (4) it includes the influence of the number of storeys; (5) it recognizes the influence of the type of the lateral load resisting system (moment resisting frame (MRF) or concentrically braced frame (CBF)); (6) it recognizes the influence of geometrical (setbacks) or mass irregularities. A recent comparison of the HFD, FBD and DDBD procedures for plane regular steel MRF yielded favorable results for HFD [1], while recent research work explores the extension of the method to the case of pulse-like (near-fault or soft-soil sites) earthquake ground motions [12].

15.2 Steps of Proposed Design Method

The proposed hybrid force/displacement (HFD) seismic design procedure can be summarized in the following steps:

(1) *Definition of the basic building attributes.* With reference to the types of frames depicted in Fig. 15.1, definition of the number of storeys, n_s, number of bays, n_b, bay widths and storey heights, presence of setbacks (geometrical irregularity), different use of a specific floor compared to the adjacent ones (mass irregularity) and limits on the depth of beams and columns due to possible architectural requirements.

(2) *Definition of the design performance level.* For example, immediate occupancy (IO) under the frequently occurred earthquake (FOE), life safety (LS) under the design basis earthquake (DBE) or collapse prevention (CP) under the maximum considered earthquake (MCE).

(3) *Definition of the input parameters (performance and irregularity metrics).*

(i) *Performance metrics.* Definition of the acceptable values of the maximum *IDR* and maximum local ductility (rotation ductility μ_θ for beams/columns and cyclic elongation ductility μ_{cb} for braces) along the height of the frame.
(ii) *Irregularity metrics.* For MRF with setbacks, quantification of the geometrical irregularity through the indices Φ_s and Φ_b which, with reference to Fig. 15.2, are given by the formulae [10]

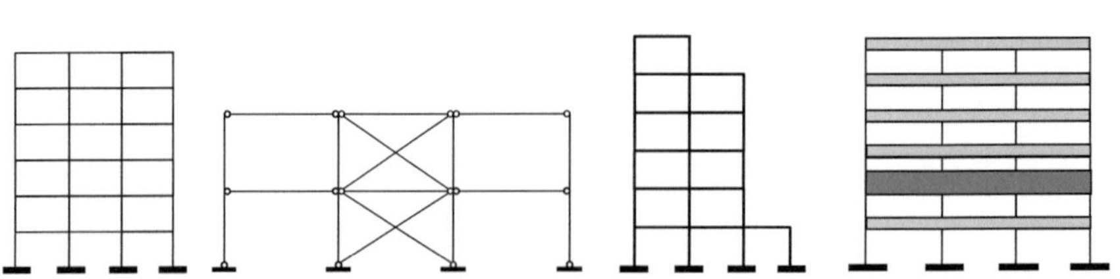

Fig. 15.1 Types of planar steel building frames considered in the HFD seismic design method

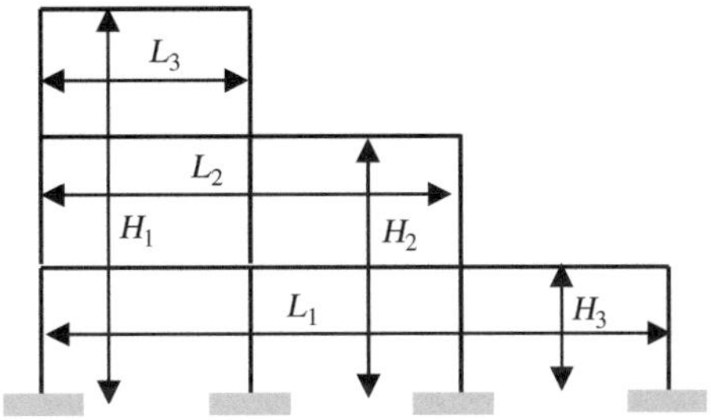

Fig. 15.2 Geometry of frame with setbacks for definition of geometrical irregularity indices

$$\Phi_s = \frac{1}{n_s - 1} \cdot \sum_{i=1}^{i=n_s-1} \frac{L_i}{L_{i+1}} \quad \Phi_b = \frac{1}{n_b - 1} \cdot \sum_{i=1}^{i=n_b-1} \frac{H_i}{H_{i+1}} \tag{15.1}$$

For MRF with mass irregularity, definition of the location (bottom, mid-height or top of the building) of the location (bottom, mid-height or top of the building) of the mass discontinuity [11].

(4) *Estimation of the input variables (yield roof displacement and mechanical characteristics).*

(i) *Estimation of the yield roof displacement,* u_{ry}. This displacement corresponds to the formation of the first plastic hinge (for MRF) or to the initiation of buckling (for CBF).
(ii) *Estimation of mechanical characteristics.* For MRF, estimation of the column-to-beam strength ratio, *a*, and beam-to-column stiffness ratio, ρ [9]

$$a = \frac{M_{RC,1,av}}{M_{RB,av}} \quad \rho = \frac{\sum (I/l)_b}{\sum (I/l)_c} \tag{15.2}$$

where $M_{RC,1,av}$ is the average of the plastic moments of resistance of the columns of the first storey, $M_{RB,av}$ is the average of the plastic moments of resistance of the beams of all the storeys of the frame, and *I* and *l* are the second moment of inertia and length of the steel member (column c or beam b). For CBF, estimation of the fundamental period of vibration, *T*, brace slenderness, λ, and ratio, a (contribution of the columns over that of the diagonals to the storey stiffness) [7]

$$\lambda = \frac{l}{\pi \cdot r} \cdot \sqrt{\frac{f_y}{E}} \quad \alpha = \frac{\sum_{i=1}^{N} EI}{k \cdot h^3} \quad k = \frac{V_{base}}{\Delta_{1y}} \tag{15.3}$$

where *l* is the buckling length, *r* is the radius of gyration of the cross section, f_y is the yield strength of the material, *E* is the Young's modulus, *I* is the second moment of inertia of the columns, *h* is the storey height, *k* the storey stiffness, V_{base} is the frame shear strength at the base and Δ_{1y} is the yield displacement of first storey.

Good initial estimates of the aforementioned input variables may be obtained by designing the frame only for strength requirements under the FOE earthquake by assuming elastic behaviour, i.e., with $q = 1$. The capacity design rules and the gravity load combination should be also considered in order to improve the initial estimation of the input variables of the proposed method.

The characteristics λ, a and ρ vary along the height of a steel frame and therefore, their nominal values are taken equal to those of the storey closest to mid-height of the building.

(5) *Transformation of local performance metrics to target roof displacement.* Transformation of the $IDR_{\max}$ to target maximum roof displacement, u_{rmax}, by employing the relation

$$u_{\mathrm{rmax}(IDR)} = \beta \cdot IDR_{\max} \cdot H \tag{15.4}$$

where H is the building height from its base and β is a coefficient depending on building properties and calculated through

$$\beta = 1 - 0.19 \cdot (n_s - 1.0)^{0.54} \cdot \rho^{0.14} \cdot \alpha^{-0.19} \tag{15.5}$$

for regular MRF [9];

$$\beta = 1 - 0.42 \cdot (n_{\mathrm{s}} - 1.0)^{-0.04} \cdot \lambda^{-0.706} \cdot \alpha^{-0.038} \cdot T^{0.13} \cdot \gamma^{-0,029} \tag{15.6}$$

for regular CBF [7], where γ is the distribution of strength up the height of the frame according to MacRae et al. [8];

$$\beta = 1 - 0.13 \cdot (n_{\mathrm{s}} - 1.0)^{0.52} \cdot \Phi_{\mathrm{s}}^{0.38} \cdot \Phi_{\mathrm{b}}^{0.14} \tag{15.7}$$

for irregular MRF with setbacks [10] and finally;

$$\beta = 1 - 0.18 \cdot (n_{\mathrm{s}} - 1)^{0.4} \cdot a^{-0.13} \tag{15.8a}$$

$$\beta = 1 - 0.17 \cdot (n_{\mathrm{s}} - 1)^{0.64} \cdot a^{-0.47} \tag{15.8b}$$

$$\beta = 1 - 0.12 \cdot (n_{\mathrm{s}} - 1)^{0.66} \cdot a^{-0.28} \tag{15.8c}$$

for irregular MRF with mass discontinuities located near to bottom, midheight and top of the building, respectively [11].

Transformation of local ductility to target roof displacement by employing the relation

$$u_{\mathrm{rmax}(\mu)} = \mu \cdot u_{\mathrm{ry}} \tag{15.9}$$

where the roof displacement ductility, μ, is associated with local ductility through

$$\mu = 1 + 1.35 \cdot (\mu_\theta - 1)^{0.86} \cdot \alpha^{0.43} \cdot n_s^{-0.31} \quad (15.10)$$

for regular MRF [9];

$$\mu = 1 + 0.84 \cdot (\mu_b - 1)^{0.56} \cdot n_s^{0.11} \cdot \lambda^{0.49} \cdot a^{0.119} \quad (15.11)$$

for regular CBF [7];

$$\mu = 1 + 0.44 \cdot (\mu_\theta - 1)^{1.26} \cdot \alpha^{0.26} \quad (15.12)$$

for irregular MRF with setbacks [10] and finally;

$$\mu = 1 + 0.50 \cdot (\mu_\theta - 1)^{1.12} \cdot \alpha^{0.36} \quad (15.13a)$$

$$\mu = 1 + 0.78 \cdot (\mu_\theta - 1) \quad (15.13b)$$

$$\mu = 1 + 0.50 \cdot (\mu_\theta - 1)^{1.18} \cdot \alpha^{0.39} \quad (15.13c)$$

for irregular MRF with mass discontinuities located near to the bottom, midheight and top of the building, respectively [11].

The design target roof displacement, $u_{\mathrm{rmax(d)}}$, is defined as the minimum value of $u_{\mathrm{rmax(IDR)}}$ and $u_{\mathrm{rmax}(\mu)}$.

(6) *Calculation of the behavior (or strength reduction) factor.* Calculation of the design value of the roof displacement ductility

$$\mu_d = \frac{u_{r\,\mathrm{max(d)}}}{u_{\mathrm{ry}}} \quad (15.14)$$

and then, calculation of the required strength reduction factor

$$q = 1 + 1.39 \cdot (\mu_d - 1) \quad \text{for } \mu_d \leq 5.8 \quad (15.15a)$$

$$q = 1 + 8.84 \cdot (\mu_d^{0.32} - 1) \quad \text{for } \mu_d > 5.8 \quad (15.15b)$$

for regular MRF [9];

$$q = 1 + 0.28 \cdot (\mu - 1)^{0.41} \cdot \lambda^{-1.06} \cdot \alpha^{-0.19} \cdot n_s^{0.9} \cdot T^{0.18} \quad (15.16)$$

for regular CBF [7];

$$q = 1 + 1.92 \cdot (\mu_d - 1)^{0.85} \cdot \Phi_s^{-0.17} \quad (15.17)$$

for irregular MRF with setbacks [10] and finally;

$$q = 1 + 2.26 \cdot (\mu_d - 1)^{0.69} \tag{15.18a}$$

$$q = 1 + 2.42 \cdot (\mu_d - 1)^{0.68} \tag{15.18b}$$

$$q = 1 + 2.45 \cdot (\mu_d - 1)^{0.60} \tag{15.18c}$$

for irregular MRF with mass discontinuities located near to the bottom, midheight and top of the building, respectively [11].

(7) *Design of the structure.* Divide the ordinates of the elastic design spectrum with the q factor and design the building on the basis of an elastic response spectrum analysis by taking into account the capacity and ductile design rules of seismic codes [2]. The design is strength-based, i.e., the designer does not need to impose limits on the required stiffness (or period of vibration) of the frame. The required stiffness (or period for given mass) of the frame is controlled by the assumed value of the yield displacement (see step 4), while the required strength is imposed by the value of the strength reduction factor.

(8) *Iteration.* Iterate with respect to the input variables u_{ry}, ρ, α, T, λ, a. The sufficient number of iterations for achieving convergence depends on the initial estimations of the input variables (see step 4 of the method). Good initial estimates of the aforementioned input variables can be easily obtained by designing the frame only for strength requirements under the FOE earthquake by assuming elastic behaviour, i.e., with $q = 1$ (see design example in the next section of the paper).

15.3 Application of the Proposed HFD Design Method

15.3.1 Description of Building and Design Assumption

The proposed HFD and the already used in practice FBD method are applied to the seismic design of the 5-storey office S275 steel building shown in Fig. 15.3. The building has storey heights equal to 3 m and bay widths equal to 6 m. Lateral load moment resisting frames are located only at the perimeter of the building, while

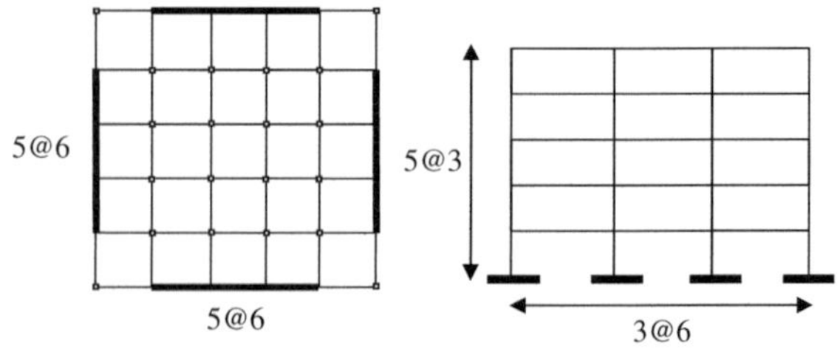

Fig. 15.3 Five storey steel office building structure

gravity load resisting frames are arranged in the interior of the building. Only one perimeter lateral load resisting frame was considered in analysis, while a "dummy" column was used for simulating the gravity load columns and P-Δ effects. Braced and irregular MRF frames are not examined herein since their treatment is similar with that of regular MRF, while simple design examples for braced and irregular MRF can be found in Karavasilis et al. [10, 11] and Zotos and Bazeos [7].

15.3.2 Definition of Seismic Performance Levels

For the office building considered herein, it is assumed that immediate occupancy (IO) under the frequently occurred earthquake (FOE), life safety (LS) under the design basis earthquake (DBE) and collapse prevention (CP) under the maximum considered earthquake (MCE) are the appropriate performance levels for seismic design. The FOE, DBE and MCE earthquakes are expressed through the elastic design spectrum of EC8 for soil class B and damping ratio equal to 3%, while assumed seismological data provide the corresponding values of the peak ground acceleration (*PGA*) as $PGA_{\mathrm{DBE}}=0.35g$, $PGA_{\mathrm{FOE}}=0.3 \times PGA_{\mathrm{DBE}}=0.1g$ and $PGA_{\mathrm{MCE}}=1.5 \times PGA_{\mathrm{DBE}}=0.53g$ (Fig. 15.4). The design criteria (target drift and local ductility values) for special moment resisting frames (SMRF) are adopted from FEMA-273 [3] and provided in Table 15.1.

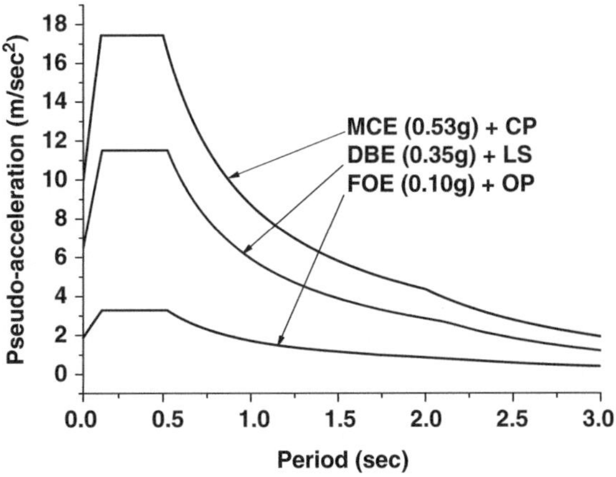

Fig. 15.4 EC8 elastic design spectra

Table 15.1 Target values of performance metrics

IO		LS		CP	
IDR	μ_{local}	*IDR*	μ_{local}	*IDR*	μ_{local}
0.7%	2.00	2.5%	7.00	5%	9.00

15.3.3 Moment Resisting Frames (MRF)

An initial elastic design for the FOE earthquake yields "HEB400-IPE400, HEB400-IPE500, HEB360-IPE400, HEB360-IPE400, HEB340-IPE300" sections for the five storeys of the frame, respectively. According to Eq. (15.2), the values of the parameters ρ and a of the frame are equal to 0.2 and 2.4, respectively, while roof displacement under the FOE response spectrum equals 0.1 m. The above values serve as initial estimates for the input variables of the HFD. The target values of the $IDR_{\max}$ and μ_θ for the LS performance level are equal to 2.5% and 7.0, respectively [3]. The value of the ratio β is calculated on the basis of Eq. (15.5) and found to be equal to 0.73. By employing Eq. (15.4), the target roof displacement $u_{\mathrm{rmax(IDR)}}$ is calculated equal to $0.73 \times 2.5\% \times (5 \times 3) = 0.27$ m, while by employing Eq. (15.10) the displacement ductility μ is found to be equal to $1 + 1.35 \times (7–1)^{0.86} \times 2.4^{0.43} \times 5^{-0.31}$=6.6 and therefore, the target roof displacement $u_{\mathrm{rmax}(\mu)}$ becomes equal to 0.1×6.6=0.66 m. The design roof displacement is equal to min (0.27, 0.66) = 0.27 and therefore, drift controls the LS performance level design. The design value of the roof displacement ductility is 0.27/0.1=2.7. The required strength reduction factor is calculated on the basis of Eq. (15.15) and found to be equal to 3.4. The DBE response spectrum is reduced by this factor and the design yields "HEB450-IPE400, HEB450-IPE450, HEB400-IPE450, HEB400-IPE400, HEB360-IPE360" sections for the five storeys of the frame, respectively. The new values of the input variables of HFD are u_{ry}=0.095, ρ=0.22 and a=3. A second execution of the hand calculations of HFD provides a value of q equal to 3.5, which does not change the sections obtained with response spectrum analysis/design and therefore, the design with respect to the LS performance level is finalized. Apparently, this frame remains elastic under the IO response spectrum since it has larger sections than the frame which serves to provide initial estimates of the input variables. Elastic analysis under the IO response spectrum provides a value of the *IDR* equal to 0.65% and roof displacement equal to 0.09 m and therefore, the frame which satisfies the LS performance level satisfies also the IO performance level. The same initial values (u_{ry}=0.1, ρ=0.2 and a=2.4) are used for designing the frame with respect to the CP damage state under the MCE earthquake. The first execution of the hand calculations of HFD prove again that drift controls design with a corresponding value of the q factor equal to 7.2 which is used to divide the MCE spectrum. A response spectrum analysis/design under the reduced MCE spectrum provides a lighter frame than the one obtained by designing with respect to the LS performance level. Therefore, the LS performance level controls the design of the frame. Strictly speaking, the designer may want to obtain the expected values of the u_{r}, *IDR* and μ_θ under all earthquake intensities. Under the FOE earthquake, the frame remains elastic and thus μ_θ=1.0, while the u_{rmax} and $IDR_{\max}$ were calculated equal to 0.65% and 0.09 m, respectively. Under the DBE earthquake, the $IDR_{\max}$ and u_{rmax} were calculated equal to 2.5% and 0.27 m, respectively, while by using the values q=3.5 and a=3 in Eq. (15.15) and Eq. (15.10), the μ_θ is calculated equal to 2.41. The q factor of the frame under the MCE earthquake is easily obtained as $(PGA_{\mathrm{MCE}} / PGA_{\mathrm{LS}}) \times q_{\mathrm{LS}} = 1.5 \times 3.5 = 5.25$. This value is used in order to estimate the response of the frame under the MCE earthquake, i.e., u_{rmax}=0.41 m, IDR_{max}=3.65% and μ_θ=3.67.

According to EC8 [2], the FBD of the frame starts by performing strength-based (no drift control) design under the DBE earthquake on the basis of a constant value of the q factor which equals to 6.5 (= 5 × 1.3, where 1.3 is the recommended overstrength factor for MRF) which yields "HEB240-IPE300, HEB240-IPE300, HEB220-IPE270, HEB220-IPE270, HEB200-IPE240" sections for the five storeys of the frame, respectively. Then, EC8 demands drift control for the IO earthquake. The latter frame behaves inelastically under the IO earthquake, while the equal-displacement rule provides a value of the IDR_{max} equal to 1.23% which is larger than the 0.7% drift limit recommended by FEMA [3]. Thus, an iterative design process is needed in order to find larger sections of the frame that satisfy this drift limit. Finally, the sections which are obtained are "HEB400-IPE400, HEB400-IPE500, HEB360-IPE400, HEB360-IPE400, HEB340-IPE300". This frame remains elastic under the IO earthquake and experiences u_{rmax}=0.10 m, IDR_{max}=0.7% and μ_θ=1.0. According to FBD, the IO performance level controls the design. Under the DBE earthquake, the frame will experience u_{rmax}=3.5 × 0.1=0.35 m and IDR_{max}=3.5 × 0.7=2.45%, while under the MCE earthquake the frame will experience u_{rmax}=5.3 × 0.10=0.53 m and IDR_{max}=5.3 × 0.7=3.7. These drifts are calculated in order to check (see next section of the paper) the validity of the equal-displacement rule and they are not used to check the performance of the frame under the FOE and DBE earthquake, i.e., they should not be compared with the target drift values of Table 3.2. Even if the target drifts of Table 3.2 were smaller (for instance, IDR=1.5% under the DBE earthquake), the design product of FBD would be the same since it satisfies drifts under the FOE earthquake and strength under the DBE earthquake. Strictly speaking, FBD may be used for drift control for any performance level by employing the iterative process which was employed for the FOE earthquake.

Data of the MRF frames designed according to the HFD and FBD procedures are presented in Table 15.2. The MRF obtained by using FBD is slightly lighter than the one obtained by using HFD.

Table 15.2 Data pertinent to the designed frames

	HFD	FBD (EC8)
Sections	HEB450-IPE400 HEB450-IPE450 HEB400-IPE450 HEB400-IPE400 HEB360-IPE360	HEB400-IPE400 HEB400-IPE500 HEB360-IPE400 HEB360-IPE400 HEB340-IPE300
T(s)	1.70	1.75
Steel weight	147	144
Controlling performance level	LS	IO

15.3.4 Evaluation of the Design Through Nonlinear Dynamic Analyses

Ten semi-artificial accelerograms, compatible with the frequency content of the EC8 [5] design spectrum, were generated via a deterministic approach. The ordinates of these ground motions were scaled in order to match the three design levels of seismic intensity considered here and then, nonlinear time-history analyses were executed by using the program ANSR-PC [13].

The mean values of the maximum response quantities from time-history analyses (TH) are compared with the estimations (EST) of the HFD and FBD design methods in Table 15.3. The results reveal the consistency of HFD to accurately estimate inelastic deformation demands and the tendency of FBD to overestimate the maximum roof displacement and to underestimate the maximum interstorey drift ratio along the height of the frame.

Table 15.3 Time history analyses results and comparison with design estimations (values in parentheses)

	HFD			FBD (EC8)		
	FOE	DBE	MCE	FOE	DBE	MCE
$IDR_{max}(\%)$	0.63	2.4	3.75	0.67	3.00	4.20
	(0.65)	(2.5)	(3.65)	(0.70)	(2.45)	(3.70)
$u_{max}(m)$	0.11	0.26	0.43	0.12	0.31	0.46
	(0.09)	(0.27)	(0.41)	(0.11)	(0.35)	(0.53)
μ_θ	1.0	2.7	3.6	1.0	2.8	3.8
	(1.0)	(2.4)	(3.7)	(1.0)	–	–

15.4 Conclusions

A performance-based seismic design methodology for steel building frames which combines the advantages of the well-known force- and displacement-based seismic design methods in a hybrid force/displacement (HFD) design scheme has been proposed. The method has been applied to the case of a realistic 5-storey office building structure and compared with the force-based design (FBD) procedure of the EC8 seismic code. The advantages of the HFD method over the FBD method were illustrated by exploring (a) the ability of the methods to identify the performance level which truly controls the design and (b) by comparing the inelastic deformation estimates of both methods with the results of the rigorous nonlinear dynamic analysis. More specifically, the HFD method identified the life safety performance level as the critical one, while the FBD design was controlled by the immediate occupancy performance level. The results of the nonlinear time history analyses for three earthquake intensities (FOE, DBE and MCE) revealed the consistency of the HFD to accurately estimate inelastic deformation demands and the tendency of the FBD to overestimate the maximum roof displacement and to underestimate the maximum interstorey drift ratio along the height of the frame.

Acknowledgments The first two authors are grateful for the support provided to them through the "K. Karatheodoris" research program of the University of Patras, Greece. All the authors acknowledge helpful discussions with Professors S.A. Anagnostopoulos, M.N. Fardis and N. Makris.

References

1. Bazeos N (2009) Comparison of three seismic design methods for plane steel frames. Soil Dyn Earthq Eng 29(3):553–562
2. CEN (2004) EN 1998-1:2004 Eurocode 8. Design of structures for earthquake resistance, part 1: general rules, seismic actions and rules for buildings. European Committee for Standardization, Brussels
3. FEMA (1997) Federal Emergency Management Agency. NEHRP guidelines for the seismic rehabilitation of buildings, Report FEMA-273, Washington, DC
4. SEAOC (1995) Structural Engineers Association of California. Vision 2000 – A Framework for performance based earthquake engineering, Vol. 1. SEAOC, Sacramento, CA
5. Karavasilis TL, Bazeos N, Beskos DE (2006a) A hybrid force/displacement seismic design method for plane steel frames. In: Mazzolani F, Wada A (eds), Behavior of Steel Structures in Seismic Area, Proceedings of STESSA Conference, pp 39–44, Yokohama, Japan, August, Taylor & Francis
6. Karavasilis TL, Bazeos N, Beskos DE (2006b) A hybrid force/displacement seismic design method for plane steel frames. Proceedings of 1st European conference on earthquake engineering and seismology (1st ECEES), Geneva, Switzerland, 3–8 September, Paper No 1013
7. Zotos PC, Bazeos N (2009) Estimation of seismic response in planar x-braced multi-storey steel frames. Proceedings of 2nd international conference on computational methods in structural dynamics and earthquake engineering, Island of Rhodes, Greece, 22–24 June, Paper CD364
8. MacRae GA, Kimura Y, Roeder C (2004) Effect of column stiffness on braced frame seismic behavior. J Struct Eng ASCE 130:381–391
9. Karavasilis TL, Bazeos N, Beskos DE (2008a) Drift and ductility estimates in regular steel MRF subjected to ordinary ground motions: a design-oriented approach. Earthq Spectra 24(2):431–151
10. Karavasilis TL, Bazeos N, Beskos DE (2008b) Seismic response of plane steel MRF with setbacks: estimation of inelastic deformation demands. J Constr Steel Res 64(6):644–654
11. Karavasilis TL, Bazeos N, Beskos DE (2008c) Estimation of seismic inelastic deformation demands in plane steel MRF with vertical mass irregularities. Eng Struct 30(11):3265–3275
12. Karavasilis TL, Makris N, Bazeos N, Beskos DE (2010) Dimensional response analysis of multi-storey regular steel MRF subjected to pulse-like earthquake ground motions. J Struct Eng (ASCE), in press
13. Maison BF (1992) PC-ANSR. A computer program for nonlinear structural analysis

Chapter 16
Significance of Modeling Deterioration in Structural Components for Predicting the Collapse Potential of Structures Under Earthquake Excitations

Helmut Krawinkler, Farzin Zareian, Dimitrios G. Lignos, and Luis F. Ibarra

16.1 Introduction

A primary objective of good earthquake engineering is to provide an adequate margin of safety against collapse. Collapse prevention is not an absolute goal, because ground motion uncertainties and economic constraints will always necessitate the acceptance of a small probability of collapse. Assessment of this small probability of collapse requires the ability to predict, with sufficient confidence, the response of structures all the way to collapse. This aspect of performance-based earthquake engineering has posed major challenges, some of which are discussed in this paper.

Observations of collapsed buildings in past earthquakes show that two modes of collapse can be envisioned for a building: sidesway collapse and vertical collapse. Sidesway collapse is caused by large story drifts that lead to a successive reduction of load carrying capacity of structural components that are part of the building's lateral load resisting system, to the extent that second order (P-Delta) effects overcome gravity load resistance. In contrast, vertical collapse is the consequence of direct loss of gravity load carrying capacity in one or several structural components. This paper is concerned with sidesway collapse.

Analytical tools devised by researchers to predict the collapse of a building subjected to a ground motion record range from SDOF representation of the building (e.g., [8]) to sophisticated Finite-Element-based MDOF models that can predict the successive failures of structural components up to collapse (e.g., [9]). Paramount throughout this spectrum of analytical tools developed to predict building collapse is the ability to model strength and stiffness deterioration of structural components when subjected to cyclic loading (Lignos and Krawinkler, [7]).

The collapse capacity of a building is defined here as the maximum ground motion intensity (often represented by the spectral acceleration at the first mode period) at which the structural system still maintains dynamic stability. Different

H. Krawinkler (✉)
Department of Civil and Environmental Engineering, Stanford University, Stanford, CA 94305, USA
e-mail: krawinkler@stanford.edu

M.N. Fardis (ed.), *Advances in Performance-Based Earthquake Engineering*, Geotechnical, Geological, and Earthquake Engineering 13, DOI 10.1007/978-90-481-8746-1_16,

ground motions will lead to different collapse capacities because of the inherent record-to-record (RTR) variability. A collapse fragility curve that incorporates aleatory uncertainty due to RTR variability can be obtained by ordering the collapse capacities for a representative set of ground motions. Additional variability in the collapse potential is caused by epistemic uncertainties due to modeling assumptions and variability in the parameters on which analytical predictions are based. Probabilistic approaches have been introduced that integrate possible sources of variability in the process of collapse prediction and present the buildings collapse potential in the form of probability of collapse (e.g., [11]).

In the context of seismic performance assessment, the collapse potential of a building can be expressed as the probability of collapse at a discrete hazard level or the mean annual frequency of collapse, both of which can be computed from the collapse fragility curve and the hazard curve for the site of the structure. Assessment of this collapse potential has been the goal of focused research at Stanford University and other institutions for the last 10 years. An attempt is made here to present, to various degrees of detail, the process utilized to accomplish this goal. The essential components of this process are as follows:

- Modeling of strength and stiffness deterioration in structural components
- Modeling of the structure so that its response can be predicted all the way to collapse through nonlinear dynamic analysis
- Representation of the seismic input and prediction of the collapse capacity
- Incorporation of measures of uncertainty in expressing safety against collapse.

16.2 Modeling of Strength and Stiffness Deterioration in Structural Components

16.2.1 Observations on Component Behavior

Figure 16.1a shows the monotonic and cyclic moment-chord rotation response of two "identical" steel beam specimens. It is quite evident that the strength and stiffness properties deteriorate, and that this deterioration is a function of the applied loading history. Various deterioration modes are illustrated in Fig. 16.1a. The monotonic test shows that strength is "capped" and is followed by a negative tangent stiffness. Thus, from a certain deformation on there is evident strength deterioration under monotonic loading. Cyclic loading causes additional modes of deterioration, with the following three shown in Fig. 16.1a: (1) basic strength deterioration, (2) post-capping strength deterioration, and (3) unloading stiffness deterioration (Ibarra et al., [6] 2005).

The consequence of cyclic deterioration is that hysteresis loop boundaries move towards the origin and the so-called "capping point" (the point at which maximum strength is attained) moves continuously as a function of the loading history. The consequence is that there is usually a large difference between an initial backbone curve and a cyclic envelope curve obtained by enveloping the peaks of a cyclic

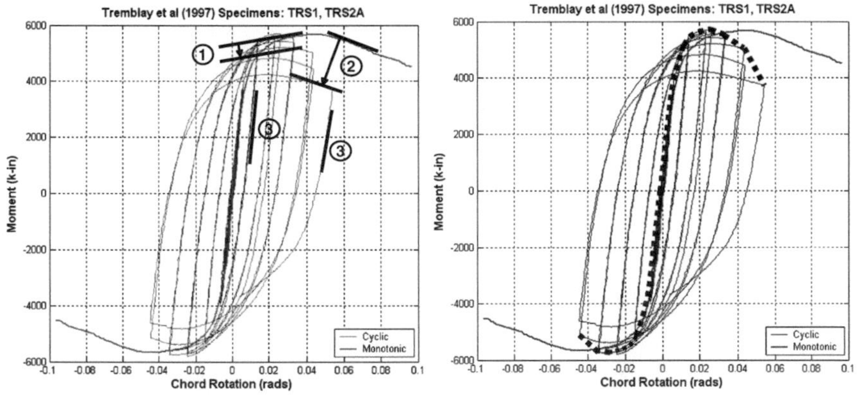

(a) (b)

Fig. 16.1 Monotonic and cyclic experimental response of a steel beam; (**a**) modes of deterioration; (**b**) backbone curve and cyclic envelope curve (ATC-72-1 [1]; data from Prof. Tremblay)

response. Both the initial (monotonic) backbone curve and a cyclic envelope curve are shown in Fig. 16.1b. The cyclic envelope curve is shown with bold dots. The following observations, which hold true in general, are made from these graphs:

- The cyclic envelope curve, with an exception, falls below the monotonic loading curve.
- The exception is evident at relatively small deformations at which the cyclic envelope curve exceeds the monotonic loading curve because cyclic hardening effects exceed cyclic deterioration effects, particularly for steel components.
- The cyclic envelope curve may be strongly loading-history dependent. If any one of the intermediate cycles would have been executed with larger amplitude, then the envelope curve would be enlarged. Thus, there is a nearly unique initial backbone curve but no unique cyclic envelope curve.
- Using the monotonic loading curve as a stable (unchanging) boundary surface for cyclic loading is inappropriate, because it ignores the effect of cyclic deterioration.

16.2.2 The Ibarra-Krawinkler Deterioration Model

The following concepts apply to force-deformation or moment-rotation relationships that can be represented by translational or rotational springs (concentrated plastic hinge models). The modified Ibarra-Krawinkler model [6, 7] is used to illustrate these concepts. For many realistic cases of steel, reinforced concrete, masonry, and wood components this deterioration model provides a good match of experimental results with analytical calibrations.

In the modified Ibarra-Krawinkler model deterioration is modeled in two modes: monotonic and cyclic. Monotonic deterioration is modeled through the initial

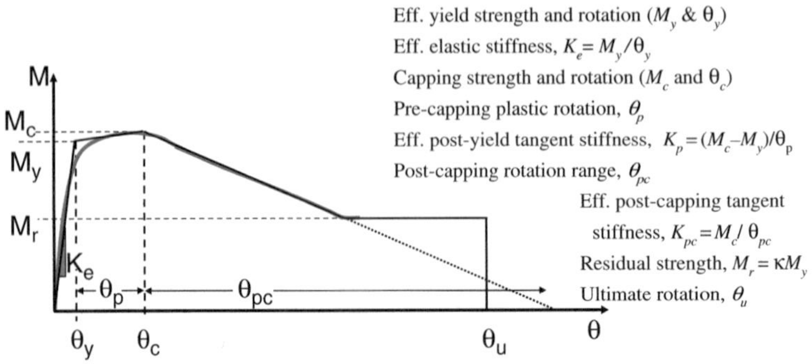

Fig. 16.2 Parameters of the initial backbone curve of the modified Ibarra-Krawinkler model

backbone curve, which is a reference force-deformation curve relationship that defines the bounds within which the hysteretic response of the component is confined (from here on, moment and rotation are being used in lieu of force and deformation). If no cyclic deterioration occurs, then the initial backbone curve is close to the monotonic loading curve. Once cyclic deterioration sets in, the branches of the backbone curve move towards the origin and are continuously updated (they may translate and/or rotate). The initial backbone curve might contain compromises made in order to simplify response description. For instance, it might account for an average effect of cyclic hardening (which likely is small for RC components, but may be significant for steel components). A typical initial backbone (moment – rotation) curve and necessary definitions are illustrated in Fig. 16.2.

Refinements (e.g., more accurate multi-linear descriptions) can be implemented as deemed necessary. It is important to note that the initial backbone curve incorporates monotonic strength deterioration for deformations exceeding the so-called capping point (point of maximum strength under monotonic loading). Residual strength may or may not be present. The ultimate deformation capacity usually is associated with a sudden failure mode such as ductile tearing in a steel component.

Rules defining cyclic response without consideration of cyclic deterioration are adapted to the mode of deformation that dominates behavior of the component. When appropriate, basic hysteresis rules may follow well established concepts such as linearized bilinear, peak-oriented, or pinching hysteretic models.

In the Ibarra-Krawinkler model it is postulated that every component possesses an inherent reference hysteretic energy dissipation capacity, regardless of the loading history applied to the component [6]. Cyclic deterioration in excursion i is defined by the parameter β_i, which is given by the following expression:

$$\beta_i = \left(\frac{E_i}{E_t - \sum_{j=1}^{i} E_j} \right)^c \tag{16.1}$$

where

β_i = parameter defining the deterioration in excursion i
E_i = hysteretic energy dissipated in excursion i
E_t = reference hysteretic energy dissipation capacity, expressed as M_y times a multiple λ of θ_p, i.e., $E_t = M_y \cdot \lambda \cdot \theta_p$
ΣE_j = hysteretic energy dissipated in all previous excursions
c = exponent defining the rate of deterioration (typically a value of $c = 1.0$ is used)

This deterioration parameter can be applied to any of the aforementioned deterioration modes in the form shown in Eq. (16.2)

$$Q_i^+ = (1 - \beta_{Q,i})Q_{i-1}^+ \quad \text{and} \quad Q_i^- = (1 - \beta_{Q,i})Q_{i-1}^- \tag{16.2}$$

where

$Q_i^{+/-}$ = deteriorated quantity (i.e., stiffness or strength) after excursion i.
$Q_{i-1}^{+/-}$ = deteriorated quantity (i.e., stiffness or strength) before excursion i.
$\beta_{Q,i}$ = based on an appropriate λ value

The modified Ibarra-Krawinkler model was tested on about 700 cyclic load-deformation histories obtained from experiments on steel, reinforced concrete components [7]. Adequate simulations were obtained in practically all cases by tuning the model parameters to the experimental data.

16.3 Modeling of Structures for Collapse Prediction

It is a noble objective to accomplish detailed finite element analysis that accounts for all nonlinear phenomena in all parts of a structure, and ultimately accounts also for the variabilities in material properties and seismic input in the process of collapse prediction. At this time such attempts are for the research domain, but hardly can be implemented in a global seismic collapse assessment process. Due to challenges such as modeling of confinement, bond slip, rebar buckling and rebar fracture in reinforced concrete, and modeling of crack propagation, fracture, and local and lateral torsional buckling in steel, we use relatively simple phenomenological concentrated plasticity models of the type discussed in Section 16.2. Such simple component models make it possible at this time to focus attention on modeling of full structural systems and on the large effect that ground motion uncertainties (record to record variability) will have on collapse. From the vast number of collapse analyses performed by the authors, the following general observations are made in regard to structure modeling:

- The component parameters θ_p and θ_{pc} have the largest effect on collapse capacity of a structure
- Realistic representation of structure P-Δ effects is critical
- Large displacement effects do not become important, until very large story drifts at which structures are close to incipient collapse anyway
- The story drift at which structures approach collapse depends on the deformation characteristics of the components, but even for frame structures that consist of very ductile components (e.g., $\theta_p = 0.04$), collapse is imminent at a maximum story drift of clearly less than 10%.

The experience gained over the last 10 years is based mostly on collapse prediction of moment resisting frame structures and wall structures in which inelastic behavior can be represented by flexural plastic hinging. Little experience exists to date on collapse prediction of structures whose behavior is controlled by shear failures or axial load failures in columns (P-M interaction, column buckling).

16.4 Representation of Seismic Input and Prediction of Collapse Capacity

While attempts have been made to relate static pushover results to the collapse capacity of structures [10], the primary means for the determination of the seismic collapse capacity of structures is nonlinear dynamic analysis, whereby an analytical model of the structure is subjected to a ground motion acceleration record. In the following discussion it is assumed, for simplicity, that the ground motion record has only one component and that the structural model is two-dimensional.

It is customary to describe the intensity of a ground motion by a single intensity measure (*IM*). More often than not, the spectral acceleration of the record at the fundamental period of the structure ($S_a(T_1)$) is used as the *IM*. The collapse capacity of a structure is defined here as the intensity of a given ground motion at which the structure is at the "verge of collapse". This implies that the structure is still stable, but that for a small increment in intensity of that ground motion the structure would collapse. In the analytical prediction, collapse is associated with numerical instability, i.e., the solution process no longer converges and the story drift in one or several stories grows without bounds. Thus, given a ground motion, the collapse capacity, ($S_{ac}(T_1)$), can be determined by performing incremental dynamic analyses (IDAs), incrementing the intensity of the ground motion in small enough steps to "catch" the intensity at which the structure is at the verge of collapse, i.e., when a very small increment in intensity will cause a very large increment in lateral displacement [4].

This definition of the collapse capacity presumes that the employed *IM* is representative of the frequency content of the ground motion associated with a seismic event that indeed will cause collapse of the structure. It is well established that the frequency content of ground motion changes with the hazard level. Baker and

Cornell [2] have shown that an adequate measure of this frequency (spectral shape) effect is the parameter epsilon (ε), which is a by-product of standard seismic hazard analysis. The parameter ε quantifies the difference between the selected ground motion's spectral acceleration at a specific period and the median of the ground motion spectral acceleration obtained from attenuation relationships. Epsilon is expressed in terms of a number of logarithmic standard deviations of the attenuation function. A positive ε indicates a peak in the spectral acceleration of a ground motion, compared to the median spectral acceleration obtained from the attenuation function. Baker and Cornell [2] have shown that for rare hazard levels (e.g., 2% in 50 years hazard) ground motions tend to have positive ε values.

The fact that ground motions have different ε characteristics at different hazard levels contradicts the assumption made in IDAs that a single suite of ground motions can represent all hazard levels once the records are scaled to the proper *IM* value. It has been shown by various researchers (e.g., [3, 12]) that ignoring the effect of ε will result in an over-estimation of the demand on the structure for rare hazard levels and under-estimation of the seismic demands for frequent hazard levels. Methods for accounting for the ε effect in collapse prediction are proposed in those references.

16.5 Assessment of Probability of Collapse

In the present approach the collapse capacity for a given structure subjected to a given ground motion is defined as the ground motion intensity, IM_c, at which the structure experiences dynamic instability. Collapse capacity values are computed for a sufficiently large number of ground motions so that a statistical evaluation of the collapse capacity values (IM_c) can be performed. Ordering these collapse capacity values and fitting an appropriate distribution function to the data provides a collapse fragility curve for the specific structure that accounts explicitly for the record-to-record (RTR) variability. For good reasons [5] a lognormal distribution is being used to represent the collapse fragility.

The process for obtaining collapse capacities and the collapse fragility curve for a 4-story frame structure subjected to the set of 40 records is illustrated in Fig. 16.3a. The projection of the last point of each IDA curve (last stable solution) on the vertical axis, illustrated with a solid gray circle, shows the collapse capacity of this building for an individual record. The cumulative distribution function, assuming a lognormal distribution, of these spectral acceleration values that correspond to structural collapse is defined as the "collapse fragility curve" and is shown with a black line in Fig. 16.3a. A more conventional representation of the collapse fragility curve for this moment-resisting frame is shown in Fig. 16.3b.

Fragility curves can be employed directly to evaluate the probability of collapse at specific hazard levels (*IM* values) or to evaluate the mean annual probability of collapse by integrating the fragility curve over the hazard curve. The assessment of probability of collapse and methods for incorporation of aleatory (mainly record-to-record variability) and epistemic (modeling) uncertainties are discussed in detail in Ibarra and Krawinkler [5] and Zareian and Krawinkler [11].

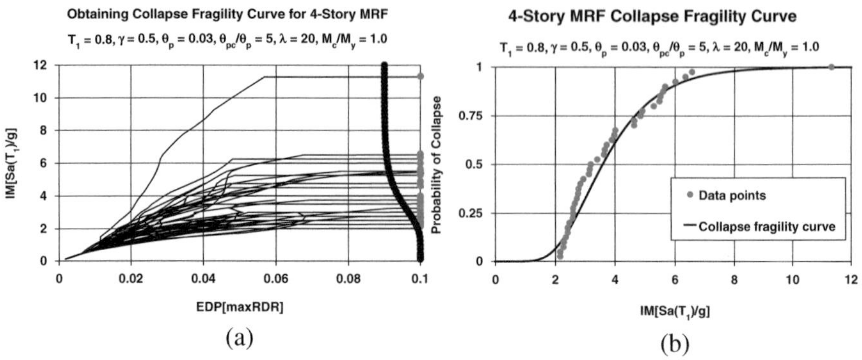

Fig. 16.3 (**a**) Determination of collapse capacities for a set of ground motions; (**b**) corresponding collapse fragility curve [12]

The aleatory variability reflects the random nature of ground motions, whereas the epistemic variability is mainly due to lack of knowledge about the building's real model (inability to incorporate all elements that may contribute to lateral strength and stiffness in the structural model) and real element properties (stiffness, strength, deterioration properties, etc.). Simultaneous consideration of the effects of both variabilities on the collapse capacity necessitates executing a Monte Carlo simulation or using simplified approaches, such as inflating the collapse fragility curve dispersion due to record to record variability, β_{RC}, to the square-root-of-the-sum-of-the-squares of β_{RC} and β_{UC}, with β_{UC} being the dispersion due to epistemic uncertainties.

Once aleatory and epistemic variability is incorporated in the description of the collapse fragility curve – and presuming that the effect of hazard dependence of the spectra shape can be accounted for (the previously discussed epsilon issue) – it becomes a straight forward process to evaluate the collapse probability for a given hazard (S_a) level or the mean annual frequency of collapse. To accomplish the latter, the collapse fragility curve has to be integrated over the hazard curve associated with the *IM* used to describe the fragility curve.

16.6 Conclusions

This paper provides a summary of structural modeling procedures for collapse assessment of structural systems. Putting aside all issues associated with ground motion and hazard modeling, and looking at collapse prediction from a computational analysis perspective only, we can expect realistic results for collapse capacity predictions provided we are capable of

- modeling deterioration characteristics of all important structural components (and in many cases this capability needs still to be acquired)
- modeling all collapse modes (local, story, global)

- predicting propagation of local collapse
- incorporating all "intangible" contributions that delay collapse, such as slab action, exterior cladding, stair cases, interior partitions, infill walls, etc.
- quantifying all important modeling uncertainties, including those attributable to human errors.

Acknowledgments Much of this research was supported by the NSF sponsored Pacific Earthquake Engineering Research (PEER) Center. Additional support was provided by the National Science Foundation (NSF) through Grant No. CMS-0421551 as part of the George E. Brown, Jr. Network for Earthquake Engineering Simulation Consortium Operations, and by a grant of the CUREE-Kajima Phase VI research program. All support is gratefully appreciated. Any opinions, findings, and conclusions or recommendations expressed in this paper are those of the authors and do not necessarily reflect the views of the sponsors.

References

1. ATC-72-1 (2009) Interim guidelines on modeling and acceptance criteria for seismic design and analysis of tall buildings (95% draft). Applied Technology Council, Redwood City, CA
2. Baker JW, Cornell CA (2005) Vector-valued ground motion intensity measure consisting of spectral acceleration and epsilon. Earthq Eng Struct Dyn 34(10):1193–1217
3. Haselton CB, Deierlein GG (2007) Assessing seismic collapse safety of modern reinforced concrete frames. PEER Report 2007/08, Pacific Engineering Research Center, University of California, Berkeley, CA
4. Ibarra L, Medina R, Krawinkler H (2002) Collapse assessment of deteriorating SDOF systems. Proceedings of the 12th European conference on earthquake engineering, Elsevier Science Ltd, London, September 2002, paper # 665
5. Ibarra LF, Krawinkler H (2005) Global collapse of frame structures under seismic excitations. Report No. PEER 2005/06. Pacific Earthquake Engineering Research Center, University of California at Berkeley, Berkeley, CA
6. Ibarra LF, Medina RA, Krawinkler H (2005) Hysteretic models that incorporate strength and stiffness deterioration. Earthq Eng Struct Dyn 34(12):1489–1511
7. Lignos DG, Krawinkler H (2009) Sidesway collapse of deteriorating structural systems under seismic excitations. Report No. 172, The John A. Blume Earthquake Engineering Center, Stanford University, Stanford, CA
8. Takizawa H, Jennings P (1980) Collapse of a model for ductile reinforced concrete frames under extreme earthquake motions. Earthq Eng Struct Dyn 8(2):117–144
9. Talaat M, Mosalam K (2007) Towards modeling progressive collapse in reinforced concrete buildings. Proceedings of ASCE Structures Congress, Long Beach, CA, paper ID 116
10. Vamvatsikos D, Cornell CA (2005) Direct estimation of seismic demand and capacity of multi-degree-of-freedom systems through incremental dynamic analysis of single degree of freedom approximation. ASCE J Struct Eng 131(4):589–599
11. Zareian F, Krawinkler H (2007) Assessment of probability of collapse and design for collapse safety. Earthq Eng Struct Dyn 36(13):1901–1944
12. Zareian F, Krawinkler H (2009) Simplified performance-based earthquake engineering. Report No. 169, The John A. Blume Earthquake Engineering Center, Stanford University, Stanford, CA

Chapter 17
Enhanced Building-Specific Seismic Performance Assessment

Eduardo Miranda

17.1 Introduction

Performance-based earthquake engineering (PBEE) is aimed at designing and building structures that will meet the performance expectations of their owners, stakeholders and society. Inherent in performance-based design is the ability to estimate the performance of the structure at all possible levels of ground motion intensity. Traditionally, a building was judged to have an acceptable performance provided that collapse was prevented. However, recent earthquake events have indicated that for most owners and for society a certain level of control of economic loss is also necessary.

The Pacific Earthquake Engineering Research (PEER) Center recently developed a rational and comprehensive methodology to estimate earthquake damage and economic losses. In particular, approaches developed by Porter and Kiremidjian [12], Krawinkler and Miranda [5], Miranda et al. [8], Aslani and Miranda [1], Mitrani-Rieser and Beck [9], compute economic losses as a sum of losses resulting from cases in which the structure does not collapse, plus economic losses in which the structure is estimated to have collapsed. However, previous work has inherently assumed that if the structure does not collapse, it would be repaired. Furthermore, damage is estimated only based on peak interstory drift demands and peak floor accelerations.

Residual deformations, although often ignored, are of utmost importance in defining the seismic performance of a structure. In particular, the amplitude of residual deformations is critical in determining the technical and economical feasibility of repairing damaged structures. For example, many damaged reinforced concrete buildings in Mexico City had to be demolished after the 1985 Michoacan, Mexico earthquake because of the technical difficulties to straighten and repair buildings with large lateral residual drifts [14]. Similarly, many reinforced

E. Miranda (✉)
Department of Civil and Environmental Engineering, Stanford University,
Yang&Yamazaki Bldg. Rm 281, Stanford, CA 94305-4020, USA
e-mail: emiranda@stanford.edu

M.N. Fardis (ed.), *Advances in Performance-Based Earthquake Engineering*, Geotechnical, Geological, and Earthquake Engineering 13, DOI 10.1007/978-90-481-8746-1_17,

concrete bridge piers were demolished in the city of Kobe in Japan after the 1995 Hyogo-Ken-Nambu, Japan earthquake due to the technical difficulties and elevated costs that would be required to straighten and repair piers with large permanent lateral deformations [4].

Recent analytical and experimental studies [7, 6, 10, 15, 16] have shown that structures subjected to large inelastic deformations have a very high probability of experiencing residual deformations (see Fig. 17.1). This suggests that ductile lateral force resisting systems which are designed and detailed to be able to sustain large lateral displacements without collapse, unless they have a self-centering capability, are especially susceptible to experiencing residual deformations when subjected to intense seismic ground motions. Large residual displacement can thus lead to a total loss of stakeholders' investment on the building despite having avoided collapse.

The objective of this manuscript is to summarize an improved loss estimation methodology that explicitly incorporates economic losses resulting from the possibility of having to demolish buildings that have experienced large residual interstory drifts. The improved methodology is illustrated by computing economic losses in two reinforced concrete moment-frame buildings. In each case economic losses are estimated using both the existing loss estimation methodologies and the proposed approach in order to compare the results.

Fig. 17.1 Examples of buildings with residual displacements that lead to their demolition (Photos courtesy of NISEE, UC Berkeley)

17.2 Improved Loss Estimation Methodology

In recently proposed loss estimation methodologies [5, 8, 1, 9] expected losses at a given level of ground motion intensity are computed as

$$E\left[L_{\mathrm{T}}|IM\right] = E\left[L|NC, IM\right] \cdot P\left(NC|IM\right) + E\left[L|C\right] \cdot P\left(C|IM\right) \tag{17.1}$$

where $E[L|NC, IM]$ is the expected value of the economic loss associated with necessary repairs due to the damage sustained in the building given that it has not collapsed when subjected to a ground motion with intensity $IM{=}im$; $E[L_T|C]$ and $E[L_T|C]$ is the expected value of the loss when the building collapses; $P(NC|IM)$ and $P(C|IM)$ are the probabilities of not collapsing and of collapse when the building is subjected to a ground motion with intensity $IM{=}im$, respectively.

In Eq. (1) $E[L|NC, IM]$ is an increasing function that describes the increment in losses with increasing ground motion intensity, but fails to recognize that the building may have to be demolished and therefore lead to a total loss even if it has survived the earthquake without collapse. In the enhanced approach the expected value of the loss in the building conditioned on the ground motion intensity is computed as:

$$E\left[L_{\mathrm{T}}|IM\right] = L_{\mathrm{R}} + L_{\mathrm{D}} + L_{\mathrm{C}} \tag{17.2}$$

where L_{R} are the contributions to the total expected loss from losses associated with cases in which collapse does not occur (non-collapse, NC) and damage in the building is repaired, R, (i.e., $NC \cap R$); L_{D} are the contributions to the total expected loss from losses associated with cases in which collapse does not occur (non-collapse, NC) but the building subsequently demolished, (i.e., $NC \cap D$); and L_{C} are the contributions to the total expected loss from losses associated with cases in which building collapse (C) occurs.

In Eq. (17.2) the loss contribution due to damage repairs is computed as

$$L_{\mathrm{R}} = E\left[L|NC \cap R, IM\right] \cdot P\left(NC \cap R|IM\right) \tag{17.3}$$

where $E[L|NC \cap R, IM]$ is the expected value of the loss in the building given that the building does not collapse and the damage is repaired knowing that it has been subjected to earthquakes with a ground motion intensity $IM{=}im$. $P(NC \cap R|IM)$ which is the probability that the building will not collapse and that it will be repaired given that it has been subjected to earthquakes with a ground motion intensity $IM{=}im$. Similarly, the loss contribution due to building demolition is computed with:

$$L_{\mathrm{D}} = E\left[L|NC \cap D\right] \cdot P\left(NC \cap D|IM\right) \tag{17.4}$$

where $E[L|NC \cap D]$ is the expected loss in the building when there is no collapse but the building is demolished and $P(NC \cap D|IM)$ the probability that the building will not collapse but it will have to be demolished given that it has been subjected to earthquakes with a ground motion intensity $IM=im$. Finally, the collapse contribution is computed as:

$$L_C = E\left[L|C\right] \cdot P\left(C|IM\right) \tag{17.5}$$

where $E[L|C]$ is the expected value of the loss in the building when collapse occurs and $P(C|IM)$ is the probability that the structure will collapse under a ground motion with a level of intensity, im. Comparing Eqs. (17.1) and (17.2) it is then clear that previous building-specific loss estimation investigations [8, 1] neglected the intermediate term, Eq (17.4). Given that, in general, this term is larger than zero, a systematic underestimation of losses was produced.

The probability that the building will not collapse and that it will be repaired given that it has been subjected to earthquakes with a ground motion with a level of intensity, im is given by:

$$P\left(NC \cap R|IM\right) = P\left(R|NC, IM\right) \cdot P\left(NC|IM\right) \tag{17.6}$$

Similarly, the probability that the structure will not collapse but that will need to be demolished when subjected to an earthquake ground motion with intensity level $IM=im$ is computed as:

$$P(NC \cap D|IM) = P(D|NC, IM) \cdot P(NC|IM) \tag{17.7}$$

where $P(D\,|NC,IM)$ is the probability that the structure will be demolished given that it has not collapsed, when subjected to an earthquake ground motion with intensity level $IM=im$ and $P(NC|IM)$ is the probability of no collapse when the building is subjected to an earthquake ground motion with intensity level $IM=im$.

Since repair and demolition events conditioned on non collapse are mutually exclusive (i.e., if the structure survives the earthquake without collapse you either demolish it or not) then:

$$P\left(R|NC, IM\right) = 1 - P\left(NR|NC, IM\right) \tag{17.8}$$

Similarly, collapse and non-collapse are also mutually exclusive events (i.e., the structure will either collapse or not collapse during an earthquake); then:

$$P\left(NC|IM\right) = 1 - P\left(C|IM\right) \tag{17.9}$$

Substituting Eqs. (17.8) and (17.9) into Eq. (17.7) we obtain

$$P(NC \cap D|IM) = P(D|NC, IM) - P(D|NC, IM) \cdot P(C|IM) \tag{17.10}$$

Estimating the probability that the structure will need to be demolished given that it has not collapsed is particularly challenging, because of the many factors that may be involved in arriving to such decision. In the proposed methodology we estimate such a probability as a function of residual lateral deformations. Experience after the 1985 Mexican earthquake, the 1995 Hyogo-ken-Nambu (Kobe) earthquake and other earthquakes indicates that permanent (residual) lateral deformation was the primary factor driving the decision to demolish buildings and other structures even when damage was relatively small. In the proposed enhanced approach the probability of having to demolish a building that has not collapsed given that it has been subjected to an earthquake ground motion with intensity *IM=im* is computed as a function of the peak residual interstory drift ratio as follows:

$$P(D|NC, IM) = \int_0^\infty P(D|RIDR)\mathrm{d}P(RIDR|NC, IM) \qquad (17.11)$$

where *P(D|RIDR)* is the probability of having to demolish the structure conditioned on the peak residual interstory drift in the building (maximum from all stories in the building) and *P(RIDR|NC,IM)* is the probability of experiencing a certain level of residual interstory drift ratio in the building given that it has not collapsed and that it has been subjected to a ground motion with intensity *IM=im*. Equation (17.11) considers that there is not a single residual interstory drift that triggers demolition, but rather that there is variability in the decision to demolish a building for a given level of residual interstory drift. This probability may be interpreted as the percentage of engineers that would recommend demolition of the building for a given residual interstory drift. Based on limited information and on engineering judgment *P(D|RIDR)* is assumed to be lognormally distributed with a median of 0.015 and a logarithmic standard deviation of 0.3. The resulting cumulative probability distribution is shown in Fig. 17.2. As shown in this figure, residual drifts that would lead to demolition range from about 0.7 to 3%. In particular, buildings with a residual

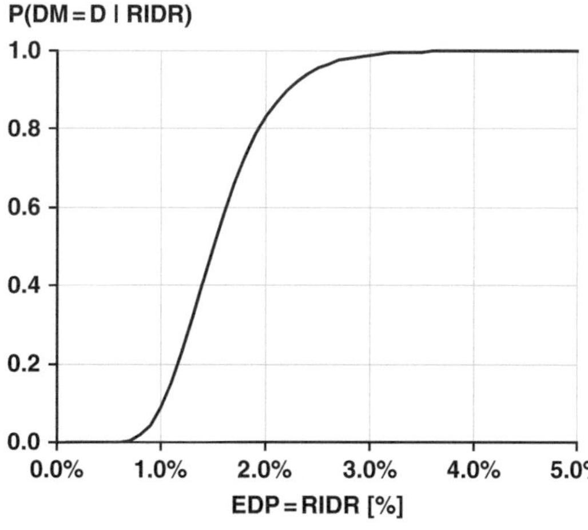

Fig. 17.2 Probability of having to demolish a building that has not collapsed as a function of the peak residual interstory drift in the building

interstory drift ratio of 1% would have approximately a 10% probability of having to be demolished; buildings experiencing residual interstory drift ratios larger than 3% would practically be certain that they would be demolished.

17.3 Illustrative Examples

In order to illustrate its use, the proposed enhanced loss estimation methodology was used to evaluate economic losses in two reinforced concrete frame buildings whose seismic response was previously studied by other investigators [2]. The first building is a 4-story building and the second one a 12-story building, both with ductile detailing. Both structures were assumed to be located at a site in Los Angeles, CA, south of the city's downtown area, representative of a typical urban California site with high seismicity. The two buildings were designed by Haselton and Deierlein [2] according to the 2003 International Building Code and related ACI and ASCE provisions. For detailed information on the designs and modeling parameters of these structures the reader is referred to Haselton and Deierlein [2] .

The buildings were modeled in OpenSees [11] using a two-dimensional, three-bay model of the lateral resisting system and a leaning (P-Δ) column. Beams and columns were modeled with concentrated hinge (lumped plasticity) elements and employ a material model developed by Ibarra et al. [3]. The nonlinear simulation models of the reinforced concrete frames were analyzed by Haselton and Dierlein using the incremental dynamic analysis technique, by analyzing each model using a large set of ground motions scaled at increasing levels of ground motion intensity. Subsequently economic losses were computed using the story-based approach suggested by Ramirez and Miranda [13]. In each case economic losses were computed either considering or not considering the intermediate term in Eq. (17.2), in order to evaluate its influence on economic losses. The influence of this term was evaluated for each building by comparing expected losses at increasing levels of ground motion intensity, by comparing expected annual losses and by comparing the probability of exceedance of large economic losses.

Figure 17.3 compares the expected economic losses with and without considering losses due to the possibility of having to demolish in the two buildings for three different levels of seismic hazard. For each building, the first pair of results corresponds to expected losses for a service level earthquake with a 50% probability of exceedance in 50 years. The middle pair of bars corresponds to the expected economic losses at the Design Basis Earthquake (DBE), defined as the one with a 10% of exceedance in 50 years and the third pair corresponds to the losses due to seismic event that has probability of exceedance of 2% in 50 years (often referred to as the Maximum Credible Earthquake, MCE). The values of the seismic ground motion intensity that correspond to all three hazard levels are indicated at the bottom of the figure. Expected values are normalized by the replacement cost of the structure.

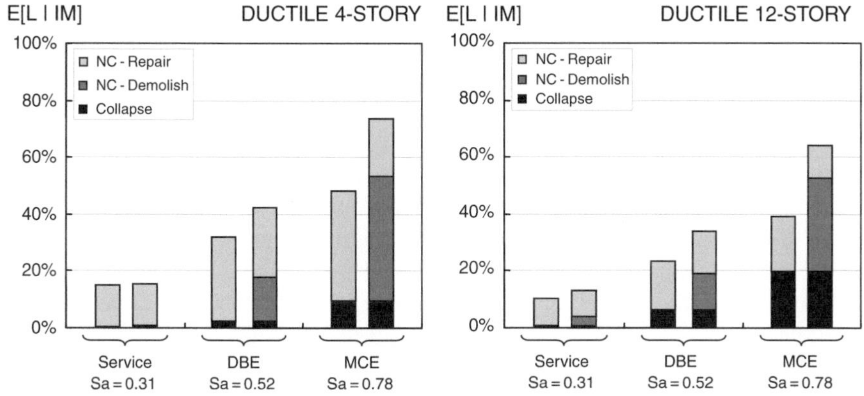

Fig. 17.3 Comparison of expected losses in the 4 and 12-story buildings computed without the inclusion of the possibility of demolition and including the possibility of demolition

For each hazard level, the left bar corresponds to losses that do not consider losses due demolition and the right bar to losses that considering losses due to demolition. It can be seen that at the service-level earthquake, the effect of losses due to building demolition does not have an influence on the overall normalized loss for the four-story building and a relatively small influence for the 12-story one. On the other hand, for the four story building the normalized economic losses increase from 31 to 42% at the DBE and from 48 to 73% at the MCE, corresponding to increments in expected losses of 35 and 52% and the DBE and MCE levels, respectively. The relative increase is the difference between the two values of expected loss, with and without considering losses due to demolition, divided by the expected loss considering losses due to demolition, multiplied by 100. This means that losses due to demolition have a large influence in the overall loss estimate, and that neglecting them can lead to significant underestimation of economic loss.

To gain further insight of the influence of the possibility of having to demolish a building after an earthquake even though it has not collapsed, loss results at these levels were disaggregated following the approach proposed by Aslani and Miranda [1]. In Fig. 17.3 each bar in the figure is divided up into collapse losses, non-collapse (NC) losses due to building demolition and non-collapse losses due to repair costs. The proportions of each bar are equal to how much each type of loss contributes to the overall loss. As shown in this figure, demolition losses have the largest contributions to the overall loss at the MCE. At this intensity level, losses conditioned on non-collapse due to demolition dominate the expected loss. In particular, the losses due to demolition are significantly larger than those of collapse, even though both lead to total loss of the initial investment. This is because at the MCE the probability of demolition is much higher (45%) than the probability of collapse (8%). That is, at this level of ground motion intensity the structure is more likely to experience large residual deformations that will lead to demolition, than collapsing. Similar results were computed for the 12-story building.

17.4 Summary and Conclusions

The Pacific Earthquake Engineering Research (PEER) Center methodology for seismic performance assessment has been extended to explicitly account for the possibility of having to demolish a building that did not collapse during an earthquake. In the proposed framework the probability of having to demolish the building given that it has not collapsed is computed as a function of the peak residual interstory drift in the building conditioned on the ground motion intensity. The latter is computed by conducting an incremental dynamic analysis in which peak residual interstory drifts are computed at increasing levels of ground motion intensity. By doing so, the record-to-record variability of residual drift demands is explicitly taken into account.

Results indicate that neglecting the probability of demolition due to excessive residual lateral deformations, as typically done presently, leads to significant underestimation of economic losses. Underestimations are typically larger in ductile buildings than in non-ductile ones. This is because ductile structures are very effective in reducing the probability of collapse when subjected to intense ground motions, but they have a significant probability of having to be demolished due to residual drifts. Meanwhile, when non-ductile structures are subjected to intense ground motions, they typically have a relatively large probability of collapse and the probability of surviving the earthquake with large permanent deformations that will lead to demolition is much smaller.

The proposed framework provides an ideal tool to assess the tradeoffs and benefits of various design alternatives. In particular it provides a framework to properly account for the economic benefits of incorporating self-centering technologies which significantly reduce or even eliminate residual drifts.

Acknowledgements The work described herein was supported by the Pacific Earthquake Engineering Research (PEER) Center with support from the Earthquake Engineering Research Centers Program of the National Science Foundation under Award No. EEC-9701568 and by the John A. Blume Earthquake Engineering Center at Stanford University. Incremental dynamic analyses were conducted by Curt Haselton. The loss estimates computed in the example were computed by Dr. Marc Ramirez.

References

1. Aslani H, Miranda E (2005a) Probabilistic earthquake loss estimation and loss disaggregation in buildings, report no. 157. John A. Blume Earthquake Engineering Center, Stanford University, Stanford, CA
2. Haselton CB, Deierlein GG (2007) Assessing seismic collapse safety of modern reinforced concrete frame buildings, technical report no. 156. John A. Blume Earthquake Engineering Center, Stanford University, Stanford, CA
3. Ibarra LF, Medina RA, Krawinkler H (2005) Hysteretic models that incorporate strength and stiffness deterioration. Earthq Eng Struct Dyn 34(12):1489–1511
4. Kawashima K (2000) Seismic design and retrofit of bridges. Proceedings of the 12th world conference on earthquake engineering, paper 2828, Auckland

5. Krawinkler H, Miranda E (2004) Performance-based earthquake engineering. In: Borzognia Y, Bertero V (ed) Earthquake engineering: from engineering seismology to performance-based engineering, 1st edn. CRC Press, Boca Raton, FL, pp 9-1–9-59
6. MacRae GA, Kawashima K (1997) Post-earthquake residual displacements of bilinear oscillators. Earthq Eng Struct Dyn 26:701–716
7. Mahin SA, Bertero VV (1981) An evaluation of inelastic seismic design spectra. J Struct Eng 107(ST9):1777–1795
8. Miranda E, Aslani H, Taghavi S (2004) Assessment of seismic performance in terms of economic losses. Proceedings of an international workshop on performance-based seismic design: concepts and implementation, Bled, Slovenia, pp 149–160
9. Mitrani-Reiser J, Beck J (2007) An ounce of prevention: probabilistic loss estimation for performance-based earthquake engineering. Department of Civil Engineering and Applied Mechanics, California Institute of Technology, Pasadena, CA
10. Pampanin S, Christopoulos C, Priestley MJN (2002) Residual deformations in the performance-seismic assessment of frame structures. Research Report No. ROSE-2002/02. European School for Advanced Studies in Reduction of Seismic Risk, Pavia, Italy
11. PEER (2006) OpenSEES (Open System for Earthquake Engineering Simulation). Retrieved May 2007, from Pacific Earthquake Engineering Research Center. opensees.berkeley.edu.
12. Porter KA, Kiremidjian AS (2001) Assembly-based vulnerability of buildings and its uses in seismic performance evaluation and risk management decision-making, Technical Report No. 309. John A. Blume Earthquake Engineering Center, Stanford University, Stanford, CA
13. Ramirez CM, Miranda E (2009) Building-specific loss estimation methods & tools for simplified performance-based earthquake engineering, Technical Report No. 173. John A. Blume Earthquake Engineering Center, Stanford University, Stanford, CA
14. Rosenblueth E, Meli R (1986) The 1985 Mexico earthquake: causes and effects in Mexico City. Concrete International 8(5):23–34
15. Ruiz-Garcia J, Miranda E (2005) Performance-based assessment of existing structures accounting for residual displacements, Technical Report No. 153. John A. Blume Earthquake Engineering Center, Stanford University, Stanford, CA
16. Ruiz-Garcia J, Miranda E (2006). Residual displacement ratios for assessment of existing structures. Earthq Eng Struct Dyn 35:315–335

Chapter 18
A Damage Spectrum for Performance-Based Design

Ahmed Ghobarah and Mahmoud Safar

18.1 Introduction

In performance-based earthquake engineering (PBEE) displacement and drift are often used as performance criteria. The rationale is that there is a correlation between drift and damage. The challenge in the design process is to ensure that damage will be limited to an acceptable level. Within a performance-based design framework the multiple performance limit states reflecting the building damage levels should be satisfied for the defined hazard levels. The process requires the development of a generic damage-based assessment tool that is sensitive to various performance characteristics and applicable to different structural materials.

The use of constant ductility response spectra (CDRS) is based on the assumption that seismic damage is a function of ductility. Ductility as a performance parameter does not reflect the damage accumulated due to the cycles of repeated load reversals. To establish generic design and assessment procedures for PBEE a general definition of damage, independent of the structural material type, is needed. For an integrated PBEE, damage performance objectives need to be considered at the beginning of the design cycle. Thus, there is interest in including a quantitative damage measure for the construction of strength demand spectra independent of the type of the structural material. The objective of this study is to develop a seismic demand spectrum based on the constant damage concept.

18.2 Quantification of Damage

In the design process, the level of damage associated with a given hazard can be specified as a performance objective. Damage indices have been developed on the local element or the global levels. Recent reviews of the trends in damage quantification were presented by Williams and Sexsmith [10] and Ghobarah et al. [5].

A. Ghobarah (✉)
Department of Civil Engineering, McMaster University, Hamilton, ON L8S 4L7, Canada
e-mail: ghobara@mcmaster.ca

M.N. Fardis (ed.), *Advances in Performance-Based Earthquake Engineering*,
Geotechnical, Geological, and Earthquake Engineering 13, DOI 10.1007/978-90-481-8746-1_18,

Some of the damage indices that account for both excessive deformation and damage accumulation under cyclic loading use overall response parameters such as stiffness, period or frequency. These damage indices are applicable at both local and global levels [5, 4]. Park and Ang [6] assumed a linear combination of two terms accounting for excessive deformation and the damage accumulation through hysteretic energy dissipation. The method of combining the two components of damage depends on the structural material. A recent study proposed a general damage index that is independent of the structural material [3].

18.3 Demand Representation

Inelastic response spectra formed the basis for establishing capacity-demand index relations to be used for design. Empirical relationships based on statistical analysis were proposed for strength reduction factor, ductility and period (R-μ-T relations).

The use of R, μ and T as the performance parameters for design is generally accepted. From a damage-based design point of view, this implicitly identifies R as the parameter limiting the design force to be applied to the structure; μ as the damage limiting parameter and T is the period as the input parameter representing the dynamic property of the system for design.

18.4 Ductility as the Damage Limiting Parameter

Constant ductility response spectra (CDRS) are based on the assumption that seismic damage is a function of ductility. However, ductility as a performance parameter does not account for the cumulative damage resulting from hysteretic response. Thus, the margin of safety against collapse for structures designed on the basis of CDRS may not be reliable. New design methods incorporating the effect of cumulative damage in the computation of strength demand spectra, can be classified into: (a) methods based on equivalent ductility factor using a weighted factor accounting for ductility reduction due to dissipated hysteretic energy; and (b) methods based on the direct use of damage indices. However, all these methods are dependent on the type of structural material.

18.5 Period as the Input Design Parameter

Traditionally, the adopted load-deformation models for the lateral force resisting elements are based on the assumption that the strength is independent of the stiffness. This was the reason that the period was adopted as a main design parameter. Recent investigations indicated that the load deformation models of several lateral force resisting elements are characterized by strength and stiffness interdependency [7]. Figure 18.1 compares strength-stiffness independent and dependent

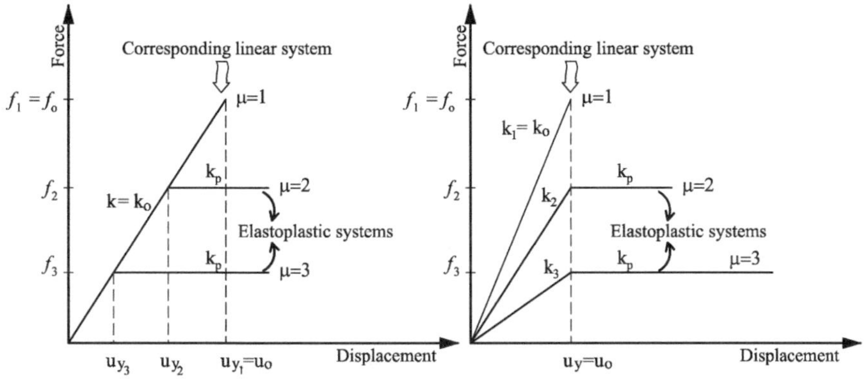

a) Strentgh-Stiffness independant model. b) Strentgh-Stiffness dependant model.

Fig. 18.1 Load-deformation models

load-deformation models. For the strength-stiffness dependent models, the yield displacement was considered to be a stable parameter that is suitable for seismic applications [7, 1]. Safar and Ghobarah [8] addressed the conceptual basis and methods for adopting constant yield displacement in the response spectrum framework.

18.6 Performance Point Index

Damage, as a performance parameter, depends on several local and global response characteristics. It is desirable to maintain the advantage of simplicity associated with spectral representations and propose a generic seismic demand parameter reflecting damage as a performance parameter. The damage index in [3] is:

$$D = 1 - \frac{M_{ac}}{M_{yo}} \tag{18.1}$$

$$\frac{M_{ac}}{M_{yo}} = f(\mu, \int dE) \tag{18.2}$$

where,

D is the damage index
M_{ac} is the actual (deteriorated) value of the yield moment (force),
M_{yo} is the value characterizing the yield point in a theoretical skeleton curve,
μ is the deformation ductility factor,
$\int dE$ is the energy dissipated through hysteretic behaviour,
$f(\mu, \int dE)$ is the evolution equation for the yield strength,

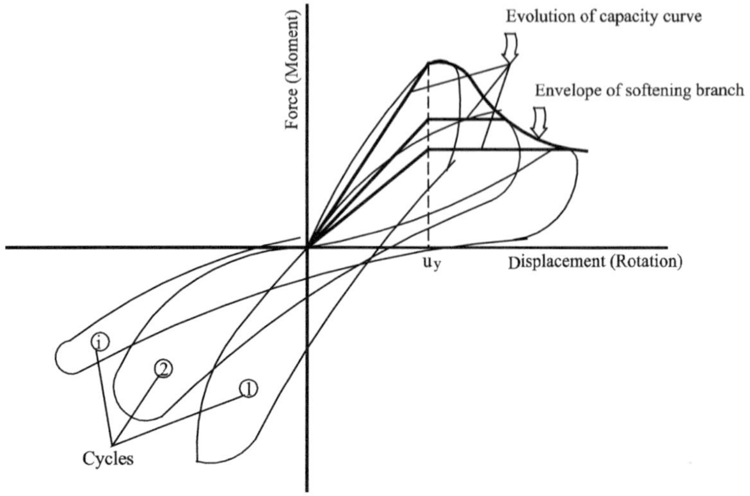

Fig. 18.2 Evolution of capacity curves for strength dependent stiffness model

The index relies on strength deterioration as the main indicator of damageability. The yield strength as given by Eq. (18.2) represents a general definition of damage. This is established based on the observation that the softening branch of hysteretic curves depends on the type of failure, ductile or non-ductile, not on the material.

The same concept of damage modeling is used as basis to formulate a damage index within the response spectrum framework. The index is termed Performance Point Index (PPI). The strength dependent stiffness (SDS) load-deformation model is used to represent the evolution of the capacity curve for a hysteretic behaviour of a structure subjected to earthquake ground motion. Figure 18.2 shows a typical hysteresis curve with a typical representation for a SDS model superimposed. The load-deformation model is used to track the envelope of the softening branch for the hysteretic behaviour. The beginning of the envelope curve represents the yield point and is modeled using the equivalent linear system for the load-deformation model. In formulating the PPI, modifications are introduced to the index presented in [3] as described by the following subsections.

18.6.1 Dissipated Energy

For a strength dependent stiffness model (SDS) as shown in Fig. 18.1b, the dissipated energy is the difference in the energy capacity of the equivalent linear system and the inelastic system. This difference can be calculated for different ductility levels of the inelastic system throughout the hysteretic behaviour. The dissipated energy for the SDS model $\int dE_{SDS}$, at a ductility level μ_{max}, can be defined as:

$$\int \mathrm{d}E_{SDS} = u_y f_e \left(\frac{1}{2} - \mu_{max} \bar{f_y} + \frac{1}{2} \bar{f_y} \right) \tag{18.3}$$

where, u_y is the yield displacement, f_e is the strength level of the equivalent linear system and the normalized yield strength $\overline{f_y}$ is defined as:

$$\overline{f_y} = \frac{f_y}{f_e} \tag{18.4}$$

where f_y is the strength level of the inelastic system with ductility level μ_{max}. The term $\frac{1}{2} u_y f_e$ defines the energy capacity of the equivalent linear system whereas the term $u_y f_y (\mu_{max} - \frac{1}{2})$ defines the energy capacity of the inelastic system.

18.6.2 Ductility Capacity

For a SDS model as described in Fig. 18.1b, the ultimate ductility capacity for a system with hysteretic behaviour μ_u, can be determined based on an explicit definition of the minimum inelastic cyclic deformation performance of structures required target ductility ratio μ_t. The definition given by the New Zealand code [9] is used, where the structure should be capable to undertake four cycles in horizontal displacement to $\pm\mu_t u_y$ without the lateral load carrying capacity reducing by more than 20%. At this displacement it was assumed that the damage index of a structure designed for μ_t would reach 1.0 (total collapse).

This assumption can be applied to the damage model presented in [3]. Assuming that the system has a bi-linear hysteretic behaviour with a post-elastic stiffness of 3% of the elastic stiffness, the following relation is obtained:

$$\begin{aligned} &\left(1 - \frac{\mu_t}{\mu_u}\right)^{1/\beta_1} . \frac{1}{2}\left(1 - \tanh\left(\beta_2 \frac{(15.5\mu_t - 16)}{(\mu_u - 1)} - \pi\right)\right) \\ &. \exp\left(-\beta_3 \frac{(15.5\mu_t - 16)}{(\mu_u - 1)}\right) = 0 \end{aligned} \tag{18.5}$$

Knowing the value of the target ductility μ_t and the parameters β_1, β_2, and β_3 [3], the ultimate ductility capacity μ_u can be obtained by solving Eq. (18.5) iteratively. The definition of target ductility values for different performance objectives is still an ongoing challenge. However, proposed values were given by ATC-40 [2].

18.6.3 Ultimate Hysteretic Energy

The ultimate hysteretic energy for a SDS model ($E_{uSDS}{}^*$) can be defined as:

$$E_{uSDS}{}^* = u_y f_e \left(\frac{1}{2} - \mu_u \overline{f_{yu}} + \frac{1}{2}\overline{f_{yu}}\right) \tag{18.6}$$

There is a similarity between Eqs. (18.3) and (18.6). However, the ultimate ductility μ_u and the ultimate normalized yield strength $\overline{f_{yu}}$ were used in Eq. (18.6). This is different from Eq. (18.3), where the attained ductility μ_{max} and the attained $\overline{f_y}$ were used. For Eq. (18.6), μ_u is obtained using the procedure described above. The value of $\overline{f_{yu}}$ is taken equal to 0.8. This value is inferred by the assumption made by the New Zealand code [9] requirement: a reduction of the lateral load carrying capacity equal to 20% for a structure designed to target ductility μ_t and subjected to four cycles of horizontal displacement to $\pm\mu_t\mu_y$, defines collapse.

18.7 Proposed Damage Index

Applying the modifications described in the preceding subsections to the index in [3] results in the following expression for the Performance Point Index (PPI):

$$PPI = 1 - \left(\left(1 - \frac{\mu_{max}}{\mu_u}\right)^{1/\beta_1} \cdot \frac{1}{2}\left(1 - \tanh\left(\beta_2 \frac{\left(\frac{1}{2} - \mu_{max}\overline{f_y} + \frac{1}{2}\overline{f_y}\right)}{\left(\frac{1}{2} - \mu_u\overline{f_{yu}} + \frac{1}{2}\overline{f_{yu}}\right)} - \pi\right)\right) \cdot \exp\left(-\beta_3 \frac{\left(\frac{1}{2} - \mu_{max}\overline{f_y} + \frac{1}{2}\overline{f_y}\right)}{\left(\frac{1}{2} - \mu_u\overline{f_{yu}} + \frac{1}{2}\overline{f_{yu}}\right)}\right)\right) \quad (18.7)$$

18.8 Constant Damage Spectrum

Constant values of PPI are used as the performance limiting parameter instead of the ductility to construct a constant damage spectrum (CDS). The proposed procedure for constructing CDS can be summarized by the following steps:

(1) According to the type of the structure, determine the damage index parameters β_1, β_2, and β_3 [3]. Define the ground motion acceleration $\ddot{u}_g(t)$.
(2) Select the damping ratio ζ for which the spectrum is to be plotted.
(3) Select a value for T_n, where T_n is the undamped natural period.
(4) Determine the response $u(t)$ of the linear system with T_n and ζ values selected. From $u(t)$ determine the maximum displacement response for the elastic system u_o and the peak force $f_o = k\,u_o$, where k is the stiffness. For a linear system k is equal to the pre-yield stiffness of the system k_o. The spectral ordinates for the linear system can be calculated as:

$$D_y = u_o, \quad V_y = \omega_o u_o, \quad \text{and} \quad A_y = \omega_o^2 u_o$$

where D_y is the yield deformation spectral ordinate, V_y is the pseudo-velocity spectral ordinate, A_y is the pseudo-acceleration spectral ordinate and ω_o is the pre-yield frequency.
(5) Determine the response $u(t)$ of an elastoplastic system with the same ζ, period T_y and yield force $f_y = \overline{f_y} f_o$, with selected $\overline{f_y} < 1$, where $\overline{f_y}$ is the normalized yield strength, f_y is the post-yield system strength. And T_y can be obtained

from the relation $\overline{f_y} = (f_y/f_o) = (k_y/k_o) = (T_n^2/T_y^2)$, where k_y is the post-yield system stiffness. From $u(t)$ determine the peak deformation u_m and the associated ductility factor $\mu_{max} = u_m/u_o$, where u_m is the maximum displacement response of the inelastic system. Calculate the damage index value (PPI) using Eq. (18.7). Repeat the analysis for enough values of $\overline{f_y}$ to develop data points $(\overline{f_y}, PPI)$ covering the ductility range of interest.

(6) (a) For a selected damage index, *PPI* determine the $\overline{f_y}$ value from the results of step 5 using an interpolative procedure. If more than one $\overline{f_y}$ value corresponds to a particular value of *PPI*, the largest value of $\overline{f_y}$ is chosen.
(b) Determine the corresponding value of T_y using the value of $\overline{f_y}$ determined in step (6a) from the relation $\overline{f_y} = (T_n^2/T_y^2)$. Accordingly the value of the post-yield frequency ω_y can be calculated as $\omega_y = 2\pi/T_y$.
(c) Determine the spectral ordinates corresponding to the value of $\overline{f_y}$ determined in step (6a). The spectral ordinates can be calculated as:

$$D_y = u_o, V_y = \omega_y u_o, \quad \text{and} \quad A_y = \omega_y^2 u_o$$

where D_y is the yield deformation spectral ordinate, V_y is the pseudovelocity spectral ordinate and A_y is the pseudoacceleration spectral ordinate. This data provides one point on the response spectrum plot.

(7) Repeat steps (4)–(6) for a range of T_n to determine the spectrum that is valid for the μ value chosen in step (6a).
(8) Repeat steps (4)–(7) for different values of the ductility factor *PPI*.

18.9 Damage-Based Assessment of a RC Frame

The developed spectrum can be used for spectral damage-based assessment. The capacity of the structure to be assessed can be established by a static pushover analysis. To compare the capacity and demand in a spectral framework, the structure is represented as an equivalent single degree of freedom system (SDOF).

To illustrate the procedure, a 6-storey reinforced concrete frame is considered. The frame was designed according to a current code. The frame configuration, member sizes and reinforcement details are shown in Fig. 18.3. The Northridge time-history is considered as the demand. The developed CDS for the time-history is presented in Fig. 18.4. Linear scale is used for a clear superposition of the capacity curve. A static pushover analysis is performed for the frame and the equivalent SDOF formulation [2] is used to establish the capacity curve. The modeling and pushover analysis for the frame was conducted with OpenSees. Since the spectrum is plotted with the yield displacement as the system-representative parameter, the performance point is defined by the yield point on the capacity curve. The location

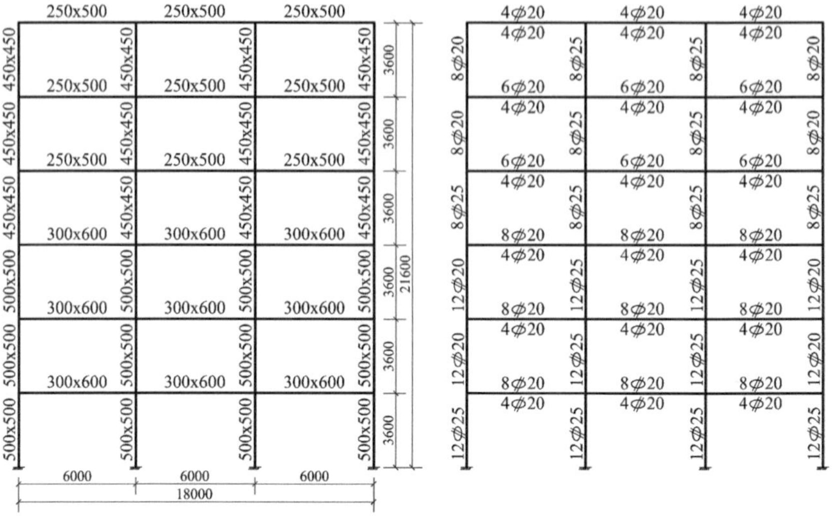

a) Cross section dimensions in mm b) Steel reinforcement details

Fig. 18.3 Details of 6-storey reinforced concrete ductile frame

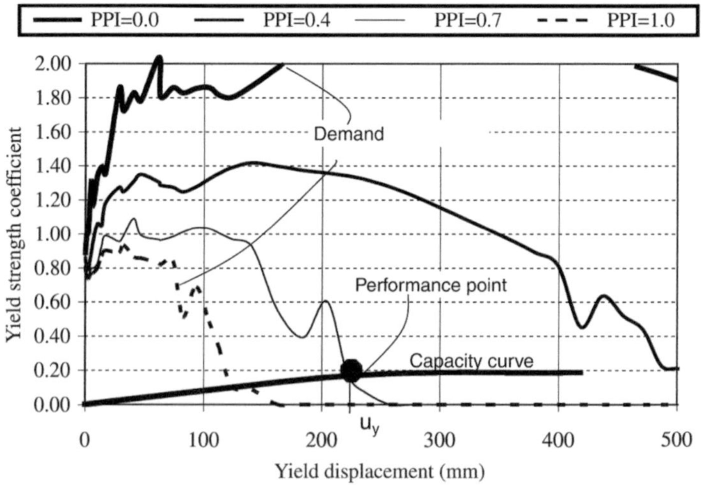

Fig. 18.4 Spectral damage-based assessment of the 6-storey frame

of the performance point with respect to the demand curves determines the performance whether elastic or inelastic. In addition, the performance point location can be used to estimate the damage index of the system under the effect of the considered demand. In the case presented in Fig. 18.4, the *PPI* can be estimated as 0.70. In terms of performance, this reflects a severe (irreparable) damage state.

To verify the estimated damage index value, a static pushover analysis is conducted before and after a dynamic analysis of the frame under the effect of the Northridge time-history. The change in the fundamental period of the structure is determined from the results of the pushover analyses.

The Final softening index *FSI* [4] is calculated. T_{initial} and T_{final} are the initial and final fundamental periods of the structure before and after the earthquake, respectively. Based on this analysis procedure, the *FSI* value for the 6-storey frame subjected the Northridge earthquake was determined to be 0.61, which is comparable to the estimated *PPI* value of 0.7 from spectral assessment. The *FSI* results confirm the reliability of the proposed procedure.

18.10 Conclusions

The proposed approach for constructing strength demand spectra results in a simple and rational framework for spectral damage-based assessment, independent of the structural material type. This generalization is a necessary step in developing codified performance-based provisions. The developed spectrum represents a large class of structures which have load-deformation behaviour with interdependent strength and stiffness.

References

1. Aschheim M, Black EF (2000) Yield point spectra for seismic design and rehabilitation. Earthq Spectra 16:317–335
2. ATC-40 (1996) Seismic evaluation and retrofit of concrete buildings. Applied Technology Council, Report ATC 40, Volumes 1 and 2, Redwood City, CA
3. Colombo A, Negro P (2005) A damage index of generalised applicability. Eng Struct 27:1164–1174
4. Dipasquale E, Cakmark AS (1987) Detection and assessment of seismic structural damage, NCEER 87-0015, National Center for Earthquake Research Center, State University of New York at Buffalo, NY
5. Ghobarah A, Aboul-Elfath H, Biddah A (1999) Response-based damage assessment of structures. Earthq Eng Struct Dyn 28(1):79–104
6. Park YJ, Ang H-S (1985) Mechanistic seismic damage model for reinforced concrete, J Struct Eng ASCE 111(4):722–739
7. Priestley MJN (2000) Performance-based seismic design, 12th world conference of earthquake engineering, paper no. 2831, New Zealand Society of Earthquake Engineering, Auckland, New Zealand
8. Safar M, Ghobarah A (2008) Inelastic response spectrum for simplified deformation-based seismic vulnerability assessment. J Earthq Eng 12(2):222–248
9. SANZ (1992) Standards Association of New Zealand, Commentary on code of practice for general structural design and design loading for buildings, (NZS 4203-Part 2), Wellington
10. Williams MS, Sexsmith RG (1995) Seismic damage indices for concrete structures: a state-of-the-art review. Earthq Spectra 11(2):319–349

Chapter 19
Construction of Response Spectra for Inelastic Asymmetric-Plan Structures

Jui-Liang Lin and Keh-Chyuan Tsai

19.1 Introduction

The conventional response spectra of the inelastic single-degree-of-freedom (SDOF) system have been extensively studied [5, 6] and widely applied in engineering practice. The force-deformation relationship of the inelastic SDOF system represents the roof translation versus base shear relation of a physical multi-degree-of-freedom (MDOF) building. For inelastic structures, the vibration "modes" may be viewed as varying in time. Nevertheless, the conversion of the inelastic force-deformation relationship of roof translation versus base shear for each vibration "mode" into a SDOF system is well developed [2]. The capacity-spectrum method [1] is one of the well-known applications of the SDOF systems to estimate the seismic response of the original MDOF structure. In order to compute the seismic demands of asymmetric-plan structures, not only the translational response at the center of mass (CM), but also the rotational response is required. However, the conventional SDOF inelastic response spectra, representing the roof translation versus base shear relationship, give no insight into the rotational demands of asymmetric-plan structures. Moreover, the SDOF systems are not able to consider the translation-rotation interaction of the inelastic asymmetric-plan structures. Thus, the conventional SDOF inelastic response spectra can neither accurately estimate the rotational demands nor properly calculate the translational demands for asymmetric-plan structures.

There are two types of pushover curves simultaneously obtained for an asymmetric-plan building under the modal inertia forces [3]: one for the roof translation versus base shear and another for the roof rotation versus base torque relationship of the original MDOF building. Figure 19.1a illustrates the typical one-cycle push-pull curves representing these two force-deformation relationships in acceleration-displacement-response-spectrum (ADRS) format. Figure 19.1a shows

J-L. Lin (✉)
National Center for Research on Earthquake Engineering, 200, Sec.3, Xinhai Rd., Taipei, 10668, Taiwan
e-mail: jllin@ncree.org

M.N. Fardis (ed.), *Advances in Performance-Based Earthquake Engineering*, Geotechnical, Geological, and Earthquake Engineering 13, DOI 10.1007/978-90-481-8746-1_19,

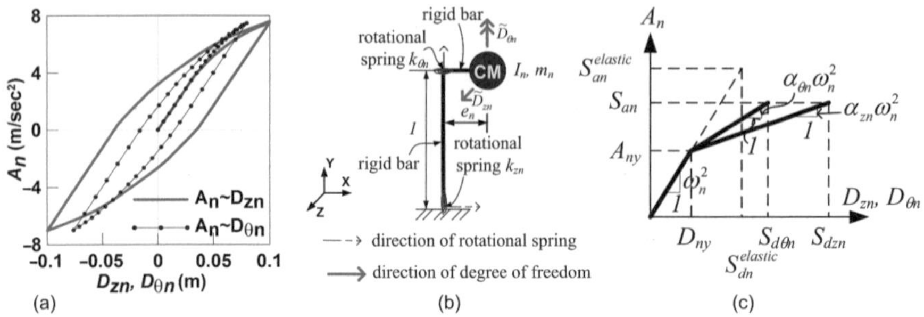

Fig. 19.1 (**a**) Typical one-cycle push-pull curves for an asymmetric-plan building under the *n*th modal inertia force; (**b**) a 2DOF modal system; (**c**) the two bilinear pushover curves for one-way asymmetric-plan structures

that the two curves bifurcate after the original asymmetric-plan building becomes inelastic. Clearly, the bifurcation is due to the non-proportionality between the modal translation and modal rotation of the inelastic asymmetric-plan structure. The 2DOF modal equations of motion and the associated 2DOF modal system (Fig. 19.1b) have been developed to simultaneously simulate these two force-deformation relationships [3].

Although the 2DOF modal system has been developed, the inelastic response spectra constructed using it is not just a standard procedure. Before the associated inelastic response spectra can be constructed, three key tasks should be completed. First, the independent elastic parameters for the 2DOF modal systems have to be determined. Second, the relationships between the inelastic 2DOF modal parameters and the strength ratio have to be established. Third, the ranges of values of the 2DOF key modal parameters should be clarified. This paper describes in detail how these three tasks are carried out. As this new type of response spectra includes both translation (T) and rotation (R), for the purpose of discussion it is termed the T-R response spectra. The T-R inelastic response spectra of asymmetric-plan buildings under an ensemble of seismic ground motion are compared with the corresponding SDOF inelastic response spectra in the example in this paper.

19.2 Theoretical Background

For convenience, the development of the 2DOF modal systems is briefly presented herein. More details can be found in Lin and Tsai [3].

19.2.1 Two-Degree-of-Freedom Modal Systems

The two horizontal axes are *X* and *Z*. The *Y*-axis is vertical. The one-way asymmetric-plan building is symmetric about the *X*-axis. The seismic ground motions are applied at right angles to the *X*-axis. One-way asymmetric-plan buildings with proportional damping are considered. The equation of motion for an N-storey one-way asymmetric-plan building, with floors simulated as rigid diaphragms, is

$$\mathbf{M}\ddot{\mathbf{u}} + \mathbf{C}\dot{\mathbf{u}} + \mathbf{K}\mathbf{u} = -\mathbf{M}\iota\ddot{u}_g(t) = -\sum_{n=1}^{2N} \mathbf{s}_n \ddot{u}_g(t) \tag{19.1}$$

where

$$\mathbf{u} = \begin{bmatrix} \mathbf{u}_z \\ \mathbf{u}_\theta \end{bmatrix}_{2N\times 1} = \sum_{n=1}^{2N} \Gamma_{zn}\boldsymbol{\varphi}_n D_n,\ \boldsymbol{\varphi}_n = \begin{bmatrix} \boldsymbol{\varphi}_{zn} \\ \boldsymbol{\varphi}_{\theta n} \end{bmatrix}_{2N\times 1} \quad \Gamma_{zn} = \frac{\boldsymbol{\varphi}_n^T \mathbf{M}\iota}{\boldsymbol{\varphi}_n^T \mathbf{M}\boldsymbol{\varphi}_n}$$
$$\mathbf{M} = \begin{bmatrix} \mathbf{m} & \mathbf{0} \\ \mathbf{0} & \mathbf{I}_0 \end{bmatrix}_{2N\times 2N},\quad \mathbf{K} = \begin{bmatrix} \mathbf{k}_{zz} & \mathbf{k}_{z\theta} \\ \mathbf{k}_{\theta z} & \mathbf{k}_{\theta\theta} \end{bmatrix}_{2N\times 2N},\quad \mathbf{s}_n = \Gamma_{zn}\mathbf{M}\boldsymbol{\varphi}_n \tag{19.2}$$

In Eqs. (19.1), and (19.2) $\mathbf{M}$ and $\mathbf{K}$ are the mass matrix and stiffness matrix, respectively, $\mathbf{u}$ and $\mathbf{s}_n$ are the displacement vector and the nth modal inertia force distribution, respectively, Γ_{zn} and $\boldsymbol{\varphi}_n$ are the nth modal participation factor and the mode shape, respectively, and D_n and ι are the generalized nth modal coordinate and the influence vector, respectively. Only the nth modal displacement response, $\mathbf{u}_n = \Gamma_{zn}\boldsymbol{\varphi}_n D_n(t)$, of the elastic one-way asymmetric-plan building is excited under the n-th modal inertia force distribution, $\mathbf{s}_n$ [2]. Thus, the equation of motion can be presented in the following form:

$$\mathbf{M}\ddot{\mathbf{u}}_n + \mathbf{C}\dot{\mathbf{u}}_n + \mathbf{K}\mathbf{u}_n = -\mathbf{s}_n\ddot{u}_g(t),\ n = 1 \sim 2N \tag{19.3}$$

We redefine $\mathbf{u}_n$ as:

$$\mathbf{u}_n(t) = \Gamma_{zn}\begin{bmatrix} \boldsymbol{\varphi}_{zn} D_{zn}(t) \\ \boldsymbol{\varphi}_{\theta n} D_{\theta n}(t) \end{bmatrix}_{2N\times 1} = \Gamma_{zn}\begin{bmatrix} \boldsymbol{\varphi}_{zn} & 0 \\ 0 & \boldsymbol{\varphi}_{\theta n} \end{bmatrix}_{2N\times 2}\begin{bmatrix} D_{zn}(t) \\ D_{\theta n}(t) \end{bmatrix}_{2\times 1} \tag{19.4}$$

When D_{zn} is equal to $D_{\theta n}$, the new definition of $\mathbf{u}_n$ in Eq. (19.4) is the same as the customary definition: $\mathbf{u}_n = \Gamma_{zn}\boldsymbol{\varphi}_n D_n(t)$. D_{zn} and $D_{\theta n}$ are called the nth modal translation and modal rotation respectively. Substituting Eq. (19.4) into Eq. (19.3) and pre-multiplying both sides of Eq. (19.3) by $\begin{bmatrix} \boldsymbol{\varphi}_{zn} & 0 \\ 0 & \boldsymbol{\varphi}_{\theta n} \end{bmatrix}^{\mathrm{T}}$, yields:

$$\mathbf{M}_n\ddot{\mathbf{D}}_n + \mathbf{C}_n\dot{\mathbf{D}}_n + \mathbf{K}_n\mathbf{D}_n = -\mathbf{M}_n\mathbf{1}\ddot{u}_g(t) \tag{19.5}$$

where

$$\mathbf{D}_n = \begin{bmatrix} D_{zn} \\ D_{\theta n} \end{bmatrix}_{2\times 1},\quad \mathbf{M}_n = \begin{bmatrix} \boldsymbol{\varphi}_{zn}^T \mathbf{m}\boldsymbol{\varphi}_{zn} & 0 \\ 0 & \boldsymbol{\varphi}_{\theta n}^T \mathbf{I}_0 \boldsymbol{\varphi}_{\theta n} \end{bmatrix}_{2\times 2},\quad \mathbf{1} = \begin{bmatrix} 1 \\ 1 \end{bmatrix}_{2\times 1}$$
$$\mathbf{C}_n = \begin{bmatrix} \boldsymbol{\varphi}_{zn}^T \mathbf{c}_{zz}\boldsymbol{\varphi}_{zn} & \boldsymbol{\varphi}_{zn}^T \mathbf{c}_{z\theta}\boldsymbol{\varphi}_{\theta n} \\ \boldsymbol{\varphi}_{\theta n}^T \mathbf{c}_{\theta z}\boldsymbol{\varphi}_{zn} & \boldsymbol{\varphi}_{\theta n}^T \mathbf{c}_{\theta\theta}\boldsymbol{\varphi}_{\theta n} \end{bmatrix}_{2\times 2},\quad \mathbf{K}_n = \begin{bmatrix} \boldsymbol{\varphi}_{zn}^T \mathbf{k}_{zz}\boldsymbol{\varphi}_{zn} & \boldsymbol{\varphi}_{zn}^T \mathbf{k}_{z\theta}\boldsymbol{\varphi}_{\theta n} \\ \boldsymbol{\varphi}_{\theta n}^T \mathbf{k}_{\theta z}\boldsymbol{\varphi}_{zn} & \boldsymbol{\varphi}_{\theta n}^T \mathbf{k}_{\theta\theta}\boldsymbol{\varphi}_{\theta n} \end{bmatrix}_{2\times 2} \tag{19.6}$$

Equation (19.5) is the nth 2DOF modal equation of motion. The elastic properties of the 2DOF modal system (Fig. 19.1b) are determined as

$$k_{zn} = \boldsymbol{\varphi}_{zn}^T \mathbf{k}_{zz} \boldsymbol{\varphi}_{zn}, \quad k_{\theta n} = \boldsymbol{\varphi}_{\theta n}^T \mathbf{k}_{\theta\theta} \boldsymbol{\varphi}_{\theta n} - \frac{\left(\boldsymbol{\varphi}_{zn}^T \mathbf{k}_{z\theta} \boldsymbol{\varphi}_{\theta n}\right)^2}{\boldsymbol{\varphi}_{zn}^T \mathbf{k}_{zz} \boldsymbol{\varphi}_{zn}}, \quad e_n = \frac{\boldsymbol{\varphi}_{zn}^T \mathbf{k}_{z\theta} \boldsymbol{\varphi}_{\theta n}}{\boldsymbol{\varphi}_{zn}^T \mathbf{k}_{zz} \boldsymbol{\varphi}_{zn}} \tag{19.7}$$

$$m_n = \boldsymbol{\varphi}_{zn}^T \mathbf{m} \boldsymbol{\varphi}_{zn}, \quad I_n = \boldsymbol{\varphi}_{\theta n}^T \mathbf{I}_0 \boldsymbol{\varphi}_{\theta n}$$

The inelastic properties of the 2DOF modal system are derived from the two pushover curves of the original asymmetric-plan building under the nth modal inertia force distribution. These two pushover curves, representing the relationships of roof translation to base shear and of roof rotation to base torque are idealized as bi-linear curves in ADRS format. The values of the yielding accelerations, A_{zny} and $A_{\theta ny}$, and of the post-yielding stiffness ratios, α_{zn} and $\alpha_{\theta n}$, are available from these two bi-linear curves (Fig. 19.1c). The inelastic 2DOF modal parameters are determined from the four known values, A_{zny}, $A_{\theta ny}$, α_{zn} and $\alpha_{\theta n}$, as:

$$M_{yzn} = A_{zny} m_n, \quad M_{y\theta n} = A_{\theta ny} I_n - A_{zny} m_n e_n \tag{19.8}$$

$$k'_{zn} = \frac{m_n}{\dfrac{\dfrac{m_n}{k_{zn}} - \dfrac{(I_n - m_n e_n)\, e_n}{k_{\theta n}}}{\alpha_{zn}} + \dfrac{(I_n - m_n e_n)\, e_n}{k'_{\theta n}}}, \quad k'_{\theta n} = k_{\theta n} \cdot \alpha_{\theta n} \tag{19.9}$$

where M_{yzn}, $M_{y\theta n}$ and k'_{zn}, $k'_{\theta n}$ are the yielding moments and the post-yielding stiffness of the two rotational springs of the 2DOF modal system (Fig. 19.1b), respectively. Thus, the total elastic and inelastic 2DOF modal parameters are those obtained from Eqs. (19.7), (19.8), and (19.9).

19.2.2 Independent Elastic 2DOF Modal Parameters

The mass and stiffness in the conventional SDOF modal equation of motion depend on the modal period. There are five elastic modal parameters, Eq. (19.7), in the 2DOF modal equation of motion. In order to construct the 2DOF inelastic response spectra, the key independent elastic 2DOF modal parameters have to be identified first. Derivations in Lin and Tsai [4] give:

$$\frac{k_{\theta n}}{k_{zn}} = \left(\frac{I_n}{m_n} - e_n\right)(1 + e_n) \tag{19.10}$$

In this study, the corresponding uncoupled 2DOF modal system is defined as the 2DOF modal system with the lumped mass located at the top of column. That is, e_n equals 0. The ratio of the rotational frequency to the translational frequency of the corresponding uncoupled 2DOF modal system is defined as

$$\Omega_{\theta n} = \frac{\omega_{\theta n}}{\omega_{zn}} = \sqrt{\frac{k_{\theta n} m_n}{k_{zn} I_n}} \tag{19.11}$$

It can be obtained [4] that

$$\boldsymbol{\Omega}_{\theta n} = \sqrt{\left(1 - \tfrac{\boldsymbol{m}_n \boldsymbol{e}_n}{\boldsymbol{I}_n}\right)(1 + \boldsymbol{e}_n)}, \ \boldsymbol{m}_n + \boldsymbol{I}_n = 1 \tag{19.12a}$$

$$\boldsymbol{T}_n = \frac{\boldsymbol{T}_{zn}}{\sqrt{1 + \boldsymbol{e}_n}} \tag{19.12b}$$

where $T_{zn} = 2\pi\sqrt{m_n/k_{zn}}$ is the translational vibration period of the corresponding uncoupled 2DOF modal system. Equation (19.12b) implies that the nth vibration period of a one-way asymmetric-plan structure, T_n, is elongated compared to the translational vibration period of the corresponding uncoupled 2DOF modal system if e_n is negative and shortened if e_n is positive. Given the values of vibration period, T_n, frequency ratio, $\Omega_{\theta n}$, and modal eccentricity, e_n, all five elastic 2DOF modal parameter values can be obtained from Eq. (19.12). Therefore, T_n, $\Omega_{\theta n}$ and e_n can be conveniently viewed as the three key parameters for computing the five elastic 2DOF modal parameters.

19.2.3 Relationships Between the Inelastic 2DOF Modal Parameters and the Strength Ratio

The yield force of a SDOF modal system can be easily obtained from the strength ratio and the associated pseudo-spectral acceleration. Given the yield forces of the SDOF modal systems, the conventional SDOF inelastic response spectra, such as the constant-strength response spectrum, can be constructed accordingly. However, to construct the 2DOF inelastic response spectra, the yield moment versus the strength ratio relationships for the two rotational springs in a 2DOF modal system have to be established.

For most asymmetric-plan buildings, the onset of translational yielding should accompany the onset of rotational yielding. Thus, it is reasonable to assume that the two pushover curves of each vibration mode for the asymmetric-plan structure yield at the same time, i.e. $A_{zny} = A_{\theta ny} = A_{ny}$. This implies that the yield modal translation is equal to the yield modal rotation, as shown in Fig. 19.1c. In [4] the yield moments, M_{yzn} and $M_{y\theta n}$ of the two springs of the 2DOF modal system are derived in terms of the strength ratio as:

$$\boldsymbol{M}_{yzn} = \frac{\boldsymbol{S}_{dn}^{\text{elastic}}}{\boldsymbol{R}} (1 + \boldsymbol{e}_n)\, \boldsymbol{k}_{zn} \tag{19.13}$$

$$\boldsymbol{M}_{\mathbf{y}\theta\mathbf{n}} = \frac{\boldsymbol{S}_{dn}^{\text{elastic}}}{\boldsymbol{R}} \boldsymbol{k}_{\theta n} \tag{19.14}$$

Thus, the inelastic modal strength parameters (M_{yzn} and $M_{y\theta n}$) versus the strength ratio R relationships of the 2DOF modal systems have been established as in Eqs. (19.13) and (19.14).

19.3 Parametric Study

Based on the results in the previous section, the inelastic response spectra for one-way asymmetric-plan buildings can be constructed using the 2DOF modal systems, if the values of e_n, $\Omega_{\theta n}$, α_{zn}, $\alpha_{\theta n}$, R and the damping ratio, ξ_n, are given. The ranges of translational post-yield stiffness ratio, α_{zn}, strength ratio, R, and damping ratio, ξ_n, are well established. However, the range of the rotational post-yield stiffness ratio, $\alpha_{\theta n}$, modal eccentricity, e_n, and frequency ratio, $\Omega_{\theta n}$, are not entirely clear. The study of the conventional SDOF response spectra became meaningful only after the ranges of the relevant parameter values have been well understood. Therefore, the goal of this section is to study the ranges of e_n, $\Omega_{\theta n}$ and $\alpha_{\theta n}$.

Since extensive parametric studies on elastic one-storey one-way asymmetric-plan buildings have been carried out in the past, the ranges of the physical system parameter values and their effects on the physical system are already available. In order to understand the ranges of both the elastic and the inelastic 2DOF modal system parameter values, their relationships associated with the physical system parameters are investigated in this paper.

19.3.1 Ranges of Elastic 2DOF Modal Parameters

It has been shown [4] that the relationships between the stated one-storey whole-structure parameters and the associated 2DOF modal parameters are

$$\Omega_{\theta n} = \Psi_\theta, \quad n = 1,2 \tag{19.15}$$

$$e_n = \frac{1}{2}\left[-1 + \left(\bar{e}\Big/\sqrt{\frac{1+\beta^2}{12\beta^2}}\right)^2 + \Psi_\theta^2 \mp \sqrt{\left(1 + \left(\bar{e}\Big/\sqrt{\frac{1+\beta^2}{12\beta^2}}\right)^2 + \Psi_\theta^2\right)^2 - 4\Psi_\theta^2}\right] \tag{19.16}$$

where $\omega_z^2 = k_z/m$ and β is the aspect ratio of the floor plan, i.e. $\beta = a/b$; a and b are the floor plan dimensions perpendicular and parallel to the ground motions, respectively. The notation $\bar{e}$ is defined as the normalized eccentricity, i.e. $\bar{e}=e/a$; Ψ_θ is the ratio of the rotational to the translational vibration frequency of the corresponding uncoupled one-storey building, i.e. $\Psi_\theta = \left(k_\theta m/k_z I_0\right)^{0.5}$. Figure 19.2a and b illustrate the varying ranges of the first and the second modal eccentricities e_1 and e_2, respectively, for the one-storey one-way asymmetric-plan buildings with aspect

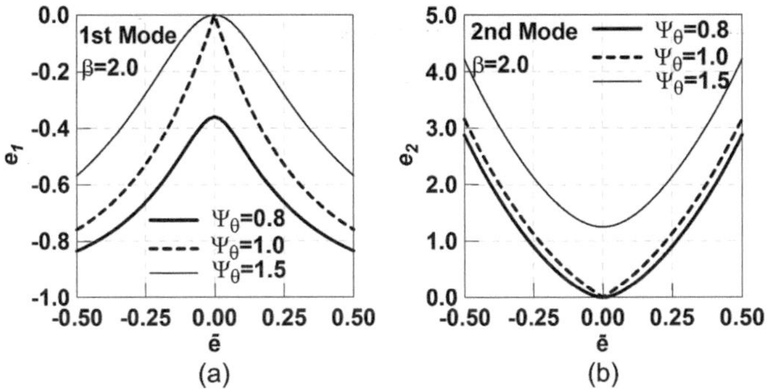

Fig. 19.2 The modal eccentricities of (**a**) the first and (**b**) the second modes for the one-storey one-way asymmetric-plan buildings with aspect ratio $\beta = 2.0$

ratio β=2.0. From all the negative values of e_1 shown in Fig. 19.2a and Eq. (19.12), the first modal vibration period of a one-storey system has been clearly elongated compared with the translational period of the corresponding uncoupled one-storey system. By contrast, the second modal vibration period of a one-storey system has been shortened, compared to the translational period of the uncoupled one-storey system.

19.3.2 Ranges of Inelastic 2DOF Modal Parameters

The strength ratio R and the translational post-yield stiffness ratio α_{zn} have been extensively investigated by other researchers. In those studies, values of R of 2, 4 and 8 were often chosen, while the value of α_{zn} is often taken as 2%. The only new inelastic 2DOF modal parameter in addition to those of the traditional SDOF modal systems is the rotational post-yielding stiffness ratio, $\alpha_{\theta n}$. Two very different example systems, Systems I and II, are used for discussing the range of $\alpha_{\theta n}$. In System I, most of the rotational stiffness is contributed by the lateral-load resisting elements perpendicular to the Z-direction ground motion. Thus, the rotational stiffness of this system is less likely to be influenced by yielding of the lateral-load resisting elements parallel to the ground motion. In other words, the rotational post-yield stiffness ratio, $\alpha_{\theta n}$, of System I is close to 1.0. In System II, by contrast, most of the rotational stiffness is contributed by the lateral-load resisting elements parallel to the Z-direction ground motion. Thus, the reduction of the rotational stiffness of System II will be similar to that of its translational stiffness when the lateral-load resisting elements parallel to the ground motion yield. That is, $\alpha_{\theta n}$ would be approximately equal to α_{zn}. So, in this study the values of $\alpha_{\theta n}$ have been chosen as 5%, 45% and 95%, to cover a wide range of possible plan-asymmetric systems, for which the proposed T-R constant-strength response spectra are constructed in the following section.

19.4 T-R Constant-Strength Response Spectra

The 10% in 50 years set of ground motion records, including 20 historical ground motions, suggested for the SAC buildings in Los Angeles were applied to construct the inelastic response spectra. The averages of the T-R inelastic response spectra and the associated conventional SDOF response spectra under this set of ground motions are shown in Fig. 19.3. The figure shows that the difference between μ_z and μ, obtained using 2DOF and SDOF modal systems, respectively, increases, as the value of R increases. Furthermore, Fig. 19.3 clearly demonstrates that the SDOF modal system is not capable of estimating the rotational demand μ_θ, which is significantly different from μ_z when R equals 4 and 8.

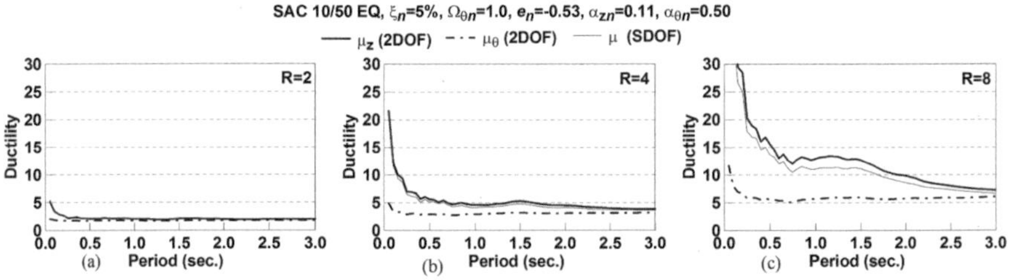

Fig. 19.3 The averaged T-R constant-strength response spectra (e_n=−0.53, $\Omega_{\theta n}$=1.0, α_{zn}=0.11, $\alpha_{\theta n}$=0.50) compared to the corresponding conventional SDOF constant-strength response spectra (α_{zn}=0.11) under the 10% in 50 years set of ground motion records used for the SAC buildings in Los Angeles. (**a**) R=2; (**b**) R=4; (**c**) R=8

19.5 Conclusions

The conventional SDOF inelastic response spectra are widely applied in engineering practice. As this type of response spectra is computed only from the roof translation versus base shear relationship, it lacks the capability to accurately capture the rotational seismic demands of asymmetric-plan structures. In addition, the conventional response spectra do not consider the possible translation-rotation interaction developed in each coupled vibration mode of inelastic asymmetric-plan structures. As the 2DOF modal systems can simultaneously consider translational and rotational force-deformation relationships, the T-R inelastic response spectra for one-way asymmetric-plan structures are proposed in this study. The T-R inelastic response spectra are able to give not only the translational but also the rotational seismic demands.

In order to construct the T-R inelastic response spectra, the independent elastic 2DOF modal parameters have been identified as the vibration period T_n, the frequency ratio $\Omega_{\theta n}$ and the modal eccentricity e_n. The relationships of the inelastic 2DOF modal strength parameters and the strength ratio have been also established. In order to understand the ranges of the 2DOF modal parameter values, a parametric study in terms of general one-storey one-way asymmetric-plan buildings has been carried out. The T-R constant-strength response spectra have been constructed to illustrate the typical T-R inelastic response spectra. It has been shown that the difference between the SDOF modal ductility μ and the 2DOF modal translational ductility μ_z increases, as the strength ratio R increases. Moreover, the SDOF modal ductility μ is often very different from the 2DOF modal rotational ductility, μ_θ.

References

1. Applied Technology Council (1996), Seismic evaluation and retrofit of concrete buildings, volumes 1 and 2, Report No. ATC-40, ATC, Redwood City, CA
2. Chopra AK, Goel RK (2002) A modal pushover analysis procedure for estimating seismic demands for buildings. Earthq Eng Struct Dyn 31:561–582
3. Lin JL, Tsai KC (2007) Simplified seismic analysis of asymmetric building systems. Earthq Eng Struct Dyn 36:459–479
4. Lin JL, Tsai KC (2009) Modal parameters for the analysis of inelastic asymmetric-plan structures. Earthq Spectra 25:821–849
5. Nassar AA, Krawinkler H (1991) Seismic demands for SDOF and MDOF systems. Report No. 95, The John A. Blume Earthquake Engineering Center, Stanford University, Stanford, CA
6. Vidic T, Fajfar P, Fischinger M (1992) A procedure for determining consistent inelastic design spectra. Proceedings of workshop on nonlinear seismic analysis of RC structures, Bled, Slovenia

Chapter 20
Multi-Mode Pushover Analysis with Generalized Force Vectors

Halûk Sucuoğlu and M. Selim Günay

20.1 Introduction

Considering the simplicity and conceptual appeal of conventional pushover analysis with a single mode, several researchers have attempted to develop multi-mode pushover analysis procedures in order to replace nonlinear response history analysis with an "inelastic" response spectrum analysis [3, 5]. Adaptive lateral force distribution schemes have further been proposed for overcoming the limitations of conventional pushover analysis arising from an invariant lateral static load distribution [1, 2, 4] which however require rigorous computations in the implementation. All multi mode pushover analysis procedures published in literature so far have two common features. First, they are adaptive except MPA [3] hence require an eigenvalue analysis at each loading increment. Moreover, an adaptive algorithm cannot be implemented with a conventional nonlinear structural analysis programming code. Second, all procedures combine modal responses statistically by SRSS, which is an approximate rule developed for combining linear elastic modal responses. Internal forces should be checked at each load increment and be corrected if they exceed the associated capacities.

A practical nonlinear static procedure is developed herein which accounts for the contribution of all significant modes to inelastic seismic response. The procedure consists of conducting a set of pushover analyses by employing different generalized force vectors. Each generalized force vector is derived as a different combination of modal lateral forces in order to simulate the effective lateral force distribution when the interstory drift at a selected story attains its maximum value during seismic response. Hence the proposed procedure is called generalized pushover analysis (GPA). Target seismic demands for interstory drifts at the selected stories are calculated from the associated generalized drift expressions where nonlinear response is considered in the first mode only. Finally, the maximum value of a response parameter is obtained from the envelope values produced by the set of generalized

H. Sucuoğlu (✉)
Department of Civil Engineering, Middle East Technical University, 06531 Ankara, Turkey
e-mail: sucuoglu@ce.metu.edu.tr

M.N. Fardis (ed.), *Advances in Performance-Based Earthquake Engineering*, 213
Geotechnical, Geological, and Earthquake Engineering 13, DOI 10.1007/978-90-481-8746-1_20,

pushover analysis conducted separately for each interstory drift. GPA can be implemented with any structural analysis software capable of performing displacement controlled nonlinear incremental static analysis. Seismic response of a twelve story reinforced concrete frame structure under twelve ground motion records are estimated by GPA in this study, and compared with the results obtained from NRHA as well as from the conventional pushover analysis.

20.2 Generalized Force Vectors

Different response parameters attain their maximum values at different times during seismic response. An effective force vector acts on the system instantaneously at the time when a specific response parameter reaches its maximum value. This effective force vector is in fact a generalized force since it has contributions from all modal forces at the time of maximum response for the specified response parameter. Accordingly, if this force vector can be defined, then it can be applied either directly or incrementally to the investigated structural system in order to produce the maximum value of this response parameter.

The derivation of generalized effective force vectors is based on the dynamic response of linear elastic MDOF systems to earthquake ground excitation $\ddot{u}_g(t)$, by employing the modal superposition procedure. The effective force vector $\mathbf{f}(t_{max})$ at time $t_{max,}$ when an arbitrarily selected response parameter reaches its maximum value, can be expressed as the superposition of modal forces $\mathbf{f}_n$ (t_{max}):

$$\mathrm{f}\,(t_{max}) = \sum\nolimits_n \mathrm{f}_n(t_{max}) \tag{20.1}$$

The n'th mode effective force in Eq. (20.1) at time t_{max} is given by

$$\mathrm{f}_n(t_{max}) = \Gamma_n \mathbf{m} \varphi_n A_n(t_{max}) \tag{20.2}$$

Here, $\Gamma_n = L_n / M_n$; $L_n = \boldsymbol{\varphi}_n^{\mathbf{T}}\ \mathbf{ml}$; $M_n = \boldsymbol{\varphi}_n^{\mathbf{T}} \mathbf{m} \boldsymbol{\varphi}_n$; $\boldsymbol{\varphi}_n$ is the n'th mode shape, $\mathbf{m}$ is the mass matrix, $\mathbf{l}$ is the influence vector, and $A_n(t_{max})$ is represented with Eq. (20.3).

$$A_n(t_{max}) = \omega_n^2 D_n(t_{max}) \tag{20.3}$$

Here ω_n^2 is the n'th mode vibration frequency and $D_n(t_{max})$ is the modal displacement amplitude at t_{max} which satisfies

$$\ddot{D}_n(t_{max}) + 2\varsigma_n \omega_n \dot{D}_n(t_{max}) + \omega_n^2 D_n(t_{max}) = -\ddot{u}_g(t_{max}) \tag{20.4}$$

$D_n(t_{max})$ cannot be determined from Eq. (20.4) unless t_{max} is known. t_{max} is the time when the selected response parameter becomes maximum, which depends on all modal responses. This response parameter is selected as the interstory drift Δ_j at the j'th story. Then,

$$\Delta_{j,\max} = \Delta_j(t_{\max}) \tag{20.5}$$

Its associated modal expansion is

$$\Delta_j(t_{\max}) = \sum\nolimits_n \Gamma_n\, D_n(t_{\max})\,(\varphi_{n,j} - \varphi_{n,j-1}) \tag{20.6}$$

where $\varphi_{n,j}$ is the j'th element of the mode shape vector $\boldsymbol{\varphi}_n$. Equation (20.6) can be normalized by dividing both sides with $\Delta_j(t_{\max})$, which yields

$$1 = \sum\nolimits_n \left[\Gamma_n \frac{D_n(t_{\max})}{\Delta_j(t_{\max})}(\varphi_{n,j} - \varphi_{n,j-1})\right] \tag{20.7}$$

Each term on the right hand side of Eq. (20.7) under summation expresses the contribution of n'th mode to the maximum interstory drift $\Delta_j(t_{\max})$ at the j'th story in a normalized form.

The maximum value of interstory drift at the j'th story in Eq. (20.5) can also be estimated by RSA through SRSS of the related spectral modal responses.

$$(\Delta_{j,\max})^2 \approx \sum\nolimits_n \left[\Gamma_n D_n(\varphi_{n,j} - \varphi_{n,j-1})\right]^2 \tag{20.8}$$

D_n in Eq. (20.8) is the spectral displacement of the n'th mode, which is directly available from the displacement response spectrum of the ground excitation $\ddot{u}_g(t)$. Equation (20.8) can also be normalized similarly, by dividing both sides with $(\Delta_{j,\max})^2$.

$$1 = \sum\nolimits_n \left[\Gamma_n \frac{D_n}{\Delta_{j,\max}}(\varphi_{n,j} - \varphi_{n,j-1})\right]^2 \tag{20.9}$$

Accordingly, the respective terms on the right hand sides of Eqs. (20.7) and (20.9) are the normalized contributions of the n'th mode to the maximum interstory drift at the j'th story. Equating them and assuming the equality $\Delta_{j,\max} = \Delta_j(t_{\max})$ from Eq. (20.5) leads to

$$D_n(t_{\max}) = \frac{D_n}{\Delta_{j,\max}}\left[\Gamma_n D_n(\varphi_{n,j} - \varphi_{n,j-1})\right] \tag{20.10}$$

Since the term in the parentheses in Eq. (20.10) is equal to $\Delta_{j,n}$ from Eq. (20.8),

$$D_n(t_{\max}) = \frac{\Delta_{j,n}}{\Delta_{j,\max}} D_n \tag{20.11}$$

It should be noted that $\Delta_{j,n}$ in Eq. (20.11) is the n'th mode contribution to the maximum interstory drift at the j'th story determined from RSA, and $\Delta_{j,\max}$ in Eq. (20.11) is the quadratic combination of the $\Delta_{j,n}$ terms as given by Eq. (20.8). The generalized force vector is obtained by first calculating $A_n(t_{\max})$ by substituting

$D_n(t_{\max})$ from Eq. (20.11) into Eq. (20.3), then substituting $A_n(t_{\max})$ from Eq. (20.3) into Eq. (20.2), and finally substituting $\mathbf{f}_n(t_{\max})$ from Eq. (20.2) into Eq. (20.1).

$$\mathbf{f}_j(t_{\max}) = \sum\nolimits_n \left(\Gamma_n \mathbf{m} \boldsymbol{\varphi}_n A_n \frac{\Delta_{j,n}}{\Delta_{j,\max}} \right) \tag{20.12}$$

A_n is the pseudo spectral acceleration of the n'th mode in Eq. (20.12). Since the formulation that is developed in Eqs. (20.1)–(20.11) is employed for obtaining the generalized force vector which acts on the structural system when the interstory drift at the j'th story becomes maximum, the associated generalized force vector in Eq. (20.12) is identified with the subscript j.

20.3 Target Seismic Deformation Demand

In linear elastic response spectrum analysis the target drift demand at the j'th story Δ_{jt} is calculated from the SRSS combination of modal drifts $\Delta_{n,j}$ expressed by

$$\Delta_{jt}^2 = \sum\nolimits_n \left[\Gamma_n \, (\varphi_{n,j} - \varphi_{n,j-1}) D_n \right]^2 \tag{20.13}$$

The displacement shape of a nonlinear system during seismic response can be expanded in terms of the linear elastic mode shapes if the modal amplitudes (coordinates) can be calculated appropriately. It has been observed by Chopra and Goel [3] that the coupling between modal coordinates due to yielding of the system is negligible. Therefore Eq. (20.13) may be employed for estimating the inelastic drift demands, provided that linear elastic modal spectral displacement demands D_n in Eq. (20.13) are replaced by the inelastic modal spectral displacement demands $D_n{}^*$.

Replacing only D_1 in Eq. (20.13) with $D_1{}^*$ while retaining the linear elastic modal spectral displacements for the second and higher modes improves the target displacement demand significantly. $D_1{}^*$ can then be estimated from either the NRHA of the nonlinear SDOF system representing the first mode contribution, or from the associated R-μ-T relations. Then the target drift demand Δ_{jt} in GPA becomes;

$$\Delta_{jt} = \left(\left[\Gamma_1 \, (\varphi_{1,j} - \varphi_{1,j-1}) D_1^* \right]^2 + \sum_{n=2}^{N} \left[\Gamma_n \, (\varphi_{n,j} - \varphi_{n,j-1}) D_n \right]^2 \right)^{1/2} \tag{20.14}$$

The contribution of higher modes to a maximum interstory drift parameter is more significant than their contribution to a maximum displacement parameter. GPA uses the interstory drift parameters as target demands. Accordingly, interstory drift is not obtained from GPA, but from an independent response spectrum analysis. When the associated generalized force vector pushes the system to this target drift, the system adopts itself in the inelastic deformation range while the further higher order deformation parameters (rotations, curvatures) take their inelastic values as in modal pushover, but by receiving more appropriate contributions from the higher modes.

20.4 Generalized Pushover Algorithm

The GPA algorithm is composed of the five basic steps summarized below.

1. *Eigenvalue analysis:* Natural frequencies ω_n (natural periods T_n), modal vectors $\boldsymbol{\varphi}_n$ and the modal participation factors Γ_n are determined from eigenvalue analysis.
2. *Response spectrum analysis:* Modal spectral amplitudes A_n, D_n are obtained from the corresponding linear elastic spectra and modal interstory drift ratios at the j'th story, $\Delta_{j,n}$ are determined from RSA. The maximum interstory drift ratio at the j'th story, $\Delta_{j,\max}$ is obtained by SRSS.
3. *Generalized force vectors:* Generalized force vectors $\mathbf{f}_j$ which produce the maximum response Δ_j are calculated from Eq. (20.12).
4. *Target interstory drift demands:* Maximum inelastic modal displacement demand $D_1{}^*$ for the first mode under an earthquake excitation is obtained from either NRHA or inelastic response spectrum of the inelastic SDOF system idealized with a bi-linear force–displacement relation. For the higher modes $n = 2$–N, D_n values are obtained from the linear elastic response spectrum. Finally, $D_1{}^*$ and D_n $(n = 2, N)$ are substituted into Eq. (20.14) for calculating the target interstory drift demands Δ_{jt}.
5. *Generalized pushover analysis:* A total number of N generalized pushover analyses are conducted. In the j'th GPA ($j = 1$–N), the structure is pushed in the lateral direction incrementally with a force distribution proportional to $\mathbf{f}_j$. At the end of each loading increment i, the interstory drift Δ_{ji} obtained at the j'th story is compared with the target interstory drift Δ_{jt} calculated from Eq. (20.14). Displacement controlled incremental loading ($i = 1, 2\ldots$) at the j'th GPA continues until Δ_{ji} reaches Δ_{jt}.

All member deformations and internal forces are directly obtained from the j'th GPA at the target interstory drift Δ_{jt}. Once all GPA is completed for $j = 1$–N, the enveloping values of member deformations and internal forces are registered as the maximum seismic response values.

20.5 Test Case: 12 Story RC Symmetrical Plan Building

The proposed GPA procedure is tested on a 12 story reinforced concrete building with symmetrical plan, where the floor plan is shown in Fig. 20.1. The building is designed according to the regulations of 2007 Turkish Earthquake Code in accordance with the capacity design principles. An enhanced ductility level is assumed for the building. The design spectrum is shown in Fig. 20.1. Concrete and steel characteristic strengths are 25 and 420 MPa, respectively. Slab thickness for all floors is

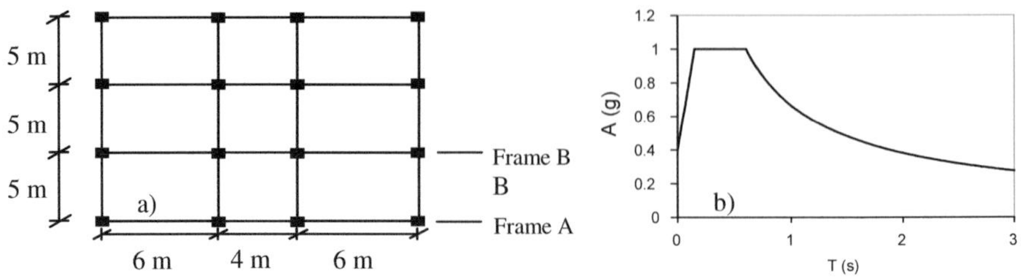

Fig. 20.1 (**a**) Typical plan of the 12 story building, (**b**) design spectrum

140 mm and live load is 3.5 kN/m^2. Dimensions of the beams at the first four, the second four and the last four stories are 300 × 550, 300 × 500 and 300 × 450 mm^2 respectively, whereas dimensions of the columns at the first four, the second four and the last four stories are 500 × 500, 450 × 450 and 400 × 400 mm^2 respectively. There is no basement; height of the ground story is 4 m while the height of all other stories is 3.2 m.

Plane frame models consisting of Frames A and B are constructed for the analysis. They are analyzed by using the nonlinear analysis software Drain-2DX. Cracked section stiffness is employed for the initial linear segment of the moment-curvature relations. Gross moments of inertia are multiplied with 0.6 and 0.4 for the columns and beams, respectively, in order to represent cracking. Free vibration periods for the first three modes of the frame are 2.38, 0.86 and 0.50 s.

20.5.1 Strong Ground Motions

The building is analyzed under twelve different strong motion components. These ground motions are selected from a larger set in accordance with the FEMA 356 criteria limiting the applicability of conventional nonlinear static procedure. They produce significant higher mode effects on the investigated 12 story frame; hence FEMA 356 does not allow using standard NSP. The basic characteristics of the twelve ground motion components are presented in Table 20.1. The first nine contain a significant pulse whereas the last three are without a pulse. All ground motions were downloaded from the PEER strong motion database.

20.5.2 Implementation of the GPA algorithm

The generalized pushover analysis algorithm is presented herein with an implementation to the 12 story building, subjected to the ground motion 7 in Table 20.1. The resulting force distributions $\mathbf{f}_j$ along height, obtained from Eq. (20.12) are presented in Fig. 20.2. It can be observed that the second mode contribution is significant on the force distributions which produce maximum interstory drifts at all 12 stories.

Table 20.1 Characteristics of strong ground motions

#	GM code	Earthquake $\overline{M}_w$	Station-component	CD (Km)[1]	Site Geol.[2]	PGA (g)	PGV (cm/s)	PGD (cm)	GM Type
1	BOL090	Duzce, 11/12/99 (7.1)	Bolu-090	12.0	D	0.822	62.1	13.6	Pulse
2	ERZ-EW	Erzincan, 03/13/92 (6.9)	Erzincan-Ew	4.4	D	0.496	64.3	21.9	Pulse
3	H-E04140	Imp. Valley, 10/15/79 (6.5)	El Centro Array #4-140	7.1	D	0.485	37.4	20.1	Pulse
4	PR1090	Kobe, 01/16/95 (6.9)	Port Island (0 m)-090	3.3	E	0.278	54.2	24.9	Pulse
5	CLS090	Loma Prieta, 10/18/89 (7)	Corralitos-090	3.9	A	0.479	45.2	11.3	Pulse
6	LEX000	Loma Prieta, 10/18/89 (7)	Los Gat. - Lex. Dam-000	5.0	A	0.420	73.5	20.0	Pulse
7	SPV270	Northridge, 01/17/94 (6.7)	Sepulveda VA-270	8.9	D	0.753	84.5	18.7	Pulse
8	PCD254	San Fer., 02/09/71 (6.6)	Pacoima Dam-254	2.8	B	1.160	54.1	11.8	Pulse
9	CHY006-E	Chi-chi, 09/20/99 (7.6)	CHY006-E	9.8	B	0.364	55.4	25.6	Pulse
10	BOL000	Duzee, 11/12/99 (7.1)	Bolu-000	12.0	D	0.728	56.4	23.1	Ordinary
11	ORR090	Northridge, 01/17/94 (6.7)	Cast.-Old Rdg Rotue-090	20.7	B	0.568	51.8	9.0	Ordinary
12	ORR360	Northridge, 01/17/94 (6.7)	Cast.-Old Rdg Rotute-360	20.7	B	0.514	52.0	15.3	Ordinary

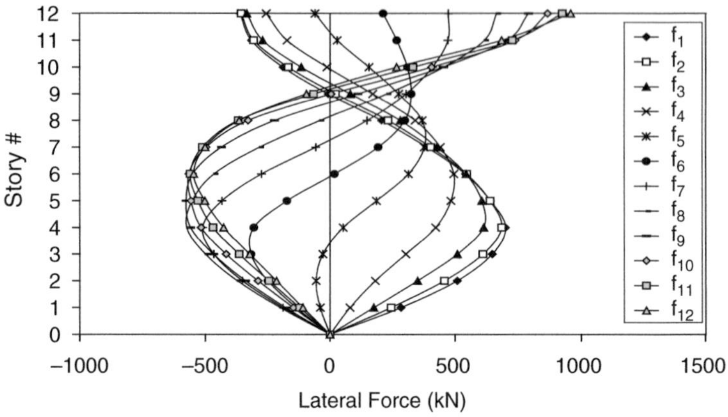

Fig. 20.2 Generalized force distributions $\mathbf{f}_j$ for the 12 story building

20.5.3 Results

Beam plastic rotations and member internal forces are calculated respectively under the 12 ground motions in Table 20.1. Beam plastic rotations at member ends calculated with NRHA, GPA and PO-1 (conventional pushover analysis) are shown in Fig. 20.3. Beam plastic rotations are calculated for each story, as the average value of all beam-end plastic rotations in that story. Plastic actions did not develop at the columns of the building under any of the twelve ground motion excitations except the ground story column bases.

The performance of GPA in predicting beam plastic rotations from NRHA is satisfactory. PO-1 is not able to capture plastic deformations in any of the beams at the upper five stories since first mode is not sufficient by itself for developing plastic rotations. GPA is quite successful in predicting the plastic beam deformations at both upper and the lower stories. Simultaneous combination of all modes in GPA leads to realistic estimations of plastic deformations. There are some unsuccessful cases at the lower stories however for GPA, such as GM2 and GM9, where higher mode contributions at the lower stories do not develop apparently during the NRHA contrary to the GPA prediction.

20.6 Conclusions

The generalized pushover analysis (GPA) procedure presented here has three basic advantages over the other multi-mode pushover procedures available in literature.

- GPA (like MPA) is not adaptive. It can be implemented conveniently by employing a general purpose nonlinear static analysis tool. There is no need for developing a special programming code.

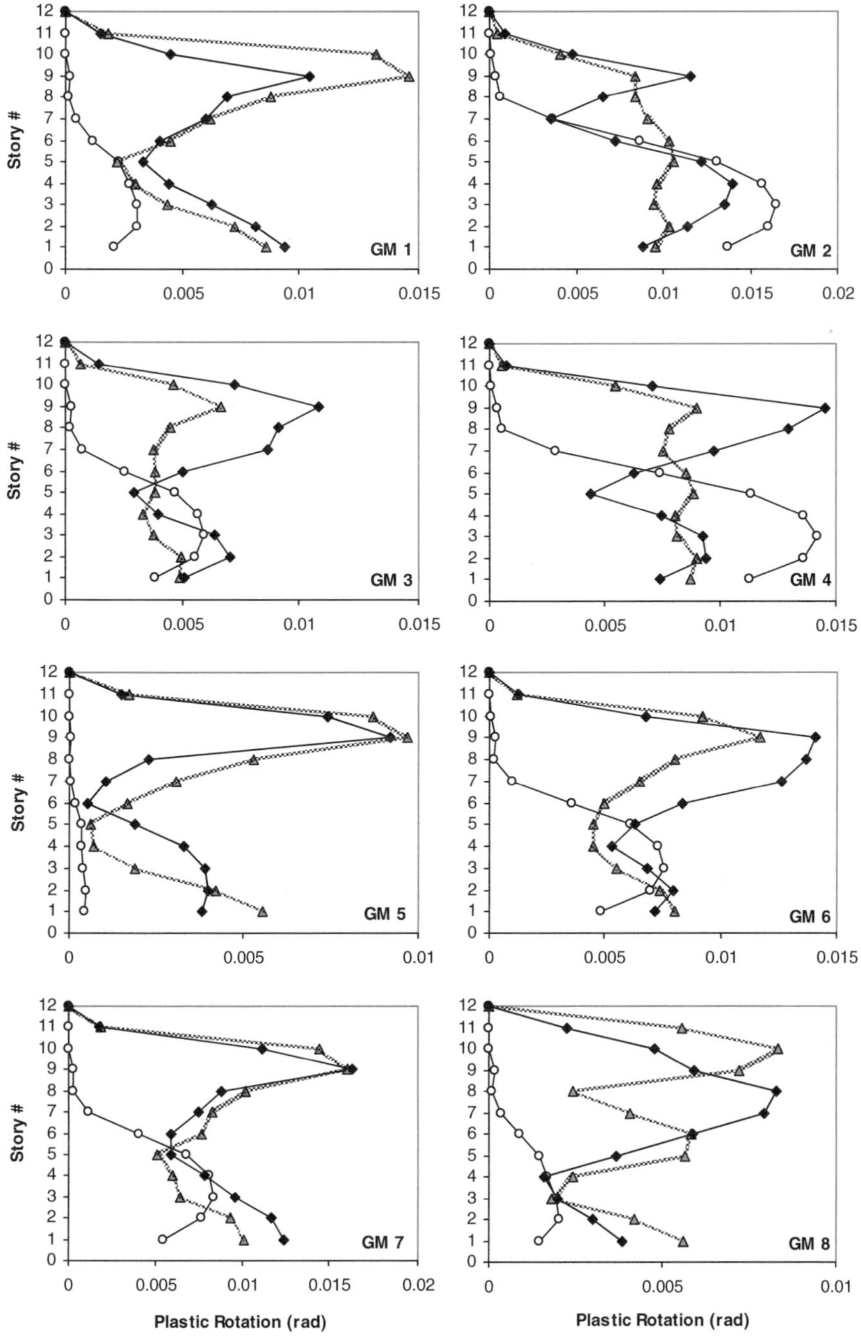

Fig. 20.3 Beam plastic rotations

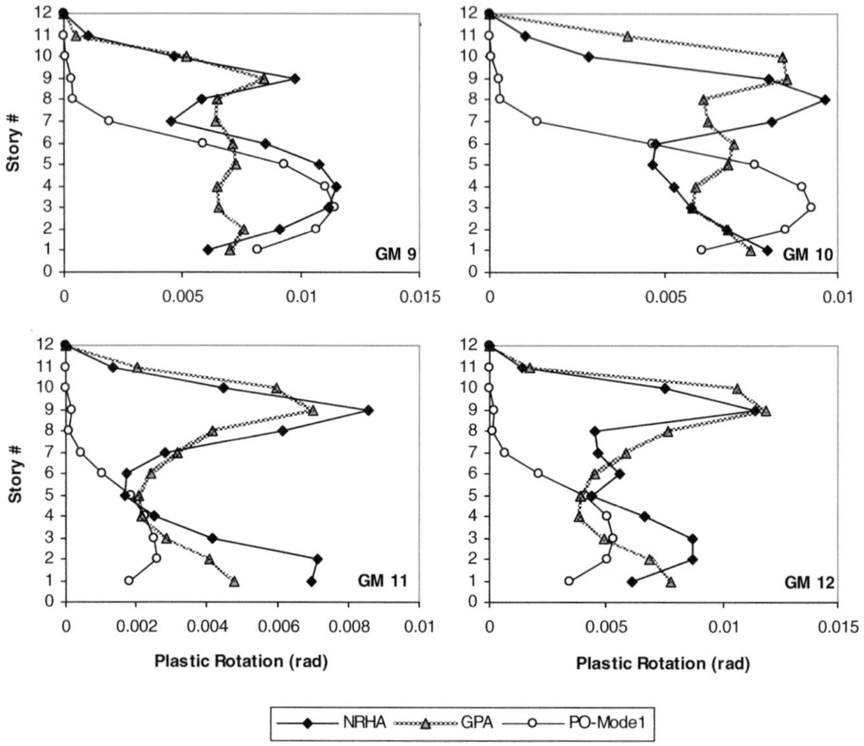

Fig. 20.3 (continued)

- The computational effort required by GPA is much less compared to the adaptive pushover procedures.
- GPA does not suffer from the statistical combination of individually calculated inelastic modal responses. It activates all modes of a multi degree of freedom system simultaneously. So, all response parameters are obtained directly from a generalized pushover analysis at the associated target drift demand.

The number of pushover analysis required in GPA under a single ground excitation expressed by its response spectrum is equal to one plus the number of stories in symmetrical plan buildings (1+*N*), where *N* is the number of lateral dynamic degrees of freedom, or the number of stories. However 1+*N* pushover in GPA is only repetitive.

References

1. Antoniou S, Pinho R (2004) Development and verification of a displacement-based adaptive pushover procedure. J Earthq Eng 8(5):643–661
2. Aydınoğlu MN (2003) An incremental response spectrum analysis procedure based on inelastic spectral displacements for multi-mode seismic performance evaluation. Bull Earthq Eng 1(1):3–36

3. Chopra AK, Goel RK (2002) A modal pushover analysis procedure for estimating seismic demands for buildings. Earthq Eng Struct Dyn 31(3):561–582
4. Gupta B, Kunnath SK (2000) Adaptive spectra-based pushover procedure for seismic evaluation of structures. Earthq Spectra 16(2):367–391
5. Sasaki F, Freeman S, Paret T (1998) Multi-mode pushover procedure (MMP). Proceedings of 6th U.S. NCEE, Seattle, WA, USA

Chapter 21
A Practice-Oriented Approach for Probabilistic Seismic Assessment of Building Structures

Peter Fajfar and Matjaž Dolšek

21.1 Introduction

One of the methods developed for the seismic risk evaluation of structures is the SAC-FEMA method, which permits probability assessment in closed form [3], and represents a part of a broader PEER probabilistic framework [4]. Within the framework of the SAC-FEMA method, the relationship between the seismic intensity measure and the engineering demand parameter is usually determined by Incremental Dynamic Analysis (IDA) [13]. IDA requires a large number of inelastic response-history analyses (and corresponding detailed data on the ground motion time-histories and hysteretic behaviour of structural elements), and is thus very time-consuming. It is often possible to create summarized IDA curves with less input data and less effort, but with still acceptable accuracy. One possible approach is to determine the seismic demand for multiple levels of seismic intensity using the N2 method [9] which is a practice-oriented nonlinear method based on pushover analysis and inelastic response spectrum. The N2 method has been included in Eurocode 8 (EC8) [1]. Such an approach yields the Incremental N2 (IN2) curve [5, 6], which is intended to approximate a summarized IDA curve. By using the IN2 curve together with default values for dispersion measures the probabilistic approach can be substantially simplified [6, 8]. In many cases, additional simplifying assumptions can be adopted, which are consistent with seismic code procedures. As a result, a very simple equation for a quick estimation of the annual probability of failure of a structure can be derived, which is appropriate for practical application. In the paper, the simplified approach is very briefly summarized and applied to the probabilistic seismic assessment of two variants of a three-storey reinforced concrete (RC) frame building.

P. Fajfar (✉)
Faculty of Civil and Geodetic Engineering, University of Ljubljana, Jamova 2, Ljubljana, SI-1000, Slovenia
e-mail: pfajfar@ikpir.fgg.uni-lj.si

M.N. Fardis (ed.), *Advances in Performance-Based Earthquake Engineering*, Geotechnical, Geological, and Earthquake Engineering 13, DOI 10.1007/978-90-481-8746-1_21,

21.2 Framework for Probabilistic Performance Assessment

A probabilistic assessment in closed form, upon which the SAC-FEMA steel moment frame guidelines are based [11], is defined by Eqs. (21.1), (21.2), (21.3), and (21.4). The x confidence level estimate of the annual probability (mean annual frequency) of the exceedance of a given limit state P_{LS}, x, can be determined as [3]

$$P_{\mathrm{LS},x} = \tilde{H}\left(s_{\mathrm{a},\tilde{C}}\right) C_{\mathrm{H}} C_{\mathrm{f}} C_x \tag{21.1}$$

$$C_{\mathrm{H}} = \exp\left[\frac{1}{2}\beta_{\mathrm{H}}^2\right], \quad C_{\mathrm{f}} = \exp\left[\frac{k^2}{2b^2}\left(\beta_{\mathrm{DR}}^2 + \beta_{\mathrm{CR}}^2\right)\right],$$
$$C_x = \exp\left[K_x \sqrt{\frac{k^2}{b^2}\left(\beta_{\mathrm{DU}}^2 + \beta_{\mathrm{CU}}^2\right)}\right] \tag{21.2}$$

$\tilde{H}\left(s_{\mathrm{a},\tilde{C}}\right)$ is the median value of the hazard function at the seismic intensity $s_{\mathrm{a},\tilde{C}}$, which causes a selected limit state. It provides a median estimate of the annual probability that the seismic intensity will be equal to or exceed the level $s_{\mathrm{a},\tilde{C}}$, i.e. the seismic intensity "corresponding" to the median displacement capacity $\tilde{D}_{\mathrm{C}}$. The seismic intensity measure (*IM*) is typically expressed by means of spectral acceleration or of peak ground acceleration. In this paper it will be denoted as s_{a}. k is a parameter of the hazard function, idealized in the form:

$$\tilde{H}\left(s_{\mathrm{a}}\right) = k_o \cdot \left(s_{\mathrm{a}}\right)^{-k} \tag{21.3}$$

b is a parameter of the function relating the engineering demand parameter (*EDP*, typically expressed in terms of top displacement or storey drift, denoted as D) to the intensity measure, i.e. of the so-called IDA curve. IDA curve is idealized as:

$$\tilde{D}\left(s_{\mathrm{a}}\right) = a \cdot \left(s_{\mathrm{a}}\right)^{b} \tag{21.4}$$

K_x is the standardized normal variate associated with the probability x of not being exceeded. For example, the values $K_x = 0$, 1 and 1.28 are associated with 50, 84 and 90% confidence levels, respectively. β_{H} is the dispersion measure for hazard. The product $\tilde{H}\left(s_{\mathrm{a},\tilde{C}}\right)$. C_{H} represents the mean value of the hazard function $\tilde{H}\left(s_{\mathrm{a},\tilde{C}}\right)$. Other β parameters represent the dispersion of the engineering demand parameter due to ground motion variability (randomness) and due to variability related to structural modeling and analysis (uncertainty). β_{DR} and β_{CR} are the dispersion measures for randomness in EDP demand and capacity, respectively, and β_{DU} and $\beta_{\mathrm{CU}}$are the dispersion measures for uncertainty in EDP demand and capacity, respectively. For practical applications, predetermined default values for dispersion measures, based on statistical studies of typical structural systems, are needed.

21.3 Simplified Procedure

An important simplification of the probabilistic seismic assessment can be introduced by employing the pushover-based N2 method instead of the response-history based IDA analysis. In such a case, the IDA curve is substituted by an IN2 curve. The whole IN2 curve can be determined by repeating the N2 approach for increasing ground motion intensity until "failure" occurs. In general, the shape of the IN2 curve depends on the relation between the reduction factor, ductility and period (the R-μ-T relation), which defines the inelastic spectra to be used in the N2 method for the determination of seismic demand. Knowing the IN2 curve, the engineering demand parameter can be easily linked to the corresponding seismic intensity measures. An IN2 curve is intended to approximate a summarized IDA curve. The term "summarized", when related to IN2 curves, applies only to mean or median curves, since the proposed simplified approach is not intended for the determination of dispersion. Default values for the dispersion measures have therefore to be used in order to determine the probability of exceedance of a given limit state.

For practical applications, further simplifications are introduced which allow a very simple application of the probabilistic approach. The analyst may decide to apply these simplifications only partly, i.e. only for determination of some parameters, whereas the other parameters are determined more accurately if appropriate data exist.

It is assumed that the spectral shape does not change with the intensity of ground motion. This assumption is used in the majority of seismic codes and is also consistent with the approach used for the determination of IDA curves developed in [13]. Using this assumption, there is a constant factor between the spectral acceleration s_a at any period and the peak ground acceleration PGA. So, instead of s_a, PGA can be used as the seismic intensity measure.

The parameter k in the hazard function (Eq. (21.3)) can be, in some parts of the world, reasonably approximated by $k = 3.0$. For the determination of the parameter k_0 at least one value of PGA (or s_a) corresponding to a return period should be known for the location under consideration. This value can be typically obtained from the seismic hazard map which applies to a specified probability of exceedance of ground motion $H(PGA)$, e.g. 10% in 50 years or, alternatively, to a specified return period (e.g. 475 years). Knowing $H(PGA)$ and PGA or s_a, the parameter k_0 can be obtained from Eq. (21.3).

Usually, there is not much difference between median and mean hazard curves. For some seismic hazard maps there is no indication if they apply to mean or median values. A typical value of the factor C_H (in Eq. (21.2)) which defines the ratio between the mean and median values of the hazard function is about 1.05. Strictly, a distinction between median and mean values should be made also when determining the structural demand. In practice, this distinction is seldom made. The N2 method and other pushover-based methods are basically expected to provide mean values of results, because they use simplified expressions for mean inelastic spectra. However, again, mean and median values are typically not very different. So, it is reasonable that, in a simplified practice-oriented method, no distinction is made between mean and median values, neither in the case of hazard curves, nor in the case of demand and capacity related to structures. Consequently, the value $C_{\mathrm{H}} = 1.0$ will be used in the proposed approach.

For the majority of medium- and long-period structures, i.e. the structures with the fundamental period longer than the characteristic period of ground motion T_{C}, the "equal displacement rule" applies, i.e. the inelastic displacement is assumed to be equal to the elastic displacement of the system with the same stiffness and mass, but with unlimited strength. In such a case the IN2 curve is a straight line (with its origin at the point (0, 0)) until "failure" occurs. The IN2 curve is defined by the point corresponding to "failure", which can be determined by the N2 method. It is conservatively assumed that the structure "fails" after the NC limit state is attained. Thus the IN2 curve after the NC limit state is horizontal. Equation (21.4) with b=1.0 defines the first part of the IN2 curve (up to the "failure" point). The "equal displacement rule", which has been widely accepted and at least implicitly adopted in many seismic codes, is used also in the proposed approach because it greatly simplifies the assessment procedure and provides fairly accurate results for a large number of structures.

At the level of the structure, there is a lack of a generally accepted definition about the NC limit state. One conservative possibility is to assume that the most critical important vertical element (column or structural wall) controls the state of the structure, i.e. the NC limit state at the level of the structure corresponds to the NC limit state ("failure") at the most critical vertical element.

In a practice-oriented approach, default values for the dispersion measures have to be used. Reliable data are not yet available. Such values will have to be determined in the future, based on extensive parametric studies. In the test examples in this paper the following value were used for the total dispersion measure for randomness: $\beta_{\mathrm{TR}} = \surd\left(\beta_{\mathrm{DR}}{}^2 + \beta_{\mathrm{CR}}{}^2\right) = 0.45$.

For practical applications which require a clear comprehension, it is convenient to express probability of exceedance with 50% confidence. In such a case, C_x in Eq. (21.1) is equal to 1.0 and the dispersion measures for uncertainty are not needed.

Considering the assumptions in this Section, simplified forms of Eqs. (21.1) and (21.2) are obtained. In the simplest case, when all assumptions are taken into account (no distinction between mean and median, k=3, fixed spectral shape, b=1, $\beta_{\mathrm{TR}} = 0.45$, the NC limit state represents "failure", the NC limit state of the structure corresponds to the NC limit state of the most critical important element, 50% confidence), Eq. (21.1) can be written as

$$P_{\mathrm{NC,50}} = 2.5\; H\,(PGA_{\mathrm{C}}) = 2.5\; k_0\, PGA_{\mathrm{C}}^{-3} \tag{21.5}$$

where $P_{\mathrm{NC,50}}$ is the annual probability that the "failure" will occur. PGA_{C} is the seismic intensity "corresponding" to the displacement capacity D_{C} (at "failure") determined by the N2 method. $H(PGA_{\mathrm{C}})$ is the median or mean value of the hazard function at the seismic intensity PGA_{C}. It provides an estimate of the annual probability that the seismic intensity will be equal to or exceed the level PGA_{C}. k_0 is a parameter of the hazard curve (Eq. (21.3)) which depends on the location of the structure. Equation (21.5) suggests that the probability of "failure" is equal to the probability of the exceedance of the intensity of ground motion corresponding to "failure", multiplied by a factor which is larger than 1. The value of the factor will change if more accurate values are available for k and dispersion measures β.

21.4 Probabilistic Assessment of the Example Structures

To demonstrate the proposed simplified probabilistic performance assessment approach, the mean annual frequency of exceedance of "failure" (defined as the NC limit state of the most critical element – see Chapter 21.3) will be determined for two variants of an asymmetric three-storey RC frame building (referred to in the following as the SPEAR building, Fig. 21.1). The building was conceived as representative of older construction in Southern European countries, but without engineered earthquake resistance. It was designed for vertical loads only. In addition to the original SPEAR building (denoted as "Test"), a variant of the building was designed in compliance with the EC8 (denoted as "EC8 H"). Realistic values were used for the permanent loads. As a result, the total mass of this variant was 45% larger than the total mass of Test structure. The geometry of the whole structure remained the same, but the dimensions of individual load-bearing elements were changed. In order to comply with the EC8 requirements, the dimensions of the columns were increased (from 25/25 to 35/35 cm for typical columns) and the dimensions of the beams were adjusted. The structure was designed for a design ground acceleration of 0.25g (on rock). Ground type C (soil factor $S = 1.15$) was assumed, resulting in a peak ground acceleration $PGA = 0.29g$. The high ductility class (DCH) was selected. For analysis the most common mathematical model was used. For the elements of the RC frame, one-component lumped plasticity elements were used, consisting of an elastic beam and two inelastic rotational hinges (defined by a moment-rotation relationship). The ultimate rotation Θ_u in the columns and beams at the NC limit state, which corresponds to a 20% reduction in the maximum

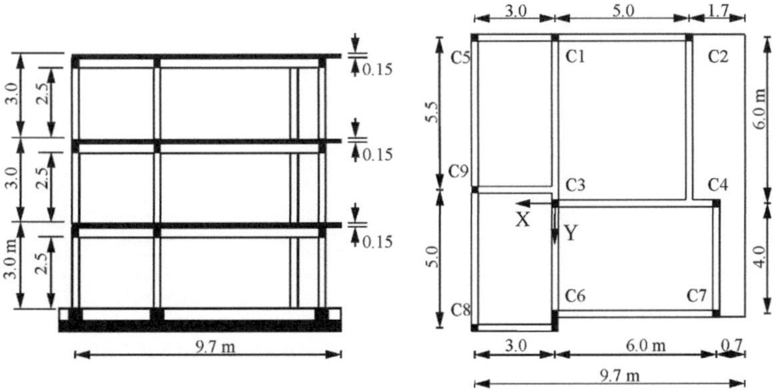

Fig. 21.1 The elevation and the plan view of the SPEAR building

moment, was determined by using the Eurocode 8-3 [2] formulas, the parameter γ_{el} being assumed equal to 1.5. This value corresponds to primary elements and represents about two/thirds of the estimated mean value of the plastic chord rotation capacity. In the case of the "Test" building, due to the absence of seismic detailing, the plastic parts of the ultimate chord rotations were multiplied by a factor of 0.825. The calculated ultimate rotations in columns in the first two stories varied from 18 to 27 mrad for the Test structure and from 24 to 34 mrad for the EC8 H structure. The values for beams varied from 15 to 30 mrad for the Test structure and from 27 to 38 mrad for the EC8 H structure. Note that the computed difference between the rotation capacities of columns in the two structures is relatively small. Details of both variants of the building, of the modelling and of the analysis procedure are provided in [12].

The mean annual frequency of exceedance of "failure" will be evaluated for both variants of the structure. First, the seismic intensity "corresponding" to the displacement capacity (at "failure"), PGA_C, will be determined by the N2 method. The procedure is presented in [8, 12], here only the main results are summarized for the X-direction which proved to be the critical one.

The lateral loads for the pushover analyses for the X-direction are determined based on the first mode shape, which is relevant for the X-direction. The pushover curves are presented in Fig. 21.2a. Marked are the points corresponding to the NC limit state, i.e. the deformation at which the NC limit state (expressed in terms of the ultimate chord rotation) is attained at the first column. The torsional influences were taken into account with correction factors based on elastic modal analysis using the extended N2 method [10]. The pushover curves clearly demonstrate larger stiffness, strength and ductility of the structure designed according to EC8. The top displacements at the NC limit state, D_C, amount to 11.1 cm (drift =1.2%) and to 22.4 cm (drift =2.5%) for the Test and EC8 structures, respectively. The critical element is the column C1 at the top of the second storey for the Test structure and the same column at the base for the EC8 H structure. In the case of the Test structure, a plastic

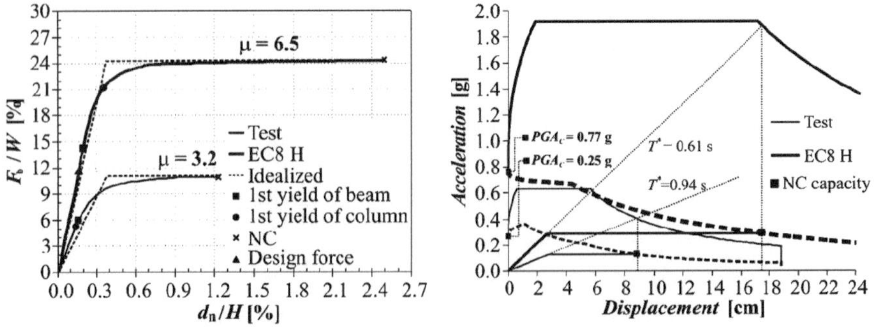

Fig. 21.2 (a) The normalized base shear – roof displacement diagrams for the X direction. (Height of the structure $H = 9$ m, weight $W = 1{,}900$ kN for Test structure and $W = 2{,}900$ kN for EC8 H structure). (**b**) Elastic demand spectra corresponding to the NC limit state and capacity diagrams for idealized SDOF systems in AD format

mechanism is formed in the lower two storeys. Almost all of the columns in the first and second storey yield at both joints, whereas most of the beams remain in the elastic region. A favourable global plastic mechanism, where all of the beams yield, as well as columns at their fixed base, occurs in the case of the EC8 H structure. The maximum inter-storey displacements at the NC limit state occur at the second storey. They amount to 6.2 cm (storey drift = 2.1%) and to 8.2 cm (storey drift = 2.7%) for the Test and EC8 H structure, respectively.

The idealized pushover curves are shown in Fig. 21.2a. Figure 21.2b shows the capacity diagrams and elastic spectra (EC8 shape, soil type C) corresponding to the NC limit state. Using the equal displacement rule, the elastic spectral acceleration corresponding to the NC capacity is defined as the crossing point of the diagonal line representing the period of the structure (in the acceleration – displacement format) and the vertical line at the displacement capacity. Knowing this point and the spectral shape, the complete elastic spectrum is defined. From Fig. 21.2b it can be seen that the Test structure would "fail" at a ground motion with $PGA_C = 0.25g$, whereas the EC8 H structure is able to resist a much larger ground motion ($PGA_C = 0.77g$). This indicates that the ground motion corresponding to "failure" is 2.7 times larger than the design ground motion. Inelastic spectra (ductility demand μ=3.2 and μ=6.5 for The Test and EC8 H structure, respectively) according to EC8 (based on the N2 method) are also plotted in Fig. 21.2b for illustration. Note, however, that they are not needed for analysis. A graphical presentation of capacity and demand allows a visualization of important structural and ground motion characteristics and is beneficial for the comprehension of the analysis procedure.

Analytically, the spectral acceleration corresponding to the NC limit state, $S_{a,C}(T^*)$, can be determined as

$$S_{a,C}\left(T^*\right) = 4\pi S_{d,C}/T^{*2} \tag{21.6}$$

where $S_{d,C}$ and T^* are the displacement at the NC limit state and the period of the equivalent SDOF system, respectively. $T^* = 0.94$ s and $T^* = 0.61$ s correspond to the Test and EC8 H structure, respectively. From the EC8 spectrum for soil type C and for $T^* = 0.94$ s or $T^* = 0.61$ s for the Test or EC8 H structure, respectively, the corresponding peak ground accelerations $PGA_C = 0.25g$ and $PGA_C = 0.77g$ are obtained.

An IN2 curve represents the relation between the seismic intensity measure (in our case *PGA*) and the engineering demand parameter (in our case top displacement *D*). As a consequence of the equal displacement rule, the IN2 curve is in the investigated case a straight line from the origin up to the "failure" point, which is defined by PGA_C and D_C (Fig. 21.3). After the "failure" point, the IN2 curve is a horizontal line. As in the case of inelastic spectra, the IN2 curve is not needed for calculations, but it helps to comprehend the procedure.

In the next step of analysis, the coefficient k_0, which defines the seismic hazard function (Eq. (21.3)), will be determined. It is assumed that both variants of the building are located at the location which was considered for the design of the EC8 variant of the building (the EC8 spectrum for ground type C normalized to $PGA =$

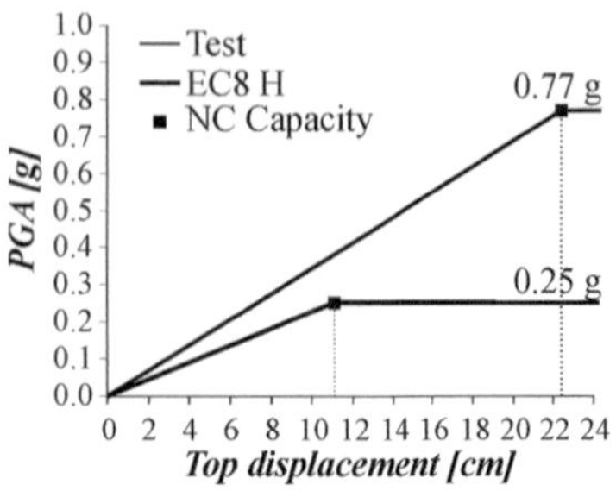

Fig. 21.3 IN2 curves for Test and EC8 structures

0.29*g*). According to the seismic hazard map for 475 year return period, the peak ground acceleration amounts to 0.25*g*, and the soil factor for soil type C amounts to 1.15. Consequently, the seismic demand for 475 year return period amounts to $PGA_{475} = 0.29g$. The 475 year return period corresponds to 10% probability of exceedance in 50 years or to 0.2% annual probability of exceedance. Assuming k=3, Eq. (21.3) yields $k_0 = 0.002 / (0.29g)^{-3} = 0.046$ (the unit for acceleration is m/s^2, the same unit should be used in Eq. (21.5)).

The annual probabilities of "failure" (50% confidence), determined by using Eq. (21.1), and the probabilities of "failure" in 50 years (average lifetime of the structures), which are determined as $1–(1–P_{NC,50})^{50}$, amount to 0.78×10^{-2} and 0.32, respectively, for the Test structure, and to 2.67×10^{-4} and 0.013 for the EC8 H structure. The probability of "failure" of the structure which has not been designed for seismic resistance is very large. For the structure designed according to a modern code it is much smaller, but still quite considerable. It would increase for a factor of 3.9 if 90% confidence was taken into account according to Eqs. (21.1) and (21.2) (assuming $\beta_{DU} = \beta_{CU} = 0.25$). On the other hand, it should be noted that in our analyses the "failure" was defined in a quite conservative way.

21.5 Conclusions

The presented simplified method is basically the SAC-FEMA approach, in which the most demanding part, i.e. the Incremental Dynamic Analysis, is replaced by the incremental N2 method. Using a number of simplifying assumptions, which are consistent with assumptions used in usual code analyses, a very simple formula for the annual probability of failure can be developed.

The proposed approach was applied to two variants of a RC frame building. The results show that the building, which has not been designed for earthquake resistance, has a large probability of failure in its lifetime if located in a moderate seismic region, in spite of a considerable deterministically determined intensity of ground motion associated with its ultimate capacity. The ultimate capacity of the building designed according to a modern standard (EC8) is, of course, much larger. A deterministic analysis (using some conservative assumptions) indicates that the structure designed for $PGA=0.29g$ would "fail" if subjected to a ground motion with $PGA= 0.77g$. However, the probability of failure in the lifetime is still considerable (1.3% for 50% confidence).

The proposed method has, like other simplified methods, limitations, which are basically the same as those which apply to the basic N2 method and to the SAC-FEMA method. The "equal displacement rule" is used which is not valid for short period structures. Some other simplifying assumptions have also been adopted which are consistent with seismic code procedures. Reliable default dispersion measures for randomness and uncertainty are not yet available. Nevertheless, it is hoped that such simple approaches will facilitate the introduction of probabilistic considerations in practical applications.

References

1. CEN (2004) Eurocode 8: design of structures for earthquake resistance. Part 1: general rules, seismic action and rules for buildings, European standard EN 1998-1. CEN, Brussels
2. CEN (2005) Eurocode 8: design of structures for earthquake resistance. Part 3: assessment and retrofitting of buildings, European standard EN 1998-3. CEN, Brussels
3. Cornell CA, Jalayar F, Hamburger RO, Foutch DA (2002) Probabilistic basis for 2000 SAC Federal Emergency Management Agency steel moment frame guidelines. J Struct Eng ASCE 128(4):526–533
4. Deierlein G (2004) Overview of a comprehensive framework for earthquake performance assessment. In: Fajfar P, Krawinkler H (eds) Proceedings of the international workshop on performance-based seismic design – concepts and implementation, Bled, Slovenia, PEER Report 2004/05:15–26, Berkeley, CA
5. Dolšek M, Fajfar P (2004) IN2 – a simple alternative for IDA. Proceedings of the 13th world conference on earthquake engineering, Vancouver, Canada, paper 3353
6. Dolšek M, Fajfar P (2007) Simplified probabilistic seismic performance assessment of plan-asymmetric buildings. Earthq Eng Struct Dyn 36:2021–2041
7. Dolšek M, Fajfar P (2008a) Effects of masonry infills on the seismic response of a four storey reinforced concrete frame – a deterministic assessment. Eng Struct 30:1991–2001
8. Dolšek M, Fajfar P (2008b) Effects of masonry infills on the seismic response of a four storey reinforced concrete frame – a probabilistic assessment. Eng Struct 30:3186–3192
9. Fajfar P (2000) A nonlinear analysis method for performance-based seismic design. Earthq Spectra 16(3):573–592
10. Fajfar P, Marušić D, Peruš I (2005) Torsional effects in the pushover-based seismic analysis of buildings. J Earth Eng 9(6):831–854
11. FEMA 350 (2000) Recommended seismic design criteria for new steel moment frame buildings, SAC Joint Venture, Federal Emergency Management Agency, Washington (DC)
12. Rozman M, Fajfar P (2009) Seismic response of a RC frame building designed according to old and modern practices. Bull Earthq Eng 7:779–799
13. Vamvatsikos D, Cornell C A (2002) Incremental dynamic analysis. Earthq Eng Struct Dyn 31:491–514

Chapter 22
Direct Probability-Based Seismic Design of RC Buildings

Paolo Franchin and Paolo Emilio Pinto

22.1 Introduction

The literature on performance based earthquake engineering has been steadily increasing over the past ten years. It has to be noted, however, how the largest proportion of papers and reports deals with seismic (performance) assessment rather than with design. On this latter aspect the number of contributions and approaches is rather limited, due to the more difficult nature of the problem. The list of references reports a non exhaustive but representative sample of the available proposals. Most approaches employ the concepts and tools of optimisation theory to arrive at a design solution that is optimal in some sense [3, 8, 13, 11, 10, 14, 15].

In optimisation the stationary point of an objective function (minimum or maximum, possibly absolute) in the space of design variables is sought. This is usually done under some constraints on performance and/or geometric dimensions of the structural members. The corresponding problem is called a constrained optimisation problem. From this starting point several possibilities branch out: there can be a single or multiple objectives, the latter can be combined again by weights into a single one or can give rise to a set of optimal solutions rather than a single optimum (what is called a Pareto set); the design variables can be continuous or discrete, and the optimisation problem changes correspondingly, together with the solution algorithms. All of the above can be recovered into the available works on performance-based seismic design, highlighting a still unsettled situation.

Amongst the several references following the optimisation approach, one recent work that can be taken as a good example of the results obtainable is that in Fragiadakis and Papadrakakis [6]. The design optimisation procedure makes use of *evolutionary algorithms* (the optimisation problem is made discrete, by preliminarily lumping possible cross-sections and reinforcement in a list, as it is the case for structural steel design). The objective function is in terms of initial construction cost, taken proportional to the concrete and steel weight. The constraints are of three

P. Franchin (✉)
Department of Structural Engineering and Geotechnics, University of Rome La Sapienza, Via Gramsci 53, 00197, Rome, Italy
e-mail: paolo.franchin@uniroma1.it

M.N. Fardis (ed.), *Advances in Performance-Based Earthquake Engineering*, 235
Geotechnical, Geological, and Earthquake Engineering 13, DOI 10.1007/978-90-481-8746-1_22,

types: (1) compliance with Eurocodes 1, 2 and 8 design rules (2) additional geometric constraints (e.g. symmetry of some elements, dimensional limitations) (3) limits on mean annual frequencies (MAF) of exceedance for different performance levels. It is noted that the first two types of constraint are of deterministic nature, while the third is of probabilistic nature. Satisfaction of constraints is imposed in a sequential fashion, starting with those that do not require analyses (2), going to linear analyses (1), and finally to incremental dynamic analysis (IDA) to check probabilistic constraints (3). It can also be observed that initial cost and cost of damage incurred in the event of an earthquake are treated differently, the former as an objective to be minimised, the latter as a constraint on the MAF of the response parameter related to damage.

The review of the available optimisation-based approaches shows that, though conceptually appealing, since they allow satisfaction of both deterministic and probabilistic performance constraints, while adding the search for an optimum on cost (initial or life-cycle), they are still associated with a prohibitive computational burden, preventing their application in real-sized situations.

To the knowledge of the authors only one, non optimisation-based and practice-oriented approach is available, namely that in Krawinkler et al. [9]. The design procedure iteratively enforces satisfaction of two performance targets in terms of cost, associated with 50/50 and 2/50 hazard levels, respectively. The procedure makes use of the median IDA curves to relate the hazard levels with the corresponding demand parameter, as well as of average loss curves, for structural and non structural damage, to relate response with damage/cost.

The design parameters are the period T (initial stiffness) and the base shear ratio γ (strength). The procedure requires a prior production of so-called "design aids" in the form of alternative median IDA curves for different values of the design parameters.

22.2 Proposal

The current proposal takes as a starting point the approach in Krawinkler et al. [9] and attempts to: (a) working directly in terms of MAF of exceedance of chosen limit-states; (b) eliminating the need for design-aids by making use of structure-specific analysis.

In line with [9] are the underlying ideas of: (a) renouncing the concept of optimal design in favour of the simple compliance with given constraints, thus reducing the computational burden, (b) working in terms of global design variables, like the period T (stiffness) and the base shear ratio γ (strength) of the building, leaving the task of modifying the design in each iteration to the designer. The resulting procedure would not be completely automatic, governed by an algorithm, but it would leave more control and flexibility to the designer. The latter is regarded as a desirable feature, more in line with current design practice. The proposed method, however, guides the designer in choosing target stiffness and strength in each iteration. The fact that the constraints are expressed probabilistically in terms of their respective MAFs is regarded as a step towards a more rational design approach, since all possible seismic intensities leading to a given performance level can be accounted for.

As it is well-known, a closed-form expression for the MAF of exceedance of a limit state is available in the following form [4]:

$$\lambda = \lambda_{\mathrm{IM}}\left(\mathrm{IM}_{\hat{D}=\hat{C}}\right)\exp\left[\frac{1}{2}\frac{k^2}{b^2}\left(\beta_{\mathrm{D}}^2+\beta_{\mathrm{C}}^2\right)\right] = k_0\left(\frac{\hat{C}}{a}\right)^{-\frac{k}{b}}\exp\left[\frac{1}{2}\frac{k^2}{b^2}\left(\beta_{\mathrm{D}}^2+\beta_{\mathrm{C}}^2\right)\right] \tag{22.1}$$

where $\mathrm{IM}_{\hat{D}=\hat{C}} = \left(\hat{C}/a\right)^{1/b}$ is the intensity measure (IM) value that induces a median demand equal to the median limit state capacity, while $\beta_{\mathrm{D}}, \beta_{\mathrm{C}}$ are the demand and capacity dispersions. The above expression holds under the assumption that demand and capacity are lognormally distributed with median demand varying with the IM as a power law $\hat{D} = a\mathrm{IM}^b$, and the MAF of the IM (hazard) is approximated as $\lambda_{\mathrm{IM}}(x) = k_0 x^{-k}$.

The parameters a and b of Eq. (22.1) are established by linear regression, usually on the results of inelastic time-history analyses carried out with unscaled recorded motions selected to cover the IM range of interest (i.e. a "cloud" analysis). A more expensive alternative relies on IDA. There are also proposals to replace these computationally intensive techniques with simplified nonlinear methods such as nonlinear static analysis. In particular two such ways are the so-called Incremental N2 method (IN2) [5], and SPO2IDA [12]. The use of pushover techniques appears a good compromise for design and is therefore pursued herein.

The transformation from a static pushover curve (SPO) to the median IDA curve can be carried out for instance as shown in Dolšek and Fajfar [5], with b (measuring the degree of nonlinearity) being equal to unity unless a degrading pushover curve is obtained (i.e. degrading constitutive laws and/or P-Δ effects are considered in the model). The process is sketched in Fig. 22.1.

The performance constraints that the structure must comply with, as anticipated, are expressed in terms of the MAF λ, or, conversely, of the mean return periods $\tau = 1/\lambda$ of the limit-state violations for a number of limit-states of interest, e.g. a damage related limit-state such as light damage (LD), and a safety-related limit-state such as collapse prevention (CP). For example (the limits being arbitrary):

$$\lambda_{\mathrm{LD}}(T,\gamma) \le \lambda_{LD}^* = 1/72 \text{ years} \tag{22.2}$$

$$\lambda_{\mathrm{CP}}(T,\gamma) \le \lambda_{CP}^* = 1/2{,}475 \text{ years} \tag{22.3}$$

Starting from an initial design (T_i,γ_{i}) the compliance with the above constraints is checked and if any of them is not respected a new set (T_{i+1},γ_{i+1}) of period/strength values is found by a Newton method, as the point closest to (T_i,γ_i) lying on the intersection between the plane tangent to the MAF surface in (T_i,γ_i) and the T-γ plane (see Fig. 22.2):

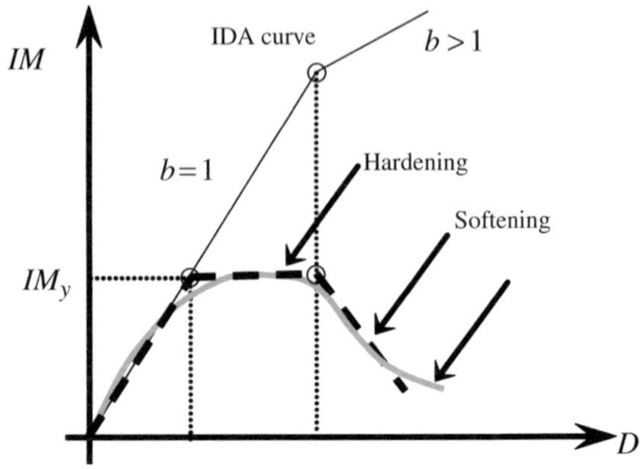

Fig. 22.1 Derivation of median IDA curve from a pushover curve according to Dolšek and Fajfar [5]

$$(T_{i+1}, \gamma_{i+1}) = \arg\min \left(d_{i,i+1} \middle| \lambda(T_i, \gamma_i) + \left.\frac{\partial \lambda}{\partial T}\right|_i (T_{i+1} - T_i) + \left.\frac{\partial \lambda}{\partial \gamma}\right|_i (\gamma_{i+1} - \gamma_i) = \lambda^* \right) \tag{22.4}$$

where $d_{i,i+1} = \sqrt{(T_{i+1} - T_i)^2 + (\gamma_{i+1} - \gamma_i)^2}$.

If more than one constraint is violated the new design point will be found by using the governing constraint, defined as that having the largest value of the quantity $\tilde{\lambda} = \lambda/\lambda^* - 1$. At the end of the process only one of the constraints is satisfied in equality, while the remaining ones are satisfied with more or less wide margins.

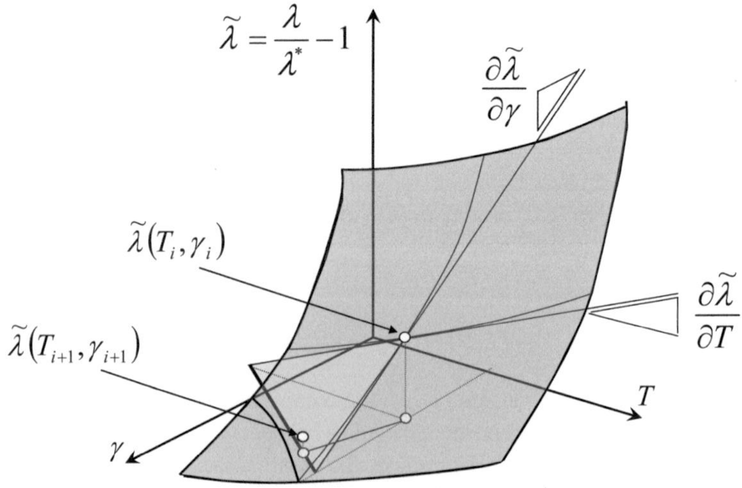

Fig. 22.2 Updated period and base shear ratio based on MAF expansion at the current design

Given the global nature of the chosen design variables T and γ, they cannot be controlled directly. The designer, or alternatively an algorithm, will modify local variables (reinforcement, cross-section dimensions, adding/removing elements) so as to approximate the new target $(T,\gamma)_{i+1}$ pair: the *actual* new point will be in most cases somewhat distant from the *target* $(T,\gamma)_{i+1}$ resulting in slower convergence, not impairing, however, the method.

Derivatives with respect to the design variables are carried out numerically. Attention must be given to the fact that T and γ cannot be independently modified, hence the partial derivatives with respect to the design variables need to be obtained as the solution of the linear system:

$$\begin{bmatrix} \Delta T_1 & \Delta\gamma_1 \\ \Delta T_2 & \Delta\gamma_2 \end{bmatrix} \begin{Bmatrix} \partial\lambda/\partial T \\ \partial\lambda/\partial\gamma \end{Bmatrix} = \begin{Bmatrix} \Delta\lambda_1 \\ \Delta\lambda_2 \end{Bmatrix} \tag{22.5}$$

where $(\Delta T_1,\Delta\gamma_1,\Delta\lambda_1)$ and $(\Delta T_2,\Delta\gamma_2,\Delta\lambda_2)$ are the variations corresponding to two separate perturbations of the local/basic variables.

22.3 Application to an Elasto-Plastic SDOF Oscillator

The procedure is first checked with reference to the simplest inelastic system consisting of a SDOF oscillator with bilinear (no hardening) force-displacement relationship. In this case the two design variables can be directly and independently controlled. Further, in this case the partial derivatives can be analytically derived without the need for perturbing the structural analysis (no analysis is required for this simple system).

The initial properties of the system are T=1.0 s and γ=0.1. The following two constraints are considered:

- Damage limitation: the MAF of the response (expressed as the ductility $\mu=d/d_y$) exceeding the (deterministic, i.e. β_C=0) limit of 1 (yielding), must be lower or equal to λ^*=1/100 years (β_D=0.3)
- Collapse prevention: the MAF of the response (expressed as the maximum displacement d) exceeding the capacity d_C (lognormal variable, median 0.1 and dispersion β_C=0.3), must be lower or equal to λ^*=1/2,475 years (β_D=0.3)

The hazard curves for several closely spaced structural periods were derived for the site of interest inverting the information provided by the Italian design code in terms of uniform hazard spectra for return periods ranging between 30 and 2,475 years. The hazard curves where then linearised in the log-log space to obtain k and k_0 as a function of T. This same hazard is also used in the next example.

Figure 22.3 shows the trajectory of the design point towards satisfaction of both constraints. It can be observed how at the initial design the governing constraint is the safety-related one (black), then the two constraints more or less alternate until, after 10 iterations, they are both satisfied for the new values T=0.52 s and γ=0.28.

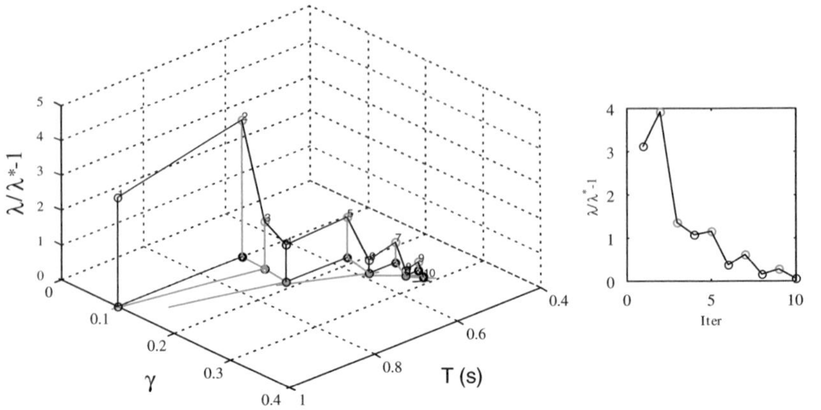

Fig. 22.3 Path of the design point towards satisfaction of the two constraints. *Black* and *gray* colours denote the collapse and damage performance constraints, respectively

22.4 Application to a Five-Storey RC Plane Frame

The proposed design procedure is illustrated in the following with reference to the 5-storey 3-bay plane RC frame shown in Fig. 22.4 with dimensions and gravity (dead plus live) loads.

The initial design of the frame is carried out as follows. Girders have a constant cross-section of width $b_w = 0.30$ m and height $h_b = 0.60$ m. Reinforcement is determined from linear elastic gravity load analysis. Bottom reinforcement at supports is set equal to half the top reinforcement. Columns are designed according to capacity design (with a minimum 1% reinforcement ratio, enforced also in the

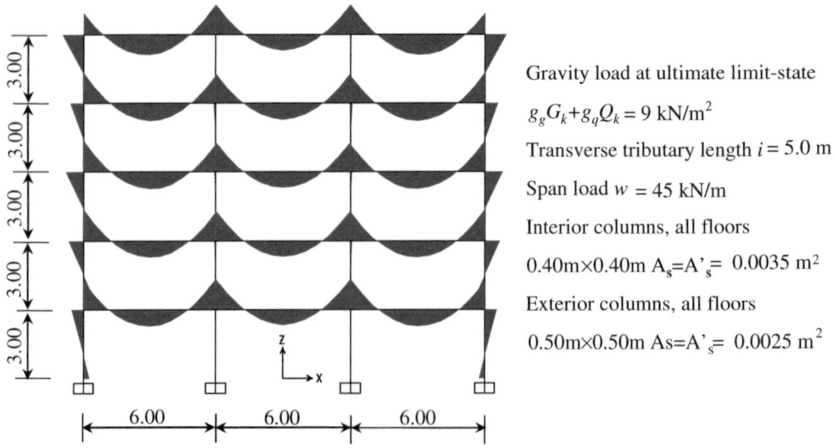

Fig. 22.4 Dimensions and loads of the 5-storeys 3-bays plane RC frame

following design modifications). Shear reinforcement is not designed until the final configuration is obtained.

The initial properties of the system are T=0.95 s and γ=0.165. The following two constraints are considered:

- Damage limitation: the MAF of the response (expressed as the top displacement ductility $\mu = d/d_y$) exceeding the (deterministic, i.e. β_C=0) limit of 1 (yielding), must be lower or equal to λ^*=1/100 years (β_D=0.3)
- Collapse prevention: the MAF of the response (expressed as the peak interstorey drift ratio θ_{max}) exceeding the capacity θ_C (lognormal variable, median 0.02 and dispersion β_C=0.3), must be lower or equal to λ^*=1/2,475 years (β_D=0.3)

The procedure allows the designer to select his own strategy in modifying the stiffness and strength in order to meet the MAFs constraints. As an example, in this case the preference is given to modifying the exterior column size, if needed, leaving interior ones, as well as the beams' size unchanged.

As far as the base-shear ratio is concerned, the strength of the structure is proportional to the bending strength of its members. For this reason the increase in reinforcement in all beams and columns is taken equal to $\rho_\gamma = \gamma^{(i+1)}/\gamma^{(i)}$:

$$\rho_{A_s} = \frac{A_s^{(i+1)}}{A_s^{(i)}} = \rho_\gamma \tag{22.6}$$

In this way the capacity design enforced on the initial design is preserved.

As far as the period is concerned the required change $\rho_T = T^{(i+1)}/T^{(i)}$ is translated into a change in stiffness, since the mass of the structure can be considered practically constant:

$$\rho_T = \frac{T^{(i+1)}}{T^{(i)}} = \sqrt{\frac{k^{(i)}}{k^{(i+1)}}} = \rho_k^{-0.5} \rightarrow \rho_k = \rho_T^{-2} \tag{22.7}$$

It is observed that the above stiffness should refer to the secant stiffness obtained from the bilinear approximation of the pushover curve, which actually depends on the geometry as well as the reinforcement (through the strength of the structure). Well aware of the fact that by so doing the convergence rate of the procedure will be slowed, in this preliminary application this dependence is neglected and ρ_k is applied to modify only the initial stiffness. The latter is modified by changing only the external column cross-section height h_c, leaving the width unchanged. The required change in height can be then written as:

$$\rho_{k\,\text{ext}} = \frac{k_{\text{ext}}^{(i+1)}}{k_{\text{ext}}^{(i)}} = \frac{I_{\text{ext}}^{(i+1)}}{I_{\text{ext}}^{(i)}} = \left(\frac{h_{c,\text{ext}}^{(i+1)}}{h_{c,\text{ext}}^{(i)}}\right)^3 = \rho_h^3 \rightarrow \rho_h = \rho_{k\,\text{ext}}^{1/3} \tag{22.8}$$

where $\rho_{k\,\text{ext}} = 2\rho_k - 1$.

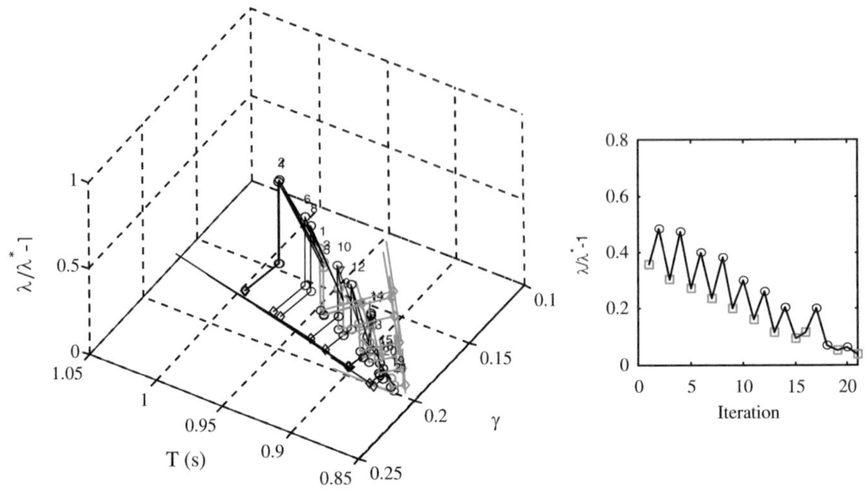

Fig. 22.5 Five-storeys three-bays plane RC frame: design point path towards satisfaction of the two constraints. *Black* and *gray* colours denote collapse and damage constraints, respectively

Figure 22.5 shows the trajectory of the design point towards satisfaction of both constraints. It can be observed how at the initial design the governing constraint is the safety-related one (black), then the two constraints more or less alternate until, after 21 iterations, they are both satisfied for the new values T=0.87s and γ=0.199: the pushover curve for this iteration (continuous), with its bilinearisation (dashed), are shown in Fig. 22.6, left, while on the right there are the corresponding median IDA curve (continuous)and its power-law approximation (dashed).

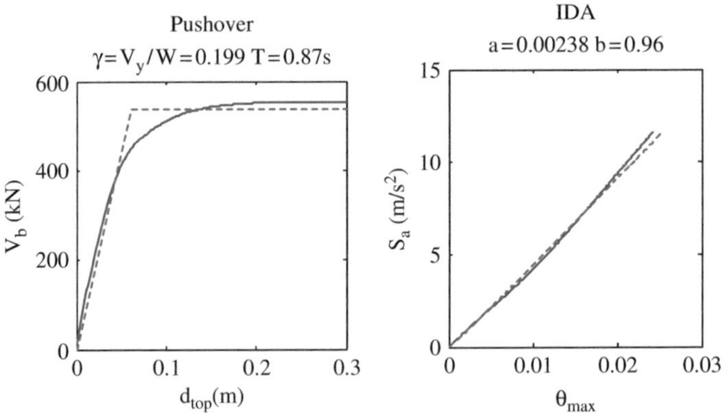

Fig. 22.6 Five-storeys three-bays plane RC frame: *left*, pushover curve at satisfaction of constraints, together with bilinear approximation; *right*, median IDA curve for θ_{max} versus S_a as derived from the pushover bilinearisation, together with (*dashed*) power-law approximation

22.5 Conclusions

This paper presents a first step towards the ambitious goal of designing a structure to comply with a number of probabilistically-defined performance constraints. The method governs the adjustments of an initial design in terms of two global design variables of clear engineering significance: the fundamental period and the base shear ratio. It employs structure-specific analysis, in the form of pushover analysis, to determined the median IDA curve for the performance measures of interest and, by using the closed-form expression for risk by Cornell and co-workers, it considers the contributions to limit-state exceedance from all seismic intensities.

Several aspects require progress. Amongst these there are the applicability of the procedure to structures of general (3D) geometry, e.g. by means of a modal pushover technique, and the use of constraints on the MAF of non-engineering performance measures, such as cost of damage to structural and non-structural components [1, 2, 7].

Finally, a robust procedure is needed to modify of the structural properties in order to match as closely as possible with the target variations in T and γ.

References

1. Aslani H, Miranda E (2004) Probabilistic damage assessment for building specific loss estimation, PEER report
2. Bachman RE (2004) The ATC-58 project plan for nonstructural components. In: Fajfar P, Krawinkler H (eds) Proceedings of the international workshop on Performance-based seismic design concepts and implementation (Report PEER 2004/05), Bled, Slovenia, June 28– July 1, 2004
3. Beck JL, Chan E, Irfanoglu A, Papadimitriou C (1999) Multi-criteria optimal structural design under uncertainty. Earthquake Engng Struct Dyn 28:841–761
4. Cornell CA, Jalayer F, Hamburger RO, Foutch DA (2002) The probabilistic basis for the 2000 SAC/FEMA steel moment frame guidelines. J Struct Engng 128:526–533
5. Dolšek M, Fajfar P (2004) IN2 – a simple alternative for IDA. Proceedings of 14th world conference on earthquake engineering, Vancouver, Canada
6. Fragiadakis M, Papadrakakis M (2008) Performance-based optimum seismic design of reinforced concrete structures. Earthq Eng Struct Dyn 37:825–844
7. Hamburger RO (2004) Development of next generation performance based seismic design guidelines. In: Fajfar P, Krawinkler H (eds) Proceedings of the international workshop on Performance-based seismic design concepts and implementation (Report PEER 2004/05), Bled, Slovenia, June 28– July 1, 2004
8. Haukaas T (2008) Unified reliability and design optimization for earthquake engineering. Probabilistic Eng Mech 23:471–481
9. Krawinkler H, Zareian F, Medina RA, Ibarra L (2006) Decision support for conceptual performance-based design. Earthq Eng Struct Dyn 35:115–133
10. Lagaros ND, Fotis AD, Stilianos AK (2006) Assessment of seismic design procedures based on the total cost. Earthq Eng Struct Dyn 35:1381–1401
11. Lagaros ND, Papadrakakis M (2007) Seismic design of RC structures: a critical assessment in the framework of multi-objective optimization. Earthq Eng Struct Dyn 36:1623–1639
12. Vamvatsikos D, Cornell CA (2005) Direct estimation of seismic demand and capacity of multidegree-of-freedom systems through incremental dynamic analysis of single degree of freedom approximation. Proceedings of ICOSSAR'05, Rome, Italy

13. Vamvatsikos D, Papadimitriou C (2005) Optimal multi-objective design of a highway bridge under seismic loading through incremental dynamic analysis. Proceedings of ICOSSAR'05, Rome, Italy
14. Zou XK, Chan CM (2005a) An optimal resizing technique for seismic drift design of concrete buildings subjected to response spectrum and time history loadings. Comput Struct 83: 1689–1704
15. Zou XK, Chan CM (2005b) Optimal seismic performance-based design of reinforced concrete buildings using nonlinear pushover analysis. Eng Struct 27:1289–1302

Chapter 23
Probabilistic Models for Visual Damage

Terje Haukaas, Shahrzad Talachian, and Kenneth J. Elwood

23.1 Introduction

The ultimate objective in this work is to predict the consequences of earthquakes in urban areas. In particular, the cost of damage to buildings is addressed in this paper. This objective is aligned with the workshop's focus on performance-based earthquake engineering. The growing concern with performance of civil infrastructure signals a need for structural engineers to consider broader consequences than potential loss of structural integrity. While life safety has been the principal concern in the building codes since seismic provisions were rigorously included some 40 years ago, it is now apparent that both direct and indirect consequences of damage should influence the design decisions.

The modeling of direct and indirect consequences of damage must start with the prediction of actual damage. This objective is not new to the structural engineering community. A number of damage indices have been proposed [9, 11]. The numerical value of a damage index is typically correlated with generic damage states such as "slight damage" or "severe damage." Recently, conditional probability curves, sometimes referred to as fragility curves, have emerged as an alternative means of assessing damage to structural and non-structural damage [8, 12]. The abscissa axis of these curves is a measure of the demand on the structural component, while the ordinate axis provides the probability that a discrete damage state is attained. In turn, the damage state is linked with a cost estimate. Usually, there are several damage states and each probability curve represents the probability that the damage is equal to or greater than a specific damage state, for the given value of the demand.

A new methodology is outlined in this paper. In this approach, it is first recognized that the *repair action* is the central ingredient in determining the cost and downtime associated with damage. Hence, the focus is shifted to predicting the visual damage that determines the repair action. Once the repair action is determined, the direct repair cost is obtained from material and labor cost estimates.

T. Haukaas (✉)
Department of Civil Engineering, University of British Columbia, Vancouver, BC, V6T 1Z4, Canada
e-mail: terje@civil.ubc.ca

M.N. Fardis (ed.), *Advances in Performance-Based Earthquake Engineering*, Geotechnical, Geological, and Earthquake Engineering 13, DOI 10.1007/978-90-481-8746-1_23,

Similarly, the indirect costs associated with evacuation of the building and business disruption are assessed from the time of the specific repair action.

Clearly, estimates of damage, repair cost, and repair time are associated with significant and unavoidable uncertainty. In this paper, this is addressed by developing explicit probabilistic models to predict the visual damage and the ensuing time and cost of the repair action. The development of such models is the central theme in this paper. From these models, fragility curves can be extracted if needed. However, an appealing aspect of the probabilistic modeling approach is that it facilitates the use of reliability methods, such as FORM, to carry out the probabilistic analysis.

In the following, the reliability approach is described and contrasted with an alternative approach based on conditional probabilities. Next, the development of generic probabilistic models is described. This is given particular attention because the education of the engineering community on this topic is regarded as an important objective. The paper concludes with a brief discussion of the application of the methodology to the assessment of consequences of damage after earthquakes. The specific models for visual damage, repair cost, and repair time are under development and will be presented in future publications.

23.2 Probabilistic Models and Reliability Analysis

To provide context for the modeling approach adopted in this paper, the analysis framework for which the models are intended is outlined. It is referred to as unified reliability analysis. In essence, it comprises classical structural reliability methods in conjunction with a flexible framework of probabilistic models [6, 7]. It is a flexible, powerful, and transparent approach, but it is perhaps perceived as somewhat advanced for users who are unfamiliar with reliability methods. To facilitate the understanding of the approach it is contrasted with a different approach that has reached a broader audience. This approach is here referred to as the "PEER integral" approach, named after the Pacific Earthquake Engineering Research Centre based in Berkeley, California [2]. Several other research institutions have similar formulations [10], which combine conditional probabilities and the theorem of total probability.

For the purpose of comparing methods, consider an earthquake intensity measure, im, a resulting structural response measure (or engineering demand parameter), edp, a resulting damage measure dm, and a resulting decision variable (here considered to be repair cost), dv. The theorem of total probability is applied three times to obtain the cumulative distribution function of the cost:

$$\begin{aligned} F(\mathrm{dv}) &= \int_0^\infty F(\mathrm{dv}|\mathrm{dm}) \left(\frac{d}{d\mathrm{dm}} \int_0^\infty F(\,\mathrm{dm}|\mathrm{edp}) \cdot \left(\frac{d}{d\mathrm{edp}} \int_0^\infty F(\mathrm{edp}|\mathrm{im}) \cdot f(\mathrm{im}) \cdot d\mathrm{im} \right) \cdot d\mathrm{edp} \right) \cdot d\mathrm{dm} \\ &= \int_0^\infty \int_0^\infty \int_0^\infty F(\mathrm{dv}|\mathrm{dm}) \cdot dF(\mathrm{dm}|\mathrm{edp}) \cdot dF(\mathrm{edp}|\mathrm{im}) \cdot f(\mathrm{im}) \cdot d\mathrm{im} \end{aligned} \tag{23.1}$$

where $F(\ |\ )$ denotes a conditional cumulative distribution function and $f()$ denotes a probability density function. This integral form was initially formulated in PEER to coordinate the research efforts into hazard modelling, addressing $f(\text{im})$, structural modelling, addressing $F(\text{edp}|\text{im})$, damage modelling, addressing $F(\text{dm}|\text{edp})$, and consequence modelling, addressing $F(\text{dv}|\text{dm})$. The equation is also useful for quantitative evaluations in simple cases when scalar measures are sufficient and conditional probabilities are readily available. However, the approach outlined in Eq. (23.1) is sometimes misinterpreted as a general-purpose analysis approach rather than its original coordinating purpose. The formulation may also have contributed to an unnecessarily narrow focus on developing "fragility curves" for structural components rather than the underlying probabilistic models.

The reliability-based approach adopted in this paper is intended as an alternative to Eq. (23.1) to address a broader range of applications. In the proposed approach, probabilistic models replace the conditional probabilities and reliability methods replace the triple integral. To this end, consider the classical structural reliability problem. This problem comprises the computation of the probability that a function, $g(\mathbf{x})$, takes on negative values, where $\mathbf{x}$ is the vector of random variables. Mathematically, this problem is formulated as a multiple integral in the space of random variables:

$$p = \int_{-\infty}^{\infty} \cdots \int_{-\infty}^{\infty} I(\mathbf{x}) f(\mathbf{x}) \mathrm{d}\mathbf{x} \tag{23.2}$$

where $I(\mathbf{x})=1$ when $g(\mathbf{x})\leq 0$ and 0 otherwise. Consider next the specific function $g(\text{dv},\mathbf{x})=\text{DV}(\mathbf{x})\text{-}dv$, where dv is a selected threshold value and DV is the repair cost for a building subjected to an earthquake ground motion. Solving the reliability problem in Eq. (23.2) with this function yields the probability that DV is less than dv. In other words, the value of p in this case represents the point on the cumulative distribution function for the repair cost at the selected cost threshold, dv:

$$F(\text{dv}) = p(\text{dv}) = \int_{-\infty}^{\infty} \cdots \int_{-\infty}^{\infty} I(\text{dv}, \mathbf{x}) \cdot f(\mathbf{x}) \cdot \mathrm{d}\mathbf{x} \tag{23.3}$$

Comparing Eq. (23.3) to Eq. (23.1) it is recognized that reliability analysis is employed to obtain the same result. Equation (23.3) is the essence of unified reliability analysis.

Figure 23.1 illustrates the flow of information in a unified reliability analysis. The reliability analysis module repeatedly generates realizations of the random variables (e.g., earthquake rupture location, material properties, model errors, etc.) and distributes them to the models. The models are then sequentially evaluated until a total monetary loss value is computed, for the given realization of the random variables. It is reemphasized that the models for hazard, building, and consequences are linked by communication of physical quantities rather than conditional probabilities. The presence of a repair selection model is also noted.

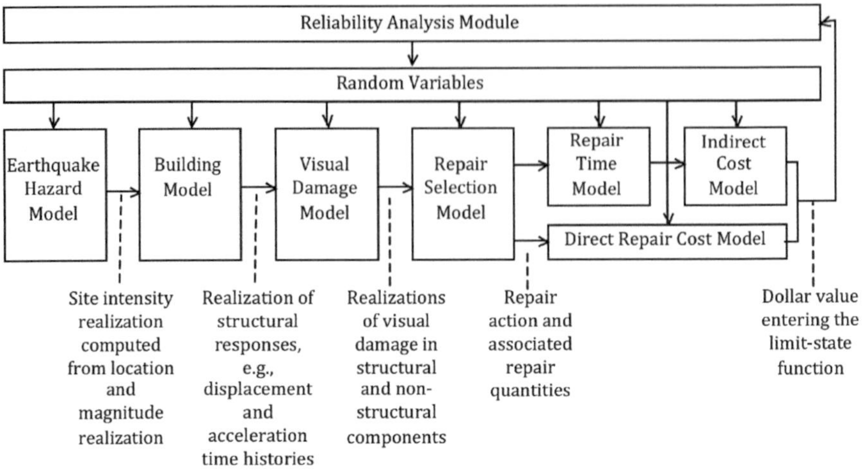

Fig. 23.1 Flow of information through models in a unified reliability analysis

Ongoing work in the authors' research group aims at developing and implementing a library of probabilistic models for hazards, infrastructure, and consequences in the context of the approach outlined in this paper. The models are implemented in the software *Rt* (www.inrisk.ubc.ca). This is a reliability analysis software that emphasizes the link with multiple models for hazards, infrastructure, and consequences. Notably, external software, such as OpenSees, SAP 2000, and Matlab can be used as a model. A screenshot of Rt is shown in Fig. 23.2, in which attention

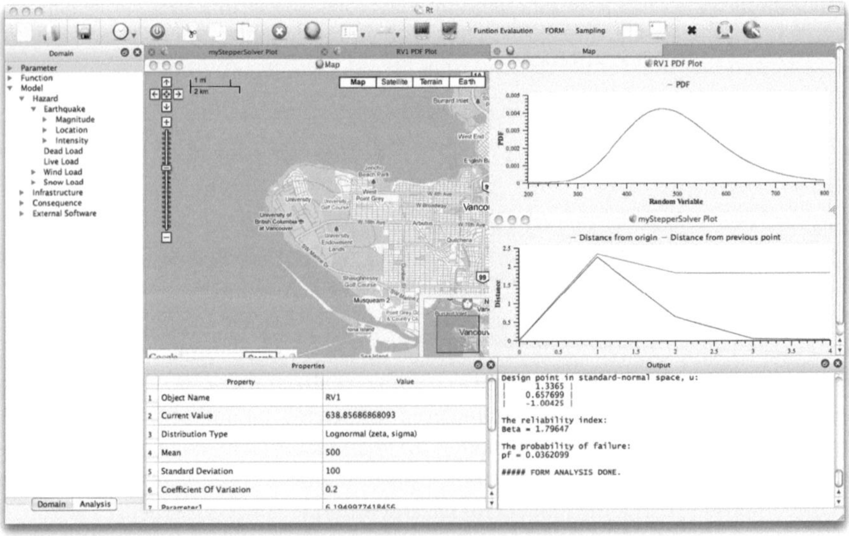

Fig. 23.2 Screenshot of the Rt software, which links the probabilistic models and carries out the reliability analyses

should be given to the leftmost Object Pane. Here, the hazard, infrastructure, and consequence categories are recognized. Further information and free downloads are available at the website given above.

From the above discussion it is understood that the prediction of performance probabilities, e.g., repair cost probabilities by means of reliability methods entails repeated evaluation of DV(**x**). That is, the value of the repair cost is sought for given realizations of the random variables. This turns the attention to the models.

23.3 Development of Probabilistic Models

The term "probabilistic model" takes on a particular interpretation in the approach adopted herein. According to this definition, a probabilistic model should not produce a probability. Rather, it should take realizations of the intervening random variables as input, along with other input parameters, and produce a unique response. This is in contrast with the fragility curve approach, in which a probability is produced. Examples of probabilistic models that yield a unique output for given values of model parameters and regressors, all being random variables, are provided later in this paper. In [7], a list of specific requirements of probabilistic models is provided. In the following, a review of the actual development of probabilistic models is provided. It is intended that this review will contribute to increased use of these methods.

The approach takes concepts from classical regression analysis as a starting point. Reference [5] and many other texts are available on this topic. In turn, those equations form the foundation for the Bayesian probabilistic models reviewed next. To this end, consider first an unknown quantity, such as the visual damage or the ensuing repair cost, denoted y. Assume that observations on y are available, accompanied by values of some always-quantifiable variables denoted by x. The variable y is called dependent variable, regressand, response, or output. The x-variables are called independent variables, predictor variables, regressors, or explanatory variables. Also assume that all information comes as n paired observations of x and y. The observations of y comes in the vector y_u with u=1,. . ., n and the observations of x comes in the matrix x_{ui} with i=1,. . ., k, where k is the number of explanatory variables. In other words, there are k regressors and n observations.

Although numerous model formulations are possible, the most common category of models, by far, is the linear model. However, this does not imply that the model must be linear in the regressors. Rather, it must be linear in the regression coefficients, which will be denoted by the symbol θ. That is, the model remains linear even if terms such as $\theta_2\sqrt{x_2}$ and $\theta_3\ln(x_3)$ appear in the model.

In the literature, the same lower-case y-notation is employed for the unknown response and its observed outcomes. Hence, the linear model is written

$$\mathbf{y} = \mathbf{X}\boldsymbol{\theta} + \boldsymbol{\varepsilon} \tag{23.4}$$

where $\mathbf{y}=y_u$ is the vector of observations, $\mathbf{X}=x_{ui}$ is the matrix of observed regressor values, $\boldsymbol{\theta}=\theta_i$ is the vector of unknown regression coefficients, and $\boldsymbol{\varepsilon}=\varepsilon_u$ is the vector of errors. In the following, uppercase **X** is used for the matrix that contains all the n observations for all the k regressors. Conversely, lowercase **x** is reserved for the vector of k regressors. In other words, the matrix **X** contains the observations of **x** as its rows.

Upon developing the model, i.e., determining the probability distribution for the model parameters, $\boldsymbol{\theta}$, predictions are made with the expression

$$y = \mathbf{x}\boldsymbol{\theta} + \varepsilon = x_1\theta_1 + x_2\theta_2 + \cdots + x_k\theta_k + \varepsilon \tag{23.5}$$

where y is the response, $\mathbf{x}=x_i$ is the vector of known regressors, $\boldsymbol{\theta}=\theta_i$ is the vector of unknown regression coefficients, and ε is the model error. The first regressor, x_1, is usually taken equal to unity and called the intercept of the model. Removal of the intercept parameter is inappropriate because it is considered as introducing additional information.

At first it may seem that the linear model form shown in Eq. (23.5) can only be empirical, i.e., calibrated by data alone and not mechanics-based equations that predict the outcome y. Models that include mechanics are preferred since they will likely be more robust when applied to conditions not considered in the database used to determine the model parameters, $\boldsymbol{\theta}$. Several approaches are available to develop such semi-empirical models. One is to use regressors (x-variables) that are obtained from first principles. In other words, use regressors that represent an amalgamation of different material and geometry parameters. Another approach is to use the purely mechanics-based model as an added deterministic term in Eq. (23.5) and let the other terms represent corrections to the mechanical model. Consequently, those corrective terms serve to introduce information gained from observations. Yet another approach is to consider non-linear models, in which the model parameters, $\boldsymbol{\theta}$, appear in non-linear terms. The last approach is outside the scope of this paper.

In classical linear regression the objective is to obtain point estimates of the θ-values. Although this does not yield a probabilistic model it finds a place in this review because it forms the basis for subsequent developments. In the linear regression approach, all uncertainty is lumped into the model error ε. It is assumed that $\boldsymbol{\varepsilon}\sim(\mathbf{0}, \sigma^2\, \mathbf{I}_n)$. In words, it is assumed that the observed errors have zero mean, standard deviation σ, and uncorrelated with each other, denoted by the n-dimensional square unit matrix $\mathbf{I}_n$. It is also assumed that the number of observations is greater than the number of regressors: $n>k$ and that the **X** matrix has rank k, i.e., full column rank. The last assumption is notable: it implies that linear dependence between the regressors, referred to as collinearity, must be avoided.

The least squares solution is obtained by minimizing $\|\varepsilon\|^2 = \varepsilon_1^2 + \varepsilon_2^2 + \cdots + \varepsilon_n^2$, i.e., the sum of the observed squared errors. Introducing the linear model from Eq. (23.4), the problem reads

$$\hat{\boldsymbol{\theta}} = \arg\min\left(\|\boldsymbol{\varepsilon}\|^2\right) = \arg\min\left(\|\mathbf{y} - \mathbf{X}\boldsymbol{\theta}\|^2\right) \tag{23.6}$$

where $\hat{\boldsymbol{\theta}}$ is the sought point estimate of the model parameters. The solution is found by setting the derivative of the objective function with respect to $\boldsymbol{\theta}$ equal to $\mathbf{0}$, with the result

$$\hat{\boldsymbol{\theta}} = \left(\mathbf{X}^{\mathrm{T}}\mathbf{X}\right)^{-1}\mathbf{X}^{\mathrm{T}}\mathbf{y} \tag{23.7}$$

The associated estimator of the variance of the error term is

$$\hat{\sigma}^2 = \frac{1}{n-k}\left(\mathbf{y} - \mathbf{X}\hat{\boldsymbol{\theta}}\right)^{\mathrm{T}}\left(\mathbf{y} - \mathbf{X}\hat{\boldsymbol{\theta}}\right) \tag{23.8}$$

Bayesian inference extends the objective from classical regression analysis. Rather than point estimates of the model parameters and associated variance and confidence measures, the objective is to establish the joint probability distribution for the model parameters and the model error, i.e., $f(\boldsymbol{\theta}, \varepsilon)$. Following classical regression analysis it is first assumed that the observed errors follow the Normal distribution. Consequently, the point estimate of the model parameters has the multivariate normal distribution, while the variance of the model error is distributed independently with the chi-squared distribution. Reference [1] employs this information in a Bayesian approach to derive the probability distributions for the model parameters, $\boldsymbol{\theta}$, and the model error, σ. The distribution of the model parameters is the multivariate t-distribution

$$f(\boldsymbol{\theta}) = \frac{\Gamma(\frac{1}{2}(\nu+k))s^{-k}\sqrt{\mathbf{X}^{\mathrm{T}}\mathbf{X}}}{(\Gamma(\frac{1}{2}))^k\Gamma(\frac{1}{2}\nu)\nu^{k/2}}\left(1 + \frac{(\boldsymbol{\theta}-\hat{\boldsymbol{\theta}})^{\mathrm{T}}\mathbf{X}^{\mathrm{T}}\mathbf{X}(\boldsymbol{\theta}-\hat{\boldsymbol{\theta}})}{\nu s^2}\right)^{-\frac{n}{2}} \tag{23.9}$$

from which the mean vector and covariance matrix read [3]

$$\mu_{\theta} = \hat{\boldsymbol{\theta}}, \qquad \Sigma_{\theta\theta} = \frac{\nu s^2}{\nu - 2}\left(\mathbf{X}^{\mathrm{T}}\mathbf{X}\right)^{-1} \tag{23.10}$$

Turning to the model error, σ, the distribution of the model variance, σ^2, is ν_s^2 χ_ν^{-2}, where χ_ν^{-2} is the inverse chi-squared distribution with ν degrees of freedom [1]. The mean and variance from this distribution is

$$\mu_{\sigma^2} = \nu s^2 \frac{1}{\nu - 2}, \qquad \sigma_{\sigma^2}^2 = \nu^2 s^4 \frac{2}{(\nu-2)^2(\nu-4)} \tag{23.11}$$

Subsequently, the mean and variance of the model standard deviation is obtained by a first-order second-moment approximation:

$$\mu_{\sigma} = \sqrt{\mu_{\sigma^2}}, \quad \sigma_{\sigma}^2 = \frac{\sigma_{\sigma^2}^2}{4\mu_{\sigma^2}} \tag{23.12}$$

In a procedure put forward in [3, 4] it is proposed to remove terms $\theta_i x_i$ that are associated with high coefficient of variation of θ_i. This elimination procedure is carried out while monitoring the mean of the model error until an appropriate balance between model parsimony and model error is achieved.

It is noted that the least square estimates appear in the probability distributions for the random model parameters. Hence, diagnostics known from classical regression maintain their importance in the Bayesian model development paradigm. Potential issues include collinearity (linear dependence between regressors), heteroskedasticity (dependence of the model error on the regressor values), error correlation (for example increasing error for every new observation), non-normality (non-normal distribution of the model error observations), outliers (extreme errors), non-linearity (inappropriate linear model form), and inappropriate variable selection (when too few or too many regressors are included). Although remedies exist for the issues, they are listed to emphasize that development of robust probabilistic models is an iterative process between inference and diagnostics.

23.4 Modeling the Consequences of Damage

The aim of the ongoing work is to develop two sets of models. First the visual damage that determines the repair action is predicted, followed by the cost/time of the repair action. In the context of reinforced concrete members, examples of visual damage include cracking, spalling of cover concrete, and buckling of longitudinal reinforcement. The repair action depends on the presence and quantity of these measures of visual damage. In other words, threshold models and quantity models for visual damage may be developed. A generic example of a threshold model for a reinforced concrete column is

$$\delta_{\text{Onset of spalling}} = \theta_1 + \theta_2 x_2 + \theta_3 x_3 + \theta_4 x_4 + \varepsilon_{\delta_{\text{Onset of spalling}}} \tag{23.13}$$

where $\delta_{\text{Onset of spalling}}$ is the drift at onset of spalling, θ_i, i=1,2,3,4 are random model parameters, x_i are material and geometry parameters, such as reinforcement ratios and strength of concrete and reinforcing steel, and ε is the model error.

Similarly, a generic example of a quantity model is

$$A_{\text{Spalling}} = \theta_5 + \theta_6 x_6 + \theta_7 x_7 + \theta_8 x_8 + \varepsilon_{A_{\text{Spalling}}} \tag{23.14}$$

where A_{Spalling} is the spalling area requiring repair, θ_i, i=5,6,7,8 are random model parameters, x_i are material and geometry parameters, and ε is the model error.

For a given repair action, the direct repair cost is a function of the cost of materials and labor. A generic model is

$$C_{\text{Direct}} = \theta_9 + \theta_{10} x_{10} + \theta_{11} x_{11} + \theta_{12} x_{12} + \theta_{13} x_{13} + \varepsilon_{C_{\text{Direct}}} \tag{23.15}$$

where C_{Direct} is the cost, θ_i, i=9,10,11,12,13 are random model parameters, x_{10} is a demand surge factor, x_{11} is an economy-of-scale factor, x_{12} is related to the material cost, x_{13} is related to the labor cost, and ε is the model error. Similarly, the time of repair will depend on the repair action and on demand surge conditions after an earthquake event, expressed by the model

$$T = \theta_{14} + \theta_{15}x_{15} + \theta_{16}x_{16} + \varepsilon_T \quad (23.16)$$

where T is the repair time, θ_i, i=14,15,16 are random model parameters, x_{15}=x_{10} is the demand surge factor, x_{16} is related to the repair action, and ε is the model error.

In the ongoing work to develop specific models of the form listed above, the availability of data is a key issue. Limited data exist with detailed description of the damage (cracking, spalling, etc.) in typical structural tests aimed at consequences beyond repairable damage. This leads to a plea for extended testing protocols, which include the recording of repairable damage during the testing of structural specimens. Such data is invaluable in the development of the probabilistic models outlined herein.

23.5 Conclusions

This paper advocates the development of probabilistic models to predict the performance of buildings subjected to ground motion. These models entail a comprehensive account of uncertainties and facilitate reliability analysis to obtain probabilistic performance predictions. A model development methodology is reviewed and its application to damage predictions is discussed. In a new approach, the focus of the damage modeling is to predict the repair action followed by prediction of the time and cost of that action. The paper concludes with a call for extended recording protocols in testing of structural and non-structural components.

References

1. Box GEP, Tiao GC (1992) Bayesian inference in statistical analysis, Wiley classics library. Wiley-Interscience, New York, 608 pp
2. Cornell CA, Krawinkler H (2000) Progress and challenges in seismic performance assessment, PEER News, Spring. http://peer.berkeley.edu/news/2000spring/index.html. Accessed Sept 2009
3. Gardoni P (2002) Probabilistic models and fragility estimates for structural components and systems, PhD Thesis, University of California, Berkeley, CA
4. Gardoni P, Der Kiureghian A, Mosalam KM (2002) Probabilistic capacity models and fragility estimates for reinforced concrete columns based on experimental observations. J Eng Mech 128(10):1024–1038
5. Gross J (2003) Linear regression, Volume 175 of Lecture Notes in Statistics. Springer, Dordrecht, 394 pp
6. Haukaas T (2008) Unified reliability and design optimization for earthquake engineering. Probabilistic Eng Mech 23(4):471-481
7. Haukaas T, Bohl A (2009) Comparison of approaches for performance-based earthquake engineering, Workshop on Performance-based Engineering, Villa Orlandi, Anacapri, Italy, 2–4 July. http://www.reluis.it/doc/EEBTB.htm. Accessed Sept 2009

8. Moehle J, Stojadinovic B, Der Kiureghian A, Yang TY (2005) An application of PEER performance-based earthquake engineering methodology, Research Digest No. 2005-1, Pacific Earthquake Engineering Research Center, University of California, Berkeley, CA
9. Park Y-J, Ang AH-S (1985) Mechanistic seismic damage model for reinforced concrete. J Struct Eng 111(4):722–739
10. Wen YK, Ellingwood BR (2005) The role of fragility assessment in consequence-based engineering. Earthq Spectra 21(3):861–877
11. Williams MS, Sexsmith RG (1995) Seismic damage indices for concrete structures: a state-of-the-art review. Earthq Spectra 11(2):319–349
12. Yang TY (2006) Performance evaluation of innovative steel braced frames, PhD Thesis, University of California, Berkeley, CA

Part III
Performance-Based Seismic Design and Retrofitting – Implementation

Chapter 24
Dual Flexural Plastic Hinge Design for Reducing Higher-Mode Effects on High-Rise Cantilever Wall Buildings

Marios Panagiotou and José I. Restrepo

24.1 Introduction

Derecho et al. [5] while examining the results of a comprehensive study on the nonlinear dynamic response of reinforced concrete cantilever walls of high-rise buildings, pointed out: *The difference between UBC-76 and 0.9 fractile normalized (bending) moments is particularly significant near mid-height. At about two-thirds of the height of the walls, the 0.9 fractile (bending) moments exceed the corresponding UBC moments by as much as 100 percent for the longer period.* Despite this observation, codes in United States have not recognized the significant effect higher modes have on the bending moment demands in cantilever walls of high-rise buildings. In contrast, codes like Eurocode 8 [3] which has similar design provisions for cantilever wall buildings to the New Zealand 3101 Concrete Design Standard [7] and the Canadian Concrete Code CSA A23.3-04 [4] do recognize the higher mode effects.

Design codes recognize the difficulties in ensuring elastic response of the lateral force resisting system in buildings. These codes recommend the use of reduced lateral forces in design. As a result, they recognize the possibility of developing nonlinear deformations in some parts of the structural system during a rare and strong intensity earthquake. Nonlinear deformations in cantilever walls occur preferably in flexure in regions defined as plastic hinges [12, 8, 10]. Traditionally, a single plastic hinge has been advocated in the seismic design of each wall in these buildings at the base of the walls in vertically regular buildings, or at the top of a podium in buildings with podiums, or at the ground floor in buildings with floors below grade. Detailing of the reinforcement in the plastic hinge regions is critical to ensure deformation demands have a low probability of exceeding the capacity in these rare events.

Seismic design codes such as EC8, NZS-3101 and CSA use Capacity Design (CD) to ensure elastic response in regions other than the plastic hinges. In these

---

M. Panagiotou (✉)
Department of Civil and Environmental Engineering, University of California at Berkeley, 705 Davis Hall, Berkeley, CA, USA
e-mail: panagiotou@berkeley.edu

M.N. Fardis (ed.), *Advances in Performance-Based Earthquake Engineering*, Geotechnical, Geological, and Earthquake Engineering 13, DOI 10.1007/978-90-481-8746-1_24,

codes the flexural design envelope varies linearly from the expected flexural overstrength at the wall base to zero at the top. The intention of such linear variation is to consider the effect of the higher modes. Recently, Panneton et al. [11], Priestley et al. [13] and Panagiotou and Restrepo [9] have found that the linear variation of bending moment with height does not always preclude the spread of plasticity into the upper regions.

Codes such as ACI-318 [1] are based on the premise that plasticity concentrates at the base of the walls only. This code does not restrict the designer to concentrate all the plasticity at the base, but at the same time, does not prompt the designer to check and detail regions above the base as probable plastic hinge regions. However, this code does not use CD, and thus does not recognize the effect of base overstrength and of higher modes. So, plasticity is likely to spread anywhere in the upper levels of the walls, as it has been recently pointed out by Moehle et al. [6]. The main problem with the design by this code is that undesirable premature modes of response in these upper regions, if they are not specially detailed for ductility.

This study proposes a dual plastic hinge design approach useful for the design of high-rise reinforced concrete wall buildings. Note that from the static viewpoint the notion of dual hinges is nearly unthinkable. However, under dynamic excitation this notion is not only conceivable but is believed to be favorable to the system's response and attractive from the design and constructability viewpoints. The second hinge at an intermediate height is specifically intended to reduce the large bending moment demands imposed by the second mode.

24.2 Dual Plastic Hinge Design Approach

Figure 24.1 shows three possible approaches as to where plasticity can develop in cantilever wall buildings. Figure 24.1a shows the first approach: plasticity develops anywhere along the height of the walls; this is termed here Extended Plasticity (EP). The second approach, shown in Fig. 24.1b, is that of a Single Plastic Hinge (SPH): this hinge develops only at the wall base. The third approach, proposed by the authors and shown in Fig. 24.1c, allows two plastic hinges in a wall: one at the base and the second one at mid-height; it is termed the Dual Plastic Hinge (DPH) design approach.

The EP and the SPH approaches have clear disadvantages. In the EP approach yielding up the height in walls would typically require special reinforcement detailing all along the height of the walls. Extended yielding, as inferred in the EP approach, is theoretical in nature. In practice the longitudinal reinforcement is detailed to show stepped bending moment capacity diagrams that envelope the code's demand. These steps in the capacity diagrams form critical flexural strength discontinuities where inelastic response concentrates. In the SPH approach the rigorous use of CD to preclude yielding above the plastic hinge region can result in longitudinal reinforcement ratios that exceed those calculated at the base of the walls. The need for these large ratios will be discussed later with the design examples chosen.

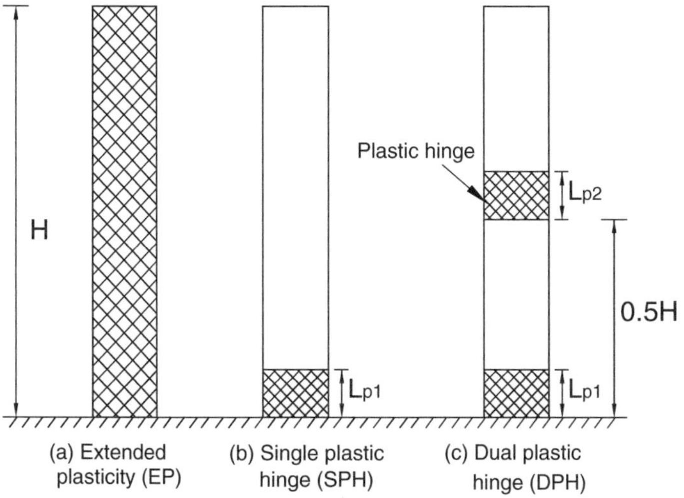

Fig. 24.1 Three different cases of plasticity location in an Euler-Bernoulli cantilever

The DPH design approach, Fig. 24.1c, overcomes the disadvantages of the EP and SPH approaches. Like the bottom plastic hinge, the mid-height plastic hinge can be designed to meet specific objectives, such as curvature ductility or strain demands for which design alternatives to the current force-based approaches may be more suitable. The base and mid-height regions of the wall where plastic hinges will develop are designed following a strength hierarchy. This hierarchy precludes the first mode of response alone from developing the mid-height plastic hinge. CD is subsequently employed to keep the remaining portion of the walls elastic and to ease the detailing of the reinforcement there. So, on one hand the performance of the building is controlled as is in the SPH design approach; on the other the ease of detailing and/or reduction in the longitudinal reinforcement along a significant portion of the walls' height in the DPH design approach brings significant optimization to construction compared with the SPH design approach.

24.3 Numerical Verification of the Design Approaches

24.3.1 Buildings Design

This section examines the design and nonlinear dynamic response of 10-, 20- and 40-story representative core-wall buildings. The buildings are designed for the response spectra of specific ground motions that have distinct near-fault characteristics. The buildings were designed with three different approaches: the ACI-318 building code, SPH according to EC8, and the proposed DPH design approach. Each building was designed for the 5%-damped response spectra of each of the three ground motions for each of the three design approaches. Their design is discussed

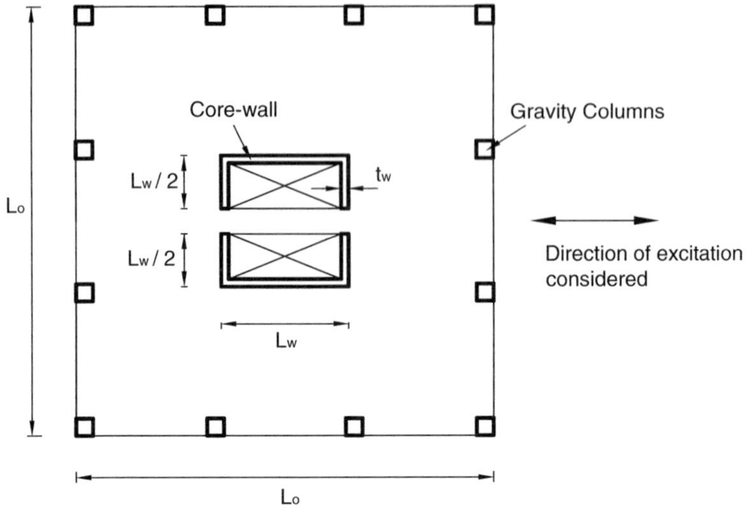

Fig. 24.2 Floor plan view of the buildings

Table 24.1 Main characteristics of buildings considered

	10-story	20-story	40-story
Floor height h (m)	3.4	3.4	3.4
Building height H (m)	33.5	67.1	134.1
Floor plan view dimension L_o(m)	12.2	24.4	45.7
Core-wall length L_w (m)	4.6	8.2	15.2

in detail in Panagiotou and Restrepo [9]. The lateral force resistance in the buildings studied was solely provided by a reinforced concrete core-wall. Figure 24.2 shows the floor plan view of the core wall buildings and Table 24.1 lists their main characteristics, including the floor height h, the seismic weight w per floor.

24.3.1.1 Designs Based on ACI-318 2005 Building Code

The bending moment envelopes for the ACI-318 design were obtained from a Modal Response Spectrum Analysis (MRSA). The longitudinal reinforcement is curtailed at four locations, ensuring the nominal flexural strength envelope is greater than the design envelope. The longitudinal reinforcement of the 40-story walls at the base is just above the minimum. Boundary elements meet ACI-318 requirements and extend a distance equal to 19, 13 and 6% of the building height from the base of the walls for the 10-, 20-, and 40-story buildings, respectively.

24.3.1.2 Single Plastic Hinge (SPH) Design Approach

The design bending moment at the base of the core-walls is identical to the ACI-318 design. The remaining portions of the core-walls are assumed elastic. This assumption is made to prove the adequacy of the current CD design recommendations in EC8 and NZS-3101 concerning the effect of higher modes.

24.3.1.3 Dual Plastic Hinge (DPH) Design Approach

For comparative purposes the expected flexural strength at the base of the core-wall in this design is the same as in the previous two approaches. To ensure the development of the base plastic hinge, the flexural design at the location of the mid-height plastic hinge follows a specific design procedure described in Panagiotou and Restrepo [9], whereas those portions of the walls away from the plastic hinges are assumed elastic.

24.3.2 Computational Model

Simple nonlinear analytical tools and simple models are used in this investigation. All floors have identical lumped masses. One-component Giberson beam elements model the core-walls. The computer program Ruaumoko [2] was used to perform the nonlinear dynamic time-history analyses (NDTHA).

24.3.3 Ground Motions

Each building was analyzed for the near-fault ground motion it was designed for. The ground acceleration time histories and the acceleration and displacement response spectra can be found in Panagiotou and Restrepo [9]. Two of them, SYLOV360 and RIN228, were recorded in the M_w 6.6 1994 Northridge earthquake. The third is the TAK090 record from the M_w 7.2 1995 Great Hanshin earthquake. The motions used in this study have distinct strong pulses with period content T_p in the period range of the second mode for the buildings considered. The destructiveness of these motions in terms of second and higher mode excitation is due to the fact that they produce not only large second mode spectral but also significant first mode spectral accelerations that are highly correlated in the time domain with the second modal accelerations.

24.3.4 Results of the Analyses

Figure 24.3 compares the bending moment envelopes obtained from the NDTHA for the ACI-318, SPH and DPH designs. Each plot in Fig. 24.3 shows three bending moment envelopes: (i) ACI; (ii) SPH; and (iii) DPH computed with NDTHA for the

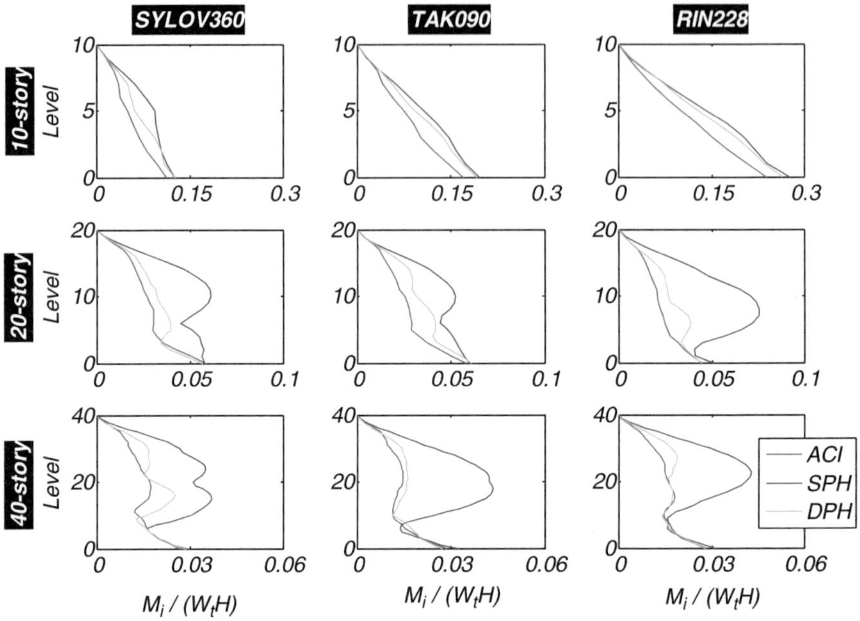

Fig. 24.3 Bending moment envelopes obtained from the NDTHA for the three design approaches

ACI, SPH and DPH designs, respectively. The bending moment envelopes have been normalized by the product of the total seismic weight and height of the structure, $W_t H$.

As expected, the SPH design approach consistently shows the greatest bending moment demands at the intermediate portion of the walls. In contrast, the DPH design approach effectively limits these demands. Limiting the intermediate height bending moments in the DPH design approach makes it possible to obtain longitudinal reinforcement ratios that are smaller than those at the wall base [9]. The bending moment envelopes of the DPH design approach for the 20- and 40-story buildings is close to this of the ACI design.

A main finding of the analysis for the SPH design approach is the practical difficulty that arises when trying to ensure elastic response in the walls except at the base. The reason for the large ratios is the larger bending moment demands combined with smaller axial forces acting on the walls at mid-height. In summary, the SPH design approach requires large amounts of longitudinal reinforcement in the intermediate portion of the walls and this is associated with significant congestion and higher cost.

The extent and magnitude of plasticity in the walls can be observed through the curvature ductility demands. Figure 24.4 shows the curvature ductility demands for the cases studied. It clearly illustrates the concentration of plasticity in regions along the wall height in the ACI-318 designs, labeled ACI in this figure. The curvature

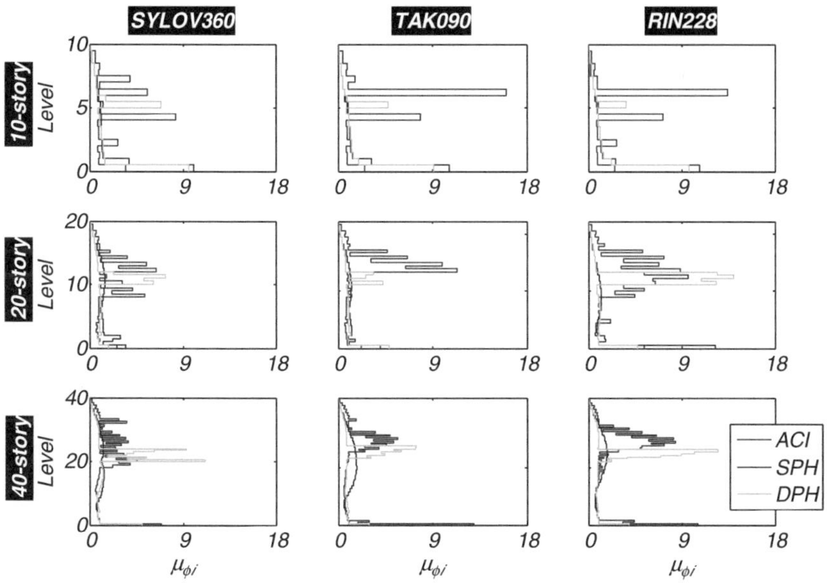

Fig. 24.4 Curvature ductility envelopes obtained from NDTHA for the three design approaches

ductility demand in the upper part of the 10-story buildings reaches 14 in one case and 17 in another, which requires special detailing to sustain. For one of the 20-story buildings the curvature ductility demand observed at 60% of the height reaches the large value of 12. For another of the 20-story and for one of the 40-story buildings the curvature ductility demand observed at 60% of the height reaches the moderate value of 9. Note that in all analyses yielding takes place in the upper portions of the walls well above the extent of the boundary elements mandated by ACI-318. The SPH design has, in seven out of the nine cases, the largest base curvature ductility demands, see Fig. 24.4. This is due to constraining plasticity to a single plastic hinge. The DPH design approach concentrates the plasticity at two specific regions along the height not exceeding 20% of the total building height, see Fig. 24.4.

For the 10-story buildings the DPH design significantly reduces the extent and magnitude of curvature ductility demand in the upper part of the building in comparison with ACI-318. In all cases the curvature ductility demands are small or modest. For the 20-story buildings the DPH design significantly reduces the extent and one case the magnitude also of curvature ductility demand in the upper part of the building in comparison with ACI-318. In all cases except for RIN228 the curvature ductility demands are small or modest. For the 20-story building and the RIN228 case a curvature ductility demand equal to 14 was computed, which can be achieved with proper detailing. For the 40-story buildings the mid-height plastic hinge does not completely spread throughout the allocated length $L_{p2} = 0.1H$. This is because of the significant effect of axial load on the flexural strength of the lightly

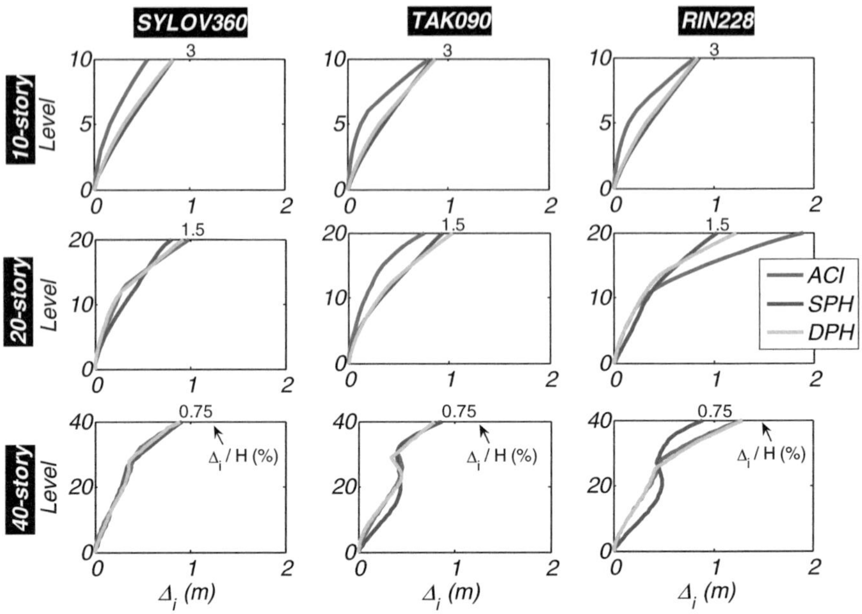

Fig. 24.5 Lateral displacement response envelopes obtained from NDTHA

reinforced 40-story walls. At the base of the walls the DPH design results in curvature ductility demands in between or even smaller than the demands obtained from the other two approaches. The curvature ductility demands in the 40-story buildings are greater for the DPH design approach than for the ACI-318 design and reaches a maximum of 12 for the RIN228 motion. In all these cases curvature ductility demands can be achieved with proper detailing.

Figure 24.5 plots the normalized lateral displacement envelopes for each of the cases analyzed. Lateral displacements have also been normalized by the total height. The maximum roof drift ratio, defined as the maximum roof lateral displacement at the top of the wall over the wall height, is practically independent of the design approach, save the cases of the 10-story building designed for and subjected to the SYLOV360 motion and the 20-, 40-story building designed for and subjected to the RIN228 motion. This is due to the similar first modal periods for the different design approaches in most cases and to the almost constant spectral displacements over the periods of interest. Differences in the lateral displacement envelopes are observed primarily for the 10-story buildings.

24.4 Summary and Conclusions

This paper discussed the effect of higher modes, and especially of the second mode, on the nonlinear dynamic response of cantilever reinforced concrete wall buildings. It proposed a dual plastic hinge approach to better control the seismic response of these buildings to strong shaking. The paper investigated numerically the seismic response of cantilever wall buildings designed using three different approaches: (i) ACI-318, (ii) a single plastic hinge concentrated at the base of the walls according to EC8, and (iii) the proposed dual hinge approach where one plastic hinge concentrated at the wall base and another develops near mid-height. Nonlinear dynamic analyses of these buildings were carried out for three strong near-fault ground motions. The investigation led to the following conclusions:

1. Near fault ground motions including strong pulses, characterized by large elastic spectral accelerations in the range of the second translational mode of high-rise cantilever buildings are likely to have a significant effect on the bending moment demands in the walls. Current design codes do not address such large demands explicitly.
2. Based on this numerical study, designs of cantilever wall buildings following the current ACI-318 building code may result in unintended concentration of nonlinear deformations higher up in the walls where elastic response is generally expected. Current detailing requirements may not ensure controlled flexural response in such regions. This study also observed large interstory drift ratios, as well as a large concentration of residual rotations, at about 60% of the walls' height in the ACI-318 designs.
3. Codes like EC8 allow the development of plastic hinges at the wall bases only. The results presented in this paper indicate that, under the near fault ground motions considered in this paper, bending moment demands at intermediate height in walls developing base plastic hinges compare closely or even exceed the base bending moments. Such intermediate height moment demands are not recognized in the code prescriptive requirements for Capacity Design. Hence, elastic response up the height in walls may not actually occur as intended. It is recommended that current design provisions be examined and appropriately revised.
4. The proposed dual plastic hinge design approach, in which plastic hinges are allowed to form at the wall base and near mid-height while ensuring elastic response elsewhere, was found to have significant advantages: reduction in the amount of longitudinal reinforcement when compared to the EC8, NZS-3101 and CSA designs, and ease of detailing along most of the height. This approach can be easily implemented in design, bringing a reduction in the amount of longitudinal reinforcement and of confinement reinforcement in a significant portion of the walls.

References

1. ACI (2005) ACI318-05. Building code requirements for structural concrete and commentary. ACI Committee 318, Farmington Hills, MI
2. Carr AJ (1998) Ruaumoko – a program for inelastic time-history analysis. Department of Civil Engineering. University of Canterbury, New Zealand

3. CEN (2004) EN 1998-1:2004 Eurocode 8: design of structures for earthquake resistance. European Committee for Standardisation, Brussels
4. CSA (2005) CSA Standard A23.3-04. Design of concrete structures, Canadian Standard Association, Rexdale, Canada, 214 pp
5. Derecho AT, Iqbal M, Ghosh SK, Fintel M, Corley WG, Scanlon A (1981) Structural walls in earthquake-resistant buildings dynamic analysis of isolated structural walls development of design procedure – design force levels. Portland Cement Association, Skokie, IL
6. Moehle J, Bozorgnia Y, Yang TY (2007) The tall buildings initiative. Proceedings of SEAOC Convention, Squaw Creek, CA, USA, pp 315–324
7. NZS (2006) NZS 3101 New Zealand Standard, Part 1 – the design of concrete structures, Standards New Zealand, Wellington, New Zealand
8. Panagiotou M, Restrepo JI (2007) Lessons learnt from the UCSD full-scale shake table testing on a 7-story residential building slice. Proceedings of SEAOC Convention, Squaw Creek, CA, USA, pp 57–73
9. Panagiotou M, Restrepo JI (2009) Dual-plastic hinge design concept for reducing higher-mode effects on high-rise cantilever wall buildings. Earthq Eng Struct Dyn 38:1359–1380
10. Panagiotou M, Restrepo JI, Conte JP (2007) Shake table test of a 7-story full scale reinforced concrete structural wall building slice phase I: rectangular wall. SSRP 07-07 Report, Department of Structural Engineering, University of California at San Diego, San Diego, CA
11. Panneton M, Léger P, Tremblay R (2006) Inelastic analysis of a reinforced concrete shear wall building according to the national building code of Canada 2005. Can J Civ Eng Vol. 3: 854–871
12. Paulay T, Priestley MJN (1992) Seismic design of reinforced concrete and masonry buildings. Wiley, Hoboken, NJ
13. Priestley MJN, Calvi GM, Kowalsky MJ (2007) displacement based seismic design of structures. IUSS Press, Pavia, Italy

Chapter 25
High Seismic Performance Systems for Steel Structures

Constantin Christopoulos

25.1 Introduction

Steel structures have long been recognized as having excellent seismic performance primarily because of the inherent ductility of steel. However, recent earthquakes such as the 1994 Northridge California earthquake and the 1995 Kobe, Japan earthquake demonstrated that achieving structures that truly meet their design intent of forming stable plastic mechanisms and developing a ductile system response required very careful, significantly more evolved detailing and fabrication quality than previously thought. Further, socio-economic realities have also shifted the emphasis of seismic performance from the life-safety goal mitigating the cost of repair or replacement of structures and occupant downtime following major earthquakes. As a result, steel structures, as designed and built following current codes, have a number of drawbacks: (i) the structural elements intended to yield and absorb seismic energy during seismic loading are part of the main structure and are therefore difficult to inspect and repair following an earthquake, (ii) ductile steel structures are expected to sustain significant residual deformations as a result of their inelastic response to seismic loading, (iii) it is nearly impossible to fully assess the remaining capacity of steel members that have undergone significant inelastic yielding unless each structure is carefully instrumented and data is collected during the earthquake and (iv) excessive residual deformations can also result in the total loss of a structure.

In this paper, first the inelastic response of ductile steel structures is discussed and the magnitude of the expected residual deformations that can lead to significant costs of repair and upgrade of the structure are highlighted. New systems that have been developed in the past few years to address the drawbacks of current steel structures are then presented with an emphasis on a new family of self-centering steel framing systems.

C. Christopoulos (✉)
Department of Civil Engineering, University of Toronto, 35 St. George Street, Toronto, M5SIA4, Canada
e-mail: c.christopoulos@utoronto.ca

M.N. Fardis (ed.), *Advances in Performance-Based Earthquake Engineering*, Geotechnical, Geological, and Earthquake Engineering 13, DOI 10.1007/978-90-481-8746-1_25,

25.2 Residual Response of Ductile Steel Structures

Residual drifts are the permanent deformations of a structure that remain at the end of a seismic excitation and they are caused by the nonlinear behavior of the yielding components in the system. Studies have shown that it is necessary to consider these residual deformations to fully characterize the performance of a structural system after a seismic excitation and the potential damage that the system has suffered [4]. Recently, McCormick et al. [11] conducted a study of one occupied building at Kyoto University in Japan and conducted a review of previous research in Japan including consideration of both physiological and psychological effects of residual drifts on occupants. They concluded that residual drifts of 0.5% are generally perceivable by occupants and as residuals approach 1.0% occupants of a building experience dizziness and nausea. More importantly, they concluded that in Japan it was generally less expensive to rebuild a structure than to repair it when an earthquake resulted in residual drifts greater than 0.5%. Pettinga et al. [12] performed nonlinear analyses of 4-story BRB frames and found mean maximum residual story drifts between 0.85 and 0.89% when subjected to New Zealand design-level earthquakes which are similar to the DBE seismic hazard for the Los Angeles area. Nonlinear analyses conducted by Tremblay et al. [15] predicted median residual drifts varying between 0.84 and 1.38% under DBE ground motions for BRB frames varying between 2 and 16 stories. Kawashima et al. [6] quantified the residual displacement response of SDOF structures with bilinear hysteretic behavior by developing residual displacement response spectra. Ruiz-Garcia and Miranda [14] evaluated the residual displacement demands in moment-resisting frames in relation to the maximum elastic displacement demand, the dynamic properties of the system and the type of hysteretic modeling.

Studies have also been conducted to determine appropriate measures that may be taken to mitigate residual drifts. Pettinga et al. [12] investigated the benefit of increasing the post-yield stiffness of traditional framed and braced systems. Residual drifts can also be reduced or completely eliminated by using systems that have self-centering capability; a number of these systems will be discussed in the following paragraphs.

In a study by Choi et al. [2] extensive 2D nonlinear dynamic analyses were carried out to determine and compare the peak and residual drift responses of a large number of special moment-resisting frame (SMRF) and buckling-restrained braced frame (BRB frame) buildings that are strictly designed according to ASCE 7-05 [1]. Six different heights of each frame type were designed and modeled. These models were subjected to both pushover and time-history analyses, where the suite of earthquakes was divided into two groups, each representing a different seismic hazard level.

The pushover analyses showed that while the BRB frames are approximately 50% stronger than the design strength, the SMRFs, are significantly more over-designed to accommodate the ASCE 7-05 story drift limit. For the design-based seismic hazard level, the SMRF systems had the largest interstory drift response for the shortest buildings, but the systems tended to converge to approximately the same peak interstory drift and variation as the building height increased. For the maximum credible seismic hazard level, the BRB frames performed slightly better than the SMRFs, except for the 12-story structures where soft-story formation and P-Delta effects in the BRB frames caused the displacement and interstory drift responses to deteriorate such that the peak drift values were significantly higher than the SMRFs.

As shown in Fig. 25.1, on average the 6- to 12-story building models slightly exceeded the prescribed code drift limits under design-based earthquakes even though they were designed to adhere to them. As shown also in this figure, both types of systems experienced significant permanent residual drifts after the excitations with an average magnitude between 0.8 and 1.8% for the BRB frames and between 0.5 and 1.2% for the SMRF, values which are both noticeable to occupants and would warrant expensive repair or replacement of the structure. The residual drifts resulting from the MCE level excitations were, on average, between 2 and 4%. These results also have a high amount of variability, meaning that for certain excitations, the actual residual could be up to double the mean value.

If a residual drift of 0.5% may represent the total loss of a building from an economic point of view, these results mean that the average BRB or SMRF building designed according to ASCE 7-05 has a high potential to be economically useless after even a design-level seismic event, since, as shown in Fig. 25.1, under that level of seismic hazard 79% of earthquakes caused a maximum residual drift greater than 0.5% in the BRB structures and 60% of earthquakes did so when applied to the SMRF structures.

In sum, if 0.5% residual drift can be considered to be the level beyond which structures are no longer practically usable, then these results suggest that BRB frame buildings with a peak drift larger than about 1.0% and SMRF buildings with a peak drift larger than about 1.5% will likely represent a total economic loss following a design-level seismic event. This in-turn suggests that for these ductile steel structures, only two relevant economic performance levels are achievable: (i) the immediate occupancy performance level where the building is practically

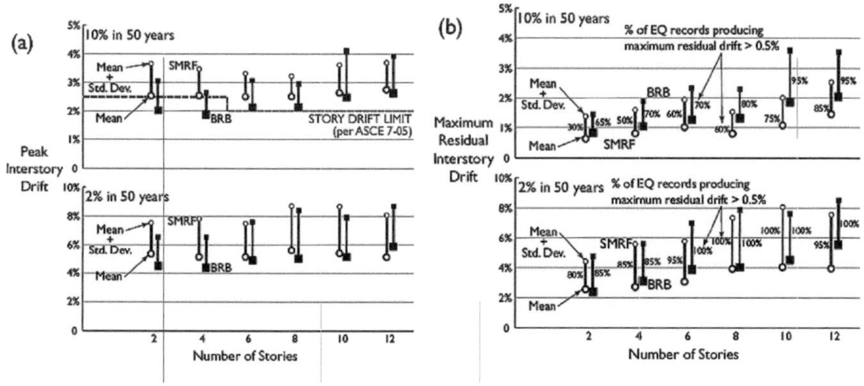

Fig. 25.1 Peak and residual interstorey drifts for BRB and MRF steel structures designed according to ASCE 7-05

undamaged because it only sustains a fraction of the design level drift and is still usable with minimal repair after the earthquake, and (ii) the complete economic loss performance level where it is less expensive to replace a building than it is to repair it because of excessive residual drifts (this later performance level being the most likely after a design level earthquake).

25.3 Steel Structures with Replaceable Nonlinear Links

While steel EBF and MRF structures designed according to the current seismic design specifications can provide life safety during a design level earthquake, they are expected to sustain significant damage through repeated inelastic deformation and localized buckling. The design of the yielding elements is also interlinked with the design of the connecting beam, which often results in a significant over-design. Because the yielding fuse is a part of the beam in current construction, strength design and drift design of the structure are interlinked, often resulting in over-designed structures. In addition, significant damage can result in the main structural elements from repeated inelastic deformation and localized buckling during a design level earthquake. As the cumulative inelastic deformations sustained by these yielding elements of the structure is unknown, it is difficult to assess the extent of damage and the structure's ability to provide adequate level of safety for any subsequent loading. Furthermore, repair the main structural elements is very difficult, disruptive, and costly. These drawbacks can be mitigated by introducing nonlinear replaceable links at the locations of expected inelastic action as illustrated in Fig. 25.2 [8, 9]. It allows for independent control of beam stiffness and required strength, resulting in more efficient structures; welding of critical elements to be done in the shop, considerably improving construction quality and reducing erection time; and for quick inspection and replacement of damaged links following a major earthquake, significantly minimizing the disruption time of the structure.

Full-scale links were tested at the University of Toronto following incremental cyclic loading protocols for verification of strength and ductility. Two link types

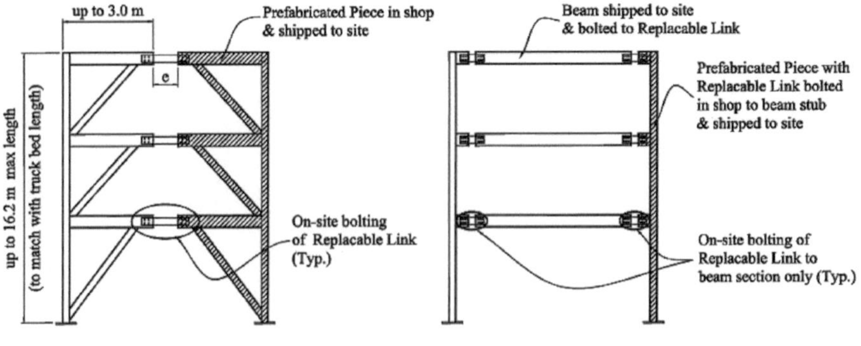

Fig. 25.2 MRF and EBF with replaceable links

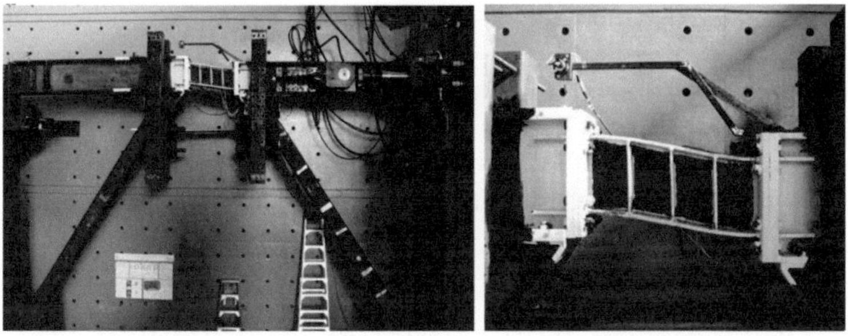

Fig. 25.3 Full-scale EBF with replaceable links at maximum drift (*left*): entire frame, (*right*): deformation of link

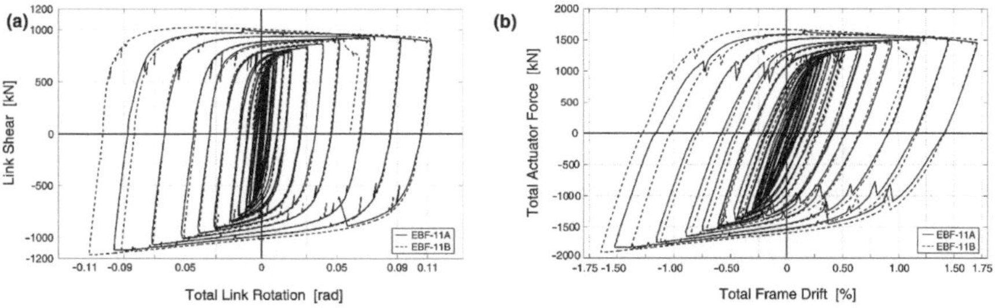

Fig. 25.4 Response of W-section specimens: (**a**) link rotation versus link shear; (**b**) frame storey drift versus actuator load

were tested for each system: (i) back-to-back channels bolted to the web of the connecting beam, (ii) W-sections with end plate connections. All end plate links provided good performance and energy dissipation. Full-scale frame tests were also carried out at Ecole Polytechnique in Montreal to validate the response of the proposed system and of the different link configurations that were developed in this research project [10]. Figure 25.3 shows the full-scale frame at the target maximum drift as well as a close-up view of the link deformation.

Figure 25.4 shows the response of two W-section link specimens that were tested in the full-scale frame setup. Superimposed in this figure are the responses of the initial link and of the repaired (or replacement link). As can be seen, the response of both links is for all practical purposes identical, indicating that a full upgrade of the structure was achieved following the replacement of the yielded link.

25.4 Steel Self-Centering Frame Systems

Structural systems possessing self-centering characteristics that minimize residual deformations represent very promising alternatives to current lateral force resisting systems. The response has more frequent stiffness changes within one nonlinear cycle than the traditional elastic-plastic hysteresis. The amount of energy dissipation is reduced compared to that of typical yielding systems but, more importantly, the system returns to the zero-force, zero-displacement point at every cycle as well as at the end of the seismic loading. This characteristic eliminates residual deformations and prevents the progressive drifting response observed with traditional elastic-plastic hysteresis, thus mitigating the potential for large P-Δ effects.

25.4.1 Post-Tensioned Self-Centering Moment-Resisting Frames

Recently, self-centering systems for moment-resisting steel framed structures have been proposed by Ricles et al. [13] and Christopoulos et al. [3]. The post-tensioning consists of high strength steel bars or strands that run along the web of the beams and are anchored to the exterior column flanges at the end of the frame. In the connections developed by Ricles et al. [13], seat and top angles are bolted to both columns and beams, and shear resistance is provided by a combination of friction at the beam-column interface and the steel angles. The system is designed so that the steel angles are the only yielding elements. Therefore, only the steel angles would need to be replaced after a major earthquake. In the steel frame concept proposed by Christopoulos et al. [3], four buckling restrained steel bars that can yield in tension and compression without buckling are symmetrically placed at each connection to provide energy dissipation under cyclic loading. The connection relies on the post-tension force to maintain contact between the beams and columns. Horizontally slotted shear tabs are welded to the column flanges and bolted to the beam web for erection. Additional benefits of these connection systems include: no field welding required, use of conventional materials and skills, and similar initial stiffness to conventional welded connections.

Recent tests by Kim and Christopoulos [7] validated new PT self-centering friction damped (SCFR) steel moment-resisting full-scale connections (see Fig. 25.5).

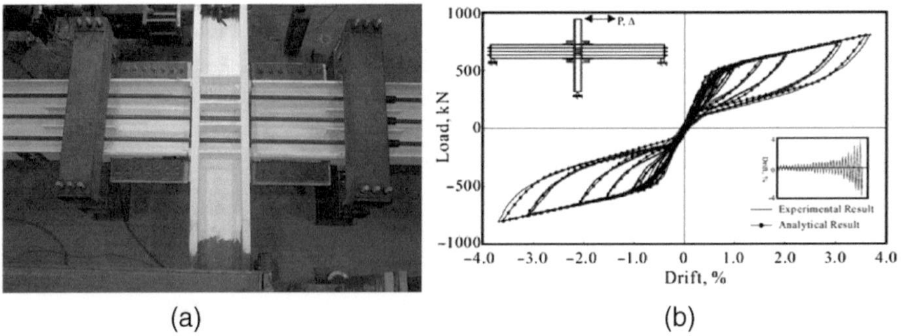

Fig. 25.5 Self-centering friction moment resisting frame: (**a**) full-scale interior connection specimen at 3% drift; (**b**) flag-shaped force-deflection response of specimen

Test results showed that self-centering moment connections with the proposed friction dampers were capable of developing stiffness and strength similar to that of welded connections. The test specimens were tested twice to demonstrate that the cyclic response of the second run was, for all practical purposes, identical to those of the first run, thus demonstrating that the proposed system is capable of a full recovery of strength, stiffness, and ductility by simply de-stressing and prestressing the bolts in the friction devices.

When tested beyond the self-centering limit (i.e., beyond the expected MCE level), the proposed connections were shown to exhibit a ductile response with the formation of flexural hinges in the beams, thus avoiding the sudden loss of strength and stiffness that occurs when the post-tensioning elements are overloaded or when the beams buckle under excessive combined axial loads and bending.

25.4.2 The Self-Centering Energy Dissipative (SCED) Bracing System

The self-centering energy dissipative (SCED) brace is an advanced cross-bracing system that has recently been developed at the University of Toronto and École Polytechnique de Montréal [5]. The advantage of the SCED brace system is that, unlike other comparable advanced bracing systems that dissipate energy, such as the buckling restrained brace system, it has a self-centering capability that reduces or eliminates residual building deformations after major seismic events. This leaves the main structural components of the building intact, undamaged and in-place, allowing the building to be re-occupied relatively soon following a seismic event. This important feature could potentially significantly reduce the financial losses due to repair and disruption time following an earthquake.

A thorough explanation of the SCED brace concept and validation can be found in Christopoulos et al. [5]. A schematic of the SCED brace showing its basic function is given in Fig. 25.6. The system consists of two rigid longitudinal members that are abutted by end plates on each end. These end plates are connected together by pre-tensioned tendons. These tendons provide the self-centering capability of the

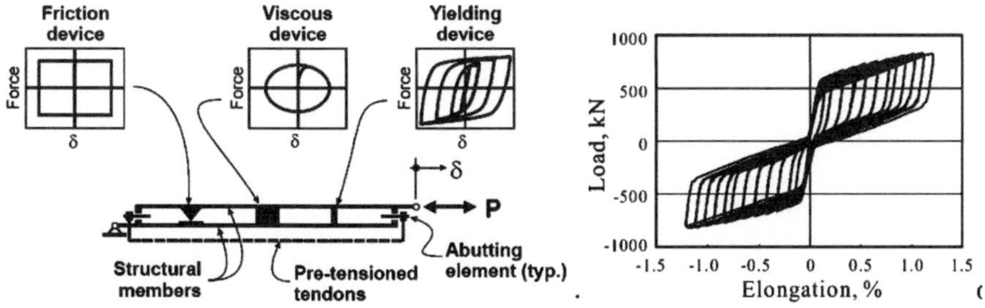

Fig. 25.6 Concept of SCED system and hysteretic response of SCED prototype under quasi-static axial loading

brace. Since the end plates are not connected to the longitudinal elements, the tendons are in tension regardless of the direction of axial deformation. Between the two longitudinal members, there exists some type of energy-dissipative mechanism either in the form of a friction, viscous, or yielding device.

Recently, Tremblay et al. [15] performed an extensive analytical study to assess the seismic response of SCED braced frames. The response of these frames was also compared to that of BRB frames. According to these results, the maximum response of the SCED frames was similar to that of the BRB frames but the residual deformations for SCED braced frame systems were negligible under low and moderate hazard levels and were reduced considerably under maximum considered earthquake level.

25.5 Conclusion

Steel structures, despite their inherent ductility capacity, are expected to be significantly damaged following a design level earthquake, with the cost of repair severely affected by residual deformations and by the fact the main structural elements sustain damage. In this paper, an overview of the residual deformation response of steel structures was first carried out. A summary of some of the recent developments in seismic resistant steel structures was then presented. Results from the validations of these new systems indicate that high-performance steel structures potentially represent an economically viable alternative to current traditional steel systems with significant benefits to occupants and owners in terms of loss mitigation following a major earthquake.

Acknowledgements The work that is summarized in this paper has been carried out in close collaboration with a number of researchers in Canada. The closest collaborators that the author wishes to acknowledge are: Professor Robert Tremblay from Ecole Polytechnique in Montreal, Dr. Hyung-Joon Kim from the University of Seoul in Korea (formerly a Ph.D. student at the University of Toronto), Dr. Hyunhoon Choi from Sungkyunkwan University in Korea (formerly a Postdoctoral Fellow at the University of Toronto), as well as Nabil Mansour and Jeff Erochko who are currently Ph.D. students at the University of Toronto. The Financial support of the Natural Sciences and Engineering Research Council of Canada, which has supported most of the work that was outlined in this paper, is also gratefully acknowledged.

References

1. American Society of Civil Engineers (ASCE) (2005) Minimum design loads for buildings and other structures, ANSI/SEI7-05 including supplement No.1. American Society of Civil Engineers, Reston, VA
2. Choi H, Erochko J, Christopoulos C, Tremblay R (2008) Comparison of the seismic response of steel buildings incorporating self-centering energy-dissipative dampers, buckling restrained braces and moment resisting frames. Report No. 05-2008, Department of Civil Engineering, University of Toronto, Toronto, ON
3. Christopoulos C, Filiatrault A, Folz B, Uang C-M (2002) Post-tensioned energy dissipating connections for moment-resisting steel frames. J Struct Eng ASCE 128(9):1111–1120

4. Christopoulos C, Pampanin S, Priestley MJN (2003) New damage index for framed systems based on residual deformations: part I. J Earthq Eng 7(1):97–118
5. Christopoulos C, Tremblay R, Kim H-J, Lacerte M (2008) Self-centering energy dissipative bracing system for the seismic resistance of structure: development and validation. ASCE J Struct Eng 134(1):96–107
6. Kawashima K, MacRae GA, Hoshikuma J-I, Nagaya K (1998) Residual displacement response spectrum and its application. J Struct Eng 124(5):523–530
7. Kim H-J, Christopoulos C (2008) Friction damped post-tensioned self-centering steel moment-resisting frames. J Struct Eng 134(11):1768–1779
8. Mansour N, Christopoulos C, Tremblay R (2006) Seismic design of EBF steel frames using replaceable nonlinear links. Proceedings of 2006 STESSA conference, Yokohama, Japan, pp 745–750
9. Mansour N, Shen Y, Christopoulos C, Tremblay R (2008) Seismic design of EBF steel frames using replaceable nonlinear links. The 14th world conference on earthquake engineering, Beijing, China
10. Mansour N, Christopoulos C, Tremblay R (2009) Experimental performance of full-scale eccentrically braced frames with replaceable shear links. International conference on the seismic behaviour of steel structures, STESSA 2009, Philadelphia, USA
11. McCormick J, Aburano H, Ikenaga M, Nakashima M (2008) Permissible residual deformation levels for building structures considering both safety and human elements, Proceedings of 14th world conference earthquake engineering, Beijing, China, Paper No. 05-06-0071
12. Pettinga D, Christopoulos C, Pampanin S, Priestley N (2007) Effectiveness of simple approaches in mitigating residual deformations in buildings. Earthquake Eng Struct Dyn 36(12):1763–1783
13. Ricles JM, Sause R, Garlock M, and Zhao C (2001) Post-tensioned seismic-resistant connections for steel frames. J Struct Eng ASCE 127(2):113–121
14. Ruiz-Garcia J, Miranda E (2006) Evaluation of residual drift demands in regular multi-story frames for performance-based seismic assessment. Earthq Eng Struct Dyn 35:1609–1629
15. Tremblay R, Lacerte M, Christopoulos C (2008) Seismic response of multistory buildings with self-centering energy dissipative steel braces. ASCE J Struct Eng 134(1):108–120

Chapter 26
Performance-Based Seismic Design and Experimental Evaluation of Steel MRFs with Compressed Elastomer Dampers

James M. Ricles, Richard Sause, Theodore L. Karavasilis, and Cheng Chen

26.1 Introduction

Passive damping systems can significantly enhance the seismic performance of buildings by reducing inelastic deformation demands on the primary lateral load resisting system and by reducing drift, velocity, and acceleration demands on non-structural components.

Among the different kinds of passive damping systems, viscoelastic dampers have been extensively studied. Karavasilis et al. [6] evaluated the hysteretic behavior of an innovative compressed elastomer damper [11] and based on the results of nonlinear dynamic time history analyses found that steel moment resisting frames (MRFs) with compressed elastomer dampers can be designed to perform better than conventional special moment resisting frames (SMRFs), even when the MRF with dampers is significantly lighter in weight than the conventional SMRF.

To demonstrate and verify the full potential of new types of dampers, damper designs and performance-based design procedures for structural systems with dampers should be experimentally validated. Full-scale testing is a reliable but, at the same time, a challenging experimental technique. In particular, full-scale testing of structural systems designed to experience inelastic deformations may be cost and time prohibitive since the damaged components of the structural system need to be repaired or rebuilt after each test.

Real-time hybrid simulation combines physical testing and numerical simulation such that the dynamic performance of the entire structural system can be considered during the simulation. When real-time hybrid simulation is utilized to evaluate the performance of structures with rate-dependent damping devices, the damping devices may be tested as experimental substructures and the remaining part of the structural system is modeled analytically. The added benefit of this experimental technique is that it enables a large number of ground motions to be applied to the

---

J.M. Ricles (✉)
Department of Civil and Environmental Engineering, ATLSS Engineering Research Center, Lehigh University, Bethlehem, PA 18015, USA
e-mail: jmr5@lehigh.edu

M.N. Fardis (ed.), *Advances in Performance-Based Earthquake Engineering*, Geotechnical, Geological, and Earthquake Engineering 13, DOI 10.1007/978-90-481-8746-1_26,

structure, resulting in various levels of damage, without the need to repair the test specimens since the damage will be contained within the analytical substructure.

This paper discusses an experimental program based on the use of real-time hybrid simulation to verify the performance-based seismic design of a two story, four-bay steel MRF equipped with compressed elastomer dampers. The experimental substructures are two individual compressed elastomer dampers with the remaining part of the building modeled as an analytical substructure. The explicit CR integration algorithm [1, 3], a robust nonlinear finite element code [5] and an adaptive compensation scheme to minimize actuator delay [2] are integrated together and used in the real-time hybrid simulation to compute the structural response based on feedback restoring forces from the experimental and analytical substructures.

26.2 Steel MRF with Compressed Elastomer Dampers

26.2.1 Prototype Building

Figure 26.1a shows the plan view of the 2-story, 6-bay by 6-bay prototype office building used for the study. The study focuses on one typical perimeter MRF, which is comprised of four bays. This MRF is designed either as a conventional steel SMRF as defined in the 2006 International Building Code [4], referred to herein as IBC 2006, or as a steel MRF equipped with compressed elastomer dampers. In

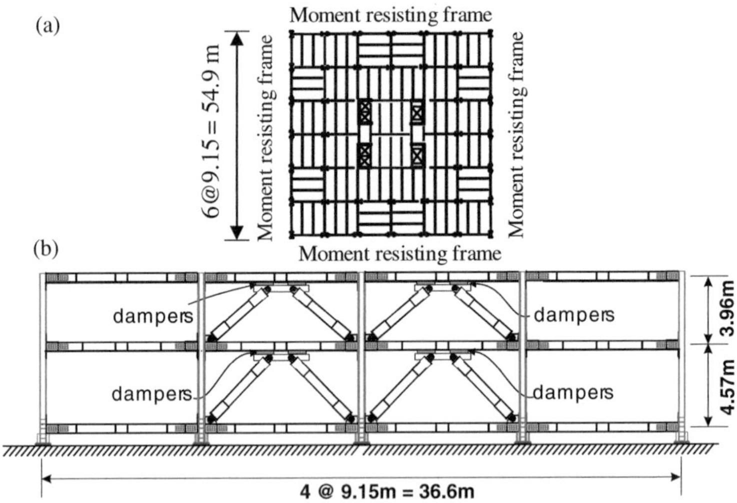

Fig. 26.1 Prototype building structure: (**a**) plan view, and (**b**) perimeter MRF with dampers and diagonal bracing

the latter case, dampers and diagonal bracing are added to the two interior bays, as shown in Fig. 26.1b.

The members of the MRF are assumed to be A992 steel with a nominal yield stress of 345 MPa. The gravity loads considered in the design are those described in IBC 2006. A smooth design response spectrum with parameters $S_{DS}=1.0$, $S_{D1}=0.6$, $T_0=0.12$ s and $T_s=0.6$ s, defined by IBC 2006, represents the Design Basis Earthquake (DBE).

26.2.2 Design of perimeter MRF as a Conventional SMRF

The perimeter MRF in Fig. 26.1b is initially designed as a conventional SMRF using the equivalent lateral force procedure in the IBC 2006. This SMRF design without dampers, referred to herein as UD100V, satisfies the member strength criteria of the IBC 2006 with a response modification factor R equal to 8 and also the 2% story drift limit of IBC 2006 with a deflection amplification factor C_d equal to 5.5.

To study whether MRFs with compressed elastomer dampers can be designed to have less strength than a conventional SMRF (without dampers) but achieve similar or better levels of seismic performance, a perimeter MRF was designed without dampers using a design base shear equal to $0.50V$, where V is the design base shear for UD100V. The resulting MRF design, referred to herein as UD50V, does not satisfy the drift criteria of the IBC 2006. This MRF design is significantly lighter than UD100V.

Table 26.1 summarizes the properties of the two MRF designs, where the properties for UD50V are without the dampers. The table lists the column section, beam sections, steel weight, fundamental period of vibration, T_1, and the predicted maximum story drift, θ_{max}, under the DBE earthquake. The maximum story drift, θ_{max}, is determined on the basis of the equal displacement principle.

26.2.3 Design of Dampers for MRF

The damper designs are based on the new generation of compressed elastomer dampers presented in Karavasilis et al. [6]. The thickness and the area of these dampers are 4 times larger than the thickness and the area of the dampers used in the real-time hybrid simulations presented herein. The mechanical properties of

Table 26.1 Properties of MRF designs

MRF	Column section	Beam sections	Steel weight (kN)	T_1 (s)	θ_{max} (%)
UD100V	W14×211	1st story: W24×84 2nd story: W21×50	200	1.08	2.40
UD50V	W14×120	1st story: W24×55 2nd story: W18×40	124	1.48	3.23

Table 26.2 Design of UD50V MRF with compressed elastomer dampers

Brace steel weight (kN)	T_1 (s)	θ_{max} (%)	No. dampers per Story 1st	2nd
17.2	1.04	1.60	8	5

these compressed elastomer dampers, namely the equivalent stiffness and loss factor, were derived from the experimental data presented in Karavasilis et al. [6]) and used to design the compressed elastomer dampers for the UD50V MRF with the aid of the simplified design procedure (SDP) developed by Lee et al. [7] . The SDP idealizes the damper hysteresis loops as linear viscoelastic ellipses and the damper design variables are the equivalent damper stiffness and the loss factor. More details on the SDP and the design of the compressed elastomer dampers for the UD50V MRF can be found in Karavasilis et al. [6].

Table 26.2 provides information for the UD50V MRF with dampers. The story drift and damper deformation demand estimates are for the DBE. It is observed that the UD50V MRF with 8 and 5 compressed elastomer dampers in the first and second stories, respectively, exhibits a significantly better anticipated performance ($\theta_{max} =$ 1.60%) than that of the conventional UD100V SMRF ($\theta_{max} = 2.40\%$). Moreover, the UD50V MRF with dampers has a steel weight equal to 124 kN (UD50V) + 17.2 kN (braces) = 141.2 kN, while the steel weight of the conventional UD100V SMRF is 200 kN.

26.3 Real-Time Hybrid Simulation

26.3.1 Real-Time Integrated Control System Architecture, Analytical Substructure Modeling and Experimental Substructure Test Setup

The performance of the MRF with compressed elastomer dampers is experimentally evaluated by conducting real-time hybrid simulations. The experimental substructures are two individual compressed elastomer dampers with the remaining part of the building modeled as an analytical substructure.

Since the dampers at a story level are placed in parallel in the prototype MRF (Fig. 26.1b), they are subjected to the same velocity and displacement. Therefore, each of the dampers setups in the laboratory can represent all of the dampers in one story. In the real-time hybrid simulation the measured restoring force from a compressed elastomer damper is multiplied by the number of dampers to obtain the total restoring force of all the dampers at a story level in the MRF.

As discussed previously, the thickness and the area of the elastomer of the dampers that are used in UD50V MRF are considered to be 4 times larger than the

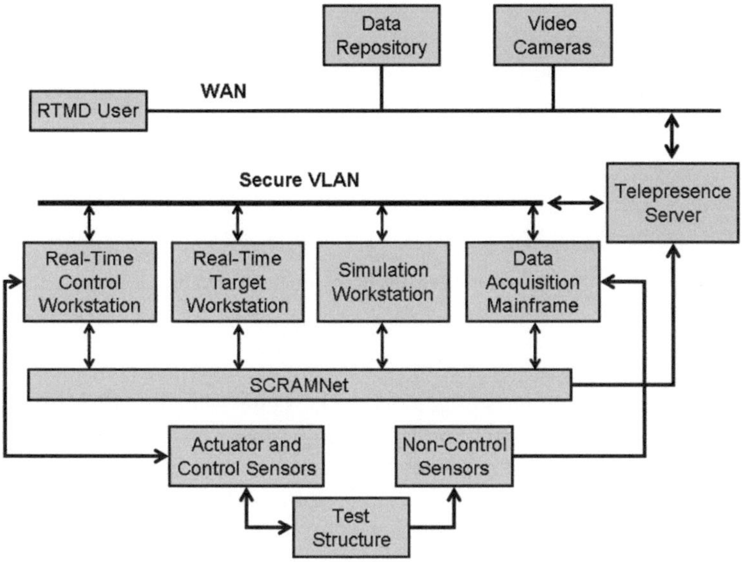

Fig. 26.2 RTMD integrated control system architecture

thickness and the area of the elastomer of the dampers in the experimental substructure. Consequently, in the real-time hybrid simulation the command displacement of the dampers was scaled down by a factor of 4 and the measured restoring force was amplified by a factor of 4.

A nonlinear finite element code [5] has been implemented into the real-time integrated control system at the NEES Real-Time Multi-Directional (RTMD) Facility at Lehigh University [8]. The architecture for the RTMD system is shown in Fig. 26.2. A digital controller (real-time control workstation) with a 1,024 Hz clock speed (sampling time $\delta t = 1/1{,}024$ s) controls the motion of the servo-hydraulic actuators and is integrated with the real-time target workstation, simulation workstation, and data acquisition workstation using a shared common RAM network (SCRAMNet). SCRAMNet has a communication rate of about 180 ns which enables the transfer of data among the integrated workstations in real-time with minimal communication delay. The nonlinear finite element code has been developed in a manner that enables the analytical substructure modeling, servo-hydraulic control law, and actuator compensation scheme to be integrated into a single SIMULINK model on the simulation workstation and then downloaded onto the target workstation using Mathworks xPC Target Software [9]).

The model of the MRF has a total of 122 degrees of freedom and 71 elements. Inelastic behavior is modeled by means of a bilinear hysteretic lumped plasticity beam-column element with 3% strain hardening and appropriate axial-moment yield surfaces. Diaphragm action is assumed at every floor level due to the presence of the

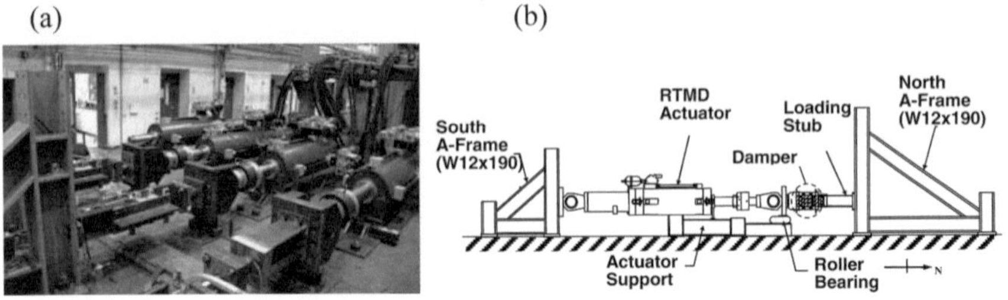

Fig. 26.3 Compressed elastomer dampers: (**a**) photograph, and (**b**) details of test setup for each damper

floor slab. A lean-on column is used to model P-Δ effects on the MRFs from gravity loads carried by gravity columns of the building.

Figure 26.3 shows the experimental setup for the real-time hybrid simulation, which consists of the experimental substructures (two large-scale compressed elastomer dampers), two servo-hydraulic actuators with supports, roller bearings, and reaction frames. The two actuators have a load capacity of 2,300 and 1,700 kN with a maximum velocity of 840 and 1,140 mm/s, respectively, when three servo valves are mounted on each actuator. The servo-controller for the actuator used in the real-time hybrid simulations consists of a digital PID controller with a proportional gain of 20, integral time constant of 5.0 resulting in an integral gain of 4.0, differential gain of zero, and a roll-off frequency of 39.8 Hz.

26.3.2 Real-Time Integration of the Equation of Motion

For the MRF structure with dampers shown in Fig. 26.1b, the temporal discretized equations of motion at the i+1th time step can be expressed as

$$\boldsymbol{M} \cdot \ddot{x}_{i+1} + \boldsymbol{C} \cdot \dot{x}_{i+1} + \boldsymbol{r}^{a}_{i+1} + \boldsymbol{r}^{e}_{i+1} = \boldsymbol{F}_{i+1} \tag{26.1}$$

where $\ddot{x}_{i+1}$ and $\dot{x}_{i+1}$ are the acceleration and velocity vectors of the structure, respectively; $\boldsymbol{r}^{a}_{i+1}$ and $\boldsymbol{r}^{e}_{i+1}$ are the restoring force vectors of the analytical and experimental substructures, respectively; $\boldsymbol{M}$ and $\boldsymbol{C}$ are the mass and damping matrices of the structure, respectively; and $\boldsymbol{F}_{i+1}$ is the excitation force.

The CR explicit integration algorithm [1, 3] is used to solve Eq. (26.1) for the structural displacement vector $\boldsymbol{x}_{i+1}$. According to the CR algorithm, the variations of the displacement and velocity vectors of the structure over the integration time step Δt are defined as

$$\dot{x}_{i+1} = \dot{x}_{i} + \Delta t \cdot \alpha_1 \cdot \ddot{x}_{i} \tag{26.2a}$$

$$\boldsymbol{x}_{i+1} = \boldsymbol{x}_{i} + \Delta t \cdot \dot{x}_{i} + \Delta t^2 \cdot \alpha_2 \cdot \ddot{x}_{i} \tag{26.2b}$$

where $\boldsymbol{x}_i, \dot{\boldsymbol{x}}_i$ and $\ddot{\boldsymbol{x}}_i$ are the displacement, velocity and acceleration vectors of the structure at the ith time step, respectively; and α_1 and α_2 are integration parameters defined as

$$\alpha_1 = \alpha_2 = 4 \cdot \left(4 \cdot \boldsymbol{M} + 2 \cdot \Delta t \cdot \boldsymbol{C} + \Delta t^2 \cdot \boldsymbol{K}\right)^{-1} \cdot \boldsymbol{M} \tag{26.3}$$

In Eq. (26.3) $\boldsymbol{K}$ is the initial stiffness matrix of the structure. It should be emphasized that this matrix includes the stiffness and damping contribution of the experimental substructures, i.e., the equivalent stiffness and damping of the two compressed elastomer dampers.

In real-time hybrid simulation, Eqs. (26.2a) and (26.2b) are used to obtain the velocity $\dot{x}_{i+1}$ and displacement $\boldsymbol{x}_{i+1}$ vectors at the i+1th time step. The displacement vector $\boldsymbol{x}_{i+1}$ is decomposed into the analytical displacement vector $\boldsymbol{x}^{\mathrm{a}}_{i+1}$ and the experimental (or command) displacement vector $\boldsymbol{x}^{\mathrm{e}}_{i+1}$, which are imposed onto the analytical and experimental substructures, respectively, to obtain the restoring force vectors $\boldsymbol{r}^{\mathrm{a}}_{i+1}$ and $\boldsymbol{r}^{\mathrm{e}}_{i+1}$. Strictly speaking, $\boldsymbol{x}^{\mathrm{e}}_{i+1}$ contains deformations, i.e., displacement differences of the nodes defining the connectivity of each of the experimental substructures. The analytical restoring force vector $\boldsymbol{r}^{\mathrm{a}}_{i+1}$ is obtained with a standard nonlinear beam-column element state-determination procedure, while the experimental restoring force vector $\boldsymbol{r}^{\mathrm{e}}_{i+1}$ is obtained from the feedback forces measured using load cells that are placed in each compressed elastomer damper test setup. The equilibrium Eq. (26.1) is then employed to calculate the acceleration response vector $\ddot{x}_{i+1}$ at the i+1th time step, and the velocity $\dot{\boldsymbol{x}}_{i+2}$ and displacement $\boldsymbol{x}_{i+2}$ vectors for the next i+2th time step are again readily available from Eqs. (26.2a) and (26.2b), this process is repeated to obtain the response over the whole duration of the earthquake ground motion.

The adaptive inverse compensation scheme by Chen and Ricles [2] is used in this research to minimize the effect of actuator delay due to inherent servo hydraulic actuator dynamics.

26.4 Real-Time Hybrid Simulation Results

An ensemble of 5 earthquake ground motions recorded on stiff soil sites (without near-fault effects) was used in the real-time hybrid simulations to evaluate the performance of the MRF with compressed elastomer dampers. The ground motions were scaled to the DBE level using the scaling procedure of Somerville [10]. Table 26.3 provides the scale factors (SF) and information for the 5 ground motions.

Time history results from the real-time hybrid simulations are presented for the HSP090 record scaled to the DBE level. Figure 26.4 shows the floor displacement time history of the MRF with dampers, UD50V MRF. Also presented in Fig. 26.4 is the floor displacement time history of the conventional UD100V SMRF from a numerical analysis. The real-time hybrid simulation results show that the

Table 26.3 Ground motions used in real-time hybrid simulations

Earthquake	Station/component	SF
Loma Prieta 1989	Hollister/HSP090	1.99
Manjil 1990	Abbar/Abbar-T	0.96
Northridge 1994	N Hollywood/CWC270	1.70
Chi Chi 1999	TCU049/TCU049-E	1.92
Chi Chi 1999	TCU105/TCU105-E	2.45

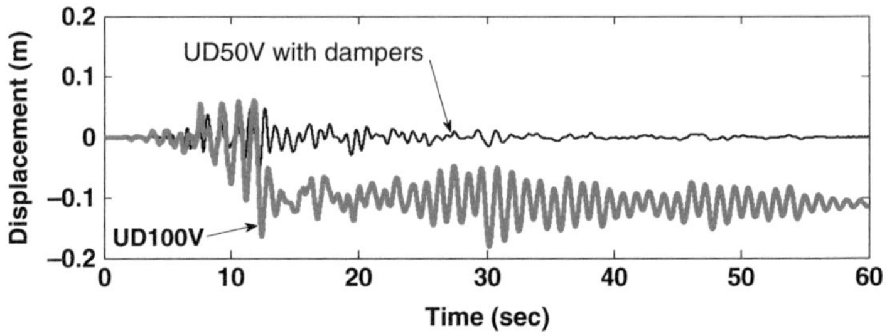

Fig 26.4 Displacement time history from real-time hybrid simulation

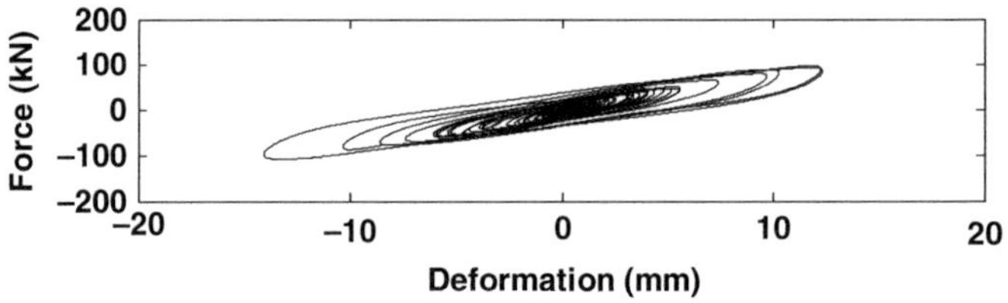

Fig. 26.5 Damper hysteresis from real-time hybrid simulation

lighter UD50V MRF with dampers experiences significantly lower transient and residual story drifts than the conventional UD100V SMRF. The hysteresis of the compressed elastomer dampers is presented in Fig. 26.5. The dampers are able to undergo numerous seismic induced deformation cycles without degradation of their behavior.

Table 26.4 presents median experimental response values for the maximum story drift, θ_{max}; maximum plastic hinge rotation $\theta_{pl.max}$; maximum 2nd floor absolute velocity v_{max}; and maximum 2nd floor absolute acceleration a_{max} of UD50V MRF from the real-time hybrid simulations. Also presented in Table 26.4 are the median values of the same response quantities of the conventional UD100V SMRF from

Table 26.4 Median values of response parameters under the DBE

Design	θ_{max} (%)	$\theta_{pl.max}$ (%)	v_{max} (m/s)	a_{max} (m/s^2)
UD50V	1.40	0.20	0.77	5.18
UD100V	2.60	1.09	1.11	5.66

numerical analysis. Table 26.4 shows that the median θ_{max} value of 1.40% for the MRF with dampers is slightly less than the θ_{max} design demand of 1.60% used in the SDP (Table 26.2), while the θ_{max} value of 2.60 for the UD100V SMRF is slightly larger than the θ_{max} design demand of 2.40% according to the equal displacement principle (Table 26.1). It is also observed that the MRF with dampers has a significantly better performance than the UD100V SMRF in terms of the maximum story drift, plastic hinge rotations, absolute floor velocities and accelerations.

26.5 Summary and Conclusions

An experimental program based on the use of real-time hybrid simulation to verify the performance-based seismic design of a steel MRF equipped with compressed elastomer dampers was presented. The experimental substructures consisted of two individual large-scale compressed elastomer dampers with the remaining part of the building modelled as an analytical substructure. Statistical experimental response results show that a steel MRF with compressed elastomer dampers can be designed to perform better than a conventional steel SMRF, even when the MRF with dampers is significantly lighter in weight than the conventional SMRF.

Acknowledgements This paper is based upon work supported by grants from the Pennsylvania Department of Community and Economic Development through the Pennsylvania Infrastructure Technology Alliance, and by the National Science Foundation (NSF) under Grant No. CMS-04002490 within the George E. Brown, Jr. Network for Earthquake Engineering Simulation Consortium Operation. The compressed elastomer dampers were manufactured and donated to the research project by Corry Rubber. Any opinions, findings, and conclusions expressed in this paper are those of the authors and do not necessarily reflect the views of the sponsors.

References

1. Chen C, Ricles JM (2008) Development of direct integration algorithms for structural dynamics using discrete control theory. J Eng Mech 134(8):676–683
2. Chen C, Ricles JM (2010) Tracking error-based servo-hydraulic actuator adaptive compensation for real-time hybrid simulation. J Struct Eng 136(4)
3. Chen C, Ricles JM, Marullo TM, Mercan O (2009a) Real-time hybrid testing using the unconditionally stable explicit CR integration algorithm. Earthq Eng Struct Dyn 38(1):23–44
4. International Code Council (2006). International Building Code, Falls Church, VA
5. Karavasilis TL, Ricles JM, Marullo T, Chen C (2009b) HybridFEM: A program for nonlinear dynamic time history analysis and real-time hybrid simulation of structures. ATLSS Engineering Research Center Report
6. Karavasilis TL, Sause R, Ricles JM (2009a) Seismic design and evaluation of steel MRFs with compressed elastomer dampers. Earthq Eng Struct Dyn, submitted for publication
7. Lee KS, Fan CP, Sause R, Ricles J (2005) Simplified design procedure for frame buildings with viscoelastic or elastomeric structural dampers. Earthq Eng Struct Dyn 34:1271–1284
8. Lehigh RTMD Users Guide (2010) http://www.nees.lehigh.edu/index.php?page=rtmd-user-s-manual
9. MATLAB (2007) The Math Works, Inc., Natick, MA

10. Somerville P (1997) Development of ground motion time histories for phase 2 of the FEMA/SAC steel project, Report No. SAC/DB-97/04, Sacramento, CA
11. Sweeney SK, Michael R (2006) Collaborative product realization of an innovative structural damper and application. Proceedings of IMECE2006, ASME international engineering congress and exposition, Chicago, IL, USA

Chapter 27
Performance-Based Design of Self-Centering Steel Frame Systems

Richard Sause, James M. Ricles, Ying-Cheng Lin, Choung-Yeol Seo, and David Roke

27.1 Introduction

Conventional moment resisting frames (MRFs) soften and dissipate energy under the design basis earthquake by developing yielding and associated damage in critical regions of the main structural members. This damage can result in significant damage and residual drift after the earthquake. To avoid this damage and residual drift, post-tensioned (PT) beam-column connections for self-centering moment-resisting frames (SC-MRFs) were developed by Ricles et al. [5] and others. The behavior of these connections is characterized by gap opening and closing at the beam-column interface. Energy dissipation occurs in special devices designed for the beam-column connection regions. Together, the PT strands and the energy dissipation devices provide the flexural resistance of the connections.

This paper presents an experimental study on an SC-MRF which uses beam web friction devices (WFDs) as energy dissipation devices. The WFDs are positioned on the beam webs to avoid interference with the floor slab and are attached to the columns at the beam-column interface. Using a performance based design (PBD) approach, the SC-MRF is designed to remain damage-free under the Design Basis Earthquake (DBE) to enable immediate occupancy, while also achieving the collapse prevention performance level under the Maximum Considered Earthquake (MCE). Experimental results of earthquake simulations on the SC-MRF are used to assess the design procedure. Related research on self-centering concentrically-braced frames (SC-CBFs) is briefly outlined.

R. Sause (✉)
Department of Civil and Environmental Engineering, ATLSS Engineering Research Center, Lehigh University, Bethlehem, PA 18015, USA
e-mail: rsause@lehigh.edu

M.N. Fardis (ed.), *Advances in Performance-Based Earthquake Engineering*, Geotechnical, Geological, and Earthquake Engineering 13, DOI 10.1007/978-90-481-8746-1_27,

27.2 SC-MRF Overview

Figure 27.1a shows an SC-MRF with PT strands and WFDs. The PT strands run across multiple bays. The WFD, shown in Fig. 27.1b, includes two "friction" channels welded to the column flange. Brass plates are sandwiched between the friction channels and beam to provide reliable friction conditions. Normal force on the friction surface between the channels and beam is provided by friction bolts shown in Fig. 27.1b. The friction channels are welded to the column flange after the friction bolts are tightened. The friction channel shape was selected to reduce the effect of weld shrinkage on the friction surface normal force. Slotted holes are used in the beam web to accommodate the travel of the friction bolts during the gap opening and closing of the connection (discussed later). The shim plates shown in Fig. 27.1b are welded to the column flange to provide good contact surfaces for the beam flanges. Reinforcing plates are welded on the outside faces of the beam flanges to avoid excessive yielding in the beam flanges.

The conceptual moment-relative rotation (M–θ_r) behavior of the connection in Fig. 27.1b is shown in Fig. 27.1c. From event 0 to 2, the connection has stiffness similar to that of a conventional welded moment connection. After the connection moment M reaches the imminent gap opening moment at event 2 (M_{IGO}), the beam tension flange loses contact with the shim plate at the column face and gap opening occurs. M_{IGO} is the sum of the decompression moment M_d due to the initial PT force (event 1) and the moment M_{Ff} due to friction in the WFD. After M_{IGO} is exceeded, M increases with increases in the PT strand force due to gap opening (event 2 to 3). Excessive gap opening will eventually yield the PT strands at event 4. During unloading between events 3 and 5, θ_r remains constant but M decreases by $2M_{Ff}$ due to the reversal in direction of the WFD friction force. Continued unloading between events 5 and 6 reduces θ_r to zero as the beam tension flange comes in contact with the shim plate at the column face. Further unloading (between events

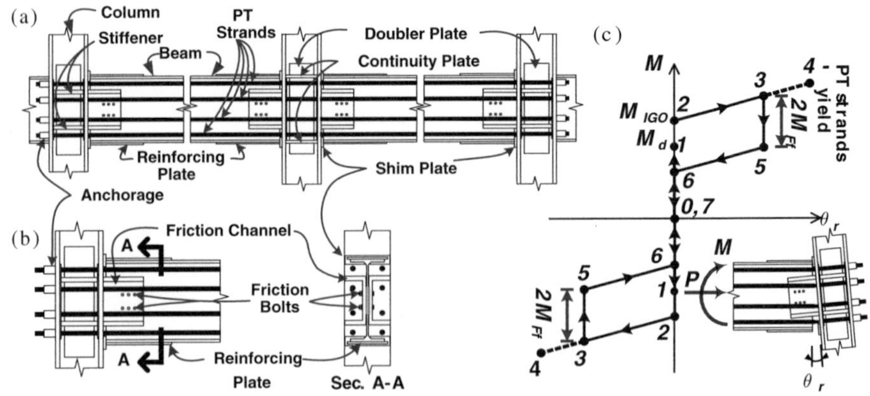

Fig. 27.1 (**a**) 2-bay SC-MRF with PT strands and WFDs, (**b**) connection details, and (**c**) conceptual M–θ_r behavior of connection

6 and 7) decreases M to zero as the beam tension flange fully compresses against the shim plate. Similar behavior occurs under load reversal.

After M_{IGO}, M is controlled by the axial force in the beam, P and the WFD friction force resultant, F_f, as follows:

$$M = Pd_2 + F_f r \tag{27.1}$$

where d_2 is the distance from the beam cross section centroid to the center of rotation (COR) of the connection, and r is the distance from F_f to the COR. The COR is at the point of contact of the beam compression flange with the column. P includes the PT force, T, and the effects of interaction between the SC-MRF and floor system (e.g., floor diaphragm forces), F_{fd}, as follows [4]:

$$P = F_{fd} + T \tag{27.2}$$

T depends on the initial PT force, T_o, and the strand elongation due to θ_r.

27.3 Performance-Based Design of SC-MRFs

The performance-based design (PBD) procedure proposed for SC-MRFs [4] considers two levels of seismic hazard, namely the DBE and MCE. The DBE has two-thirds the intensity of the MCE [3] and an approximate 10% probability of being exceeded in 50 years. The MCE has a 2% probability of being exceeded in 50 years. Under the DBE, an SC-MRF is designed to achieve immediate occupancy (IO) performance [3], with limited structural and nonstructural damage. Under the MCE, an SC-MRF is designed to achieve collapse prevention (CP) performance [3].

The design objectives and the limit states of an SC-MRF with WFDs are shown in a conceptual base shear-roof drift (V–θ_{roof}) response in Fig. 27.2. Before the IO performance limit, connection decompression and yielding at the column bases of the SC-MRF are permitted to occur. Panel zone yielding, beam web yielding, and

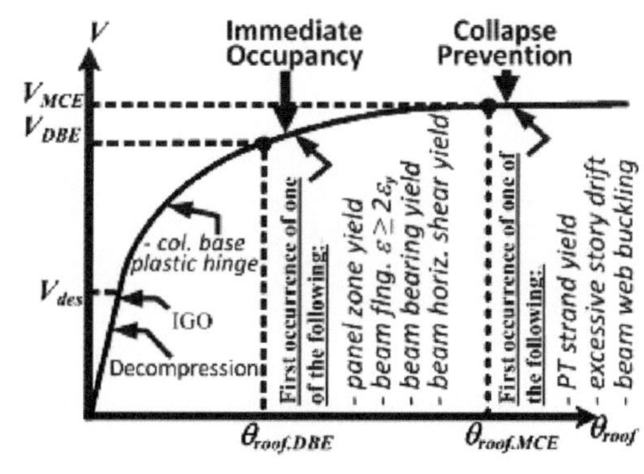

Fig. 27.2 Performance-based design objectives

a beam flange strain greater than twice the yield strain (i.e., $2\varepsilon_y$) are permitted to occur between the IO and CP performance limits. Before the CP limit, PT strand yielding, beam web buckling, and excessive story drift are not permitted.

The PBD procedure permits the use of an equivalent lateral force analysis of the SC-MRF, using an analysis model with rigid beam-column connections that is subjected to the design forces defined in ASCE7-05 [1] with a response modification factor R equal to 8. The design moment M_{des} from this analysis is used to establish an initial value of M_{IGO} approximately equal to $0.95M_{\mathrm{des}}$. The effective energy dissipation ratio of the SC-MRF connections, $\beta_{\mathrm{E}} = M_{\mathrm{Ff}}/M_{\mathrm{IGO}}$, is used to establish the proportion of M_{IGO} provided by M_{Ff}. To provide the SC-MRF with sufficient energy dissipation to achieve a satisfactory seismic response, Seo and Sause [9] recommend that $\beta_{\mathrm{E}} \geq 0.25$, however to enable connection re-centering, $\beta_{\mathrm{E}} \leq 0.40$. The PBD procedure includes numerous steps to control the limit states shown in Fig. 27.2. In this procedure, estimates of the θ_{r} demand under the DBE and MCE are critical for determining whether these limit states are reached. The details of the PBD procedure are given in Garlock et al. [4].

27.4 SC-MRF Experimental Program

The PBD procedure described above was used to design SC-MRFs for the 7 × 7-bay 4-story prototype building shown in plan in Fig. 27.3a. The building is located on a stiff soil site in the Los Angeles area. The perimeter frames include SC-MRFs. Each perimeter frame has two 2-bay SC-MRFs with WFDs. Figure 27.3a shows the floor diaphragm is attached to one bay of each SC-MRF in each perimeter frame to avoid restraining gap opening of the SC-MRF connections.

The test frame is a 0.6-scale model of one SC-MRF from the prototype building, as shown in Fig. 27.3b. The test frame has A992 steel members which were scaled

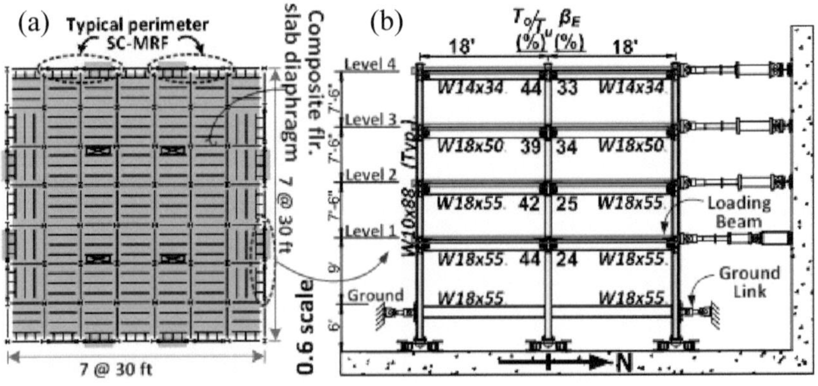

Fig. 27.3 Schematic of (**a**) plan of prototype, and (**b**) elevation of 0.6-scale 4-story 2-bay SC-MRF test frame (note: 1′ = 1 ft = 305 mm; 1″ = 1 in. = 25.4 mm)

Table 27.1 Design demands (in radians)

$\theta_{\mathrm{roof,DBE}}$	$\theta_{\mathrm{s,DBE}}$	$\theta_{\mathrm{r,DBE}}$	$\theta_{\mathrm{roof,MCE}}$	$\theta_{\mathrm{s,MCE}}$	$\theta_{\mathrm{r,MCE}}$
0.026	0.039	0.031	0.039	0.059	0.047

down from the prototype SC-MRF. Table 27.1 gives the design demands for roof drift (θ_{roof}), story drift (θ_{s}), and connection relative rotation (θ_{r}) at the DBE and MCE levels. The initial PT forces T_{o}, shown in Fig. 27.3b, are less than 45% of strand ultimate strength (T_{u}). Design values of β_{E} are shown in Fig. 27.3b.

During the experiments, lateral force is applied at each floor level by hydraulic actuators through a simulated floor diaphragm attached near the middle of the North-bay beam shown in Fig. 27.3b. The hybrid simulation method was used for earthquake simulations. In these simulations, the test frame was the experimental substructure, while the gravity load bearing system, gravity loads, and the seismic mass tributary to the test frame were included in the analytical substructure. The hybrid simulation used 2% damping in the 1st mode and 5% damping in the 3rd mode. The explicit unconditionally stable CR integration algorithm by Chen and Ricles [2] was used to solve the equations of motion.

Table 27.2 shows the test matrix of hybrid simulations. Ground motions at the frequently occurring earthquake (FOE), DBE, and MCE level were used. Three different DBE level ground motion records were used, denoted DBE-1, DBE-2 and DBE-3. Each of these records was one of a pair of recorded orthogonal components of horizontal ground acceleration. To develop a set of DBE-level ground motions, fifteen such pairs were selected from earthquake ground motion data bases, and scaled so that the geometric mean of the spectral accelerations S_{a} of the record pair was equal to the S_{a} value at a period of 1.5 s for the uniform hazard spectrum (UHS) for the site of the prototype building in the Los Angeles area. To use these records for the hybrid simulations, the time step for the records were scaled by $\sqrt{0.6} = 0.77$, where 0.6 is the scale factor of the test frame.

The selected records, DBE-1, DBE-2 and DBE-3, cause a maximum story drift that is approximately one-standard deviation lower than, similar to, and one-standard deviation higher than, respectively, the mean maximum story drift from nonlinear dynamic analyses of the test frame for the set of DBE-level ground motion

Table 27.2 Test matrix of hybrid simulations

Tests	Description	Record	Scale factor	Method
FOE-1	1979 Imperial Valley	H-CXO225	−0.70	Hybrid simulation
FOE-2			−1.41	
DBE-1	1979 Imperial Valley	H-ECC002	0.94	
DBE-2	1989 Loma Prieta	SJTE315	−2.23	
DBE-3	1994 Northridge	LOS000	1.18	
MCEs	1994 Northridge	Varied w/ tests		
Aftershock	1989 Loma Prieta	SJTE315	−2.23	

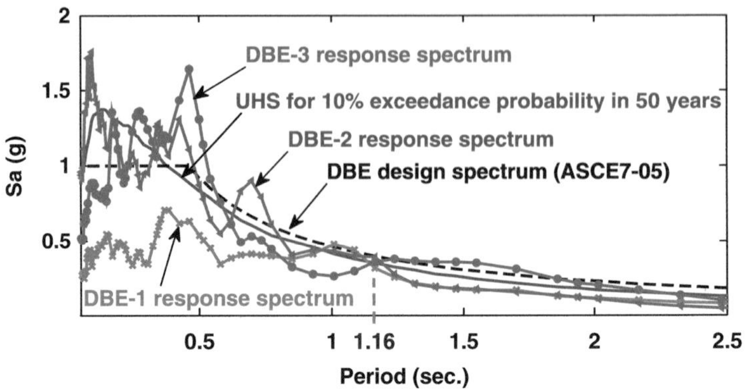

Fig. 27.4 Pseudo-acceleration response spectra

records. Figure 27.4 compares the DBE design spectrum based on ASCE7-05 [1], the uniform hazard spectrum, UHS, which has a 10% probability of exceedance in 50 years at the prototype building site, and the individual response spectra for the DBE-1, DBE-2 and DBE-3 records. In this figure, all periods are scaled by the time scale factor of $\sqrt{0.6} = 0.77$. It should be noted that S_a for the DBE design spectrum at the period of 1.16 s (which is approximately the fundamental period of the test frame) is slightly higher than the corresponding S_a of the UHS (i.e., 0.4g vs. 0.35g), because the DBE design spectrum is defined to have two-thirds the intensity of the MCE design spectrum [3], and is not directly equal to the UHS which has 10% probability of exceedance in 50 years. The S_a of the individual response spectra for DBE-1, DBE-2, and DBE-3 at 1.16 s vary from 0.32g to 0.4g because each record was not scaled individually to the target S_a value, rather, each record pair was scaled by a single scale factor so that the geometric mean of S_a for the record pair equaled the target S_a value.

27.5 SC-MRF Experimental Results

Floor displacement time histories from the DBE-2 hybrid simulation are shown in Fig. 27.5a. The residual story drifts of the test frame are given in Table 27.3 and were obtained by dividing the difference in residual displacements of adjacent floors by the story height. Table 3 shows the maximum residual story drift after DBE-2 is 0.00061 rad, which demonstrates the self-centering capability of the SC-MRF. Table 27.4 shows results from the DBE-2 simulation, and includes the maximum story drift $\theta_{s\,max}$, maximum relative rotation $\theta_{r\,max}$, maximum PT force T_{max} and loss of PT force during the simulated earthquake, ΔT, normalized by the strand tensile strength T_u. The maximum $\theta_{s\,max}$ is 2.9% radians, which is less than the expected design demand of 3.9% radians (Table 27.1) for the DBE. The maximum $\theta_{r\,max}$ is 2.7% radians, which is less than the expected design demand of 3.1% radians for the DBE. The maximum PT force in the DBE-2 simulation was $0.6T_u$. No

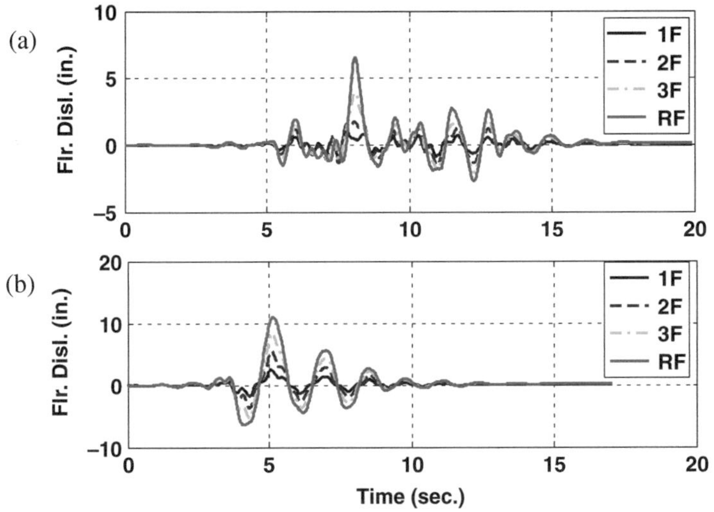

Fig. 27.5 Floor displacements time histories from (**a**) DBE-2, and (**b**) DBE-3

yielding occurred in the PT strands. The PT force decreased slightly due to seating of the PT strand anchorage, and ΔT is less than 1% of T_u.

Figure 27.5b shows floor displacement time histories from DBE-3. The maximum residual story drift after DBE-3 is 0.00074 rad, which further demonstrates the self-centering capability of the SC-MRF. The 1st story residual drift from the DBE-3 simulation is larger than that from the DBE-2 simulation due to yielding in the

Table 27.3 Residual story drift after simulations (in rad)

	Story			
Test	1st	2nd	3rd	4th
DBE-2	0.00014	0.00035	0.00061	0.00045
DBE-3	0.00074	0.00063	0.00023	0.00008

Table 27.4 Test results from DBE-2 and DBE-3

	Level	$\theta_{s\,max}$ (rad)	$\theta_{r\,max}$ (rad)	T_{max}/T_u (%)	$\Delta T_r/T_u$ (%)
DBE-2	RF	0.029	0.027	60	−0.5
	3F	0.025	0.024	54	−0.4
	2F	0.014	0.014	51	−0.0
	1F	0.008	0.005	46	−0.0
DBE-3	RF	0.039	0.038	65	−0.7
	3F	0.035	0.034	60	−0.7
	2F	0.035	0.031	61	−1.3
	1F	0.021	0.025	59	−1.0

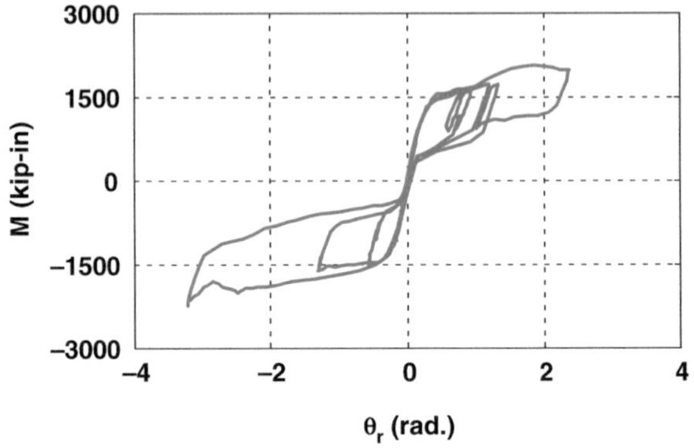

Fig. 27.6 Moment vs. θ_r of 3FSN connection from DBE-3

columns at the ground level. Table 27.4 shows the maximum $\theta_{s\,max}$ is 3.9% radians, which equals the design demand for the DBE (Table 27.1). The maximum $\theta_{r\,max}$ is 3.8% rad, which is slightly larger than the design demand of 3.1% rad for the DBE. The maximum PT force in the DBE-3 simulation was $0.65T_u$. No yielding occurred in the PT strands. ΔT is less than 1.5% of T_u. The self-centering behavior of the SC-MRF beam-column connections is illustrated by the typical M–θ_r response from the DBE-3 simulation for the North-end connection of the 3rd floor South-bay beam (denoted 3FSN), shown in Fig. 27.6. After gap opening, the stiffness differs in the positive and negative moment directions due to the floor diaphragm forces (i.e., the lateral force from the actuators) acting on the North-bay beam. When the test frame is loaded in the North direction by the floor diaphragm, a tension axial force is imposed on the South-bay beam. When test frame is loaded to the South, a compression axial force is imposed on the South-bay beam. Therefore, the axial force P in the South bay is different for the North and South loading directions. β_E estimated from the DBE-3 simulation results is around 30%. Overall, the SC-MRF achieved the DBE performance objectives.

27.6 SC-CBF Research

Steel concentrically-braced frame (CBF) systems are stiff and economical earthquake-resistant steel frame systems with limited ductility capacity and a tendency to accumulate residual drift during an earthquake. To increase the ductility and reduce the residual drift of CBFs, the self-centering concentrically-braced frame (SC-CBF) system, shown schematically in Fig. 27.7a, is being developed. The system has beams, columns, and braces in a conventional arrangement, with column base details that permit the columns to uplift at the foundation (Fig. 27.7c). Gravity loads (g) and post-tensioning (PT) forces resist column uplift and provide a restoring force after uplift. The beams, columns, and braces are intended to remain elastic under the DBE, and the column uplift controls the seismic force levels in the frame.

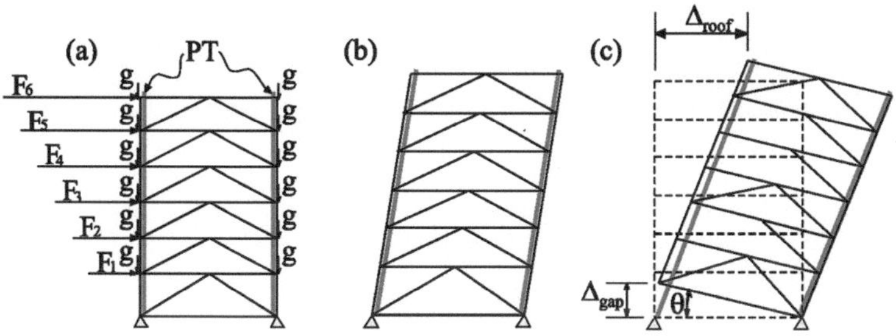

Fig. 27.7 SC-CBF: (**a**) schematic, (**b**) elastic response, and (**c**) rigid-body rotation

Idealized SC-CBF behavior under lateral load is shown in Fig. 27.7. Under low levels of lateral load, the structure deforms elastically (Fig. 27.7b). This deformation is similar to that of a conventional CBF. Under higher levels of lateral load, the base overturning moment becomes large enough for the "tension" column to decompress and uplift (Fig. 27.7c). After column decompression and uplift, the lateral displacement is dominated by rigid body rotation about the compression column base, although additional forces develop in the beams, columns, and braces of the frame. The PT steel elongates from the uplift and rotation of the frame, which increases the PT force and provides a positive stiffness to the lateral force-lateral drift behavior.

The complete behavior of a SC-CBF includes many limit states. The following limit states are considered in a performance-based design procedure developed by Roke et al. [6–8]: (1) decompression and uplift of the "tension" column, (2) yielding of the PT steel, (3) significant yielding of the beams, columns, or braces, and (4) failure of the beams, columns, or braces of the SC-CBF. The performance objectives are IO performance under the DBE and CP performance under the MCE. Decompression and uplift of the "tension" column does not cause structural damage and is permitted before the IO limit. Yielding of the PT steel and the beams, columns, and braces results in structural damage, and is not permitted before the IO limit but is permitted before the CP limit. Failure of the beams, columns, and braces is not permitted before the CP limit.

27.7 Summary and Conclusions

Results of hybrid earthquake simulations demonstrated the self-centering behavior of the SC-MRF with WFDs under the DBE. The SC-MRF performed well and satisfied the objectives of the PBD procedure. No significant damage occurred during the DBE-level tests. The beam-column connection M–θ_r behavior was as intended, and the WFDs provided reasonable levels of energy dissipation. Overall, the SC-MRF achieved the IO performance level under the DBE. The results of the MCE simulations are not shown in the paper due to space limitations, however the SC-MRF sustained modest damage during the MCE simulations and achieved at least the CP performance level. Related research on SC-CBFs was briefly outlined.

Acknowledgements The research reported in this paper was conducted at the NEES Real-Time Multi-Directional (RTMD) Earthquake Simulation Facility located at the ATLSS Center at Lehigh University. The work was supported by the National Science Foundation under Grant No. CMS-0420974, within the George E. Brown, Jr. Network for Earthquake Engineering Simulation Research (NEESR) program and Award No. CMS-0402490 NEES Consortium Operation.

References

1. ASCE 7 (2005) Minimum design loads for buildings and other structures. American Society of Civil Engineers, Reston, VA
2. Chen C, Ricles JM (2008) Development of direct integration algorithms for structural dynamics using discrete control theory. J Eng Mech 134(8):676–683
3. FEMA (2000) NEHRP recommended provisions for seismic regulations for new buildings and other structures. FEMA 450, Federal Emergency Management Agency, Washington, DC
4. Garlock M, Sause R, Ricles JM (2005) Behavior and design of post-tensioned steel frame systems. J Struct Eng ASCE 133(3):389–399
5. Ricles JM, Sause R, Garlock M, Zhao C (2001) Post-tensioned seismic-resistant connections for steel frames. J Struct Eng ASCE 127(2):113–121
6. Roke D, Sause R, Ricles J, Seo CY, Lee KS (2006) Self-centering seismic-resistant steel concentrically-braced frames. Proceedings of 8th US National Conference on Earthquake Engineering, San Francisco, Ca, EERI (CD-ROM)
7. Roke D, Sause R, Ricles JM, Gonner N (2008) Design concepts for damage-free seismic-resistant self-centering steel concentrically-braced frames. Proceedings of 14th WCEE (DVD, paper S16-01-006), Beijing
8. Roke D, Sause R, Ricles JM, Gonner N (2009) Design concepts for damage-free seismic-resistant self-centering steel concentrically-braced frames. Proceedings of ASCE Structures Congress, Austin, TX
9. Seo C-Y, Sause R (2005) Ductility demands on self-centering systems under earthquake loading. ACI Struct J 102(2):275–285

Chapter 28
Damage-Control Self-Centering Structures: From Laboratory Testing to On-site Applications

Stefano Pampanin

28.1 Introduction

Earthquake Engineering is facing an extraordinary challenging era, the ultimate target being set at increasingly higher levels by the demanding expectation of our modern society: to provide low-cost, more widely affordable, still architecturally appealing, high-seismic-performance structures capable of sustaining a design level earthquake with limited or negligible damage, minimum disruption of business (downtime) or, in more general terms, controllable socio-economical losses. These compelling requirements of cost-effectiveness and high-performance are leading to a major effort towards the development of damage-control design approaches and technologies.

In this contribution, an overview of recent developments and emerging solutions for high-performance, whilst still cost-effective, damage-control seismic resisting systems, based on the combination of traditional materials and available technology, will be given.

28.2 The Breakthrough of Jointed Ductile Articulated Systems

Current seismic design philosophies of ductile structural systems aim to ensure that the structural damage is intentionally concentrated within selected discrete regions of the structure, typical referred to as *plastic hinges*. According to the basic concepts of Performance Seismic Engineering [24], depending on the seismic intensity or ground motion return period, different (and often not negligible) levels of structural damage and, consequently, repairing costs shall be expected and accepted as unavoidable result of the inelastic behavior.

S. Pampanin (✉)
Department of Civil and Natural Resources Engineering, University of Canterbury, Christchurch, New Zealand
e-mail: stefano.pampanin@canterbury.ac.nz

M.N. Fardis (ed.), *Advances in Performance-Based Earthquake Engineering*, Geotechnical, Geological, and Earthquake Engineering 13, DOI 10.1007/978-90-481-8746-1_28,

A revolutionary alternative technological solution and its associated conceptual design philosophy have been introduced in the 1990s as outcomes of the U.S. PRESSS Program (PREcast Seismic Structural System) coordinated by the University of California, San Diego [21, 22], with the intent to create innovative solutions for precast concrete buildings, alternative to the traditional *wet* or *strong* connections based on the emulation of cast-in-situ concrete. High-performance, low-damage structural systems have been developed for both frames and walls through the use of dry jointed ductile connections, where prefabricated elements are joined together by means of unbonded post-tensioned bars/tendons (Fig. 28.1, top). During the seismic response, as per an "articulated" or segmented construction system, the precast elements are subjected to a *controlled rocking* mechanism. After the earthquake shaking, due to the elastic clamping action of the unbonded tendons the structure returns back to the original position, with negligible damage and residual/permanent deformations. Additional dissipation capability can be provided by means of internal mild steel (grouted) bars or other supplemental damping devices. Such an "hybrid" system [26] thus combines re-centering capability with self-centering properties, resulting into a peculiar "flag-shaped" hysteresis behaviour (Fig. 28.1, bottom).

The conceptual innovation of "capacity design" principles introduced by Park & Paulay in the late 1960s is universally recognized as a major milestone in the development of earthquake engineering and seismic design philosophies. Similarly, the concept of ductile connections able to accommodate high inelastic demand without

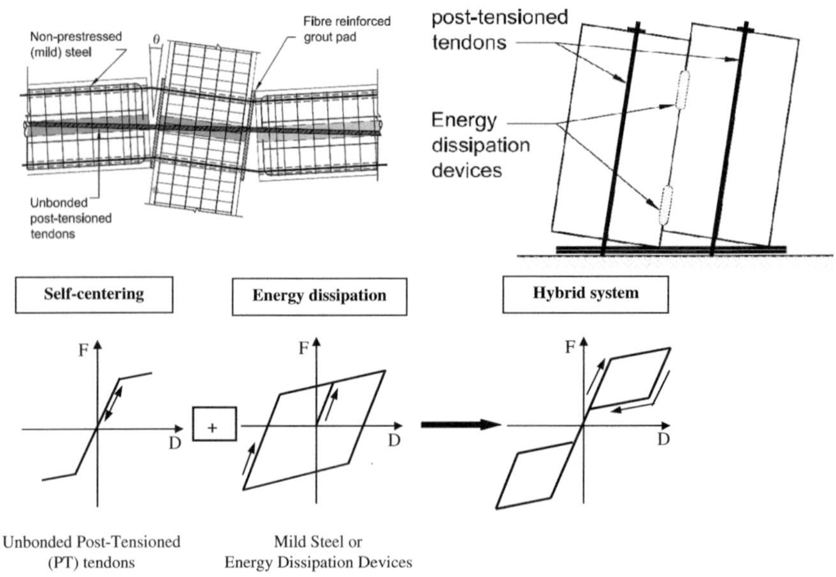

Fig. 28.1 Jointed precast "hybrid" frame and wall connections developed in the US PRESSS-Program

Fig. 28.2 Evolution of seismic resisting connections: experimental performance of beam-column joints designed according to *Left*: pre-1970 codes (shear damage in the joint); *Centre*: capacity design principles as per the NZS3101:1995 (beam plastic hinge) and *Right*: hybrid jointed ductile connections as per NZS3101:2006, Appendix B (controlled rocking) [14]

suffering extensive material damage, developed in the 1990s, can be arguably represent an equally critical milestone towards the development of the next generation of damage-resistant high-performance systems, based on the use of conventional material and techniques (Fig. 28.2).

28.3 Replaceable Fuses as "Weakest Links of the Chain"

The continuous and rapid development of post-tensioned jointed ductile connections have resulted to the validation of a wide range of alternative arrangements, under the general umbrella of "hybrid" systems, currently available to designers and contractor for practical applications. Main differentiating features are:

Longitudinal profile of the post-tensioned tendons: either straight or draped tendons or a combination of the above can be adopted in the beams depending on the ratio between gravity and lateral loads effects, as a consequence of the different levels of seismicity (target design earthquake) as well as of the assigned role of the system (i.e. gravity-load carrying system, seismic resisting system or intermediate solution). The combination of internally bonded prestressing with unbonded post-tensioning, to be implemented at different construction phases, can also provide unique flexibility in the design phase.

Type and source of energy dissipation: either internal (grouted) mild steel bars or, more recently developed, external & replaceable supplemental dissipaters can be adopted. Following the declared target to achieve a no- (or at least low-) damage system, significant effort has been dedicated in the past few years towards the development of cost-efficient external dissipaters, which, if required, can be easily demounted and replaced after an earthquake event [16]. This option would give the possibility to conceive a modular system with replaceable sacrificial fuses at the rocking connection, acting as the "*weakest link of the chain*", according to capacity design principles. Very efficient and practical dissipater systems have been developed consisting of axial, tension-compression yielding mild steel short bars, machined down to a desired "fuse" dimension, inserted and grouted (or epoxied) in a steel tube acting as anti-buckling restrainer.

These dissipaters have been developed and extensively tested within several subassemblies configurations, i.e. beam-column joint connections, wall systems, column (or bridge pier)-to-foundation connections (Fig. 28.3). As a further advantage of this type of external dissipaters, very stable flag-shape hysteresis loops can be obtained, with no stiffness degradation due to bond losses when compared to internally grouted (bonded) mild steel bars. Either metallic and/or other advanced materials (e.g. shape memory alloys, visco-elastic systems) can be used and implemented to provide alternative type of dissipation mechanisms (elasto-plastic due to axial or flexural yielding, friction, visco-elastic). Examples of application of friction and viscous devices in unbonded post-tensioned systems have been given in by Kurama [9] and Kurama and Shen [8].

A second generation of self-centering/dissipative high-performance systems, referred to as advanced flag-shape systems (AFS) has been proposed by Kam et al. [7]. AFS systems combine alternative forms of displacement-proportional and velocity-proportional energy dissipation (i.e. yielding, friction or viscous damping) in series and/or in parallel with the main source of re-centering capacity (unbonded post-tensioned tendons, mechanical springs or Shape Memory Alloys, SMA, with

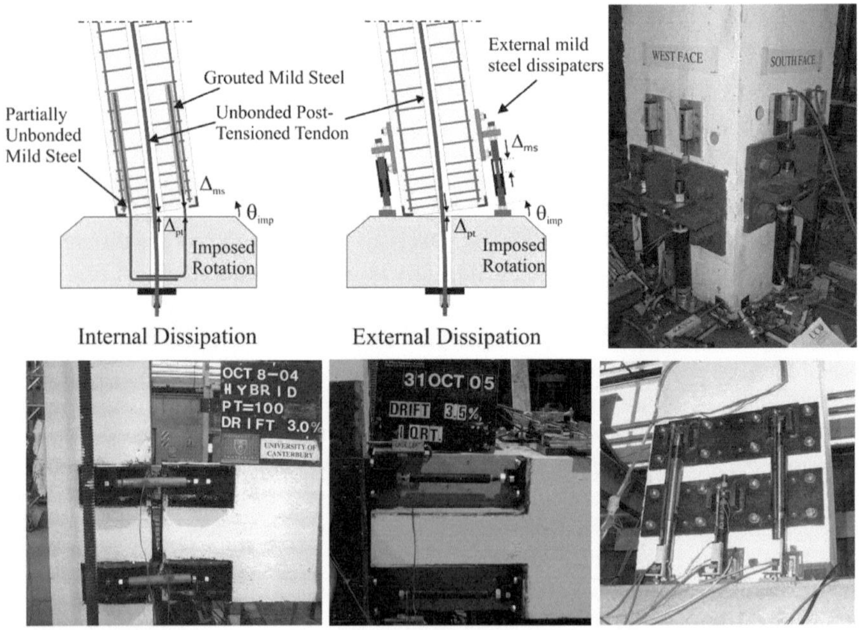

Fig. 28.3 *Top*: internal versus external replaceable dissipaters/fuses at the base-columns (or bridge pier). *Bottom*: alternative configuration of external dissipaters for columns, beam-column joints or wall systems

super-elastic behaviour). As a result, an enhanced and very robust seismic performance, under either far field and near field events (high velocity pulse) can be achieved, as proven by numerical investigations [7] as well as pseudo-dynamic and shake table testing [11, 10].

28.4 A Further Step Forward: Controlling and Reducing the Damage to the Floor

The peculiarity of a jointed ductile connection, consisting of an "articulated" assembly of precast elements can be further exploited and extended to the design of floor-to-lateral-load-resisting-system connections. The latter topic has been receiving a growing attention in the engineering community in the last decade, following the several examples of poor performance of floor-diaphragm observed in recent earthquakes. Experimental tests on 3-dimensional performance of precast super-assemblages including frames and hollowcore units ([12], Fig. 28.4, right) have further underlined issues related to the inherent displacement incompatibility between precast floor and lateral resisting systems, including beam elongation effects ([5, 6], Fig. 28.4, left).

Alternative innovative solutions have been recently developed and proposed in literature to minimize the damage to the floor system, while guaranteeing a reliable diaphragm action [1]. The first approach would consist of combining standard precast rocking/dissipative frame connections with an articulated or "jointed" floor system. According to this proposal, developed from the original concept of discrete X-plate mechanical connectors implemented in the Five-Storey PRESSS Building tested at UCSD [22], the floor (hollowcore in this case) unit is connected to the lateral beams by slider/shear mechanical connectors, acting as shear keys when the floor moves at right angles to the beam and as sliders when the floor moves parallel to the beam (Fig. 28.5). As a result, the system is able to accommodate the displacement compatibility demand between floor and frame by creating an articulated or jointed mechanism, effectively decoupled in the two directions. Also, due to the low flexural stiffness of the shear keys-connectors in the out of plane direction,

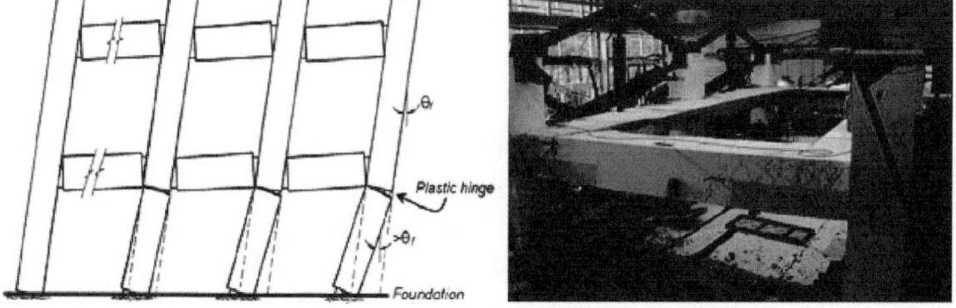

Fig. 28.4 Beam elongation effects (after [6]) and failure of precast hollowcore floor during large scale tests carried out at the University of Canterbury (courtesy of J. Matthews)

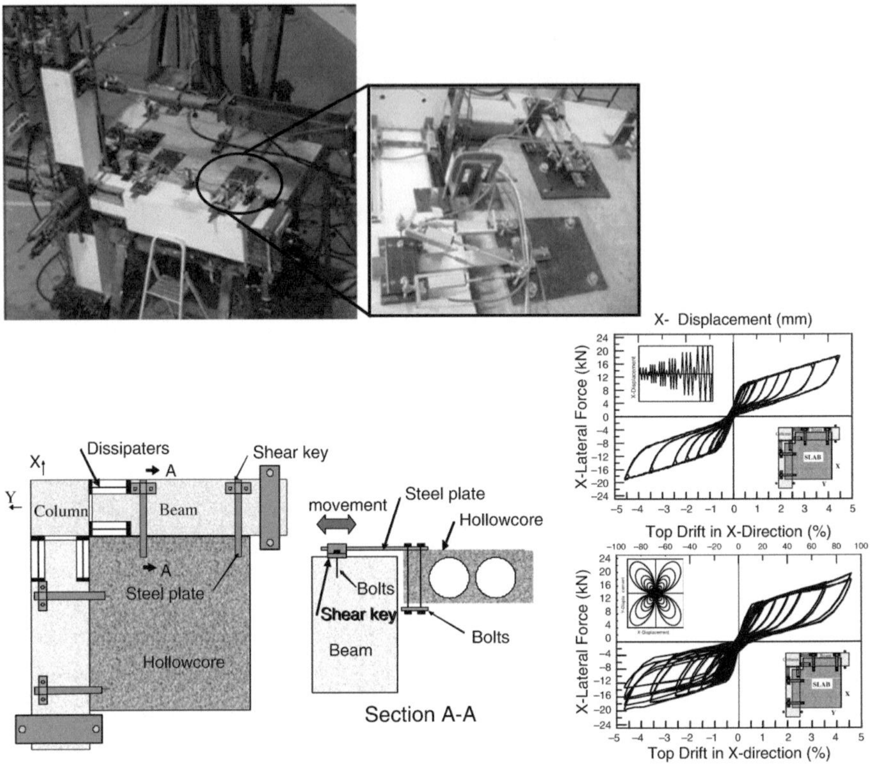

Fig. 28.5 "Articulated floor" system. Concept, connection details and response under uni-directional and bi-directional cyclic tests

torsion of the beam elements due to pull out of the floor or relative rotation of floor and edge support can be limited. A second approach to obtain a non-tearing floor system would rely on a new "top-hinge" or "top-hung" system in combination with a standard floor solution (i.e. topping and continuous starter bars). See [16, 17] for more information. A comprehensive "PRESSS Design Handbook", comprising a full design example of a five-storey building (frames and walls) as well as a software for the section analysis of jointed ductile "hybrid" connections has been recently developed under the umbrella of the NZ Concrete Society [17].

28.5 Post-Tensioned Timber Buildings: The Pres-Lam System

The concept of post-tensioned hybrid (recentering/dissipating) systems has been recently and successfully extended from precast concrete to timber frames and walls [15]. Since 2004, a series of tests (comprising quasi-static cyclic, pseudodynamic and shake-table), have been carried out on several subassemblies or larger scale systems at the University of Canterbury to develop different arrangements of prototype connections for unbonded post-tensioned timber frame and walls (Fig. 28.6). Due to

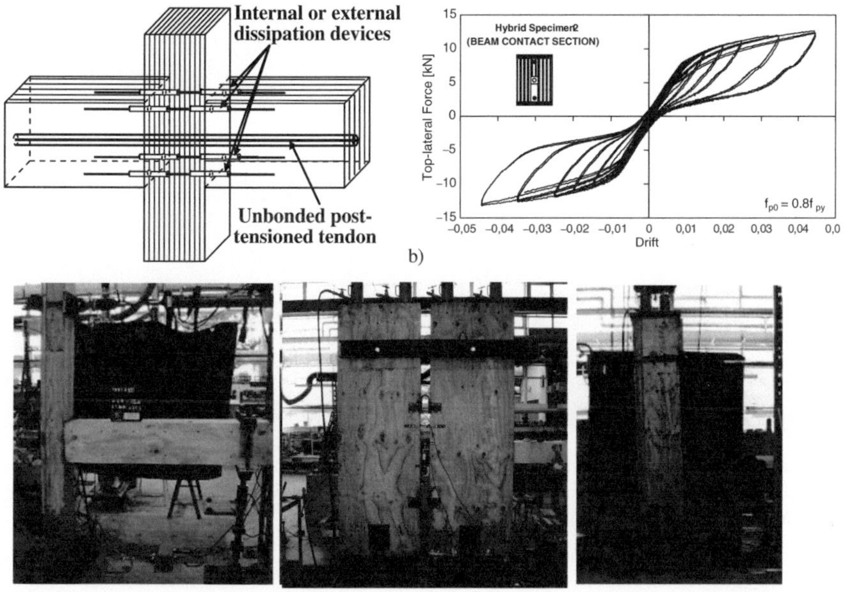

Fig. 28.6 Concept of hybrid jointed ductile connections for LVL timber frame systems and experimental campaign on beam-column joints, walll systems and column-to-foundation connections

its high homogeneity and good mechanical properties [23], laminated veneer lumber (LVL) has been selected as the preferred engineered wood material. The experimental results provided very satisfactory results and confirmation of the high potential of this new construction system, referred to as Pres-Lam, which gives opportunities for much greater use of timber and engineered wood products in large buildings, using innovative technologies for creating high quality buildings with large open spaces, excellent living and working environments, and resistance to hazards such as earthquakes, fires and extreme weather events [2]. A major multi-year R&D project has been recently initiated under the umbrella of a NZ-Australia Research Consortium, STIC Ltd (Structural Timber Innovation Company).

Based on the extension of existing procedures available in literature for jointed ductile precast concrete walls and frames and on analytical-experimental comparisons, simplified analytical/numerical modeling and (displacement-based) design procedures for these post-tensioned timber systems have been tentatively proposed and are under further refinements [15, 20, 13].

28.6 From Theory to Practice: On-Site Applications and Case Studies

Several on site applications of jointed ductile systems, adopting PRESSS-type technology have been implemented in different seismic-prone countries around the world, including U.S., South America, Europe and New Zealand. One of the first

Fig. 28.7 On site application in USA and S. America. *Left*: 39-storey Paramount Building in San Francisco (details of the design in Englerkirk [4], photo courtesy E. Miranda). *Right*: 4-storey Hotel Virgo, Mendoza (Prear Pretensados, Argentinos S.A.)

and most glamorous application of hybrid systems in high seismic regions is given by the Paramount Building in San Francisco (Fig. 28.7), consisting on a 39-storey apartment building and representing the highest precast concrete structure in a high seismic zone [4]. Several applications have followed in Italy, through the implementation of the Brooklyn system (Fig. 28.8), which adopts draped tendons and a metallic slotted & hidden corbel [19].

The first multi-storey PRESSS-building in New Zealand (Fig. 28.9), currently under completion, features some of the latest technical solutions previously described, including external replaceable dissipaters for both the frame system and dismountable steel coupling beams for the post-tensioned coupled

Fig. 28.8 Structural skeleton of a 4-storey building in Italy adopting the Brooklyn suspended (draped tendons) solution (B.S. Italia, Bergamo)

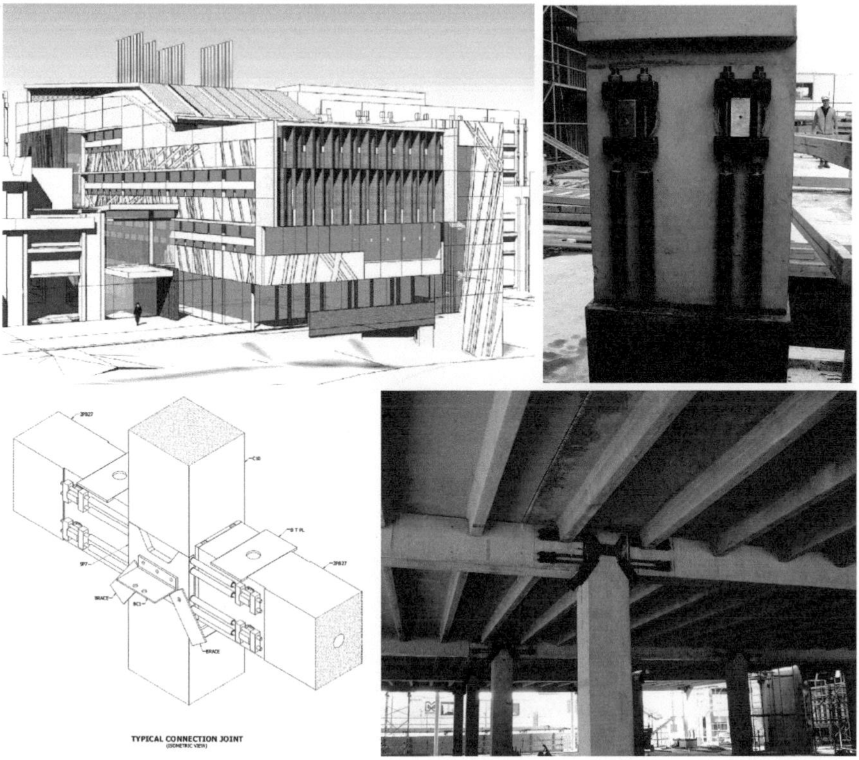

Fig. 28.9 Rendering of the first multi-storey PRESSS-Building in New Zealand (design details in Cattanach and Pampanin [3]: schematic and photo of beam-column joint and column-to-foundation connection details, implementing "plug&play" external replaceable dissipaters

walls [3]. For its innovation in implementing state-of-art design (displacement based) and technological solutions, it has been awarded the NZ Concrete Supreme Award 2008.

In the meantime, preliminary feasibility studies of virtually re-designed Pres-Lam building have been carried out. A cost-comparison with a 6-storey reinforced concrete building currently under construction at the University of Canterbury, has provided very promising and valuable information on the high potential of this "new-born" technological solution (Fig. 28.10, [25]).

As a confirmation of the growing interest on large-scale multi-storey timber construction, also promoted by an increasing sensitivity towards "sustainable and environmental-friendly" buildings, the World's first implementation of the Pres-Lam system is going to happen soon with the construction of a three storey building for the Nelson Marlborough Institute of Technology (NMIT) in New Zealand, implementing unbonded post-tensioned hybrid LVL walls coupled with U-shape Flexural Plate dissipaters (Fig. 28.10).

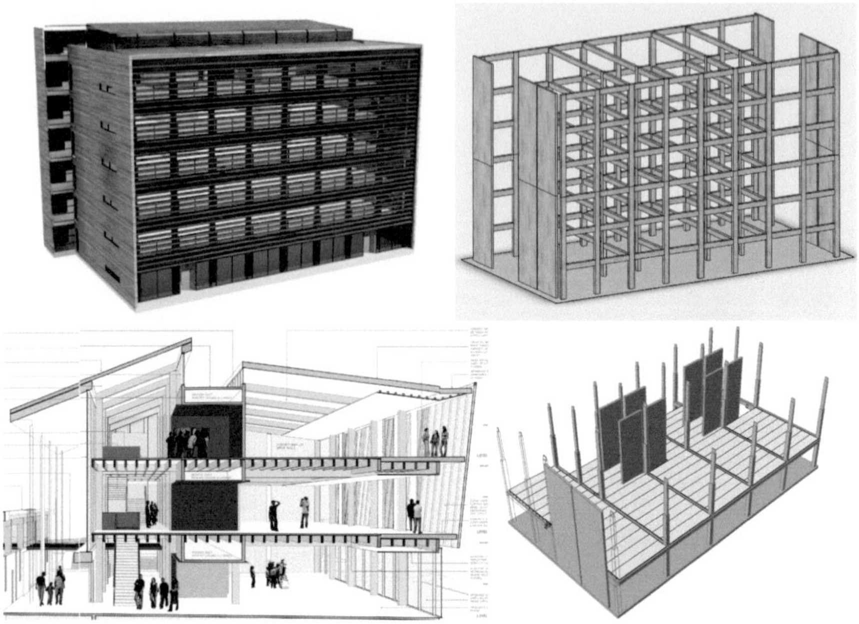

Fig. 28.10 Architectural rendering and structural skeleton of (*top*): a case-study 6-story post-tensioned timber building (virtual re-design of real concrete building at UoCanterbury, photo courtesy of T. Smith); (*bottom*): World's first Pres-Lam system application (Nelson Marborough Inst. of Technology, Nelson, NZ, photo courtesy of Irving-Smith-Jack architects and Aurecon)

28.7 Conclusions

In this contribution, an overview of recent developments and emerging solutions for high-performance, whilst still cost-effective, damage-control seismic resisting systems, either for concrete or timber structures, based on unbonded-post-tensioned techniques and a controlled-rocking mechanism has been given.

Jointed ductile connections can be implemented in the lateral load resisting systems (either frames, walls or dual systems) to sustain a design level earthquake with limited or negligible damage, costs of repairing and business interruption. External replaceable dissipaters can be adopted to provide supplemental strength and dissipation capacity to the system and designed to be the only sacrificial and replaceable fuses for the entire structure. Furthermore, research focus has been more recently given to reduce the damage to the floors, by implementing similar "jointed" or articulated solutions, based on the use of either discrete mechanical connectors between lateral resisting system and floors, or of a non-elongating (non-tearing floor) beam-column "top-hinge' or "top-hung" connection.

More importantly, the construction industry has been welcoming this new technology, supporting its refinement during the R&D phase by providing continuous and valuable feedback to improve constructability and cost-effectiveness. The recent on-site applications of such systems, featuring some of the latest technical solutions developed and extensively tested in the laboratory have to be acknowledged as an important example of successful interaction between industry and academia, resulting to the actual implementation of performance-based seismic design theory and technology.

References

1. Amaris AD, Pampanin S, Bull DK, Carr AJ, (2008) Solutions to control and minimize floor damage in precast concrete buildings under severe earthquake loading, NZ Concrete Industry Conference, Rotorua, 2–4 October
2. Buchanan A, Pampanin S, Palermo A, Newcombe M (2009) Non-conventional multi-storey timber building using posttensioning. 11th international conference on non-conventional materials and technologies, Bath, UK
3. Cattanach A, and Pampanin S (2008) 21st century precast: the detailing and manufacture of NZ′s first multi-storey PRESSS-Building, NZ Concrete Industry Conference, Rotorua
4. Englerkirk (2002) Design-construction of the paramount – a 39 story precast prestressed concrete apartment Building. PCI J 47(4):56–72
5. Fenwick RC, Megget LM (1993) Elongation and load deflection characteristics of reinforced concrete members containing plastic hinges. Bull NZ Nat Soc Earthq Eng 26(1):28–41
6. *fib* (2003) International federation for structural concrete. Seismic design of precast concrete building structures. Bulletin No. 27, Lausanne, 254 pp
7. Kam WY, Pampanin S, Palermo A, Carr A (2006) Advanced flag-shaped systems for high seismic performance. 1st European Conference on Earthquake Engineering Geneva, Switzerland
8. Kurama Y, Shen Q (2004) Posttensioned hybrid coupled walls under lateral loads. J Struct Eng ASCE 130(2): 297–309
9. Kurama YC (2001) Seismic design of unbonded post-tensioned precast concrete walls with supplementary viscous damping. ACI Struct J 97(4):648–658
10. Marriott D, Pampanin S, Bull D, Palermo A (2008) Dynamic testing of Precast, post-tensioned rocking wall systems with alternative dissipating solutions. Bull NZ Soc Earthq Eng 41(2): 90–103
11. Marriott D, Pampanin S, Palermo A (2009) Quasi-static and pseudo-dynamic testing of unbonded post-tensioned rocking bridge piers with external replaceable dissipaters. Earthq Eng Struct Dyn 38(3):331–354
12. Matthews J, Bull D, Mander J (2003) Hollowcore floor slab performance following a severe earthquake. Procedings of *fib* symposium concrete structures in seismic regions, Athens, Greece
13. Newcombe M, Pampanin S, Buchanan A, Palermo A, (2008) Section analysis and cyclic behavior of post-tensioned jointed ductile connections for multi-storey timber buildings. J Earthq Eng Special Issue 12(S1):83–110
14. NZS3101:2006 (2006) Appendix B: special provisions for the seismic design of ductile jointed precast concrete structural systems. Standards New Zealand, Wellington, NZ
15. Palermo A, Pampanin S, Buchanan AH, Newcombe M (2005) Seismic design of multi-storey buildings using laminated veneer lumber. Proceedings of the New Zealand Society of Earthquake Engineering Conference, Wairakei, New Zealand
16. Pampanin S (2005) Emerging solutions for high seismic performance of precast –prestressed concrete Buildings. J Adv Concr Technol 3(2):202–222
17. Pampanin S (ed) (2010) "PRESSS Design Hanbook", New Zealand Concrete Society, Wellington, NZ

18. Pampanin S, Amaris A, Akguzel U, Palermo A (2006) Experimental investigations on high-performance jointed ductile connections for precast frames. 1st European conference on earthquake engineering and seismology, Geneva, Switzerland
19. Pampanin S, Pagani C, Zambelli S (2004) Cable-stayed and suspended post-tensioned solutions for precast concrete frames: the Brooklyn System. NZ Concrete Industry Conference, Queenstown
20. Pampanin S, Palermo A, Buchanan AH, Fragiacomo M, Deam BL (2006) Code provisions for seismic design of multi-storey post-tensioned timber buildings, CIB W18, Florence, August
21. Priestley MJN (1996) The PRESSS program-current status and proposed plans for phase III. PCI J 41(2):22–40
22. Priestley MJN, Sritharan S, Conley JR, Pampanin S (1999) Preliminary results and conclusions from the PRESSS five-storey precast concrete test building. PCI J 44(6):42–67
23. Ranta Maunus A (1995) Laminated veneer lumber and other structural sections. In: Hans Blass et al. (eds) Timber Engineering STEP 1, 1st edn. Centrum Hout, The Netherlands
24. SEAOC Vision 2000 Committee (1995) Performance-based seismic engineering. Structural Engineers Association of California, Sacramento, CA
25. Smith T, Pampanin S, Buchanan M (2009) Post-tensioned timber building, construction, costs and business case study, ANIDIS Bologna 28 June – 2 July
26. Stanton JF, Stone WC, Cheok GS (1997) A hybrid reinforced precast frame for seismic regions. PCI J 42(2):20–32

Chapter 29
Seismic Design of Plane Steel Frames Using Modal Strength Reduction (Behaviour) Factors

George A. Papagiannopoulos and Dimitri E. Beskos

29.1 Introduction

The concept of the strength reduction factor has been the object of extensive research studies (see references in [5]). Current seismic design codes use a single constant crude value for the strength reduction factor derived on the basis of intuition, experience or approximate seismic response methods [5]. The need to implement rationality and accuracy to the strength reduction factor has led recently the authors to the development of the concept of equivalent modal damping ratios which, as it has been proven in [5, 6], (i) can essentially play the role of the strength reduction factor in a more rational and accurate way, (ii) are given for the first few modes of vibration of the structure that significantly contribute to its dynamic response and (iii) can be defined as functions of deformation and damage indices, i.e., interstorey drift and plastic hinge rotation.

In this work rationality and accuracy are introduced into the strength reduction factor in a direct way, i.e., by developing deformation and damage dependent modal strength reduction factors. This is materialized by using the equivalent damping concept and its application to seismic design of structures as developed in [5, 6] as well as by establishing a relationship between the damping reduction factor (used to construct a response/design acceleration spectrum for high amounts of damping) and the strength reduction factor. Having constructed by means of extensive numerical experiments design equations that provide equivalent damping ratios as function of period, allowable deformation and damage for the first significant modes of vibration of a steel framed structure, one makes use of the aforementioned relationship between the damping reduction and strength reduction factors and obtains strength reduction factors for the first significant modes. Thus, the design base shear of a steel frame can be determined by means of spectrum analysis utilizing different values of the strength reduction factor for each mode.

G.A. Papagiannopoulos (✉)
Department of Civil Engineering, University of Patras, Patras GR26504, Greece
e-mail: gpapagia@upatras.gr

M.N. Fardis (ed.), *Advances in Performance-Based Earthquake Engineering*, Geotechnical, Geological, and Earthquake Engineering 13, DOI 10.1007/978-90-481-8746-1_29,

The proposed modal strength reduction factor approach is used for seismic designs of a steel moment resisting frame for illustration purposes. Its validation is done by comparing it against nonlinear inelastic dynamic analysis and the usual code-based approach of employing one common value of the strength reduction factor for each mode. It is concluded that the use of modal strength reduction factors instead of a single value for that factor for all modes, leads to more accurate seismic design results in a more rational way.

29.2 Review on Seismic Design Using Equivalent Modal Damping

29.2.1 The Equivalent Modal Damping Concept

According to the developments in [5, 6], a nonlinear MDOF structure can be substituted for seismic response purposes by an equivalent linear MDOF structure having the same mass and initial stiffness with the nonlinear one and time – invariant equivalent modal damping ratios that take into account the effects of all non – linearitites. The conversion of the effects of nonlinearities into equivalent modal damping ratios is based on a balance (equivalence) between the damping work and that of non – linearities [5, 6].

The equivalent modal damping values can be viewed as playing the role of the strength reduction factor in seismic design in a more rational and accurate way. They can also be given as functions of deformation and damage and, thus, can be directly implemented to seismic design of a structure [5, 6]. Indeed in [5, 6], design equations providing equivalent damping ratios as functions of period and allowable deformation and damage for the first few significant modes have been constructed using extensive numerical data coming from a representative number of plane steel moment resisting frames excited by various seismic motions. These design equations are used in conjunction with an elastic spectrum, constructed so as to accommodate high damping values, and modal synthesis tools to calculate the design base shear and from there the corresponding forces of the structure.

29.2.2 Steel Frames, Ground Motions, Performance Levels and Highly Damped Spectra

A set of 20 steel plane moment resisting frames was used for the parametric analyses of this work in order to obtain equivalent modal damping ratios [5]. The frames are regular and orthogonal with storey heights and bay widths equal to 3.0m and 4.0m respectively. The number of storeys and bays varies taking the values of 2, 3, 4, 7, 9, 12, 15, 18 and 1, 2, 3, 4, 5, 6, respectively. The frames have been designed in accordance with the provisions of Eurocodes 3 and 8 [1, 3] assuming an acceleration design spectrum with a peak ground acceleration (PGA) equal to 0.24g, a soil of class B and a reduction (behaviour) factor equal to 3. Load on beams (dead and live loads of the floors) is constant for each frame and takes values between 25 and 30 kN/m in order to ensure different fundamental periods for the frames examined. The steel material was assumed to be elastoplastic with a yield stress of 275 MPa and 3% hardening. The frame connections were assumed to be rigid and diaphragm action was considered at every floor. Detailed data for the steel frames, including number of bays, number of storeys, beam and column sections and first and second natural periods of vibration are presented in [5].

These frames are excited by a set of 36 historical earthquake accelerograms recorded worldwide from 24 different earthquake events from 1968 to 2007 [5]. The set of accelerograms includes earthquake ground motions recorded in the proximity of faults (near fault pulse type ground motions) and motions that exhibit long duration. Reference [5] provides information regarding the location, date, magnitude and fault mechanism of every earthquake event as well as the station name, the site code, the recorded peak ground acceleration (PGA) and peak ground velocity (PGV) of the accelerograms considered and pictorial representations of the acceleration and velocity histories of these seismic motions.

Interstorey drift ratio (IDR) is used as a measure of the deformation performance of the frames and the damage performance of the frames is taken into account by using the rotational capacity of the plastic hinge [5].

The seismic design method based on equivalent modal damping ratios requires the availability of elastic response/design spectra with higher, than the usual 2–5%, amounts of damping [5]. When dealing with equivalent damping that comes from the non-linear material of a structure, then the highly damped acceleration spectra needed to perform its seismic design should correspond to absolute acceleration [5, 4, 8] and not to pseudo-acceleration. A highly damped absolute acceleration spectrum can be constructed with the aid of the damping reduction factor B_{a} [5, 4] which is defined as:

$$B_{\mathrm{a}} = |\ddot{u}_t|_{\max} \,/\, |\ddot{u}_t|_{\max,\xi_{\mathrm{eq}}=2\%} = S_{\mathrm{a},\xi_{\mathrm{eq}}} / S_{\mathrm{a},\xi_{\mathrm{el}}=2\%} \tag{29.1}$$

where ξ_{eq} is the equivalent damping ratio, $S_{\mathrm{a},\xi_{\mathrm{el}}=2\%}$ the absolute acceleration of the structure for 2% damping and $S_{\mathrm{a},\xi_{\mathrm{eq}}}$ the absolute acceleration of the structure for other than 2% damping. The value of 2% damping corresponds to the inherent damping of a steel structure in the linear range of response.

29.2.3 Design Equations for Modal Damping Ratios

The 20 steel frames of this study were analyzed to determine their response to each of the 36 earthquake ground motions considered here [5]. Equivalent modal damping values for the aforementioned seismic performance levels were obtained following [5, 6]. Curves that depict the variation of equivalent damping ξ_{eq} with period T for the first five modes of the frames and the performance levels are given for near fault pulse type and long duration seismic motions [5, 6]. Design equations

Table 29.1 Design equations for equivalent damping for long duration seismic motions as functions of IDR and damage

Mode	$IDR = 1.5\%$ and $\theta_p = \theta_y$	$IDR = 2.0\%$ and $\theta_p = 3.5\theta_y$
1	$\xi_{eq} = 0.025 \cdot (T - 0.5) + 0.10$ for $0.5 \leq T \leq 2.5$s	$\xi_{eq} = 0.47$ for $0.5 \leq T \leq 2.5$ s
2	$\xi_{eq} = 0.055$ for $0.15 \leq T \leq 0.85$ s	$\xi_{eq} = 0.11$ for $0.15 \leq T \leq 0.85$ s
3	$\xi_{eq} = 0.035$ for $0.11 \leq T \leq 0.48$ s	$\xi_{eq} = 0.10$ for$0.32 \leq T \leq 0.47$ s
4	$\xi_{eq} = 0.035$ for $0.11 \leq T \leq 0.27$ s and $\xi_{eq} = 0.8 \cdot (T - 0.27) + 0.035$ for $0.27 \leq T \leq 0.32$ s	–
5	$\xi_{eq} = 0.929 \cdot (T - 0.17) + 0.035$ for $0.17 \leq T \leq 0.24$ s	–

for equivalent modal damping are also given in [5, 6]. Thus, the seismic design of a structure can be performed by using these design equations in conjunction with a design acceleration spectrum with high levels of damping as described above. For illustration purposes, the design equations that give the equivalent modal damping ratios ξ_{eq} in the case of long duration seismic motions and for two IDR and θ_p levels are depicted in Table 29.1. In this table, a dash (–) denotes that a mode attains 100% equivalent modal damping for reasons explained in [5, 6].

29.3 The Modal Strength Reduction Factor

Each modal contribution to the seismic design base force is given as $M_k^* \cdot S_{a,k}(T_k, \xi_{eq,k})$, where M_k^* is the effective modal mass of mode k and $S_{a,k}(T_k, \xi_{eq,k})$ the corresponding acceleration spectrum ordinates computed at natural period T_k and damping $\xi_{eq,k}$. The total seismic design force is derived by combining the individual modal contributions mentioned above by using an appropriate modal combination rule, e.g., the CQC [2]. Thus, one can define the modal strength reduction factor q_k as the ratio of the modal elastic base shear $V_{el,k}$ of a structure over its corresponding modal yielding base shear $V_{y,k}$. More specifically, using the effective modal mass M_k^* as well as $S_{a,k}(T_k, \xi_{el,k})$ and $S_{a,k}(T_k, \xi_{eq,k})$ as the corresponding acceleration spectrum ordinates computed at natural period T_k and damping $\xi_{el,k}$ and $\xi_{eq,k}$ respectively, one has that

$$q_k = V_{el,k}/V_{y,k} = M_k^* S_{a,k}(T_k, \xi_{el,k})/M_k^* S_{a,k}(T_k, \xi_{eq,k}) = S_{a,k}(T_k, \xi_{el,k})/S_{a,k}(T_k, \xi_{eq,k}) \quad (29.2)$$

Using now the definition of the damping reduction factor as given in Eq. (29.1), we have:

$$q_k = S_{a,k}(T_k, \xi_{el,k})/S_{a,k}(T_k, \xi_{eq,k}) = 1/B_{a,k} \quad (29.3)$$

where $B_{a,k}$ is the modal damping reduction factor. Thus, the values of the modal strength reduction factor can be derived by inverting those of the modal damping reduction factor. In other words, the reduction of the seismic design force, as effected by the strength reduction factor, theoretically coincides with the reduction of the seismic force as effected by the damping reduction factor.

29.4 Curves and Design Equations for Modal Strength Reduction Factors

The construction of curves depicting the variation of strength reduction factor with period follows the same steps performed for the calculation of damping reduction factors in [5]. There, mean and mean plus one deviation values of the damping reduction factors B_a have been calculated on the basis of the mean and mean plus one deviation acceleration spectra of near fault pulse type as well as long duration seismic motions. Thus, using Eq. (29.3), one can obtain the variation of the strength reduction factor with period. Figure 29.1 shows the mean and mean plus one deviation values of the strength reduction factor for long duration motions. In that figure, the ratio $2\%/\xi_{eq}\%$ symbolizes the ratio $S_a(2\%)/S_a(\xi_{eq}\%)$. Combining deformation and damage dependent equivalent modal damping values from [5] with damping dependent strength reduction factor, i.e. Fig. 29.1, for the type of seismic motions

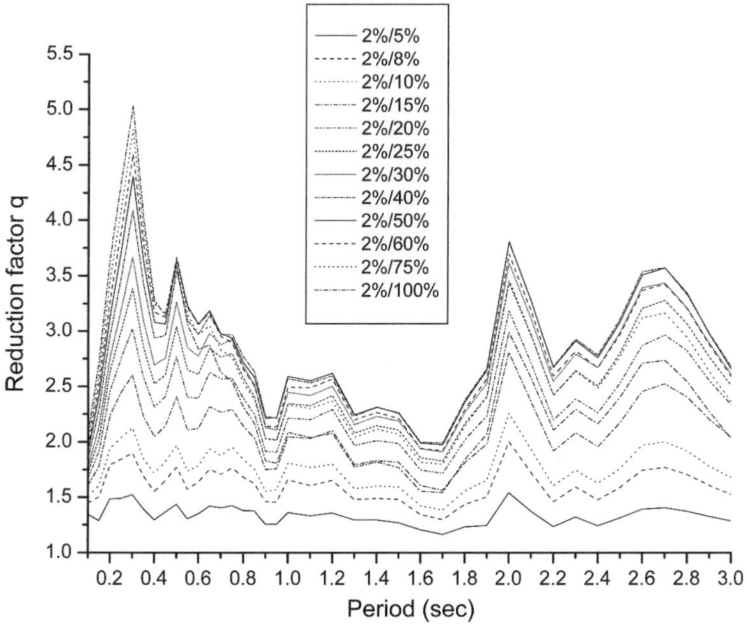

Fig. 29.1 Strength reduction factor derived from mean plus one deviation acceleration spectra of long duration motions

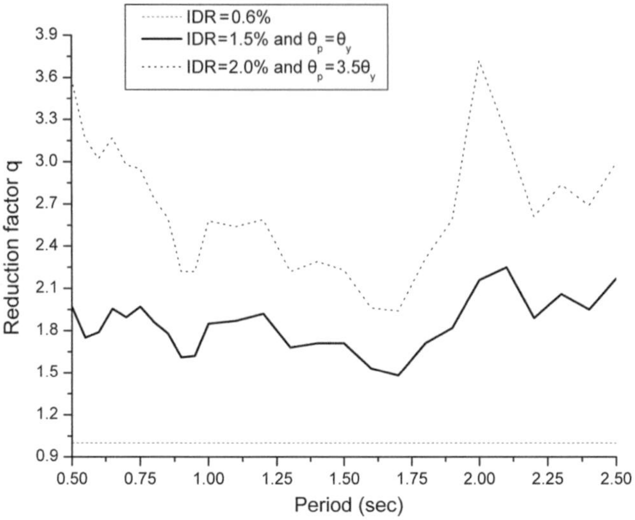

Fig. 29.2 Strength reduction factor for the first mode derived from mean plus one deviation acceleration spectra of long duration motions

considered here, one obtains the mean and mean plus one deviation values of modal strength reduction factors as functions of period, deformation and damage.

Figures 29.2 and 29.3 illustrate the mean and mean plus one deviation values of the strength reduction factor as a function of deformation and damage for the first and second modes, respectively, in the case of long duration motions. Design

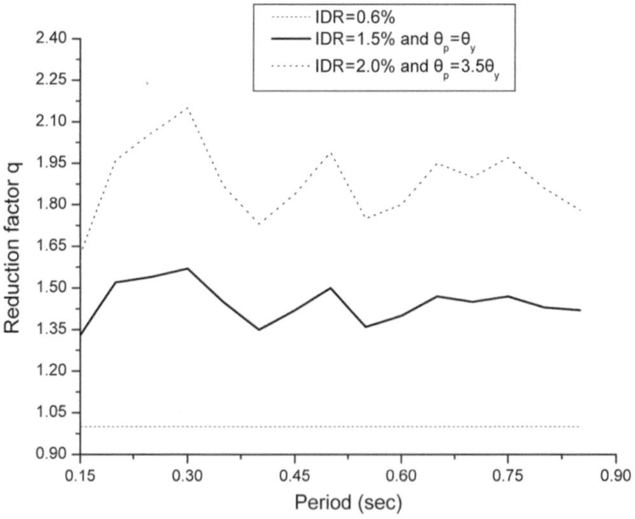

Fig. 29.3 Strength reduction factor for the second mode derived from mean plus one deviation acceleration spectra of long duration motions

Table 29.2 Design equations for modal strength reduction factors for long duration seismic motions as functions of IDR and damage

Mode	$IDR = 1.5\%$ and $\theta_p = \theta_y$	$IDR = 2.0\%$ & $\theta_p = 3.5\theta_y$
1	$q_1 = 1.5$ for $0.5 \leq T \leq 1.5$ sec and $q_1 = 1.5 + 0.3 \cdot (T - 1.5)$ for $1.5 \leq T \leq 2.5$ sec	$q_1 = 3.3 - 1.4 \cdot (T - 0.5)$ for $0.5 \leq T \leq 1.5$ sec and $q_1 = 1.9 + 0.5 \cdot (T - 1.5)$ for $1.5 \leq T \leq 2.5$ sec
2	$q_2 = 1.3$ for$0.15 \leq T \leq 0.85$ sec	$q_2 = 1.7$ for $0.15 \leq T \leq 0.85$ sec
3	$q_3 = 1.3$ for $0.11 \leq T \leq 0.48$ sec	$q_3 = 1.55 + 0.676 \cdot (T - 0.11)$ for $0.11 \leq T \leq 0.48$ sec
4	$q_4 = 1.3$ for $0.11 \leq T \leq 0.32$ sec	$q_4 = 1.9 + 10 \cdot (T - 0.11)$ for $0.11 \leq T \leq 0.32$ sec
5	$q_5 = 1.2$ for$0.10 \leq T \leq 0.24$ sec	$q_5 = 1.9 + 10 \cdot (T - 0.1)$ for $0.10 \leq T \leq 0.24$ sec

equations for modal strength reduction factors for the case of long duration seismic motions are shown for illustration purposes in Table 29.2.

29.5 Numerical Examples

A steel moment resisting plane frame of 12 storeys and 4 bays is designed using modal strength reduction factors. The dead plus live load on beams is equal to 27.5 kN/m and the expected ground motion is defined by using the mean plus one standard deviation damped spectrum that corresponds to long duration motions [5]. The design of this frame is performed according to EC3 [1] structural steel code targeting the case of $IDR = 1.5\%$ and $\theta_p = \theta_y$ with the aid of SAP 2000 [7]. HEB profiles are used for columns and IPE for beams. One can obtain from Table 29.2 the following values for the modal strength reduction factors: $q_1 = 1.51$, $q_2 = 1.30$, $q_3 = 1.30$, $q_4 = 1.30$, $q_5 = 1.20$. The sections found are 340/340/340/340/340-300 (storeys 1–5) and 320/340/340/340/320-300 (storeys 6–9) and 300/320/320/320/300-270 (storeys 10–12). In the above, expressions of the form, e.g., 340–300 (storeys 1–4) mean that storeys from 1 to 4 have columns with HEB340 sections and beams with IPE300 sections. Furthermore, expressions of the form, e.g., 300/320/320/320/300-270 (storeys 6–10) mean that for storeys from 6 to 10 there is a variation of column sections in each bay of the same storey, i.e., sections HEB300 and HEB320 for the first bay, sections HEB320 and HEB320 for the next two bays and sections HEB320 and HEB300 for the last bay and all beams have IPE270 section. Nonlinear dynamic analyses are then executed using the accelerograms of the long duration seismic motions mentioned above in order to check if the designed frame satisfies the target performance criteria of $IDR = 1.5\%$ and damage $\theta_p = \theta_y$. The results from nonlinear dynamic analyses regarding median values for interstorey drift (IDR) and plastic hinge rotation θ_p, respectively, are as follows: $IDR_{med} = 1.47\%$ and $\theta_{p,med} = 1.00\theta_y$.

In the previous example, the proposed method employed design spectra that correspond to absolute acceleration and constructed on the basis of the data of the present work. In the example that follows, the proposed method makes use of the design spectrum of EC8 [3] that corresponds to pseudo-acceleration, which as mentioned previously, cannot be used for high levels of damping. The previous steel moment resisting frame is designed for the 5% damped elastic design spectrum of EC8 [3], peak ground acceleration of $0.24g$ and soil of class B employing the same modal strength reduction factors of long duration motions case as before, i.e., $q_1 = 1.51$, $q_2 = 1.30$, $q_3 = 1.30$, $q_4 = 1.30$, $q_5 = 1.20$. The sections found are 340/340/340/340/340-300 (storeys 1–5) and 320/340/340/340/320-300 (storeys 6–9) and 300/320/320/320/300-270 (storeys 10–12). Nonlinear dynamic analyses are then executed using eight of the long duration accelerograms of [5], compatible to the aforementioned EC8 [3] design spectrum, in order to check if the designed frame satisfies the target interstorey drift value (IDR). The deformation result from nonlinear dynamic analyses of the frame designed with the aid of the modal strength reduction factors in conjunction with pseudo-acceleration design spectrum of EC8 [3] is $IDR_{\mathrm{med}} = 1.46\%$. It is observed that the IDR value obtained is close to the target value of 1.5%.

The steel frame is now designed employing the conventional method of EC8 [3] that makes use of a single value of the strength reduction factor that for the frame type considered is equal to $q = 6$. Using the 5% damped elastic design spectrum of EC8 [3] with a peak ground acceleration of $0.24g$, soil class B and a strength reduction factor $q = 6$, the following sections are obtained: 240/260/260/260/240-300 (storeys 1–5) & 240/260/260/260/240-300 (storeys 6–9) & 240/240/240/240/240-270 (storeys 10–12). With the aid of the equal displacement rule adopted by EC8 [3] for the computation of displacements, one can find after some iterations that the requirement of $IDR = 1.5\%$ is satisfied for the sections: 260/280/280/280/260-330 (storeys 1–5) & 260/260/260/260/260-330 (storeys 6-9) & 240/260/260/260/240-300 (storeys 10–12). Finally, nonlinear dynamic analyses are executed using eight of the long duration accelerograms of [5], compatible to the aforementioned EC8 [3] design spectrum, in order to check if the designed frame indeed satisfies the target $IDR = 1.5\%$ value. The results from nonlinear dynamic analyses lead to $IDR_{\mathrm{med}} = 1.65\%$. It is observed that this IDR value is not close to the target *IDR* value of 1.5% and the sections, although smaller, were obtained iteratively and not directly as it is the case of the proposed method. Obviously, to obtain the value of 1.5%, one must increase the sections (iteratively).

Similar numerical experiments for other target seismic performance levels demonstrated that large deviations appear when attempting to compare the code-based seismic design based on one single value of the strength reduction factor and the proposed one employing different values for the strength reduction factor per mode. The main reason is the difference in the values of the absolute acceleration and pseudo-acceleration spectra due to the presence of high damping. Use of pseudo-acceleration spectrum may lead to an incorrect calculation of the seismic design forces of the structure when its equivalent damping comes from its nonlinear material deformations. Moreover, the modal strength reduction factors are smaller in comparison to the proposed code-values of the strength reduction factor not only because they have been computed using absolute acceleration spectra but also because they are given as functions of deformation and damage limits.

29.6 Conclusions

On the basis of the preceding developments, the following conclusions can be stated:

(1) A seismic design method for plane steel moment resisting frames has been developed. The method employs modal strength reduction factors, spectrum analysis and modal synthesis. The modal strength reduction factor is obtained by combining equivalent modal damping ratios and damping reduction factors used to construct absolute acceleration spectra with high damping values.
(2) Curves relating modal strength reduction factor with period, deformation and damage have been constructed and corresponding design equations have been developed. These curves or equations can be used in conjunction with design spectrum and modal synthesis tools to calculate the seismic design forces of the structure.
(3) The proposed approach was applied to the seismic design of steel moment resisting framed structures and was validated using nonlinear inelastic dynamic analyses. Unlike the usual code – based approach of considering a single strength reduction factor value for all modes, the proposed approach employing different modal strength reduction factors leads to more accurate seismic response results in a more rational way.

References

1. CEN (2004) EN 1993-1-1:2003, Eurocode 3, Design of steel structures, part 1.1: general ru.les for buildings. European Committee for Standardization, Brussels
2. Chopra AK (2007) Dynamics of structures: theory and applications to earthquake engineering, 3rd edn. Prentice Hall, Englewood Cliffs, NJ
3. EC8 (2004), Eurocode 8, Design of structures for earthquake resistance, part 1: general rules, seismic actions and rules for buildings, European Standard EN 1998-1, Stage 51 Draft. European Committee for Standardization (CEN), Brussels
4. Lin YY, Chang KC (2003) Study on damping reduction factor for buildings under earthquake ground motions. ASCE J Struct Eng 129(2):206–214
5. Papagiannopoulos GA (2008) Seismic design of steel structures by using equivalent modal damping ratios or modal strength reduction factors, Ph.D. Thesis, Department of Civil Engineering, University of Patras, (in Greek)
6. Papagiannopoulos GA and Beskos DE (2009) The equivalent modal damping concept and its use in seismic design of steel structures. In: Papadrakakis M, Charmpis DC, Lagaros ND, Tsompanakis Y (eds) Progress in computational structural dynamics and earthquake engineering. CRC Press, Boca Raton, FL, pp 401–411
7. SAP2000 (2005) Static and dynamic finite element analysis of structures – version 9.1.4. Computers and Structures, Berkeley, CA
8. Weitzmann R, Ohsaki M, Nakashima M (2006) Simplified methods for design of base-isolated structures in the long-period high-damping range. Earthq Eng Struct Dyn 35:497–515

Chapter 30
Recent Advances in Seismic Isolation: Methods and Tools

Panos Tsopelas and Sashi Kunnath

30.1 Introduction

Seismic isolation has found a large number of applications all around the world and has proven to be an efficient and practical approach to achieve satisfactory seismic performance for a variety of structural systems. Having overcome a slow start in its implementation in practice during its early years, seismic isolation is finding an increasing acceptance between structural engineers in recent years. It is the research developments in the areas of experimental verification and analytical modeling of the behavior of seismic isolation hardware and structural response, together with the advances of the seismic isolation hardware itself, which have provided the fuel for this ever increasing acceptance of the technology by the structural engineering practice.

The original concept behind seismic isolation involves decoupling the structure from the horizontal ground motion, providing flexibility and energy dissipation capacity [12]. By those means the severity of structural shaking can be mitigated and damage during earthquakes can be reduced, if not eliminated. Utilization of any of the two types of conventional seismic isolation systems, elastomeric or sliding, can reduce significantly the inertial forces that are developed in a structure during a strong seismic motion. The use of the term conventional is to describe seismic isolation technology/hardware consisting of components exhibiting stiffness and energy dissipation capacity which remain constant during the system operation. Conventional seismic isolation technology cannot lead to an optimum level of performance for all possible seismic excitations, since it is inherently intended to optimally meet a single design objective. However, structural design and seismic design in particular are multi-objective processes. Therefore, in order to meet the multiple objectives of a design process and achieve benefits in performance

P. Tsopelas (✉)
Department of Civil Engineering, University of Thessaly, Pedion Areos, GR 38334, Volos, Greece
e-mail: tsopelas@uth.gr

M.N. Fardis (ed.), *Advances in Performance-Based Earthquake Engineering*, Geotechnical, Geological, and Earthquake Engineering 13, DOI 10.1007/978-90-481-8746-1_30,

which are not possible with conventional passive isolation systems, seismic isolation hardware had to evolve to exhibit an adaptive behavior.

This paper concentrates on sliding isolation systems and in particular on the evolution of the most popular isolator of this type, the Friction Pendulum (FP) isolator, which combines in a single unit flexibility and energy dissipation capacity.

The first evolution of FP isolators, the uplift restraint XY-FP isolator, comes about to prevent a potentially undesirable phenomenon, the uplift of isolators. While a reduction in inertial forces in a seismically isolated structure causes a desired reduction of the sustained overturning moments, the uplift forces may still be potentially large enough to be of concern, on account of the inherent incapacity of sliding bearings to resist uplift forces. In fact, a variety of conditions may contribute to the development of uplift. These include: slender structures with a large height-to-width ratio, certain types of bridges with large ratio of height of the centroidal axis to distance between bearings and isolators below braced columns or stiff walls. Uplift in isolation bearings may produce, under certain conditions, detrimental effects in the form of local instability and damage on sliding bearings due to large compressive forces upon impact following uplift. Loss of contact and impact on return can produce higher-mode response and large axial forces in columns.

The next advancement over the conventional FP isolators are the multi-surface (double and triple curvature) FP isolators, exhibiting adaptive behavior which can be used by design engineers to achieve benefits in performance which are not possible with other passive isolation systems. Current practice in the design of seismically isolated structures is to have a structural system resisting the base shear due to the design basis earthquake (DBE) and to design the isolation system to have sufficient displacement capacity to meet the demands of the maximum considered earthquake (MCE). These design objectives make the life of a designer rather "miserable"; the target of reducing displacement demand during the MCE (with increased stiffness and damping) results in less than optimum performance in the DBE and vice versa. This situation is exacerbated due to the substantial differences in the DBE and MCE demands (the DBE spectrum prescribed by codes is normally 2/3 of the MCE spectrum). In addition, the performance of the isolation system in more frequent events of smaller magnitude is typically not considered in the design process. Although low-level shaking is not a design issue in terms of strength or displacement capacity, it can be a performance issue. Isolation systems designed with sufficient damping and flexibility for larger earthquakes may not activate in minor events, which can adversely affect secondary system response. Even if the isolation system does activate, re-centering can be an issue.

The works by Kelly [9] and Hall [7] indicate that to control displacements in large earthquakes, while still maintaining good performance in low-to-moderate events, requires designing an isolation system which is: (a) very stiff with low damping at low-level shaking, (b) softens with increasing damping in the DBE, (c) further softens and increases damping in the MCE, and (d) stiffens beyond the MCE.

This desirable behavior can be achieved with properly designed multi-spherical sliding bearings. It may be recognized that increases in adaptability are accompanied by increased complexity and, therefore, each of these variations may be of interest to a designer. The triple FP isolator helps to overcome these challenges, since adaptive behavior permits the isolation system to be separately optimized for low intensity, design level and maximum earthquake shaking.

30.2 Uplift-Restraint Friction-Pendulum Isolator

Since the introduction of seismic isolation more than 20 years ago, seismically isolated structures have been designed so that tension in elastomeric isolators and uplift in sliding isolators are avoided. This practice is followed because, with uplift, the large bearing compressive forces, with or without impact, which are generated on engagement following uplift, might affect the integrity of the system. Tension in elastomeric bearings might also be undesirable, due to concerns for their failure in tension. Engineers are often forced to introduce extensive changes in the superstructure above the isolation level attempting to evenly distribute the reactions in order to prevent isolator uplift or tension. Such changes may be extensive, as was the case in the Oakland City Hall [8] with the construction of a large truss, and in the Corinth Canal Bridges [1] with use of counterweights at the abutments. However, often cost or architectural and functionality constraints prevent the modification of the structural system to avoid uplift or tension in bearings. Accordingly, a need arose to develop acceptable isolators with tension (or uplift restraint) capability.

The uplift-restraint FP isolator (XY-FP) belongs to the new generation of seismic isolation hardware which, contrary to the conventional FP isolator, is capable of accommodating the uplift-restraint property by allowing continuous transition of the bearing axial force from compression to tension and vice versa. Moreover, the isolator can have different frictional interface properties under compressive and tensile isolator normal force [17]. A schematic of this isolator is shown in Fig. 30.1.

The basic operation principles of the uplift-restraint FP isolator are explained below, as its behavior is compared to a regular FP isolator [17]. Consider an isolation system consisting of two conventional FP isolators supporting a rigid block weighing $2W$ as in Fig. 30.2. When the block experiences an overturning moment M that can cause uplift in one of the isolators the total weight, $2W$, is being carried by the other isolator, as shown in Fig. 30.2. Both the restoring force and the friction force

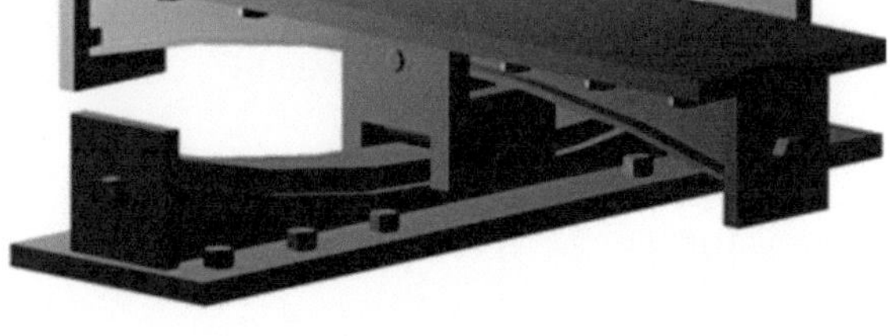

Fig. 30.1 View of the uplift-restraint FP isolator (from [17])

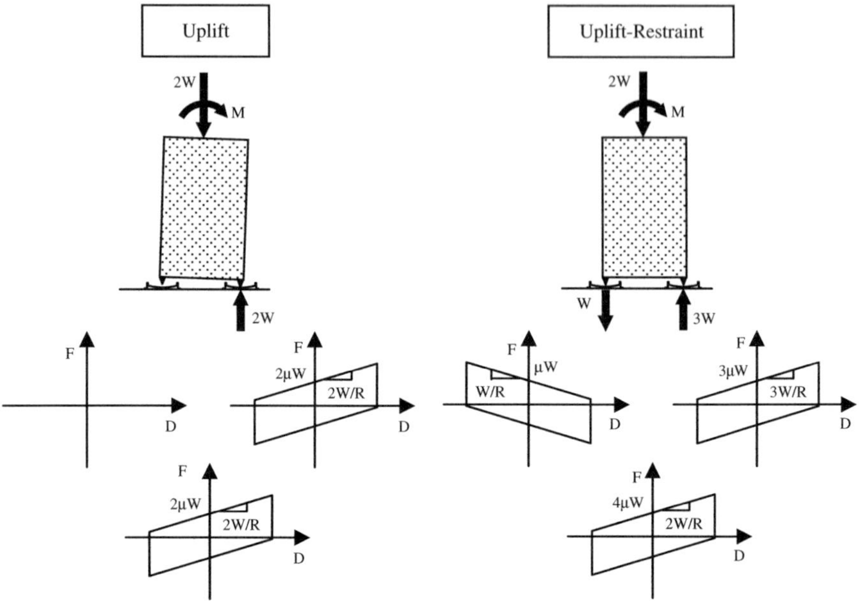

Fig. 30.2 Comparison between conventional FP and the uplift-restraint XY-FP isolators

of the total system remain unchanged. Now consider an isolation system consisting of two uplift-restraint XY-FP isolators. When the block experiences overturning moment which causes tension forces to be developed in one of the isolators the other isolator will experience additional compressive loads, so that vertical force equilibrium is satisfied. The negative stiffness of the isolator in tension will be counterbalanced by the increased stiffness of the isolator in compression, thus resulting in the same total system stiffness. However, the total friction force of the isolation system is increased by the presence of tensional forces in one of the isolators. In the simplified case presented herein, where half the isolators sustain tension with the magnitude of the tensile forces being 100% of the initial compressive load W in those isolators experiencing tension, the total friction force increases from 2 to 4 μW.

The mathematical model describing the behavior of this isolator is synthesized by two independent uniaxial hysteretic elements allowing different frictional interface properties along the principal isolator directions. The force-displacement relationship in the local coordinate system at two perpendicular directions utilized in modeling the XY-FP isolator is given by

$$\begin{Bmatrix} F_1 \\ F_2 \end{Bmatrix} = \begin{bmatrix} N/R_1 & 0 \\ 0 & N/R_2 \end{bmatrix} \begin{Bmatrix} U_1 \\ U_2 \end{Bmatrix} + \begin{bmatrix} \mu_1 |N| & 0 \\ 0 & \mu_2 |N| \end{bmatrix} \begin{Bmatrix} Z_1 \\ Z_2 \end{Bmatrix} \tag{30.1}$$

where R_1, R_2, μ_1, μ_2, and U_1, U_2 are the radii of curvature of the lower and upper concave beams, the associated sliding-friction coefficients and the displacements in bearing local axis 1 and 2, respectively; N is the normal force on the bearing (positive when compressive) and Z_1, Z_2 are hysteretic dimensionless quantities governed by differential equations [17]. The dependency of the coefficient of friction on sliding velocity and on bearing pressure, as well as the variation of the normal force on the isolation bearing owing to the effect of vertical earthquake motion and global overturning moment, are accounted for as in Roussis and Constantinou [17]. The model of this isolator has been implemented in the analysis software 3D-BASIS-ME-MB [21].

30.3 Double Curvature Friction Pendulum (DCFP) Isolators

The next step in the evolution of FPS is the double Curvature Pendulum Isolator, which has the ability, by appropriate choice of the Isolator's parameters (sliding interfaces, R radius of curvature and displacement capacity of each sliding interface), to change the dynamic properties (i.e., the *F-D* curve, stiffness and damping capacity) of the isolator, thus allowing the designer/engineer much more flexibility in design than the Single FP isolator does. Fenz and Constantinou [3], Tsai et al. [19, 18] and Kim and Yun [10] have presented the main principles of operation of these isolators, as well as the details of the mechanics involved in their behavior. The modes of operation are shown in Fig. 30.3 with sliding occurring in one and/or two interfaces, thus mobilizing the corresponding curvatures (stiffnesses) and frictional coefficients (energy dissipation). Such behavior can give the ability to a designer, as sliding commences, to start with a large stiffness (small isolation periods) for earthquakes of smaller magnitude which happen more often and transition to smaller

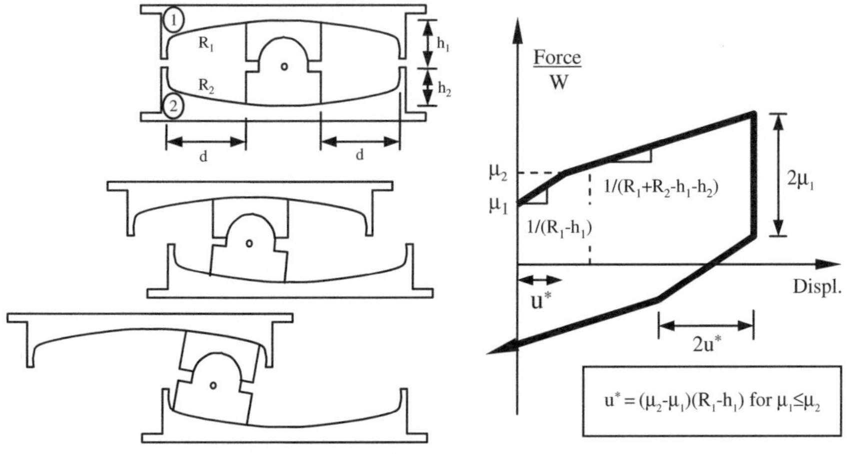

Fig. 30.3 Modes of sliding of a DCFP isolator and corresponding force displacement relation

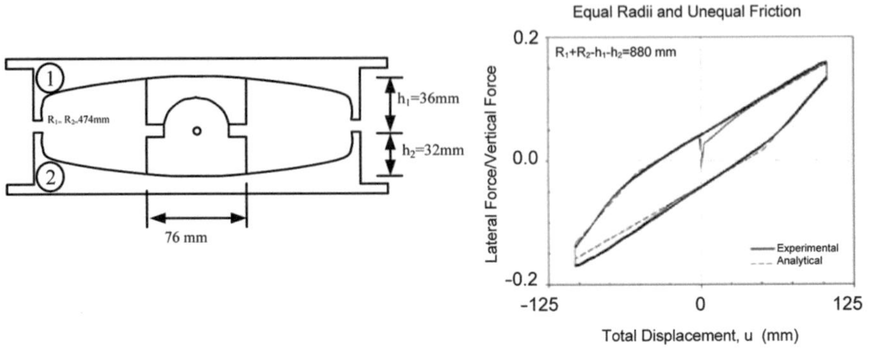

Fig. 30.4 Force displacement relation of a DCFP isolator with equal radii and un-equal friction ($\mu_1 = 0.081$, $R_{\text{eff1}} = 438$ mm, $\mu_2 = 0.012$, $R_{\text{eff2}} = 442$ mm) (from [3])

stiffness (large isolation period) when, for example, a large earthquake occurs (large displacement demand); the larger isolation period limits/reduces the level of forces in the structure and thus reduces the potential for damage to secondary systems (museum artifacts). Figure 30.4 presents the experimental results obtained by Fenz and Constantinou [3] from an isolator with equal radii and un-equal friction, as depicted in the figure.

To fully utilize the potential of such a system in the design of structural systems, it is of paramount importance to develop a computational/analytical model which can capture the detailed Force-Displacement response in all possible modes of operation of the isolator, in order to fully explore its potential. Such a model could help explore the effect of this variable/adaptive behavior on the structure, as well as on the secondary systems housed by it.

30.3.1 Smooth Hysteretic Model for DCFP Isolators

A rate rate-independent phenomenological model originally proposed by Ozdemir [14] has been shown to be similar to the Bouc-Wen model, which has been used the last two decades to model force deformation characteristics of rubber and friction based seismic isolators [21]. The following two equations describe Ozdemir's model:

$$\dot{F} = k\left[\dot{D} - \left|\dot{D}\right| \cdot \left(\frac{F-S}{F_o}\right)^n\right] \tag{30.2}$$

$$\dot{S} = k\alpha \left|\dot{D}\right| \left(\frac{F-S}{F_o}\right)^n \tag{30.3}$$

where, F is the force, D the displacement, S a back-stress term, k an elastic stiffness, F_o a yielding force, α a parameter which controls the post-hardening modulus and n an odd integer controlling the shape of the elastic-plastic transition.

To capture the general tri-linear hysteretic behavior of DCFP isolators, a second back-stress term (S_1 and S_2) should be added to the original Ozdemir model as follows [22]:

$$\dot{F} = k\left[\dot{D} - \left|\dot{D}\right| \cdot \beta_1 \left(1 - |\beta_2|\right) + \beta_2\right] \tag{30.4}$$

$$\dot{S}_1 = k\left|\dot{D}\right| \left[\alpha_1\beta_1 \left(1 - |\beta_2|\right) + \alpha_2\beta_1 \left|\beta_2\right|\right] \tag{30.5}$$

$$\dot{S}_2 = k\alpha_2 \left|\dot{D}\right| \beta_2 \tag{30.6}$$

where $\beta_i = \left[\frac{F-S_i}{F_i}\right]^n$ for i=1, 2

The presence of the second-back stress term, together with the additional parameters in the modified model, result in a generalized hysteresis loop as depicted in the schematic in Fig. 30.5. Figure 30.6 compares the predictions from Ozdemir's modified model with experimental and analytical results from Fenz and Constantinou [3] for a DCFP isolator with equal radii and equal friction (μ_1= 0.057, R_{eff1} = 442 mm, μ_2 = 0.058, R_{eff2} = 448 mm) or unequal radii and unequal friction (μ_1= 0.021, R_{eff1} = 442 mm, μ_2 = 0.038, R_{eff2} = 726 mm).

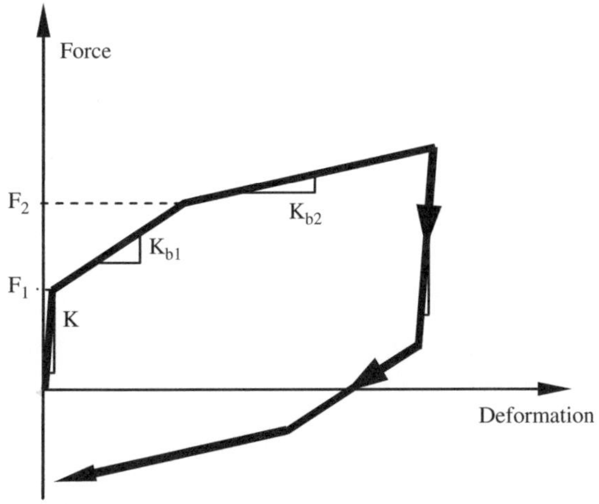

Fig. 30.5 Schematic presentation of the hysteresis loop produced by the modified Ozdemir model

30.4 Triple Friction Pendulum Isolators

Adaptive seismic isolation systems are those which can change their properties (stiffness and strength/damping) during operation. In order to achieve adaptive behavior, conventional seismic isolation hardware had to be combined with other devices active or semi-active [2, 11, 16, 15, 23, 13]. A common conclusion from

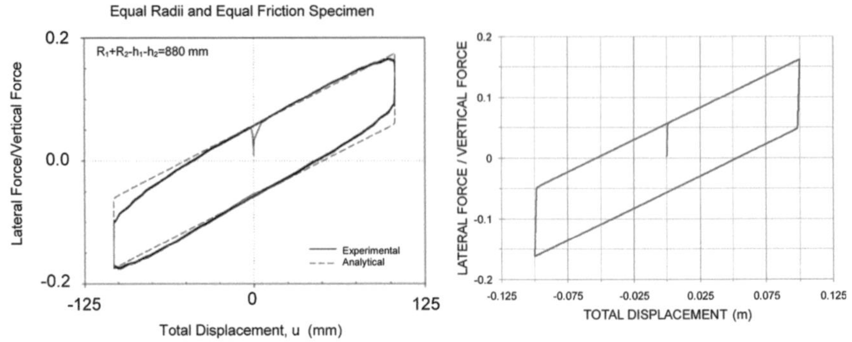

(a)

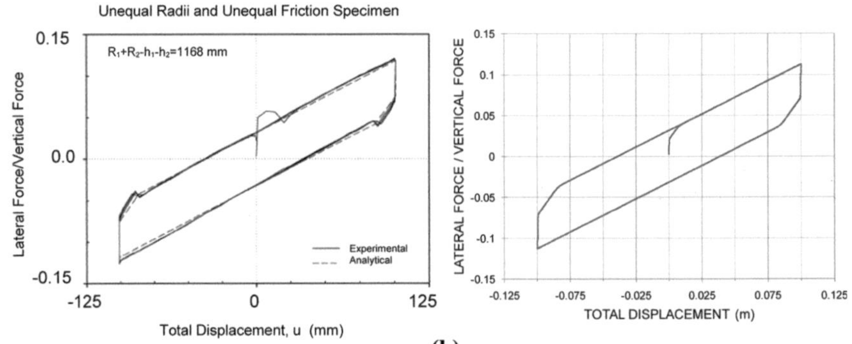

(b)

Fig. 30.6 Comparisons between experimental and analytical results from Fenz and Constantinou [3] on the *left* and the prediction of the proposed modified Ozdemir's model on the *right*; (**a**) equal radii and equal friction ($\mu_1 = 0.057$, $R_{\text{eff1}} = 442$ mm, $\mu_2 = 0.058$, $R_{\text{eff2}} = 448$ mm); (**b**) unequal radii and unequal friction ($\mu_1 = 0.021$, $R_{\text{eff1}} = 442$ mm, $\mu_2 = 0.038$, $R_{\text{eff2}} = 726$ mm)

all these studies is that properly designed active and semi-active hybrid systems can offer improved performance over passive systems for a wider range of earthquakes. However, implementation difficulties, as well as questions regarding longevity and reliability of the involved technologies, still persist.

The next step in the evolution of the FP isolators is a multi-spherical sliding isolator which expands the definition of an adaptive seismic isolation system [4, 5]. This isolator is a fully passive device which exhibits adaptive stiffness and adaptive damping behavior without assistance from another technology or device. By adaptive behavior, it is meant that stiffness and strength (friction) change to predictable values at calculable and controllable displacement amplitudes. Moreover, the isolator is a derivative of the conventional FP bearing, a mature and established seismic protection technology with demonstrated reliability. This makes practical implementation more feasible. Figure 30.7 presents the geometry and frictional parameters of the triple FP isolator.

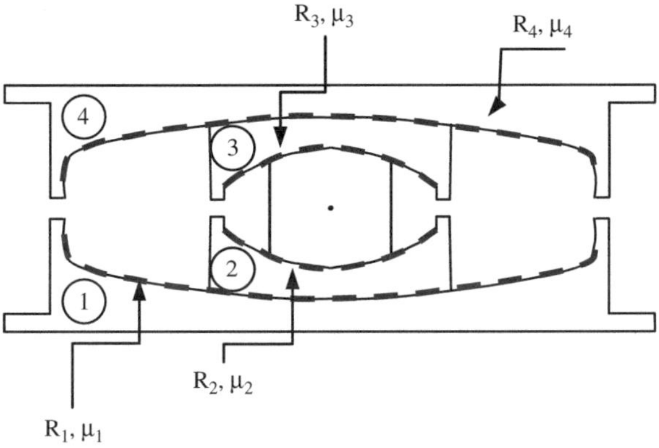

Fig. 30.7 Triple friction pendulum isolator

30.4.1 Modeling of Triple Friction Pendulum Isolators

There are currently no applicable hysteresis rules or nonlinear elements available in structural analysis software that can be used to exactly model triple FP bearings for response-history analysis. The multiphase sliding results in a distinct force-displacement relationship that is inherently more complex than any one exhibited by currently used seismic isolation devices.

Two approaches may be taken to model the behavior of this new device: (a) to develop and implement a new hysteresis rule to trace the overall behavior, or (b) to combine existing nonlinear elements in such a way that the overall behavior is captured. Series models composed of existing nonlinear elements have been proposed by Fenz and Constantinou [6]; they can be immediately implemented in currently available analysis software. However, the behavior of the Triple FP bearing is not exactly that of a series arrangement of single concave FP bearings, though it is similar.

Tsopelas [20] has developed a new hysteresis rule which traces the behavior of the triple FP isolator in every sliding interface and is capable of exactly reproducing the overall force-displacement behavior of the isolator. The model is checking equilibrium at every step in each of the sliding interfaces and decides where sliding will commence in the next step. The model based on this hysteretic rule is currently in the implementation phase in the software 3D-BASIS-ME-MB.

Figures 30.8 and 30.9 present the overall and the individual sliding interface force displacement loops of a triple FP isolator with properties: $R_{\mathrm{eff1}} = R_{\mathrm{eff4}} = 435$ mm, $R_{\mathrm{eff2}} = R_{\mathrm{eff3}} = 50$ mm, $\mu_1 = 0.05$, $\mu_2 = \mu_3 = 0.02$, $\mu_4 = 0.12$, under harmonic excitations. This isolator is the one tested by Fenz and Constantinou [5].

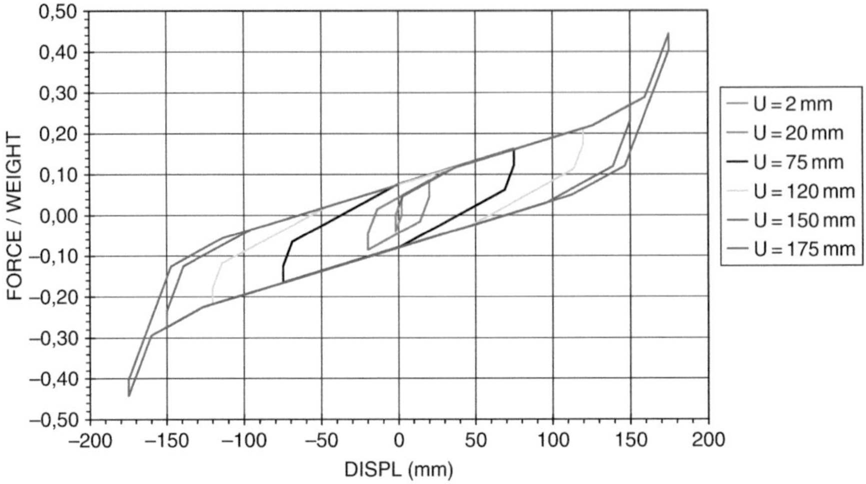

Fig. 30.8 Analytical predictions of the overall behavior of the triple FP isolator tested by Fenz and Constantinou [5] under sinusoidal displacement input of various amplitudes as obtained with the hysteretic-rule based model developed by Tsopelas [20]

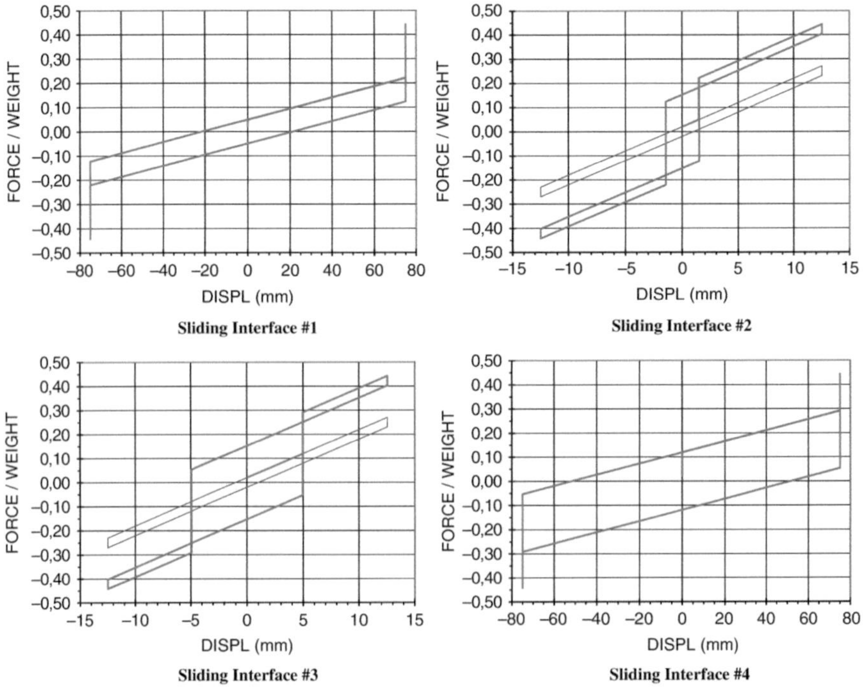

Fig. 30.9 Analytical predictions of the behavior of each sliding interface of the triple FP isolator tested by Fenz and Constantinou [5] under sinusoidal displacement input of 175 mm amplitude as obtained with the hysteretic-rule based model developed by Tsopelas [20]

References

1. Constantinou MC (1998) Application of seismic isolation systems in Greece. Proceedings of '98 structural engineers world congress, Paper T175-3, San Francisco, CA
2. Feng MQ, Shinozuka M, Fujii S (1993) Friction-controllable sliding isolation system. J Eng Mech (ASCE) 119(9):1845–1864
3. Fenz DM, Constantinou MC (2006) Behavior of the double concave friction pendulum bearing. Earthq Eng Struct Dyn 35(11):1403–1424
4. Fenz DM, Constantinou MC (2007a) Spherical sliding isolation bearings with adaptive behavior: experimental verification. Earthq Eng Struct Dyn 37(2):185–205
5. Fenz DM, Constantinou MC (2007b) Spherical sliding isolation bearings with adaptive behavior: theory. Earthq Eng Struct Dyn 37(2):163–183
6. Fenz DM, Constantinou MC (2008) Modeling triple friction pendulum bearings for response-history analysis. Earthq Spectra 24(4):1011–1028
7. Hall JF. (1999) Discussion of 'the role of damping in seismic isolation'. Earthq Eng Struct Dyn 28(12):1717–1720
8. Honeck W, Walters M, Sattary V, Rodler P (1993) Seismic isolation of the Oakland city hall. Proceedings of ATC-17-1 seminar on seismic isolation, passive energy dissipation, and active control, San Francisco, CA, 11–12 March, Vol. I, pp 221–232
9. Kelly JM (1999) The role of damping in seismic isolation. Earthq Eng Struct Dyn 28(1):3–20
10. Kim YS, Yun C-B (2007) Seismic response characteristics of bridges using double concave friction pendulum bearings with tri-linear behavior. Eng Struct 29(11):3082–3093
11. Makris N (1997) Rigidity–plasticity–viscosity: can electrorheological dampers protect base isolated structures from near-source ground motions? Earthq Eng Struct Dyn 26(5):571–591
12. Naeim F, Kelly JM (1999) Design of seismic isolated structures – from theory to practice. Wiley, New York, NY
13. Nagarajaiah S, Sahasrabudhe S (2006) Seismic response control of smart sliding isolated buildings using variable stiffness systems: an experimental and numerical study. Earthq Eng Struct Dyn 35(2):177–197
14. Ozdemir H (1976) Nonlinear transient dynamic analysis of yielding structures. Ph.D. Thesis, Department of Civil Engineering, University of California, Berkeley, CA
15. Ramallo JC, Johnson EA, Spencer Jr BF (2002) 'Smart' base isolation systems. J Eng Mech (ASCE) 128(10):1088–1100
16. Riley MA, Reinhorn AM, Nagarajaiah S (1998) Implementation issues and testing of a hybrid sliding isolation system. Eng Struct 20(3):144–154
17. Roussis PC, Constantinou MC (2005) Experimental and analytical studies of structures seismically isolated with an uplift-restraint isolation system. Technical Report MCEER005-0001, Buffalo, NY
18. Tsai CS, Chen WS, Chiang TC, Chen BJ (2006) Component and shaking table tests for full-scale multiple friction pendulum system. Earthq Eng Struct Dyn 35(11):1653–1675
19. Tsai CS, Chiang TC, Chen BJ (2005) Experimental evaluation piecewise exact solution for predicting seismic responses of spherical sliding type isolated structures. Earthq Eng Struct Dyn 34(9):1027–1046
20. Tsopelas P (2009) Exact modeling of triple friction pendulum isolators with hysteretic rule based model (under preparation)
21. Tsopelas P, Roussis P, Constantinou MC, Buchanan R, Reinhorn AM (2005) 3D-BASIS-ME-MB: computer program for nonlinear dynamic analysis of seismically isolated structures. Technical Report MCEER005-0009, Buffalo, NY
22. Ucak A, Tsopelas P (2009) A multi surface plasticity model for modeling of double curvature friction pendulum isolators (under preparation)
23. Wongprasert N, Symans MD (2005) Experimental evaluation of adaptive elastomeric base-isolated structures using variable-orifice fluid dampers. J Struct Eng (ASCE) 131(6):867–877

Chapter 31
Modal Analysis of Isolated Bridges with Transverse Restraints at the End Abutments

Nicos Makris, Georgios Kampas, and Dimitra Angelopoulou

31.1 Introduction

Seismic isolation, either with elastomeric or sliding bearings, is at present widely adopted as an effective technology for the seismic protection of highway and railway bridges [6, 3]. Traditionally, many conventionally designed bridges use elastomeric bearings (pads) between the deck and its supports to accommodate thermal movements. The long experience with this technology had a positive role on the implementation of modern seismic protection technologies in bridges [4].

The most commonly used isolation bearings are either elastomeric bearings or spherical sliding bearings. Figure 31.1 shows the view of two neighbor railway bridges currently under construction in central Greece. Both bridges are seismically isolated on identical spherical sliding bearings with radius of curvature $R = 2.2$ m. The isolation period of the spherical bearings alone, in any horizontal direction is $T_I = 2\pi\sqrt{R/g} = 2.98$ s. In both bridges the motion of the deck is restrained along the transverse direction. This restriction is nearly imperative in railway bridges in order to avoid misalignment of the rails at the deck-abutment joints during earthquake shaking; while, it is also common in highway bridges.

The eigenvalue analysis of both bridges can be conducted with a linear stick model using elastic beam elements. The stick model allows for the flexure of the center piers and the finite stiffness of pile foundations in the horizontal, vertical, rocking and cross horizontal-rocking directions [9].

Dynamic analysis with commercially available software resulted that the five-span, 200 m long bridge has as first eigenvalue the longitudinal, $T_1 = T_L = 3.07$ s, a value that is slightly larger than the spherical bearing period, $T_I = 2\pi\sqrt{R/g} = 2.98$ s, due to the finite flexibility of piers and piles connected in series with the bearings. The second eigenvalue of the five-span, 200 m-long, bridge is the first transverse eigenvalue, $T_2 = T_T = 2.17$ s, a value that is appreciably smaller from

N. Makris (✉)
Department of Civil Engineering, University of Patras, Patras GR 26500, Greece
e-mail: nmakris@upatras.gr

M.N. Fardis (ed.), *Advances in Performance-Based Earthquake Engineering*, Geotechnical, Geological, and Earthquake Engineering 13, DOI 10.1007/978-90-481-8746-1_31,

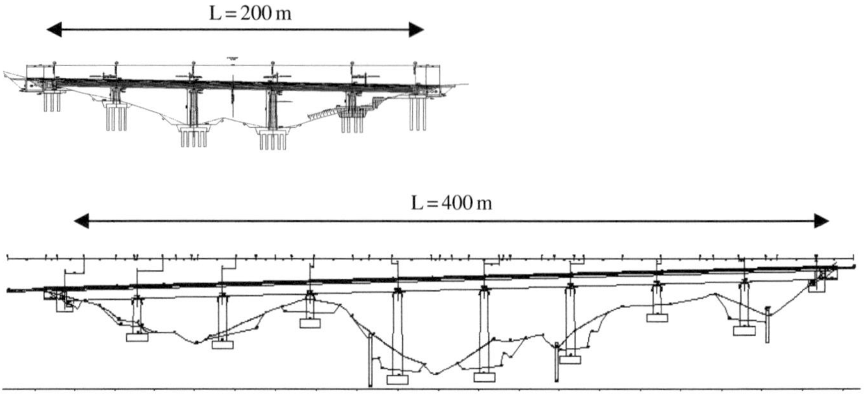

Fig. 31.1 Elevation of two seismically isolated bridges. *Top*: 5-span 200 m long bridge; *bottom*: 9-span 400 m long bridge

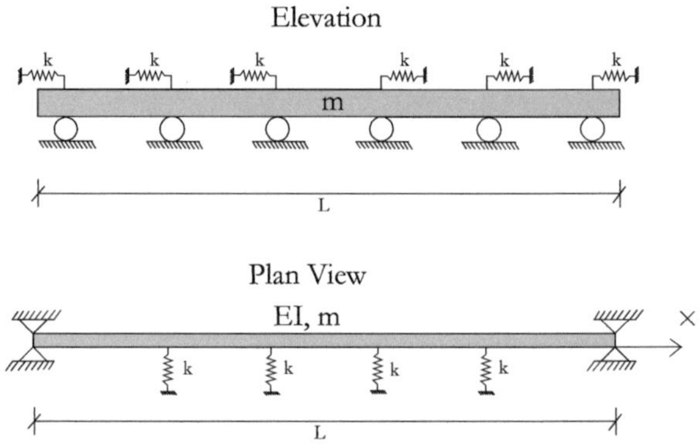

Fig. 31.2 Mechanical model of a deck fully isolated in the longitudinal direction while in the transverse direction the deck is isolated at the center supports and simply supported at the abutments

the spherical bearing period, $T_I = 2\pi\sqrt{R/g} = 2.98$ s, due to the added transverse flexural rigidity of the deck which is simple-supported at the end-abutments (see Fig. 31.2). The longer nine-span, 400 m long, bridge has as first eigenvalue, the first longitudinal eigenvalue, $T_1 = T_L = 3.16$ s and very close, yetsecond eigenvalue, the transverse eigenvalue, $T_2 = T_T = 3.13$ s. This numerical result, where the transverse period has reached so closely the longitudinal period, was the main motivation for this study in order to examine whether it is possible that the transverse period may exceed the longitudinal period.

31.2 Mechanical Idealization of Isolated Bridges

Figure 31.2 shows the mechanical model of the isolated bridge where the transverse motion of the deck is isolated with springs at the center piers and restrained at the end-abutments; while, the longitudinal motion is isolated with elastic springs

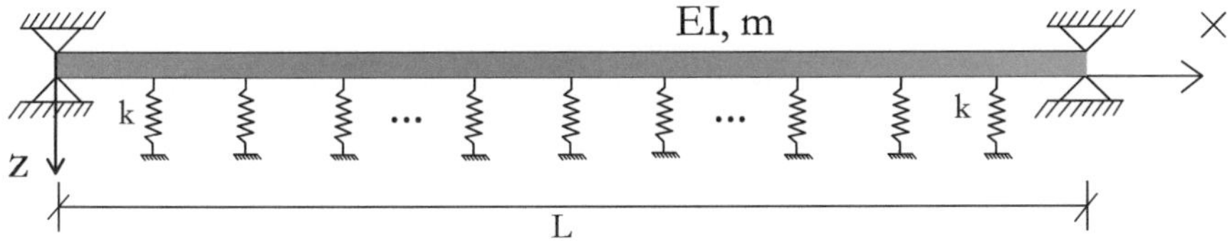

Fig. 31.3 Plan view of a beam with continuous distributed springs along its length (Winkler foundation) simply supported in the transverse direction at its ends

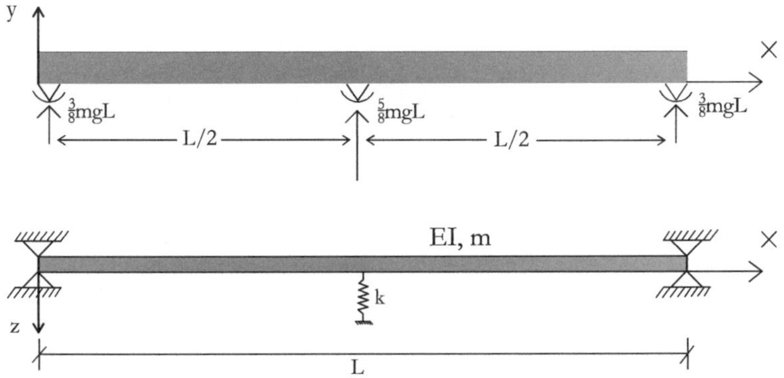

Fig. 31.4 *Top*: elevation of a two-span beam fully isolated along the longitudinal direction; *bottom*: plan view of a beam isolated at mid-span and simply supported at its ends

everywhere. In order to capture the dynamic behavior of the mechanical configuration in Fig. 31.3 we examine the mathematical solution of the two limiting cases – that of a beam fully isolated along the longitudinal direction while restrained at the end abutments along the transverse direction and having either (a) infinite distributed transverse springs along the span (beam on Winkler supports) as in Fig. 31.3, or (b) a single longitudinal and a single transverse spring at the mid-span as in Fig. 31.4.

31.3 Longitudinal and Transverse Eigenvalues of a Beam with Continuously Distributed Springs

For a beam on continuously distributed elastic supports with stiffness k $[F]/[L]^2$ and distributed mass, m $[M]/[L]$; and assuming that the axial rigidity of the beam is large compared to its flexural rigidity, its first longitudinal eigenvalue is the isolation frequency along the longitudinal direction

$$\omega_{L1} = \omega_{IL} = \sqrt{\frac{kL}{mL}} = \sqrt{\frac{k}{m}} \tag{31.1}$$

Under free vibration, the governing equation of motion along the transverse direction is ([7], among others)

$$EI\frac{d^4w(x)}{dx^4} + (k - m\omega^2)w(x) = 0 \tag{31.2}$$

It is well known in the literature ([7, 8, 1] among others) that the eigenvalues of the homogeneous Equation given by (31.2) existing only for $k - m\omega^2 < 0$ and are given by the expression

$$\omega_{Tn} = \sqrt{\frac{k}{m} + n^4\pi^4\frac{EI}{mL^4}},\ n \in \{1, 2, ...\} \tag{31.3}$$

Equation (31.3) shows that the lowest transverse eigenfrequency ($n = 1$) of the isolated deck is $\omega_{T1} = \sqrt{k/m + \pi^4 EI/mL^4}$, therefore, it will be always longer than the first longitudinal frequency $\omega_{L1} = \sqrt{k/m}$. Consequently, the limiting case model which idealizes the isolated deck on distinct bearings with a flexural beam with continuously distributed springs yields that, no matter how long the bridge is, the first transverse isolated period is always smaller than the isolated longitudinal period due to the flexural rigidity of the deck ($\pi^2\sqrt{EI/mL^4}$).

31.4 Longitudinal and Transverse Eigenvalues of a Beam with a Single Longitudinal and Transverse Spring at the Mid-Span

We now proceed with the eigenvalue analysis of the other limiting mechanical idealization – that of a beam where its transverse motion is isolated with springs at the mid-span and restrained at the end supports; while the longitudinal motion is isolated with identical elastic springs at all three supports (see Fig. 31.4).

31.4.1 Transverse Periods

Given the symmetry of the problem we can analyze half of the beam with $l = L/2$ where the right free end has zero slope ($dw(l)/dx = 0$) and the shear equals to the spring reaction. Note that this model yields only the odd eigenvalues.

The solution of the vibration of a beam with flexural rigidity, *EI* and distributed mass, *m* is [7, 1],

$$w(x) = A\sin\lambda x + B\cos\lambda x + C\sinh\lambda x + D\cosh\lambda x \tag{31.4}$$

where now

$$\lambda = 4\sqrt{\frac{m\omega^2}{EI}} > 0 \tag{31.5}$$

The boundary conditions of this configuration for vibrations along the transverse directions are zero translation and zero moment at the left end-support ($w(0) = d^2w(0)/dx^2 = 0$), while at the right end ($x = l = L/2$) $dw(l)/dx = 0$ and

$V(l) = -EI\ d^3w(l)/dx^3 = -k/2\ w(l) = 0$, where k is the transverse stiffness of the bearing(s) at the mid-span.

With the abovementioned boundary conditions the eigenvalues of the system for vibrations along the transverse direction are obtained from the solution of the homogeneous system.

$$\begin{bmatrix} 0 & 1 & 0 & 1 \\ 0 & -1 & 0 & 1 \\ \cos\lambda l & -\sin\lambda l & \cosh\lambda l & \sinh\lambda l \\ -\lambda^3\cos\lambda l - \tilde{k}\sin\lambda l & \lambda^3\sin\lambda l - \tilde{k}\cos\lambda l & \lambda^3\cosh\lambda l - \tilde{k}\sinh\lambda l & \lambda^3\sinh\lambda l - \tilde{k}\cosh\lambda l \end{bmatrix} \begin{Bmatrix} A \\ B \\ C \\ D \end{Bmatrix} = \begin{Bmatrix} 0 \\ 0 \\ 0 \\ 0 \end{Bmatrix} \tag{31.6}$$

in which $\tilde{k} = k/2EI$. The solution of the associated characteristic equation gives

$$2\tilde{k}\cosh\lambda l\cos\lambda l[\tanh\lambda l - \tan\lambda l - \frac{2\lambda^3}{\tilde{k}}] = 0 \tag{31.7}$$

Equation (31.7) is satisfied either when $\tanh\lambda l - \tan\lambda l - 2\lambda^3/\tilde{k} = 0$ or if $\cos\lambda l = 0$. The condition $\cos\lambda l = 0$ corresponds to $\lambda l = (2n+1)(\pi/2)$, which gives the eigenvalues of the simple supported beam (without springs at the mid-span). For instance the first transverse period, T_{T1}^{SS}, of the simple supported beam with length L is ($n = 0$).

$$T_{T1}^{SS} = \frac{2}{\pi}\sqrt{\frac{mL^4}{EI}} \tag{31.8}$$

while the second modal period, $T_{T2}^{SS} = T_{T1}^{SS}/4$. By setting the quantity in brackets in Eq. (31.7) equal to zero, one obtains

$$\tanh\lambda l = \tan\lambda l + \frac{4(\lambda l)^3}{\xi} \tag{31.9}$$

where $\xi = kL^3/8EI$ is a dimensionless parameter which expresses the relative contribution of the spring at the mid-span to the transverse flexural rigidity of the deck.

The solution of the transcendental equation given by Eq. (31.9) is obtained for various values of ξ with a Newton-Raphson method. The transcendental Eq. (31.9) has real and positive solutions, say $S(\xi) = \lambda(\xi)l$ with $l = L/2$ and $\lambda(\xi) = \sqrt[4]{m\omega^2(\xi)/EI} > 0$. Accordingly,

$$S(\xi) = \frac{L}{2}\sqrt[4]{\frac{m\omega_T{}^2(\xi)}{EI}} \tag{31.10}$$

or

$$\omega_T(\xi) = 4S^2(\xi)\sqrt{\frac{EI}{mL^4}} \tag{31. 11}$$

Equation (31.11) with the help of Eq. (31.8) gives

$$\frac{T_T}{T_{T1}{}^{SS}} = \frac{\pi^2}{4}\frac{1}{S^2(\xi)} \tag{31.12}$$

in which, T_T , is the first transverse period of the two-span beam supported with a bearing at the mid-span. Figure 31.5 plots the expression given by Eq. (31.12) after finding the roots of Eq. (31.9), $S(\xi) = \lambda(\xi)l$, for the entire range of ξ that is of interest. The above numerical solution serves to validate the computations of the commercially available software, SAP [2], which yields the point-circles when elastomeric bearings are used and the point-squares when friction pendulum bearings are used. Note that for $\xi \rightarrow 0$ the ratio, T_T/T_{T1}^{SS}, tends to one; while, as $\xi \rightarrow \infty$, $T_T/T_{T1}^{SS} \rightarrow 4/25 = 0.16$ [5].

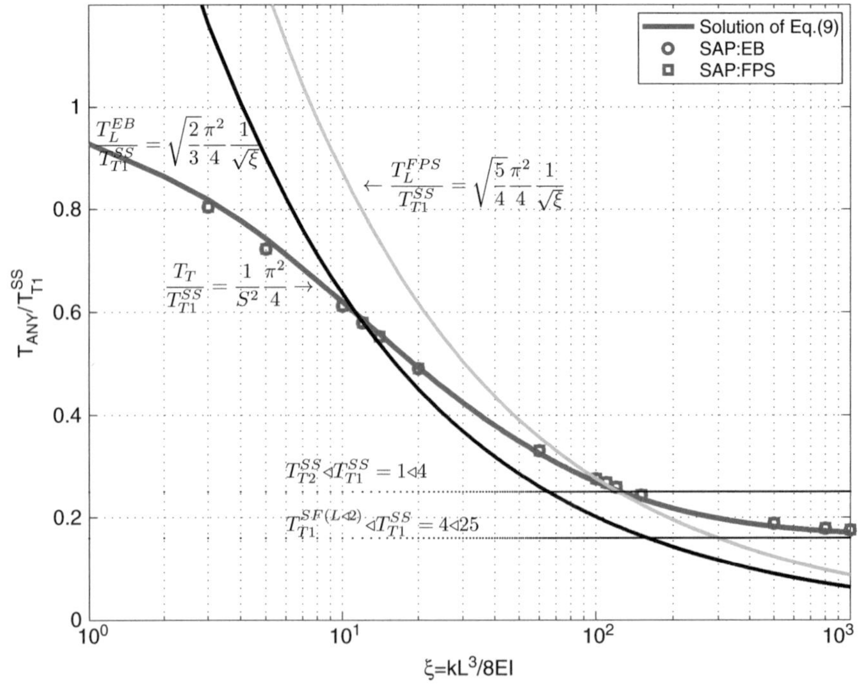

Fig. 31.5 Comparison of the normalized first transverse period T_T of a two-span isolated deck of length L (solution of Eq. (31.9)) against the longitudinal isolation periods $T_L{}^{EB}$ or $T_L{}^{FPS}$ (*EB* : elastomeric bearings, *FPS*: friction pendulum system)

Figure 31.5 also shows that regardless whether the two-span deck is isolated on elastomeric or sliding bearings, the normalized transverse period of the two-span deck, T_T/T_{T1}^{SS}, is offered by a single master curve that is only a function of the normalized stiffness of the transverse spring at the mid-span$\xi = kL^3/8EI$.

This result can be also obtained qualitatively from dimensional analysis without solving the eigenvalue problem and the associated characteristic Eq. (31.9) (see [5]).

31.4.2 Longitudinal Periods

31.4.2.1 Elastomeric Bearings

When the two-span deck of Fig. 31.4 is supported at each of the three supports (end abutments and mid-span) on identical elastomeric bearings with lateral stiffness k, the longitudinal period of the bridge is

$$T_L^{EB} = 2\pi\sqrt{\frac{mL}{3k}} \tag{31.13}$$

The dimensionless stiffness of the bearing, $\xi = kL^3/8EI$, appearing in the right-hand side of the characteristic Eq. (31.9) gives

$$k = \frac{8\xi EI}{L^3} \tag{31.14}$$

Substitution of Eq. (31.14) to (31.13) gives,

$$T_L^{EB} = \frac{1}{\sqrt{6}}\frac{1}{\sqrt{\xi}}\frac{\pi}{2}\sqrt{\frac{mL^4}{EI}} \tag{31.15}$$

From Eq. (31.8) the first transverse period of the simple supported beam with length L is $T_{T1}^{SS} = \frac{2}{\pi}\sqrt{mL^4/EI}$, and therefore Eq. (31.15) gives

$$\frac{T_L^{EB}}{T_{T1}^{SS}} = \sqrt{\frac{2}{3}\frac{\pi^2}{4}}\frac{1}{\sqrt{\xi}} \tag{31.16}$$

Figure 31.5 plots the expressions given from Eq. (31.16) next to the line given by Eq. (31.12). Note that the two lines cross at an approximate value of $\xi \approx 12$, beyond which the transverse period of the two-span bridge supported on elastomeric bearings exceeds the longitudinal isolation period, T_L^{EB}, given by Eq. (31.16). This result is quite remarkable given that the solution from the beam on elastic foundation (infinite bearings) does not predict any crossing (the transverse period of a beam on Wrinkler foundation is always shorter than the longitudinal period regardless the length of the deck: Eq. (31.3)).

31.4.2.2 Spherical-Sliding Bearings

We consider now the alternative situation where the two-span bridge of Fig. 31.4 is supported at each of the three supports (end-abutments and mid-span) on identical spherical sliding bearings with radius of curvature$R$. In this case the longitudinal period of the bridge is

$$T_L^{FPS} = 2\pi\sqrt{\frac{R}{g}} \tag{31.17}$$

Given that the two-span beam is a continuous beam, the vertical reaction at the center bearing is $N_C = 5/8\ mgL$, while the vertical reaction at the end-bearings is $N_e = 3/16\ mgL$. Accordingly, the lateral stiffness of the center spherical sliding bearing is, $k = N_C/R = 5/8\ mgL/R$ and the dimensionless stiffness of the center bearing is

$$\xi = \frac{kL^3}{8EI} = \frac{5}{64}\frac{mgL^4}{REI} \tag{31.18}$$

Substitution in Eq. (31.18) the ratio *g/R* from Eq. (31.17) gives,

$$\frac{T_L^{FPS}}{T_{T1}^{SS}} = \sqrt{\frac{5}{4}\frac{\pi^2}{4}}\frac{1}{\sqrt{\xi}} \tag{31.19}$$

Figure 31.5 also plots the expression given from Eq. (31.19) next to the line given by Eq. (31.12). Note now that the two lines cross at an approximate value of $\xi \approx 116$. This result shows that the transverse period of a two-span isolated bridge with spherical sliding bearings which is transversely restrained at the end abutments may exceed the longitudinal isolation period $T_L^{FPS} = 2\pi\sqrt{R/g}$; however, this may happen when the two-span bridge is significantly longer than the corresponding length of the two-span bridge isolated on elastomeric bearing that offer the same longitudinal period.

This remarkable result shows that while a bridge designer may use either elastomeric or spherical sliding bearings to achieve a desirable isolation period along the longitudinal direction, the transverse period of the deck when spherical sliding bearings are used will be always shorter (stiffer configuration) than the transverse period offered by elastomeric bearing which give the same isolation period along the longitudinal direction. In a recent publication [5] it has been shown using arguments from dimensional analysis that the solution of Eq. (31.9) shown in Fig. 31.5 can be used to estimate the transverse period of multispan bridges with arbitrary number, n, of equal spans.

31.5 Conclusions

This paper examines the eigenvalues of relatively long seismically isolated bridges in which the transverse displacement of the deck at the end-abutments is restricted. With this restriction the deck is fully isolated along the longitudinal direction, while along the transverse direction the deck is a simple supported beam at the end-abutments which enjoys concentrated restoring forces from the isolation bearings above the center piers. This study investigates mathematically the eigenvalue problem of a two-span isolated deck and concludes that regardless of the value of the isolation period along the longitudinal direction there is a certain length beyond which the transverse period of the deck will exceed the longitudinal isolation period. The value of this length depends on whether the deck is isolated on elastomeric or sliding bearings. This finding was obtained only after considering local springs at the deck since the beam on distributed elastic supports (Beam on Wrinkler foundation) is unable to capture this phenomenon.

References

1. Clough RW, Penzien J (1993) Dynamics of structures, 2nd edn. McGraw-Hill, New York, NY
2. Computers and Structures. SAP 2000 (2006) Documentation. University of California, Berkeley, CA
3. FHWA (1995) Seismic retrofitting manual for highway bridges. U.S. Department of Transportation, Washington, DC, USA
4. Kelly JM (1997) Earthquake-resistant design with rubber. Springer, London
5. Makris N, Kampas G, Angelopoulou D (2009) The eigenvalues of isolated bridges with transverse restrains at the end abutments. Earthq Eng Struct Dyn, in press, DOI: 10.1002/eqe.971
6. Skinner RI, Robinson WH, McVerry GH (1993) An introduction to seismic isolation. Wiley, New York, NY
7. Timoshenko SP, Young DH, Weaven WJR (1974) Vibration problems in engineering, 4th edn. Wiley, New York, NY
8. Ugural AC, Fenster SK (1995) Advanced strength and applied elasticity, 3rd edn. Prentice-Hall, Upper Saddle River, NJ
9. Zhang J, Makris N (2002) Seismic response analysis of highway overcrossings including soil-structure interaction. Earthq Eng Struct Dyn **31**:1967–1991

Chapter 32
Benefit–Cost Evaluation of Seismic Risk Mitigation in Existing Non-ductile Concrete Buildings

Gregory Deierlein and Abbie Liel

32.1 Introduction

Emerging performance-based earthquake engineering methods can offer significant new insights to the systematic evaluation of design criteria and policy-related questions for new and existing buildings. In this study, recently developed performance-based technologies are applied to assess the comparative performance of older "non-ductile" reinforced concrete (RC) buildings versus modern "ductile" buildings, which employ capacity design approaches and ductile reinforcing bar detailing. Concerns with older non-ductile designs stem from buildings constructed in the high seismic regions of the western United States, prior to the mid-1970s when major changes were instituted for seismic design of RC structures as a result of damage observed during the 1971 San Fernando earthquake. Prior to these changes, seismic design requirements for concrete frames did not require capacity design provisions to inhibit the formation of story mechanisms or column shear failures. While it is generally recognized that such buildings do not provide the same level of safety and damage control as modern buildings, there are many debates as to the safety of existing buildings and whether policies should be adopted to require detailed risk assessment and mitigation.

This study aims to improve understanding of earthquake risks in non-ductile RC buildings and the cost-effectiveness of mitigating these risks through building replacement or retrofit. Performance-based methods are applied to assess the risk of damage and collapse to a set of archetypical RC-framed office buildings that are representative of those in the high-seismic regions of California. A related objective is to illustrate the application of the performance-based cost–benefit analyses, which can be generally applied to other building types in other seismic regions.

G. Deierlein (✉)
Department of Civil and Environmental Engineering, Stanford University, Stanford, CA 94305, USA
e-mail: ggd@stanford.edu

M.N. Fardis (ed.), *Advances in Performance-Based Earthquake Engineering*, Geotechnical, Geological, and Earthquake Engineering 13, DOI 10.1007/978-90-481-8746-1_32,

32.2 Performance Assessment Methodology

The performance assessment follows an approach developed by the Pacific Earthquake Engineering Research (PEER) Center that provides a systematic formulation to characterize earthquake ground motions, structural response, building damage, and ultimately decision metrics related to economic losses and life safety risks [1, 6]. Key aspects of the implementation are summarized below, and the reader is referred to the underlying studies by Liel and Deierlein [7] and Haselton and Deierlein [3] for further details.

Structural Demand Parameter Assessment: The structural response is determined through nonlinear dynamic analyses of two dimensional RC frame models that characterize the significant design features of the archetype buildings. As shown in Fig. 32.1, the frame models utilize nonlinear springs to characterize the inelastic behavior of beams, columns, beam-column joints, and foundations. The three-bay configuration and finite joint elements reflect the important design-related aspects of interior versus exterior joints and column behavior. The analyses are conducted using the *OpenSees* simulation platform (http://opensees.berkeley.edu). The dynamic time history analyses are organized using the so-called *incremental dynamic analysis* procedure, whereby the response is calculated for specific ground motions that are scaled by increasing intensity (often spectral acceleration intensity, scaled at the first-mode period) up to the point of structural collapse. At each intensity level, statistics of structural response parameters, such as peak story drift or plastic hinge rotations, are compiled to characterize the response.

Collapse Risk Assessment: Collapse risk statistics are determined from the dynamic analyses, but considering three important adjustments. First, where certain failure modes are not directly simulated in the nonlinear analysis, these are checked through a post-processing of the data to adjust the dynamic analysis results. For example, shear failure and loss of axial load capacity of non-ductile RC columns is a mode of failure that is not simulated directly in the analysis and is incorporated through the post-processing operation. A second adjustment made is to collapse capacity to account for the spectral shape effects in extreme (rare) ground motions

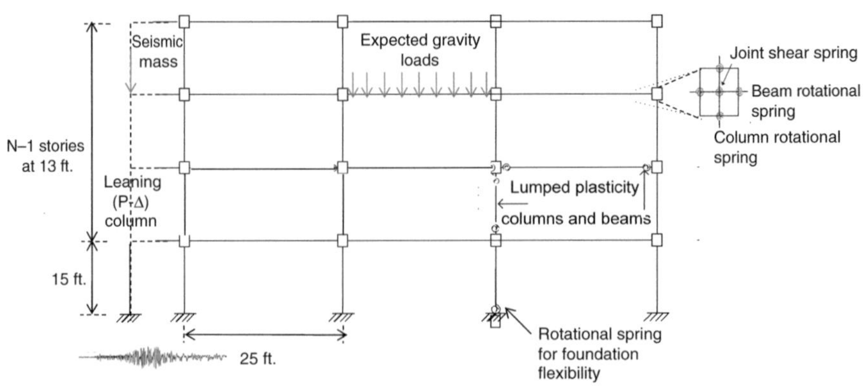

Fig. 32.1 Schematic of the RC frame structural analysis model

that are not reflected in the input ground motions used for the dynamic analyses. Following an approach described by Haselton and Deierlein [3], the median collapse capacity (described in terms of spectral acceleration) is adjusted to reflect the characteristic spectral shape of the dominant hazard spectra for the rare ground motions. Finally, the dispersion in collapse capacities from the dynamic analyses is adjusted to consider so-called modeling uncertainties. For the RC frames, the modeling uncertainty increases the total dispersion (standard deviation of the natural logarithm of collapse intensities) from about 0.4, due just to ground motion variability, to about 0.6.

Damage and Loss Assessment: The damage and associated costs for building repair or replacement are calculated considering damage to both structural and nonstructural components. Losses were determined using a toolbox of damage fragility and loss functions developed by Mitrani-Reiser [8] and applied in a related study by Goulet et al. [2]. The loss components were based on a typical architectural office building layouts, such as shown in Fig. 32.2.

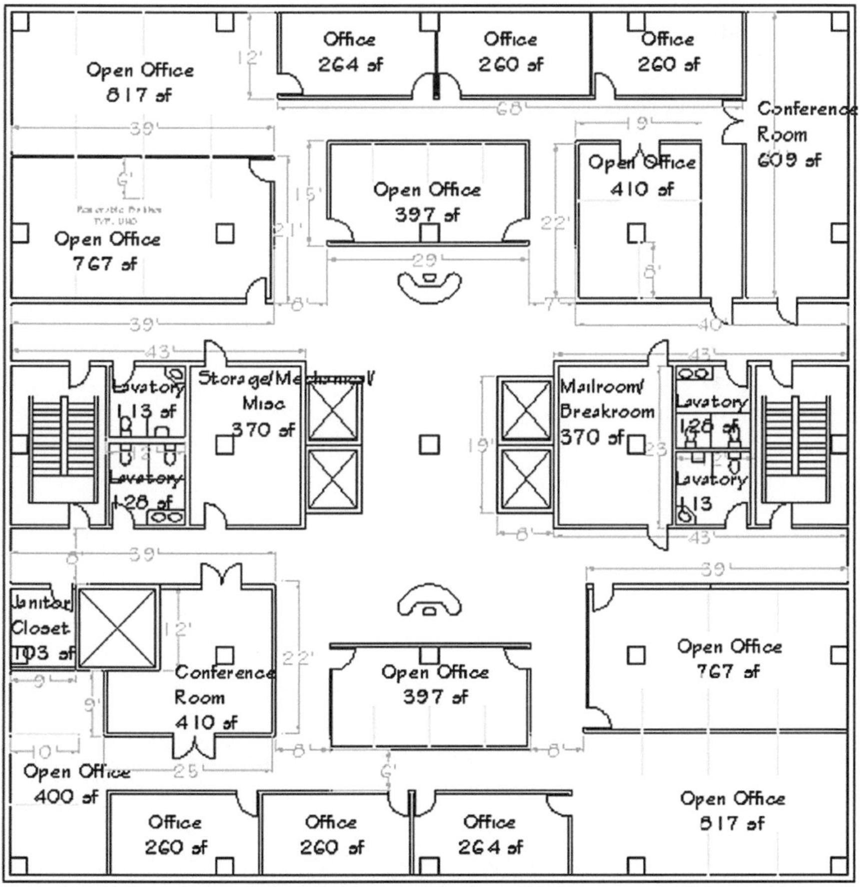

Fig. 32.2 Architectural floor plan of archetype building

Fatality Risk and Loss Assessment: The risks of human fatalities due to building collapse are calculated considering the average building occupancy, the building collapse volume ratio, and statistics on fatalities for persons trapped in collapsed buildings. Parameters assumed for this study include: occupancy density 1 person per 23 m^2 of office space, average occupancy rate of 0.33, trapped occupant rates of 0.3–0.6 for collapsed buildings, and an average fatality rate of 0.17 for trapped persons.

32.3 Comparative Assessment of RC Building Archetypes

Eight representative building designs were developed and analyzed to generalized performance of buildings with non-ductile and ductile characteristics. As summarized in Table 32.1, the buildings ranged in height from 2 to 12 stories and utilized either two-way space (S) frame or perimeter (P) frame configurations, e.g., 2S = 2-story space frame. The non-ductile and ductile RC frames were designed according to the 1967 and 2003 building code provisions, respectively, for a typical site in southern California [4, 5].

Table 32.1 Collapse performance results for (a) non-ductile RC frames (b) ductile RC frames

Structure	Ω	S_{CT} (g)	$Sa_{2/50}$ (g)	CMR	MAFc $\times 10^{-4}$	EAL (%)	EAF $\times 10^{-3}$
(a)							
2S	1.9	0.47	0.80	0.59	109	5.2	41
2P	1.6	0.68	0.79	0.85	47	3.2	24
4S	1.4	0.27	0.49	0.54	107	2.3	62
4P	1.1	0.31	0.47	0.66	100	2.3	97
8S	1.6	0.23	0.31	0.75	64	1.8	77
8P	1.1	0.29	0.42	0.68	135	2.1	141
12S	1.9	0.29	0.35	0.83	50	1.6	76
12P	1.1	0.24	0.42	0.56	119	1.6	192
(b)							
2S	3.5	3.55	1.16	3.07	1.0	1.0	0.4
2P	1.8	2.48	1.13	2.19	3.4	1.0	1.7
4S	2.7	2.22	0.87	2.56	1.7	1.1	1.3
4P	1.6	1.56	0.77	2.04	3.6	1.2	2.7
8S	2.3	1.23	0.54	2.29	2.4	1.3	3.1
8P	1.6	1.00	0.57	1.77	6.3	1.0	8.3
12S	2.1	0.83	0.44	1.91	4.7	1.1	9.4
12P	1.7	0.85	0.47	1.84	5.2	0.8	9.9

Ω: over strength, ratio of ultimate strength to design strength from pushover analysis
S_{CT}: median collapse intensity, based on Sa(T1)
$Sa_{2/50}$: ground motion intensity, Sa(T1), with 2% in 50 year chance of exceedence
CMR: Collapse Margin Ratio = S_{CT} / $Sa_{2/50}$
MAFc: Mean Annual Frequency of Collapse, collapses per year.
EAL: Expected Annual Loss as a percentage of building replacement value
EAF: Expected Annual Fatalities, fatalities per year

Summarized in Table 32.1 are the key performance assessment results for each frame, including parameters to characterize the risks associated with collapse, economic loss, and fatalities. One measure of the collapse safety is the *Collapse Margin Ratio* (CMR), which is the ratio between the median collapse spectral acceleration capacity (S_{CT}), as obtained from the dynamic analyses, to the ground motion intensity with a 2% chance of exceedence in 50 years ($Sa_{2/50}$). The resulting collapse fragilities reveal dramatic differences in collapse capacity for the non-ductile versus ductile frames, where the former have collapse probabilities on the order of 0.65– 0.85 under the $Sa_{2/50}$ ground motions, as compared to probabilities of 0.05–0.15 for the ductile frames. When the collapse fragilities are integrated with the seismic hazard curve, the resulting *Mean Annual Frequencies of collapse* (*MAFc*) for the non-ductile frames are about 35 times larger (on average) than the ductile frames. Ratios of *Expected Annual Fatality* (*EAF*) rates mirror the collapse rates since the relationships between the two are based on expected values.

In contrast to the collapse and fatality risks, the *Expected Annual Losses (EAL)* associated with building damage and repair are only about twice as large for the older non-ductile frames as compared to the modern ductile frames. As shown in Table 32.1, the *EAL* are about 0.8–1.3% of the building replacement value for the ductile frames and 1.6–5.2% for the non-ductile frames. The differences in loss behavior between the two are further illustrated in Fig. 32.3. Whereas the economic losses in the ductile frames tend to be dominated by the non-collapse condition at relatively low ground motion intensities, the losses for the non-ductile frames have a much larger contribution from building collapse. The relative performance is further evident by differences in losses under the design earthquake accelerations, indicated by the spectral intensity value S_D, in Fig. 32.3.

Whether or not the large apparent difference in safety between older versus modern buildings would warrant proactive mitigation steps, such as through mandatory

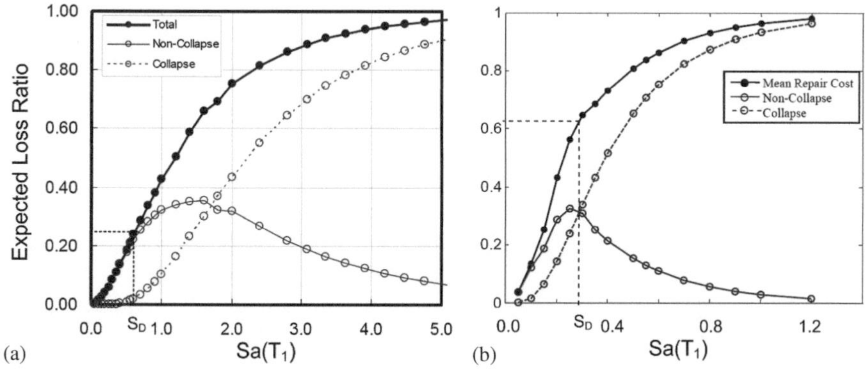

Fig. 32.3 Loss contributions for 4-story space frame buildings (**a**) ductile design and (**b**) non-ductile design

Table 32.2 Benefits and costs of replacing non-ductile RC frame structures

Building	Benefits over 50 years: Losses avoided ($, M)[a]	Lives saved[b]	Total benefits[c] ($, M)	Building replacement Cost ($, M)	Cost–benefit ratio
2S	6.5	2.0	8.6	6.1	0.7
2P	3.7	1.1	4.8	6.5	1.4
4S	3.9	3.1	7.0	12.5	1.8
4P	3.5	4.7	8.3	12.0	1.4
8S	2.7	3.7	6.5	19.9	3.1
8P	5.3	6.6	12.1	19.4	1.6
12S	4.2	3.3	7.6	29.1	3.8
12P	6.2	9.1	15.6	28.1	1.8

[a]Present value of losses avoided by building replacement, millions of US$
[b]The estimated total number of lives saved by building replacement
[c]Sum of the net present value (discounted over 50 years) of losses avoided and lives saved. The dollar value of lives saved is based on $2 million/life, discounted over 50 years based on annual discount rate of 3%

building retrofit or replacement, is an important and vexing question. To examine whether it would be worthwhile, based purely on economic considerations, to replace existing non-ductile buildings with new ductile buildings, the costs and benefits are compared in Table 32.2. Here the benefits of building replacement are considered to be the savings in economic losses associated with repair and the lives saved, calculated based on the relative risks to the non-ductile and ductile buildings. To compare the benefits on a pure economic basis, the value of life is assumed as 2 million US dollars, and the value of the benefits is calculated over an assumed 50-year building life with an annual discount rate of 3%. The cost to achieve these benefits is assumed to be the building replacement value, based on average building costs of about $1,500–$1,800 per square meter. As indicated in the last column of Table 32.2, with the exception of the 2-story space frame building, the cost–benefit ratios all violate the breakeven value of 1, indicating that the replacements are not justified on the basis of economic costs.

32.4 Cost–Benefit Assessment of Building Retrofit

Whereas the previous analysis only considered the cost–benefits of building replacement, a more likely option for risk mitigation is structural retrofit. To examine this, the following three alternative retrofit methods were examined for the 4- and 8-story space frame buildings: (1) Fiber – wrapping of columns with fiber composites to improved their ductility, (2) Column Jacketing – reinforced concrete jacketing of columns to improve both their strength and ductility, and (3) Wall Piers – addition of wall piers to add strength and ductility to the system. As described by Liel and

Table 32.3 Comparative benefits of retrofitting non-ductile frames

			Annualized losses[b]			Benefits over 50 years[b]		
Structure[a]		Ω	MAFc × 10^{-4}	EAL (% rep.)	EAF (× 10^{-3})	Losses avoided (% repl.)	Lives saved	Total benefits (% repl.)
4S	1967	1.4	107	2.3%	62	NA	NA	NA
	Fiber	1.4	75	2.1%	50	5%	0.6	10%
	CJ	3.5	7	1.2%	9	28%	2.7	49%
	WP	2.1	20	1.8%	27	13%	1.8	27%
	2003	2.7	2	1.1%	1.3	31%	3.1	56%
8S	1967	1.6	64	1.8%	77	NA	NA	NA
	Fiber	1.6	60	1.6%	68	5%	0.5	8%
	CJ	2.1	6	1.4%	9	10%	3.4	28%
	WP	2.2	7	1.5%	11	8%	3.3	25%
	2003	2.3	2	1.3%	3.1	14%	3.7	32%

[a]Structure types include the original 1967 non-ductile design and retrofits based on: fiber wrapping, column jacketing, and addition of wall piers. Results are also shown for 2003 code-conforming ductile design

[b]Definitions for annualized losses and benefits are the same as in Tables 32.1 and 32.2

Deierlein [7], the retrofit designs are fairly modest and intended to impact minimally the building architecture.

Performance assessment results of the retrofit designs are summarized in Table 32.3, using the same metrics previously discussed in Tables 32.1 and 32.2. In general, fiber wrap retrofits were only marginally effective, since they did not reduce the tendency for story collapses, and the benefits of improved column ductility were limited. On the other hand, the column strengthening (CJ) and wall (WP) retrofits, which inhibit story mechanisms, provide the largest benefits, as evidenced by increases in the static pushover index, Ω, and commensurate reductions in the risk of collapse (MAFc) and fatalities (EAF).

The relative amount of benefits gained by reduction in economic losses versus life safety risks can be inferred by comparing the "losses avoided" and "total benefits" columns of Table 32.3. For the 4-story building, the total benefits of the retrofit were due in about equal share to reductions in damage losses and lives saved. For the 8-story building, proportionately more benefit was derived from the economic value of lives saved.

Whether or not the retrofits are cost-effective will depend upon the cost of the retrofits. To achieve a favorable cost–benefit ratio less than 1, the retrofit cost would need to be less than the total benefits, as given by the last column of Table 32.3. For example, the CJ retrofit of the 4-story building could cost up to 49% of the building value and still have a favorable cost–benefit ratio. On the other hand, the least effective retrofit, fiber wrapping of columns in the 8-story building, could only cost up to 8% of the building value. Assuming that building structural retrofits generally cost in the range of \$400–\$800 per square meter of building area, and assuming a typical building value of \$1,800 per square meter, those retrofits with costs less than 20–40% of the building replacement value would have favorable cost–benefit ratios. In practice, one could imagine that thoughtful design of retrofit solutions would further improve their effectiveness and reduce their costs to improve their attractiveness.

32.5 Conclusions

As illustrated in this paper, by articulating seismic performance in terms of explicit life-safety and economic metrics, performance-based engineering approaches provide the opportunity to evaluate the relative safety of seismically deficient existing buildings with modern code-conforming buildings. For example, the comparative analyses show the older non-ductile RC buildings to have risks of collapse and fatalities that are about 35 times larger for modern buildings, and damage losses that are about 2 times larger. When applied in conjunction with cost–benefit analyses, the methods can help to inform decisions by owners and other public policy makers on the cost-effectiveness of building retrofit or replacement to mitigate earthquake risks. For the non-ductile buildings investigated in this study, retrofit is shown to be a cost-effective option for risk mitigation.

Acknowledgments: This work has been supported by the PEER Center through the Earthquake Engineering Research Centers Program of the National Science Foundation (under award number EEC-9701568). The authors would also like to acknowledge valuable input from Curt Haselton, Judith Mitrani-Reiser, Marc Ramirez, Evan Reis, Ashley Spear, and Jackie Steiner.

References

1. Deierlein GG (2004) Overview of a comprehensive framework for earthquake performance assessment, Performance-based seismic design concepts and implementation: Proceedings of the international workshop Bled, Slovenia, PEER TR 2004/05, pp 15–26
2. Goulet CA, Haselton CB, Mitrani-Reiser J, Beck JL, Deierlein GG, Porter KA, Stewart JP (2007) Evaluation of the seismic performance of a code-conforming reinforced-concrete frame building – from seismic hazard to collapse safety and economic losses. Earthq Eng Struct Dyn 36(13):1973–1997
3. Haselton CB, Deierlein GG (2007) Assessing seismic collapse safety of modern reinforced concrete frame buildings, PEER TR 2007/08
4. ICBO (1967) Uniform Building Code. ICBO, Pasadena, CA
5. ICC (2003) International Building Code. ICC, Falls Church, VA
6. Krawinkler H, Miranda E (2004) Performance-based earthquake engineering. In: Borzognia Y, Bertero V (eds) Earthquake engineering: from engineering seismology to performance-based engineering, 1st edn, pp 9-1 – 9-59). CRC Press, Boca Raton, FL
7. Liel AB, Deierlein GG (2008) Assessing the collapse risk of California's existing reinforced concrete frame structures: metrics for seismic safety decisions, Blume Earthquake Engineering Center, TR No. 166, Stanford University, Stanford, CA
8. Mitrani-Reiser J (2007) An ounce of prevention: probabilistic loss estimation for performance based earthquake engineering, Doctoral Dissertation, Caltech

Chapter 33
Seismic Retrofit of Non-ductile Reinforced Concrete Frames Using Infill Walls as a Rocking Spine

Khalid M. Mosalam and M. Selim Günay

33.1 Introduction

Reinforced concrete (RC) frames containing unreinforced masonry (URM) infill walls are a commonly used structural system around the world. URM infill walls are generally treated as non-structural elements which are used mainly for architectural purposes. However, as structural elements, they have both beneficial and detrimental effects. Infill walls contribute to the lateral force resisting capacity and damping of the structure up to a certain level of ground motion. They increase the initial stiffness and decrease the initial period of a structure, which might be beneficial depending on the frequency content of the experienced ground motion, an example of which is the recent 2009 earthquake in L'Aquila, Italy. However, the URM infill walls are prone to early brittle failure and the infill wall failure may lead to the formation of a soft story and consequent column shear failure. In addition, infill walls interact with the surrounding frame in such a way that column shear failure is made more likely. There is an interaction effect between the in-plane strength of the wall and its out-of-plane strength, with load in one direction reducing the strength in the other. Out-of-plane failure of the URM infill walls leads to life-safety hazard from falling debris. Figure 33.1 shows URM infill walls damage from 2008 Wenchuan, China and 2009 L'Aquila, Italy earthquakes.

Since URM infill walls, as structural elements, may lead to undesired damage, their realistic modeling is important. In this paper, a previously developed practical infill wall analytical model [6] considering the interaction between in-plane and out-of-plane responses is implemented in a progressive collapse algorithm [11]. The infill wall model is used to investigate the efficacy of a retrofit method which comprises of strengthening the infill walls with mesh reinforcement and a concrete layer taking advantages of the strengthened infill walls as rocking spines [4].

K.M. Mosalam (✉)
Department of Civil and Environmental Engineering, University of California, Berkeley, CA 94720-1710, USA
e-mail: mosalam@ce.berkeley.edu

M.N. Fardis (ed.), *Advances in Performance-Based Earthquake Engineering*, Geotechnical, Geological, and Earthquake Engineering 13, DOI 10.1007/978-90-481-8746-1_33,

Fig. 33.1 URM infill walls damage from earthquakes in 2008 Wenchuan, China (*top*) and 2009 L'Aquila, Italy (*bottom*)

33.2 Analytical Model for the URM Infill Wall

The analytical model employed for modeling the infill walls is a practical model previously developed in [6]. In this model, each infill wall panel is represented by a single diagonal element (Fig. 33.2), which is composed of two beam-column elements joined at the midpoint node which is assigned a lumped mass in the out-of-plane (OOP) direction. The cross-section of the beam-column elements is modeled by strategically locating a collection of nonlinear fiber elements using the structural and earthquake engineering software OpenSees [9]. The fibers are located along a line in the OOP direction (Fig. 33.2). By this way, the beam-column element acts as truss and flexural elements in the IP and OOP directions, respectively. The model considers the interaction between the in-plane (IP) axial strength and the OOP bending strength. Location of the fiber elements and the nonlinear material properties assigned to these fiber elements are set such that the intended strength interaction and the IP axial and OOP bending stiffness values are properly simulated in these elements. FEMA-356 [1] equations are used for calculating the axial stiffness and unidirectional strength in the IP direction. The OOP mass, stiffness and unidirectional bending strength are calculated such that the model has the same natural frequency as the original infill wall, it should produce the same support reactions, where it is attached to the surrounding frame, for a given support motion

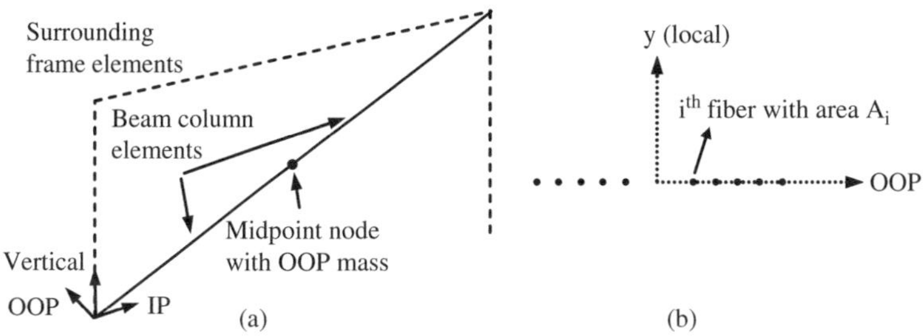

Fig. 33.2 (**a**) Utilized infill wall model, (**b**) fiber layout in the cross-section

(story acceleration) and it should exhibit initial yielding at the same level of support motion that causes the original infill wall to yield. Relevant equations and their derivation can be found in [6]. The IP axial and OOP bending strength interaction curve is accepted as a 3/2-power curve, represented with Equation 1.

$$\left(\frac{P_{IP}}{P_{IP0}}\right)^{3/2} + \left(\frac{M_{OOP}}{M_{OOP0}}\right)^{3/2} \leq 1.0 \tag{1}$$

where P_{IP} is the IP axial strength in the presence of OOP force, P_{IP0} is the IP axial strength without OOP force, M_{OOP} is the OOP bending strength in the presence of IP force, and M_{OOP0} is the OOP bending strength without IP force. Equation (1) matches quite well with the finite element results reported in [5].

Since only one diagonal element is utilized in the model, it has both tension and compression strengths. Therefore, the fibers are considered to have the same absolute value for the tensile and compression yield strengths. A bilinear relationship with a very small strain hardening is assumed for the stress-strain relationship.

33.3 Implementation into the Progressive Collapse Algorithm

The analytical infill wall model described in the previous section is implemented in a previously developed progressive collapse algorithm [11]. The mentioned progressive collapse algorithm had been developed by using element removal based on dynamic equilibrium and the resulting transient change in system kinematics, the underlying theory of which can be found in [10, 11].

The progressive collapse algorithm had been implemented for automated removal of collapsed elements during an ongoing simulation (Fig. 33.3). The implementation had been carried out as a new OpenSees module, designed so that it is called by the main analysis module after each converged load step to check each element for possible violation of its respective removal criteria. A violation of any pre-defined removal criterion triggers the activation of the element removal algorithm on the violating element before returning to the main analysis module.

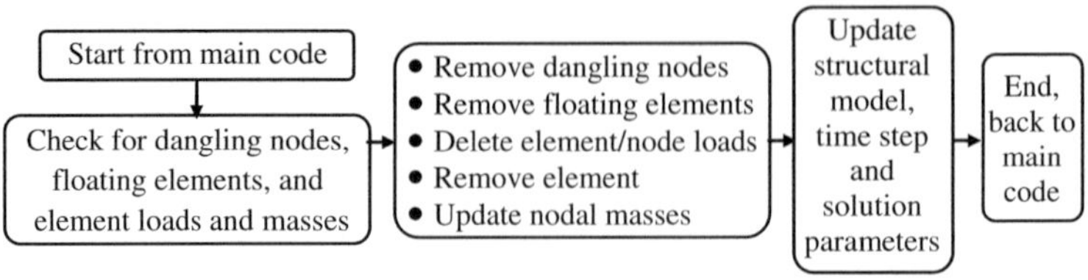

Fig. 33.3 Considered element removal algorithm

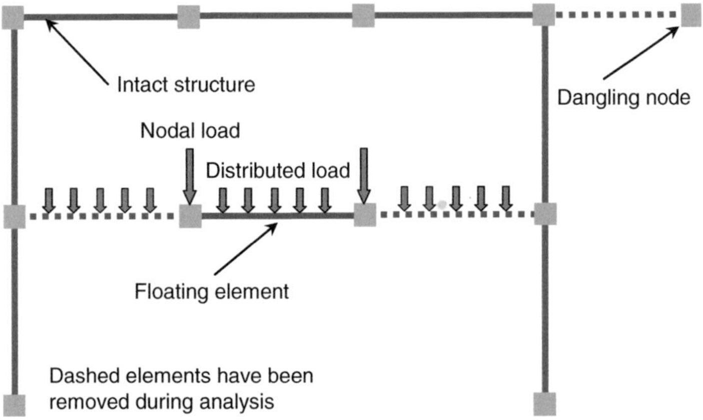

Fig. 33.4 Demonstration of dangling nodes and floating elements as a result of element removal

Activation of the element removal algorithm includes updating nodal masses, checking if the removal of the collapsed element results in leaving behind dangling nodes or floating elements, which must be removed as well (Fig. 33.4), and removing all associated element and nodal forces, imposed displacements, and constraints.

Removal criteria had been defined for force-based and displacement-based distributed plasticity fiber elements and lumped plasticity beam–column elements with fiber-discretized plastic hinges in OpenSees [9]. The criteria were based on the material-level damage indices for a confined RC cross-section model [11].

Implementation of the removal of the aforementioned infill wall analytical model in the progressive collapse algorithm is performed through defining a new criterion for the beam-column elements. The new criterion is based on the interaction between the IP and OOP displacements. IP displacement is the relative horizontal displacement between the top and bottom nodes of the diagonal element. OOP displacement is the OOP displacement of the middle node (where the OOP mass is attached) with respect to the chord which connects the top and bottom nodes in OOP direction. The same equation used for the strength interaction is accepted for the displacement interaction. When the mentioned combination of displacements from the analysis exceeds the envelope curve (Fig. 33.5), the two beam-column elements and the middle node representing the URM infill wall are removed. IP and OOP displacement capacities in the presence of zero load in the other direction

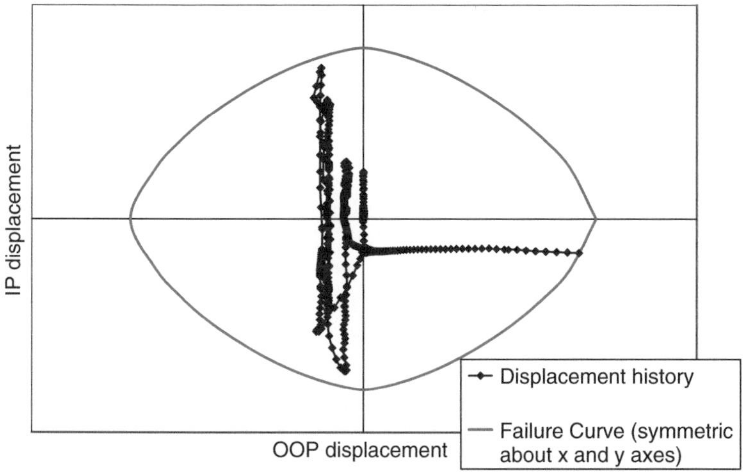

Fig. 33.5 Demonstration of the infill wall removal considering the interaction between IP and OOP displacements

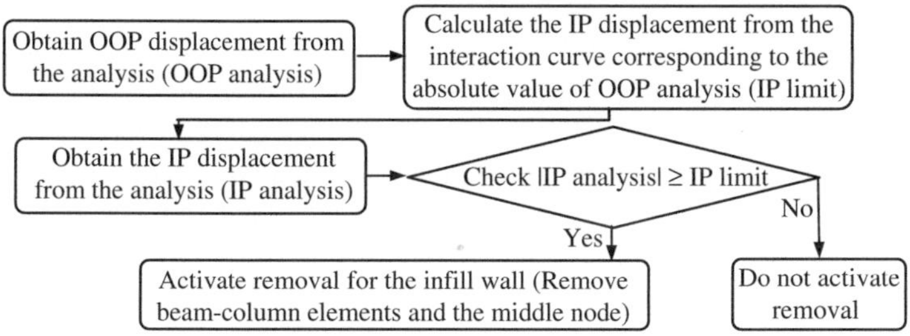

Fig. 33.6 Procedure for an infill wall removal

are obtained from FEMA-356 for collapse prevention level. The procedure for the removal of an infill wall is presented in Fig. 33.6.

An OOP stability criterion is defined in [3] for URM load-carrying walls. First, a critical acceleration value of the infill beyond which the connection between the diaphragm and the wall starts to slide is calculated. After sliding, kinetic and potential energy terms are compared to decide on the stability and failure of the wall. It is noted that this criterion is not employed in the current study since it considers OOP failure without IP interaction.

33.4 Application to a Retrofit Method

The implemented infill wall model is utilized for investigating the efficacy of a retrofit method. This method consists of strengthening the URM infill walls with mesh reinforcement and a concrete layer on one side of the infill wall. Strengthening the infill wall increases the vertical forces transferred from the infill wall to the columns and results in tensile forces in the columns. Without improving the foundation and letting it uplifts, the strengthened infill wall behaves as a rocking spine. Details of the application of this retrofit method can be found in [4].

In order to investigate the efficacy of the retrofit method, a nine story, five bay two dimensional (2D) frame is modeled using OpenSees [9], in which the removal of infill walls is implemented. The properties of the frame resemble the Bombiya Arcade Building in Karachi, Pakistan. The story height is 3.6 m; story mass is 60 tons, and bay width is 3.0 m. Cross-section sizes and reinforcement detailing of the columns and beams are indicated in Fig. 33.7.

For the as-built cases, URM infill walls with 0.15 m thickness are located at every bay of all the stories where the material is modeled with modulus of elasticity 3,450 MPa, compressive strength 6.9 MPa, and shear strength 0.34 MPa. For the retrofitted cases, infill walls of all of the stories of the middle bay are strengthened using #3 mesh reinforcement with 150 mm spacing in both directions and 100 mm thick concrete layer on one side only. As-built URM infill walls are modeled as described above and the strengthened URM infill walls are modeled as two elastic diagonal struts with only axial degrees of freedom. Stiffness of these struts are calculated by modifying the infill wall strut stiffness equation in ASCE/SEI 41-06 [2] for a composite cross-section of masonry and reinforced concrete and equally divided between the two diagonal struts.

Springs which have infinitely high stiffness and strength in compression and zero strength in tension are employed for modeling the boundary conditions in the vertical direction at the bases. In this way, uplifting of the foundation due to the rocking caused by the strengthened URM infill wall as a spine can be simulated.

Force-based beam-column element (nonlinearBeamColumn) is employed for beams and columns with five integration points. Material models designated as Concrete02 and Steel02 in OpenSees [9] are used to model concrete ($f'_c = 20\,\text{MPa}$) and steel ($f_y = 410\,\text{MPa}$) uniaxial behavior, respectively.

Nonlinear shear springs are placed at the top and bottom of the columns using zero-length elements. The envelope force-deformation relationship of these shear springs are defined by a linear relationship up to the shear force capacity, then with a negative slope equal to 1% of the initial stiffness until the residual shear force which is set as 10% of the shear capacity. The residual shear force remains constant after this point. The hysteretic rule for cyclic response is defined with moderate

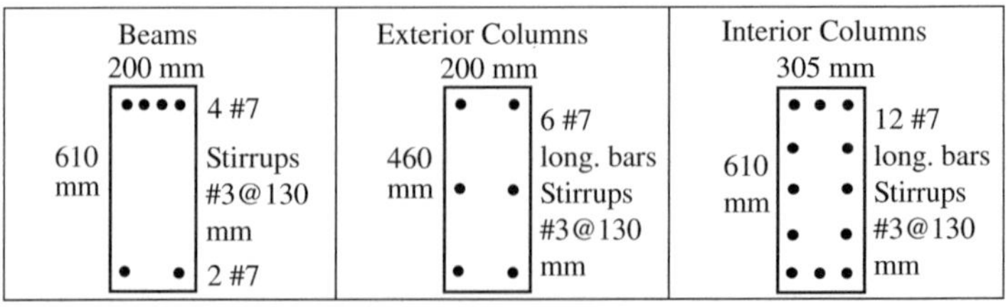

Fig. 33.7 Cross-sections and the reinforcement detailing of the columns and beams for the application 9-story, 5-bay 2D frame structure

degradation. The reason of placing shear springs both at the top and bottom of the columns is related to the local lateral forces transferred by the diagonal elements representing the different infill walls at the top and bottom of a column.

Story masses are lumped at the nodes at the story levels. Mass and tangential stiffness proportional Rayleigh damping (5% of critical) is used with constants calculated by using the first and third mode periods. It is noted that first mode periods of the as-built and retrofitted frames are 0.98 and 0.93 s, respectively.

OOP translational elastic springs are placed at the story levels to represent the frames in the transverse direction. Stiffness values of these OOP springs are adjusted such that periods in the longitudinal and transverse directions are the same.

Fragility analyses of the as-built and retrofitted frames are conducted bi-directionally using twenty ground motion pairs, which were extensively studied and carefully selected for a previous study for the seismic evaluation of a science building located on the campus of the University of California, Berkeley (UCS building) under the TestBed program of PEER [8]. A scaling based on the first mode pseudo-acceleration (PSa) is applied to the ground motions. Accordingly, six (0.25, 0.5, 0.75, 1.0, 1.25, and 1.5*g*) PSa levels are considered.

The maximum interstory drift profiles obtained from the ground motion labeled “LV_mgnp” [8] with 1.5*g* PSa are plotted in Fig. 33.8. It is observed that the retrofit method transforms the soft story drift in the as-built case to a uniform interstory drift distribution along the height. Figure 33.9a demonstrates the rocking in the retrofitted case. This figure shows that all URM infill walls (dotted) failed but the spine in the middle bay caused the system to rock. It is also concluded that the retrofit method did not prevent the failure of the URM infill walls but it prevented collapse by transforming the concentrated drifts in lower stories to a uniform drift distribution along the height. In contrast to the uniform distribution in Fig. 33.9a, the soft story mechanism in the as-built case is clearly observed in Fig. 33.9b.

For the fragility analyses, the maximum interstory drift ratio (MIDR) along the height is chosen as the response variable. It is shown in [7] that the ground motion intensity and profile are the dominant input variables affecting MIDR for a study conducted for the UCS building mentioned above. Therefore, uncertainty in other variables (e.g. mass, damping, or strength) is not taken into consideration.

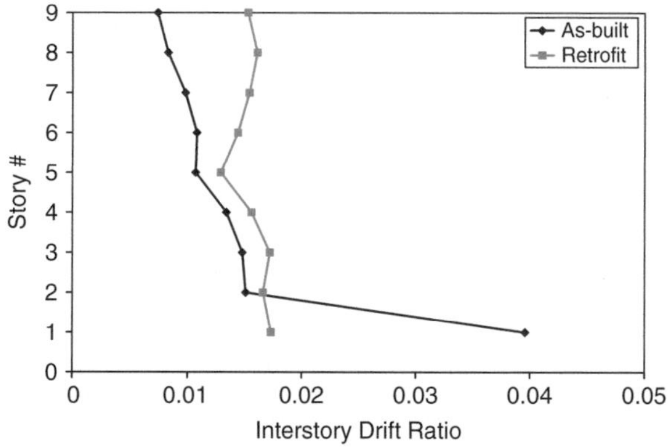

Fig. 33.8 Comparison of maximum interstory drift profiles for the as-built and retrofitted cases for the ground motion “LV_mgnp” with 1.5*g* PSa

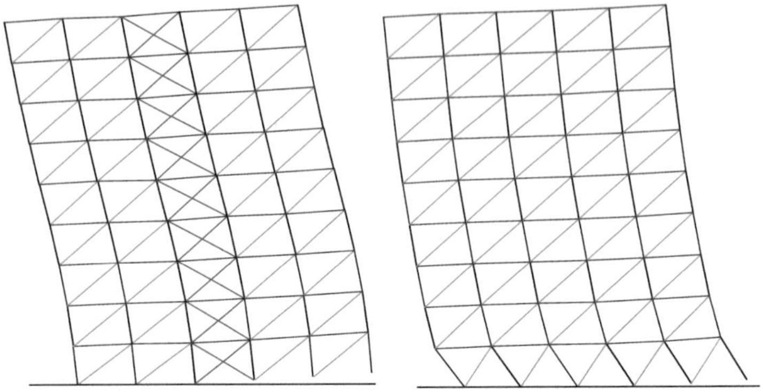

Fig. 33.9 Deformed shape at peak roof displacement under "LV_mgnp" with 1.5*g* PSa for the rocking of the retrofitted frame (*left*) and the soft story mechanism of the as-built frame (*right*)

For each PSa level, MIDR for each of the 20 ground motions is calculated. For some levels, the number of data is less than 20 because of the numerical failures due to excessive deformations. Then, a lognormal statistical distribution is fitted to these MIDR values and the lognormal parameters, the mean (λ) and the standard deviation (ξ), are calculated for each intensity level. Afterwards, the probability of exceedance of a certain MIDR value is calculated for these intensity levels (defined in terms of the PSa). Three limiting MIDR values, namely 0.01, 0.02 and 0.03, are considered in this study.

Obtained fragility relationships for the as-built and retrofitted cases are plotted in Fig. 33.10. It is observed that the retrofit method is quite successful in reducing the probability of failure in terms of the inter-story drifts (IDR).

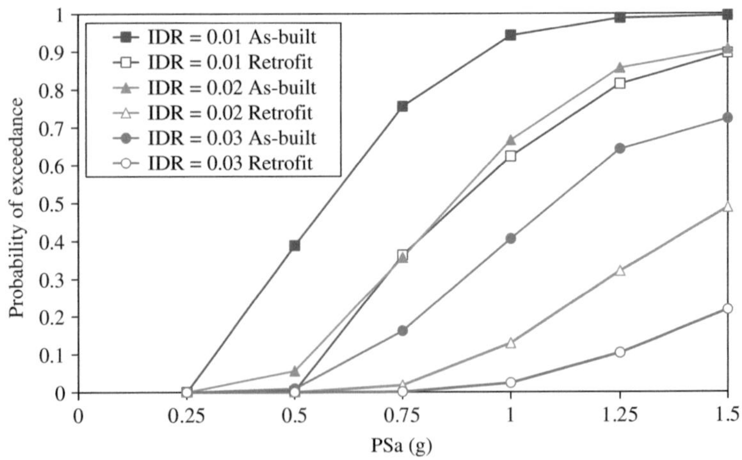

Fig. 33.10 Fragility relationships for the as-built and retrofitted frames

33.5 Summary and Conclusions

A previously developed infill wall analytical model which considers the interaction between in-plane and out-of-plane responses is implemented in a progressive collapse algorithm. The implemented infill wall model is utilized for investigating the efficacy of a retrofit method which comprises of strengthening the infill walls with mesh reinforcement and a concrete layer taking advantage of the strengthened infill walls as rocking spines. Through nonlinear dynamic analyses of a case study of nine-story, five-bay 2D RC frame with URM infill walls, it is demonstrated that the retrofit method is effective in reducing the interstory drifts and transforming the concentrated drifts due to soft story to uniform drifts along the height. Moreover, fragility analyses illustrated the effectiveness of this retrofit method in reducing the probability of failure in terms of exceeding the inter-story drift to selected threshold values.

Acknowledgements TUBITAK (Scientific and Technical Research Council of Turkey) is acknowledged for the funding provided to the second author.

References

1. American Society of Civil Engineers (2000) Prestandard and commentary for the seismic rehabilitation of buildings, FEMA-356, ASCE, Reston, VA
2. American Society of Civil Engineers (2008) Seismic rehabilitation of existing buildings. ASCE/SEI Standard 41-06, ASCE, Reston, VA
3. Erbay ÖO (2004) A methodology to assess seismic risk for populations of unreinforced masonry buildings. Mid-America Earthquake Center, MAE, Report 07-10
4. Günay S, Korolyk M, Mar D, Mosalam KM, Rodgers J (2009) Infill walls as a spine to enhance the seismic performance of non-ductile reinforced concrete frames. ATC&SEI conference on improving the seismic performance of existing buildings and other structures, 9–11 December, San Francisco, CA, USA
5. Hashemi SA, Mosalam KM (2007) Seismic evaluation of reinforced concrete buildings including effects of infill masonry walls. Pacific Earthquake Engineering Research Center, Technical Report 2007/100
6. Kadysiewski S, Mosalam KM (2009) Modelling of unreinforced masonry walls considering in-plane and out-of-plane interaction. 11th Canadian masonry symposium, 31 May–3 June, Toronto, Canada
7. Lee T-H, Mosalam KM (2003) Sensitivity of seismic demand of a reinforced concrete shear-wall building. 9th international conference on applications of statistics and probability in civil engineering, ICASP9, San Francisco, CA, USA, 6–9 July, pp 1511–1518
8. Lee T-H, Mosalam KM (2006) Probabilistic seismic evaluation of reinforced concrete structural components and systems. Pacific Earthquake Engineering Research Center, Technical Report 2006/04
9. Mazzoni S, McKenna F, Scott MH, Fenves G (2004) Opensees User's Manual. www.opensees.berkeley.edu 10 Sept 2009
10. Talaat M, Mosalam KM (2008) Computational modeling of progressive collapse in reinforced concrete frame structures. Pacific Earthquake Engineering Research Center, PEER 2007/10
11. Talaat M, Mosalam KM (2009) Modeling progressive collapse in reinforced concrete buildings using direct element removal. Earthq Eng Struct Dyn 38(5):609–634

Chapter 34
Deformation Capacity of Lightly Reinforced Concrete Members – Comparative Evaluation

S.J. Pantazopoulou and D.V. Syntzirma

34.1 Introduction

Research towards development of expressions for estimating the deformation capacity of R.C. members may be traced back to the time when strength expressions were first established [9]. However, the success in the area of deformation prediction is far more limited. For example, whereas the lateral load strength of a simple structure such as a well detailed cantilever R.C. column may be quantified with a margin of error within 10% of the actual value, the estimated drift capacity with the available tools today may be as far off as 100% of the actual value, with a generally inestimable and uncertain margin of safety.

This uncertainty is particularly relevant in displacement based assessment, where the assessment verdict explicitly rides on the ability of the member to sustain the imposed drift demand, and is reflected by the large factors of safety used to obtain design values [3, 5, 4]. A number of issues regarding behaviour of the response mechanisms in R.C. are considered responsible for the scatter. The paper reviews the mechanics of deformation of R.C. members and methods of evaluation of deformation resultants at advanced stages of inelastic response. A comparative study of consistently estimated deformation capacities with estimates obtained from Codes is conducted on a series of column-specimens tested under reversed cyclic load simulating earthquake effects; the specimens modelled former detailing practices representative of the 1950s–1970s [14, 12]. The models' performance in assessing the dependable deformation capacity is interpreted with reference to the important mechanisms controlling the observed specimen behaviour.

S.J. Pantazopoulou (✉)
Department of Civil Engineering, Demokritus University of Thrace, Vas. Sofias Str., No. 12, Xanthi, 67100, Greece
e-mail: pantaz@civil.duth.gr

M.N. Fardis (ed.), *Advances in Performance-Based Earthquake Engineering*, 359
Geotechnical, Geological, and Earthquake Engineering 13, DOI 10.1007/978-90-481-8746-1_34,

34.2 Deformation Mechanisms in R.C. Members

Behavior of R.C. frame members under combined axial load, and cyclic shear – flexure, such as occurring during earthquakes, is usually interpreted with the model of Fig. 34.1a. The static relationship between shear force and flexural moment in the span of the cantilever (Fig. 34.1a) is identical to that occurring over the length of the actual frame member extending from the point of contraflexure (zero moment) to the fixed end support. Deformations are owing to flexure, shear action, and pullout slip of the reinforcement from the support or lap splice; these mechanisms of behaviour are considered to act in series, therefore their effects are considered additive, as implied by the mechanical analogue of Fig. 34.1b, used in computer simulations of inelastic R.C. members. Elastic curvature over the member's length contributes to total drift, whereas all other effects (inelastic rotation over the plastic hinge region, shear deformation and pullout slip) are modelled through pertinent springs, each characterized by an individual resistance curve (Fig. 34.1c).

These mechanisms were originally assumed to act independently of each other. The total deformation obtained for any given load combination was approximated by the summation of the individual contributions (Fig. 34.2). The same concept was extended to deformation *capacity* which is a measure of total deformation that the member may undergo without significant irreversible loss of strength; for frame members this is usually quantified by the chord rotation θ_u, associated with a 20% loss of strength beyond the peak point.

The idea that deformation capacity may be estimated as the sum of contributions of the participating mechanisms was tested against hundreds of tests contained in many databases, including R.C. members with modern detailing as well as members

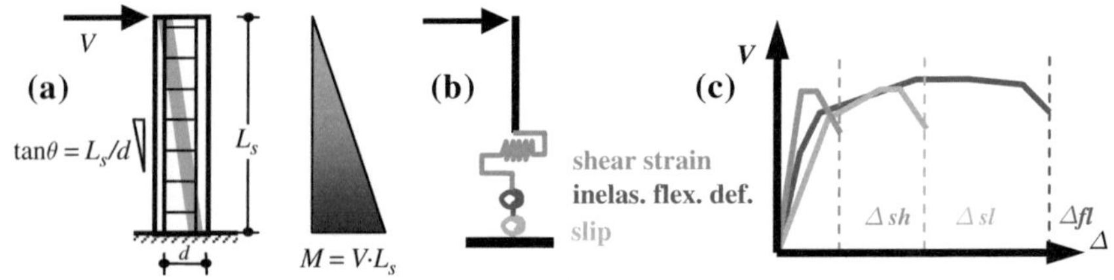

Fig. 34.1 (**a**) Idealization of shear span as a cantilever, (**b**) mechanistic analogue (**c**) resistance curves of individual springs, each having a different strength: $V_{u,\text{fl}}$, $V_{u,\text{sh}}$, $V_{u,\text{sl}}$

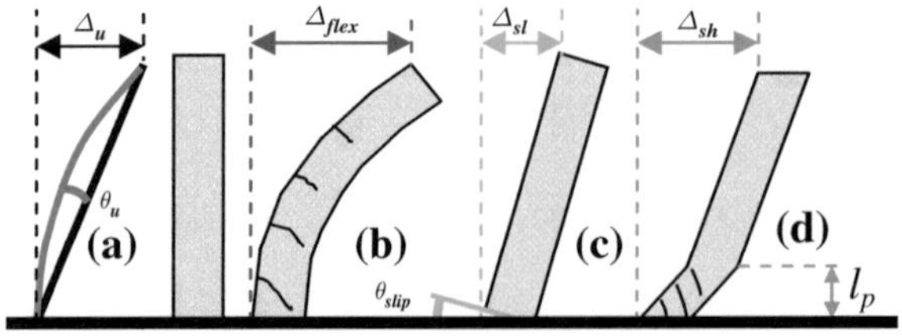

$$V = V_{fl} = \frac{M}{L_s} = V_{sh} = V_{sl};$$

$$\Delta = \Delta_{fl} + \Delta_{sh} + \Delta_{sl}$$

Fig. 34.2 Contribution of the various response mechanisms to the total drift with substandard details representative of old design practices [9, 8, 13, 7, 1]. Realistic values are obtained for well detailed members, which generally demonstrate large deformation capacity particularly when their axial load ratio is less than 0.4. The values become irrelevant when this concept is applied to members experiencing brittle failure, as often encountered with old-type frame members. Clearly, if the strength of one of the springs in the assembly of Fig. 34.1b is overcome at some value of deformation, then this event terminates the response curve of the member, well before the development of the estimated nominal deformation capacity of the other springs. This is why, of late, this approach has been retained only to describe behaviour up to the onset of yielding, i.e., $\theta_y=\theta_{y,\mathrm{fl}}+\theta_{y,\mathrm{sl}}+\theta_{y,\mathrm{sh}}$. For response beyond yielding, opinions diverge as to how to estimate deformation capacity.

Thus, the revised ASCE-41 document [4], which reflects the recommended N.A. assessment practice, evaluates directly the total inelastic drift capacity, θ_u, through empirical rules, the result being a single compound value that accounts for the various effects and design parameters through pertinent binary rules: here, the total rotation capacity is, $\theta_u=\theta_y+\theta_{\mathrm{pl}}$. Similar is the approach drafted for the revised EC8-III, which provides direct estimates for the total inelastic rotation capacity, θ_u, through calibrated expressions in terms of the relevant design variables (revised draft of [3]).

Recognizing the fundamental relevance of the model depicted in Fig. 34.1b the authors attempted to improve on its correlation with the test data, by modifying the additive expression, so that contributions of the individual contributing mechanisms were associated with the onset of occurrence of any type of premature failure [15]. This framework was referred to as *Capacity – Based Prioritizing* of failure modes (CBP):

$$\Delta_y = w_y \cdot [\Delta_{y,fl} + \Delta_{y,\mathrm{sh}} + \Delta_{y,\mathrm{sl}}] \;\; \Delta_u = \Delta_{y,\mathrm{fl}} + w_u \cdot [\Delta_{p,\mathrm{fl}} + \Delta_{p,\mathrm{sh}} + \Delta_{p,\mathrm{sl}}] \quad (34.1)$$

Factors w_y, w_u, represent strength or deformation ratios for strength-controlled or strain-controlled mechanisms of behaviour, respectively [15]. Here reference is made to the value of nominal strength terms, $V_{u,\mathrm{sh}}$, $V_{u,\mathrm{fl}}$, and $V_{u,\mathrm{sl}}$, which are considered to degrade with increasing imposed ductility, μ_Δ (i.e., even when flexural yielding is possible for $\mu_\Delta=1$, the relative hierarchy of the strength terms may be reversed for higher μ_Δ values, since the strength terms decay at different rates.) To evaluate these terms, what is needed is an immediate tool by which to transform forces and strains from the local level, where some type of material failure has been detected, to the global level, where prioritizing of strength components and addition of scaled deformation capacity contributions is needed. The necessary transformations are described in the following sections.

34.3 Local to Global Transformation of Stress Resultants

Capacity prioritizing in order to identify the weakest link of member behaviour is done on the basis of member shear forces. Thus, although any possible form of failure refers to exhaustion of some material strength or deformation capacity, it

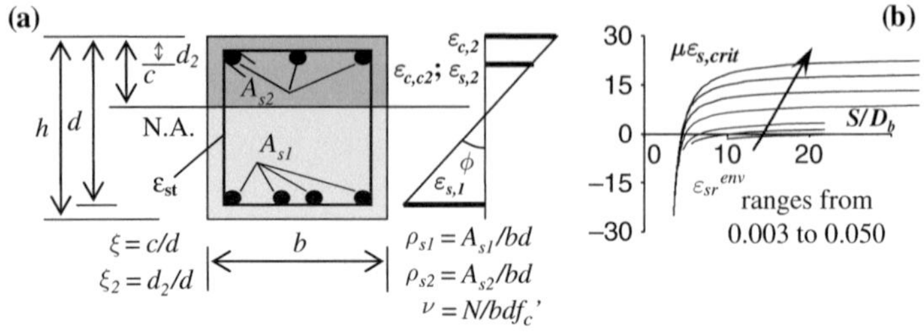

Fig. 34.3 (**a**) Definition of terms, (**b**) critical compr. strain ductility at bar buckling

is necessary that occurrence of this event be reflected in the global scale by the acting member shear force, *V*. With reference to Fig. 34.3a, possible material failure problems that would limit the strength of a column are:

(a) Cover delamination (i.e., max compr. strain $\varepsilon_{c,2}$ in the cover $\geq$ 0.0035–0.005).
(b) Confined concrete compressive strain at the extreme fiber of the confined zone, $\varepsilon_{c,c2}$, exceeds the strain capacity of the confined core, $\varepsilon_{c,\text{cu}}$.
(c) Loss of concrete contribution to lap splice strength owing to longitudinal cover cracking due to load reversals (i.e., when $\varepsilon_{c,c2} > 0.002$).
(d) Bar strain $\varepsilon_{s,1}$ exceeds the strain development capacity of the reinforcement: $\varepsilon_{s,1} \geq \min\{\varepsilon^{\text{anc - spl}}_{s,1,\max}; \varepsilon_{s,u}\}$ ($\varepsilon^{\text{anc - spl}}_{s,1,\max}$ is the tensile strain that may be sustained by the lap splice or anchorage at the critical section; $\varepsilon_{s,u}$ is the bar fracture strain.)
(e) Exhaustion of the compression strain capacity of longitudinal compression reinforcement (onset of bar buckling): $\varepsilon_{s,2} > \varepsilon_{s,\text{crit}}$.
(f) Occurrence of web diagonal tension cracking: $V \geq V_{\text{Rd1}}$ in [3] *or* V_c in [5,4].
(g) Onset of stirrup yielding, $\varepsilon_{\text{st}} = \varepsilon_{\text{st},y}$ (with regards to the nominal shear strength, V_{Rd3} in [3] or V_n in [5,4]).
(h) Occurrence of large postyielding strains in the stirrups (along the descending branch of the member response curve – associated with the degraded shear strength of the member, $V_{\text{fl}} = V_{u,\text{sh}}$, Eq. (34.3)): $\varepsilon_{\text{st}} > \varepsilon_{\text{st},y}$.

Equilibrium is used to convert from the material scale to the stress resultant of the cantilever of Fig. 34.1a (Eq. (34.2)). Normal strains over the cross section are assumed to follow a plane sections profile (Fig. 34.3a); for states of stress past flexural yielding, the normalized neutral axis depth, $\xi = c/d$, is assumed to remain approximately constant (its value may be obtained by interpolation using the gravity axial load, N_g, as a reference, average value [8]). Thus, for criteria (a) – (e) above, which refer to the occurrence of a milestone event in some component of normal strain, the corresponding shear force of the cantilever is obtained from:

$$V = M/L_s = \left[f_{s,1} A_{s1} d\,(1 - 0.4\xi) + N_g\,(0.5\,h - 0.4\xi d)\right]/L_s \tag{34.2}$$

In the above, $f_{s,1}$ is the axial stress in the steel tension reinforcement, for the corresponding axial strain, $\varepsilon_{s,1}$ (Fig. 34.3a) obtained from the steel stress-strain diagram. ($\varepsilon_{s,1}$ is the tension bar strain associated with each of the milestone events listed in (a)–(e) above: for case (a), $\varepsilon_{s,1}=\varepsilon_{c,2}(1-\xi)/\xi$; for (b), $\varepsilon_{s,1}=\varepsilon_{c,c2}(1-\xi)/(\xi-\delta_2)$, where $\delta_2=d_2/d$; for (c) and (d), the critical average bond stress $f_{b,u}$, acting over the active development length $L^u_{b,\mathrm{eff}}$, can develop a tension bar stress $f^{\mathrm{anc-spl}}_{s,1,\max}= 4f_b(L^u_{b,\mathrm{eff}}/D_b)\leq f_{s,u}$; for case (e) the critical buckling strain is obtained from pertinent interaction diagrams that relate the dependable strain ductility of compression reinforcement, $\mu\varepsilon_{c,\mathrm{crit}}$ with the normalized spacing of stirrups, S/D_b [15, 16], as shown in Fig. 34.3b – thus, the associated bar strain in the tension reinforcement, $\varepsilon_{s,1}$, is obtained from, $\varepsilon_{s,1}=\mu\varepsilon_{s,\mathrm{crit}}\, \varepsilon_{s,y}(1-\xi)/(\xi-\delta_2)$.)

For criteria (f)–(h) the stress resultant is given by the shear strength of the cantilever model. After diagonal tension cracking of the web, the shear strength, denoted by V_n in [5, 4] or as V_{Rd3} in [3] is obtained from a Mörsch type truss for (g) and (h). Shear strength degrades with increasing magnitude of imposed displacement ductility, μ_Δ, due to compression softening of the concrete struts (reduction in f'_c) and thereby of the shear strength of the equivalent truss. Expressions to model this effect have been derived from first principles with reference to deterioration of dowel action, aggregate interlock, bond and diagonal tension cracking in cracked reinforced concrete [6]. Codes model this process using a simple reduction coefficient $k(\mu_\Delta)$,according with [3, 5, 4]:

$$V_{u,\mathrm{sh}} = k(\mu_\Delta)\cdot(V_c + V_w) \tag{34.3}$$

where $k(\mu_\Delta)<1$ for $\mu_\Delta>1$ in [3], or for $\mu_\Delta>2$ in [5, 4] (i.e., after flexural yielding), whereby the process of degradation leads to a residual strength equal to 60% of V_n at $\mu_\Delta=6$ in [4], or to 75% of V_n at $\mu_\Delta=5$ in [3].

Values obtained for the milestone events listed above limit the strength of the mechanisms of resistance in series (Fig. 34.1b), namely *Flexural* ($V_{u,\mathrm{fl}}$), *Shear* ($V_{u,\mathrm{sh}}$), *Anchorage/Lap Splice* ($V_{u,\mathrm{sl}}$), or *Compression Bar Stability* ($V_{u,\mathrm{buckl}}$). Therefore, for any drift level, the above terms are organized in a hierarchy. The term with the lower strength, V_{fail}, controls the mode of damage and failure of the member:

$$V_{\mathrm{fail}} = \min\{V_{u,\mathrm{fl}}, V_{u,\mathrm{sh}}, V_{u,\mathrm{sl}}, V_{u,\mathrm{buckl}}\} \tag{34.4}$$

This value is then used to estimate the coefficient w_u in Eq. (34.1).

34.4 Strain – Displacement Transformations

Geometric relations are required to identify the magnitude of column drift or tip displacement of the model cantilever (*strain resultants*), for each of the milestone events listed in the preceding section. Deformation mechanisms participating to total drift are illustrated in Fig. 34.2: they represent flexural drift due to curvature along the member (Fig. 34.2b), rigid body rotation due to reinforcement pullout from the support anchorage or lap splice (Fig. 34.2c), and shear distortion which results in lateral offset of the member (Fig. 34.2d). Special considerations are as follows:

34.4.1 Strain Resultants Due to Flexural Curvature

34.4.1.1 Before Yielding of the Longitudinal Reinforcement

Chord rotation is $\theta = \Delta / L_s = \phi L_s / 3$. At yielding, $\theta_y = \phi_y L_s / 3$.

34.4.1.2 After Yielding of the Longitudinal Reinforcement

Inelastic flexural curvature is assumed to occur over a length of "plastic hinge", ℓ_p, measured from the face of the critical section. Inelastic tip displacement Δ_u and chord rotation θ_u or the associated plastic components Δ_{pl} and θ_{pl} are approximated by:

$$\Delta_u = \theta_u \cdot L_s; \quad \theta_u = \theta_y + \theta_{\mathrm{pl}}; \quad \theta_{\mathrm{pl}} = (\phi_u - \phi_y) \cdot \ell_p \cdot \left(1 - 0.5(\ell_p / L_s)\right) \approx \phi_u \ell_p \quad (34.5)$$

Definition of a Plastic Hinge Length, ℓ_{p}

The plastic hinge length ℓ_p was meant to account for spread of yielding along the length of the member. A consistent simple representation is obtained from the linear moment diagram along the shear span, which, for approximately constant internal lever arm after yielding, may be expressed in terms of the strain hardening ratio of main steel as per [8]:

$$\ell_p = [(f_{s,\max} - f_{s,y}) / f_{s,\max}] \cdot L_s + a_v = \beta \cdot L_s + a_v \quad (34.6)$$

where $f_{s,\max}$ is the peak stress attained by the bar at the critical section whereas a_v is the moment-shift due to shear ($\approx 0.9d$). As $f_{s,\max}$ increases approaching the fracture strength of tension reinforcement, so does the ℓ_p value; thus, there is no unique value for ℓ_p, but it increases with demand. Various interpretations are associated with practical expressions for ℓ_p. For example, to account for yield penetration, and for extensive yielding, various alternative expressions have been proposed:

$$\ell_p = \omega_1 \cdot L_s + \omega_2 \cdot D_b \cdot f_y + \omega_3 \cdot h \quad (34.7)$$

where $\omega_1 = 0.08$, $\omega_2 = 0.022$ and $\omega_3 = 0$ in [10, 11], whereas $\omega_1 = 0.1$, $\omega_2 = 0.24/\sqrt{f_c'}$ and $\omega_3 = 0.17$ in [3]. At least two issues contribute to the uncertainty about ℓ_p: The first is that the formal definition for ℓ_p given by Eq. (34.6) breaks down if the steel reinforcement is elasto-plastic (no hardening), leading to $\ell_p \to 0$, which contradicts the expectation that yield penetration will spread over the anchorage in the absence of hardening. Another is the physical significance of the D_b-dependent term: it is often associated with yield-penetration inside the support (anchorage), whereas an equal effect on θ_{pl} is also owing to yield penetration over the member length beyond the section of yielding moment (a term equivalent to a_v in Eq. (34.6)). Note that the most recent versions of the relevant assessment standards (e.g. in [4] and the revised draft of [3]) completely bypass the notion of a plastic hinge length by providing direct expressions for estimation of θ_{pl} without need for integration of inelastic curvatures. To eliminate the spurious outcome of Eq. (34.6) in practical applications, when the reinforcement exhibits little or no strain hardening, ℓ_p should be taken at least equal to $0.5d$.

34.4.2 Strain Resultants Owing to Bar Pullout/Slip

34.4.2.1 Before Yielding of Longitudinal Reinforcement

With reference to Fig. 34.2c, partial bar pullout from the support before yielding of longitudinal reinforcement causes a lumped rotation θ_{sl} at the support. This, in turn, produces a tip translation. The lumped rotation $\theta_{y,\mathrm{sl}}$ at the onset of yielding is associated with a linear attenuation of bar strains from the yield strain value $\varepsilon_{s,y}$ at the column support, to zero over the effective anchorage length, $L^y_{b,\mathrm{eff}}$, which is the length required to develop the bar yield force assuming a constant bond stress distribution equal to $f_{b,y}$ ($f_{b,y}$ is the *nominal code value* for bond strength [3, 2]).

$$\theta_{y,sl} = \phi_y L^y_{b,\mathrm{eff}}/2;\, L^y_{b,\mathrm{eff}} = D_b f_y/4f_{b,y};\, f_{b,y} = \eta_1 \cdot f'_t;\, \eta_1 = \begin{cases} 2.25 & \text{ribbed bars} \\ 1.0 & \text{smooth bars} \end{cases} \tag{34.8}$$

34.4.2.2 After Yielding of Longitudinal Reinforcement

Integration of bar strains over the part of the anchorage length where bar strains exceed yielding ($\varepsilon_{s,1} > \varepsilon_{s,y}$, ℓ_r = length of yield penetration over the anchorage) give a first order approximation of the plastic component of drift, associated with bar inelasticity over the anchorage:

$$\theta_{p,\mathrm{sl}} = 0.5(\varepsilon_{s,u} + \varepsilon_y) \cdot \ell_r/[(1-\xi)\cdot d] \tag{34.9}$$

34.4.2.3 Length of Yield Penetration in the Anchorage

To estimate the value of ℓ_r in Eq. (34.9), bond stress is assumed negligible over ℓ_r, corresponding to almost constant (post-yielding) bar stresses. (Note that if the definition of Section 34.4.2.2 is used instead, along with a nonzero strain hardening slope for the bar, the length of yield penetration may be estimated, by assuming a bi-linear distribution of bar stresses over the anchorage length. The slope of this diagram is: $4f_b/D_b$, where f_b the piecewise value of bond strength – a degraded value for the part of the anchorage that is beyond yielding, and the initial value, $f_{b,y}$, for the elastic part of the anchorage. This assumption leads to an expression for

ℓ_r Eq. (34.10) which is analogous to that obtained for the plastic hinge length, i.e., Eq. (34.6):

$$\theta_{p,\mathrm{sl}} = (\phi_u - \phi_y)\cdot\ell_r; \quad \ell_r = \frac{f_{s,\max} - f_{s,y}}{f_{s,\max}}\cdot\frac{D_b}{4}\cdot\frac{f_{s,\max}}{f_{b,u}} = \frac{\beta\cdot f_{s,\max}}{f_{b,u}}\cdot\frac{D_b}{4} = \frac{\beta\cdot f_{s,\max}}{f_{b,u}}\cdot\frac{D_b}{4} \tag{34.10}$$

Again, this breaks down for elastoplastic reinforcement without hardening, i.e., when β=0). To obtain ℓ_r for the general case, note that yield penetration cannot grow indefinitely: pullout failure of the anchorage occurs once the residual anchorage length engaged, equal to L_b–ℓ_r, exhausts the effective anchorage length, $L^u_{b,\mathrm{eff}}$ (note that $L^u_{b,\mathrm{eff}}$=$D_b f_y/4f_{b,u}$, which is the minimum length required to support the bar yield force, whereas L_b is the available anchorage length).

34.4.2.4 Limiting Strain Development Capacity After Yielding

Yield penetration limits the strain development capacity of the bar to the value: $\varepsilon^{\mathrm{anc\,-\,spl}}_{s,1,\max}$=$\varepsilon_y L_{\mathrm{anc}}/L_{\mathrm{eff},u}$. This result is a consequence of the continuity of strain requirement at the end of yield penetration. Thus, even if the available anchorage length, L_b, suffices to develop the yield force of the bar, the strain that may be sustained at the face of the anchorage is limited eventually by yield penetration: Inelastic anchorage failure will occur when the bar strain exceeds the above-set limit for $\varepsilon^{\mathrm{anc\,-\,spl}}_{s,1,\max}$.

34.4.2.5 Bond Strength in Eq. (34.10)

Ultimate bond strength, $f_{b,u}$, mobilized along a bar anchorage or lap splice is associated with the bar-concrete interface friction which depends on the clamping action provided by the surrounding concrete cover and transverse stirrups against the possible plane of splitting [15]:

$$f_{b,u} = 1.4\cdot\left(\underbrace{\frac{p}{D_b}f'_t}_{\text{cover}} + \underbrace{\frac{A_{st}\cdot f_{st,y}}{D_b\, n_b\, S}}_{\text{stirrups}}\right); \quad f'_t = 0.5\sqrt{f'_c} \tag{34.11}$$

n_b is the number of bars restrained by the stirrup legs included in A_{st} (where A_{st} is the cross sectional area of stirrups crossing the splitting plane). The concrete term depends on the critical crack path, p, required for splitting failure: p=$2.5D_b$+$2D_{st}$+$2c_o$ for lap-splices or anchorages outside the end support of the member, where c_o is the concrete cover thickness [11]. If the anchorage occurs in well confined regions (e.g. inside a column) the value of p in (11) is taken =$4c_o$ to account for plastification of the surrounding concrete prior to its pullout failure. Note that in lap-splices or anchorages that occur within the member's span (i.e. outside the end support), the cover contribution to $f_{b,u}$ is set to zero when the normal compressive strain in the cover has exceeded the value of 0.002 to account for cover separation from the bar surface due to longitudinal splitting in the compression zone (Section 34.3c); thus a significant loss of development capacity occurs in situations past the above limit.

34.4.3 Distortion Resultants

34.4.3.1 Elastic Distortion Term

Shear distortion is elastic prior to web cracking, obtained from the acting shear force to the member's shear stiffness ratio [3, 10]:

$$\theta_{y,\mathrm{sh}} = V\big/[0.4 \cdot E_c \cdot 0.8A_g] \tag{34.12}$$

34.4.3.2 Distortion in the Plastic Hinge Region

After web cracking, shear distortion is set equal to stirrup strain, $\varepsilon_{\mathrm{st}}$. From the Mörsch truss geometry it may be shown that $\gamma{=}\varepsilon_{\mathrm{st}} = (V{-}V_{c,\mathrm{cr}})/[E_s \Sigma A_{\mathrm{st},i}]$, where the numerator in this calculation represents the total force carried by the stirrups crossing a diagonal crack, V_w (Eq. (34.3)) and the denominator represents the extensional stiffness of the stirrups. Here, $V_{c,\mathrm{cr}}$ is the total shear force carried by the cracked concrete web:

$$\text{For } \frac{N}{f_c' \cdot A_g} \geq (\rho_{s1} - \rho_{s2}) \cdot \frac{f_y}{f_c'} \Rightarrow V_{c,\mathrm{cr}} = 0.5\sqrt{f_c'} \cdot \left[\frac{d}{L_s} \cdot \sqrt{1 + \frac{N}{0.5\sqrt{f_c'} \cdot A_g}}\right] \cdot A_g \tag{34.13}$$

Otherwise, $V_{c,\mathrm{cr}}{=}0$. Note that Eq. (34.13) has been derived from equilibrium of forces on the cross section: a concrete contribution is assumed to exist if there is a nonzero compressive force in the concrete (i.e. once the cracks have been closed). With particular reference to columns with distributed reinforcement on all sides of the cross section, it is necessary to establish the neutral axis location prior to estimating the effective tension and compression reinforcement ratios, ρ_{s1}, ρ_{s2}, to be used with Eq. (34.13).

Furthermore, according with the CBP model, the contribution of web reinforcement to shear strength, V_w, should be calculated from the sum of forces developed in all stirrup legs crossing the critical shear crack, while also considering the limited development capacity of inadequately anchored stirrups: $V_w = A_{\mathrm{st}} \cdot \sum_i f_{\mathrm{st},i} \neq A_{\mathrm{st}} \cdot f_{y,\mathrm{st}} \cdot (d - d_{s2})\big/S$. Thus, for old-type construction it is necessary that stirrups be accounted for discretely and not smeared through the *d/S* term, as it is essential that the least number of stirrups crossing a crack plane need be determined, rather than an average value. If it is possible to determine a shear crack path along the member that does not interrupt any stirrup legs at all, then the V_w term according to the CBP definition is zero [6], whereas the Codes [3, 5, 4] would yield a nonzero value even for excessively large stirrup spacing, *S*.

34.4.3.3 Degradation of Shear Strength

This phenomenon occurs after web cracking, as the value of the stirrup strain, ε_{st}, increases: assuming for simplicity that $V_{u,\mathrm{sh}}$ follows the pattern adopted by ASCE-41, the stirrup strain, ε_{st}, is evaluated from the stirrup force: $V_w=[V-k(\mu_\Delta)\cdot V_{c,\mathrm{cr}}]/\,k(\mu_\Delta)\rightarrow\varepsilon_{st}=V_w/E_sA_{st}$. The principal tensile concrete strain, $\varepsilon_{c1}\approx\sqrt{2}\varepsilon_{st}/2$ occurs in directions orthogonal to the concrete struts of the Mörsch truss causing the so-called compression softening of the struts according with the Modified Compression Field Theory. The compression softening coefficient is: $\lambda=1/(0.8+0.27\varepsilon_{c1}/\varepsilon_{co})$; this is responsible for the degradation of nominal shear strength, expressed empirically through $k(\mu_\Delta)$ [6].

The Angle of Sliding

An unresolved issue in calculations is the angle of inclination of the critical shear crack: a variety of tests (Fig. 34.4) demonstrate that aspect ratio and axial load ratio both affect the V_c and V_w terms; whereas this effect is considered in all alternatives for the concrete contribution term V_c (Eq. (34.3)), it is generally neglected by the Code Models and the Mechanistic expressions for V_w, which is generally obtained from a 45° crack assumption, as illustrated by the most prominent expressions for shear strength, Eqs. (34.14), (34.15), and (34.16). The model for V_n in [11] is a stark exception to this rule. Other differences between Eq. (34.14) and the N.A. code

Fig. 34.4 Angle of sliding (inclination of critical shear crack) and contribution of axial load to shear resistance (=$N\tan\alpha$)

expression [5, 4] given by Eq. (34.15) are, (a) the standalone contribution to shear strength by the applied axial load, which is also adopted by [3] (Eq. (34.16)) and (b) that the strength degradation coefficient $k(\mu_\Delta)$ is not applied on the shear strength of the equivalent truss, a feature that is at odds with experimental evidence and basic fundamentals of Compression Field Theories for concrete. The uncertainty thus introduced in shear strength estimations can more than the account for the persistent dispersion in the available data on this variable.

$$V_n = V_c + V_N + V_w = k(\phi)\cdot\sqrt{f_c'}\cdot 0.8A_g + N\cdot\tan\alpha + \frac{A_{st}f_{st,y}\,(d-d_2)}{S}\cdot\cot\alpha \quad \text{in [11]} \tag{34.14}$$

$$V_n = V_c + V_w = k(\mu_\Delta)\cdot\left[\frac{0.5\sqrt{f_c'}}{L_s/d}\sqrt{1+\frac{N}{0.5A_g\sqrt{f_c'}}}\right]\cdot 0.8A_g + k(\mu_\Delta)\cdot\frac{A_{st}f_{st,y}d}{S} \quad \text{in [5,4]} \tag{34.15}$$

$$\begin{aligned} V_{Rd3} &= V_c + V_N + V_w = k(\mu_\Delta)(0.16\max(0.5;100\rho_{tot})(1-0.16\min(5;\tfrac{L}{h})\cdot\sqrt{f_c'}\cdot 0.8A_g \\ &\quad + \min\{N;0.55A_e f_c'\}\tan\alpha + k(\mu_\Delta)A_{st}f_{st,y}\cdot\left[\tfrac{d-d_2}{S}\right] \text{[in 5]} \end{aligned} \tag{34.16}$$

34.4.4 Bar Buckling

It threatens members that do not fail prematurely, but undergo extensive flexural yielding. Due to load reversal the bar reaches instability conditions under a compression stress but with significant residual tensile strain [16]. The critical total strain ductility at buckling, $\mu\varepsilon_{s,\mathrm{crit}}$, is obtained from pertinent interaction diagrams that depend on the bar's unsupported length ratio S/D_b and the peak inelastic tension strain (envelope), attained by the reinforcement during previous displacement reversals, $\varepsilon_{sr}{}^{\mathrm{env}}$ [see Fig. 34.3(b) from [16].

34.5 The Correlation with Tests

To illustrate the effects on deformation capacity and on the controlling failure sequence imparted by important design parameters (such as the shear demand to supply ratio, sparse spacing and anchorage of stirrups, confinement in lap splice regions and pattern of imposed displacement history), an experimental study was conducted by the authors [14, 12]. The test program included 16 scaled column specimens with details representative of older practices ("non-conforming" as per [5, 4]), having a shear span $L_s = 0.9$ m, a 200 mm^2 cross section, reinforced with either eight or four longitudinal bars in order to effect a high and a low shear demand, respectively. Thus, either three or two main bars (D_b=12 mm) were arranged on each side of the cross section, respectively. Transverse reinforcement comprised smooth rectangular stirrups (D_{st}=6 mm), spaced as listed in Table 34.1.

Table 34.1 Specimen details. In all cases, clear cover = 20 mm, Long. Reinf.: D_b=12 mm, Transv. Reinf.: D_{st}=6 mm (smooth); Lines 6 and 7 list the peak shear force sustained, P_{max} (kN) and the corresp. drift capacity θ_u (%). Cases marked with * had: f_y=524 (MPa), $f_{st,y}$=271; all others had f_y=623 and $f_{st,y}$=384. Case 9b had 2 layers of stirrups at S=140 mm

Group I: Long. Reinf. = 8 bars											**Group :** L.R.= 4 bars				
1a*	1b*	2a*	2b*	3a*	3b*	4a	4aa*	4b*	5a*	5b*	7a	7b	8a	8b	9b
S=50 mm		S=70 mm (stirrup spacing in mm)							110 mm		70 mm		110 mm		140
No lap splice					L_{spl}=25D_b		L_{spl}=36D_b				No lap splice				
f'_c = 20.2 (MPa)							40.8	f'_c =25.5 (MPa)			f'_c =40.8 (MPa)				
52.8	54.1	50.2	51.1	33.0	32.4	58.1	46.0	40.8	52.6	52.4	37.8	36.7	37.1	37.3	38.9
4.4	4.5	2.1	3.7	2.25	2.0	6.3	1.9	1.6	2.25	6.15	7.25	13.2	6.05	6.9	13.3

Bars in 5 specimens were lap spliced at the base above the footing, over a length of either 25D_b or 36D_b (Table 34.1). Specimens were tested under a constant axial load $N \approx 0.1 f_c' A_g$, and lateral displacement reversals following two different displacement histories identified by index "a", "aa" or "b" in Table 34.1 [14, 12].

All specimens failed in a brittle mode. Deformation capacity and lateral strength (average value in the two directions) are listed in Table 34.1. The tests showed that flexural shear and anchorage strengths degraded at different rates with increasing displacement, confirming the basic thesis of the mechanical model regarding capacity prioritization as prerequisite in assessing deformation capacity.

The mechanistic principles outlined in this paper as well as in Code Models [3, 5, 4] were used to evaluate the strength terms (V_{flex}, $V_{u,\text{sh}}$, $V_{u,\text{sl}}$) and deformation capacities (θ_y, θ_u) of the test specimens using the actual material properties. Results are compared with the experimental values (both from tests "a" and "b" for otherwise identical specimens in each subgroup) in Fig. 34.5. Discrepancies are observed in the estimated strength of laps according with [4]; all models tend to overestimate rotation at yielding, but provide conservative estimates of rotation capacity.

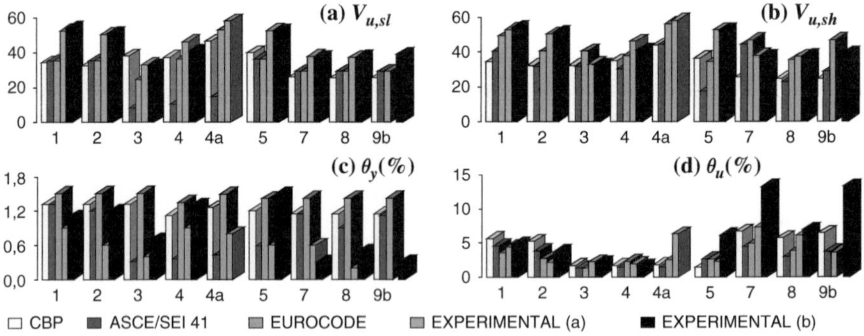

Fig. 34.5 Comparison between analytical and experimental results: $V_{u,sl}$, $V_{u,sh}$, θ_y, θ_u

Discrepancies are mostly inherited by the deviation in the values of estimated residual shear and lap strengths, which remain the least adequately understood variables from among the strength terms in the hierarchy of Eq. (34.4).

References

1. Berry M, Parrish M, Eberhard M (2004), PEER structural performance database – user's manual, PEER Center, University of California, Berkeley, CA
2. CEN (2003) EN 1992-1-1:2003 – Eurocode 2 -design of concrete structures. Part 1: general rules and rules for buildings". European Committee for Standardisation, Brussels
3. CEN (2005) EN 1998-3:2005 – Eurocode 8 Design of structures for earthquake resistance – part 3: assessment and retrofitting of buildings", European Committee for Standardisation, Brussels
4. Elwood K, Matamoros A, Wallace J, Lehman D, Heintz J, Mitchell A, Moore M, Valley M, Lowes L, Comartin C, Moehle J (2007) Update to ASCE/SEI 41 concrete provisions. Earthq Spectra 23(3):493–523
5. FEMA 356 (2000) Prestandard and commentary for the seismic rehabilitation of buildings. FEMA, Washington, DC
6. Martin-Perez B, and Pantazopoulou S (2001) Effect of bond, aggregate interlock and dowel action on shear-strength degradation of reinforced concrete. Eng Struct 23:214–227
7. Panagiotakos T, Fardis MN (2001) "Deformation of R.C. members at yielding and ultimate", ACI Struct J 98(2):135–148
8. Pantazopoulou SJ (2003) Chapter 4 in "Seismic assessment and retrofit of reinforced concrete buildings", fib Bulletin No. 24, fib, Lausanne, Switzerland
9. Park R, Paulay T (1975) Reinforced concrete structures. Wiley, New York, NY, p 769
10. Paulay T, Priestley MJN (1992) Seismic design of reinforced concrete and masonry buildings. Wiley, New York, NY, 767 pp
11. Priestley MJN, Seible F, Calvi M (1996) Seismic design and retrofit of bridges. Wiley, New York, NY
12. Syntzirma D, Thermou G, Pantazopoulou S, Halkitis G (2006) Experimental research of R.C. elements with substandard details, 1st European conference on earthquake engineering and seismology, Geneva, paper no: 819
13. Syntzirma DV, Pantazopoulou SJ (2002) Performance-based seismic evaluation of R.C. building members", 12th European conference on earthquake engineering, London, paper no:816
14. Syntzirma DV, Pantazopoulou SJ (2006) Assessment of deformability of R.C. members with substandard details. 2nd international fib congress, Naples, Italy, 5–8 June, paper no. 446
15. Syntzirma DV, Pantazopoulou SJ (2007) Deformation capacity of R.C. members with brittle details under cyclic loads. ACI Special Publ 236:1–22
16. Syntzirma DV, Pantazopoulou SJ, Aschheim M (2010) Load history effects on deformation capacity of flexural members limited by bar buckling. J Struct Eng ASCE 136:1–11

Chapter 35
The Effect of Displacement History on the Performance of Concrete Columns in Flexure

Bora Acun and Halûk Sucuoğlu

35.1 Introduction

Columns are the primary members of frame structures that dominate the frame response during earthquakes. Most of the building collapses in the past earthquakes resulted from poor column performance. Column failures in buildings are either due to insufficient shear resistance (shear failure) or due to insufficient deformation capacity (flexure-shear and flexure failure). New generation of performance-based seismic rehabilitation and design codes express the flexural performances of column in terms of total or plastic rotation capacities of the critical end regions whereas shear failure is strictly prohibited [2–4].

The deformation capacities of columns undergoing pure flexure failure are related to the reinforcement detailing as well as the imposed displacement history. Repeated number of large amplitude cycles may lead to degradation in lateral strength and stiffness, hence the exhaustion of deformation capacities of columns responding in flexure. Although the effects of longitudinal and lateral reinforcement on the deformation capacity of columns are well understood under standard displacement protocols, information on the effect of repeated severe displacement cycles is limited. Iwasaki et al. [5] tested bridge piers; Pujol et al. [7] tested small scale columns under displacement reversals and investigated the effect of the number of displacement cycles on column deformation capacity. Columns in both of these tests developed inclined cracking after flexural yielding. Verderame et al. [8] tested concrete columns with smooth reinforcing bars and substandard detailing, but failing in pure flexure, to investigate the differences in the displacement capacity under monotonic and cyclic loading.

Six full scale column specimens designed for pure flexure failure were tested in this study under repeated large displacement cycles. Typical column designs were employed in the tests that represent columns conforming to the material and special seismic detailing standards of modern concrete design codes [1]. Axial load

---

B. Acun (✉)
Department of Civil Engineering, Middle East Technical University, 06531 Ankara, Turkey
e-mail: bacun@metu.edu.tr

M.N. Fardis (ed.), *Advances in Performance-Based Earthquake Engineering*, Geotechnical, Geological, and Earthquake Engineering 13, DOI 10.1007/978-90-481-8746-1_35,

ratio was constant in all specimens. The main variable in the experiments was the imposed displacement histories. Five or seven initial cycles at different drift ratios were applied to four different specimens respectively, and then followed by cycles with increased drift amplitudes. Two specimens were tested under variable drift amplitudes. Observed rotations at the plastic hinge region are evaluated comparatively with the limits proposed by ASCE/SEI 41 and Eurocode 8, and the observed moment- rotation and lateral force-displacement behavior is assessed in view of the modeling criteria in ASCE/SEI 41.

35.2 Experimental Program

Test Specimens: The test specimens were designed for pure flexure failure. The column specimens conform to the modern code standards for seismic design. They were constructed with 25 MPa concrete, deformed bars, and transverse reinforcement with 135 degree hooks. All specimens were cast vertically. Details of the specimens are shown in Fig. 35.1; material properties are given in Table 35.1. Vp/Vn ratios for the specimens were 0.20, indicating that flexure failure was ensured.

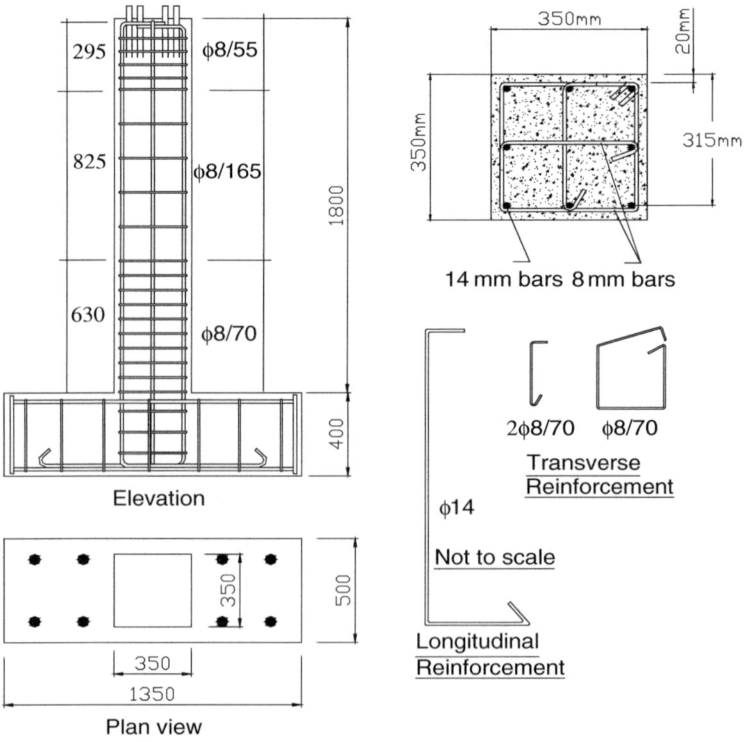

Fig. 35.1 Details of column specimens

Table 35.1 Material properties and reinforcement ratios of test specimens

Concrete	Long. reinforcement			Trans. reinforcement		
Comp. strength f'_c	Yield strength f_y	Ultimate atrength f_u	Reinf. ratio ρ_l	Yield strength f_{yw}	Ultimate strength f_{uw}	Reinf. ratio ρ_t
(MPa)	(MPa)	(MPa)	$(A_s/b_w.h)$	(MPa)	(MPa)	$(A_{sw}/b_w.s)$
25	454	604	0.01	469	685	0.0061

Test Setup and Instrumentation: Specimens were placed and tested on a mat foundation, fixed to the strong floor with post tensioned bars. A steel head was placed on top of the columns and lateral load was applied by an actuator with hinges attached at both ends, from the level of steel head. Two steel beams were placed on either side of the specimen parallel to the loading direction and a set of rollers were attached to the upper part of columns in order to prevent the out-of-plane movement of specimens.

Axial load was applied by a steel loading beam placed horizontally on the steel head, perpendicular to the loading direction. Two high strength steel rods were connected both to the steel loading beam and mat foundation and load was applied by post-tensioning of these rods. Axial load level was kept constant during the tests. The general view of test setup is shown in Fig. 35.2.

The applied loads (lateral and axial) were measured by using calibrated load cells. The response of test specimens was measured with LVDT's placed at several levels of each specimen. Four dial gages were mounted on both sides of each column at 350 mm level from the top of the specimen footing. Sixteen strain gages (12 on longitudinal, 4 on transverse bars) were installed on each specimen.

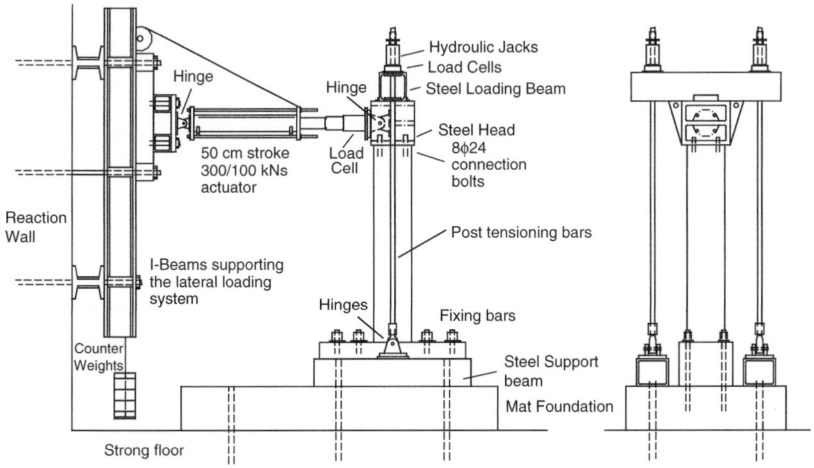

Fig. 35.2 Test setup

Table 35.2 Displacement protocols imposed on the test specimens

		Specimens					
Cycle no		1D2	2D3	3D4	4D5	5DV1	6DV2
1	Top Displacement Amp. (mm)	35	50	70	105	10	17.5
2		35	50	70	105	10	17.5
3		35	50	70	105	10	17.5
4		35	50	70	105	50	35
5		35	50	70	105	50	35
6		35	50	105	70	50	35
7		35	50	105	70	35	50
8		70	70	105	70	35	50
9		70	70		70	35	50
10		70	70		70	70	70
11		70	70			70	70
12		70	70			70	70
13		105	105			35	105
14		105	105			35	105
15		105	105			35	105
16						105	
17						105	
18						105	

Testing Program: The imposed displacement histories are shown in Table 35.2 for all specimens. Note that 35, 50, 70 and 105 mm top displacement amplitudes correspond to the drift ratios of 1.75, 2.5, 3.5 and 5.25%, respectively. The first number in the specimen labels indicate the specimen number in each group, D stands for deformed bars, and the last number is the drift ratio attained in the first set of constant displacement cycles for four specimens. V identifies a variable displacement history for two specimens where the cycle amplitudes follow a different variation pattern compared to the other four specimens. Axial load level was kept constant at $0.2f'_c\ A_g$ for all specimens.

35.3 Test Results: Deformation Capacities of Columns

The first set of test results are presented for the moment-chord rotation relationships for the bottom ends of column specimens. Chord rotation at the bottom end is equal to the drift angle, i.e. top displacement divided by the specimen height, and it represents total rotation of the plastic hinge region including the elastic and plastic components. The results are shown in Fig. 35.3. Analytical moment-chord rotation relations calculated under monotonously increasing moments, obtained from moment-curvature relations along the column height are marked on each figure. Plastic curvatures of the plastic hinge region at the column base are converted

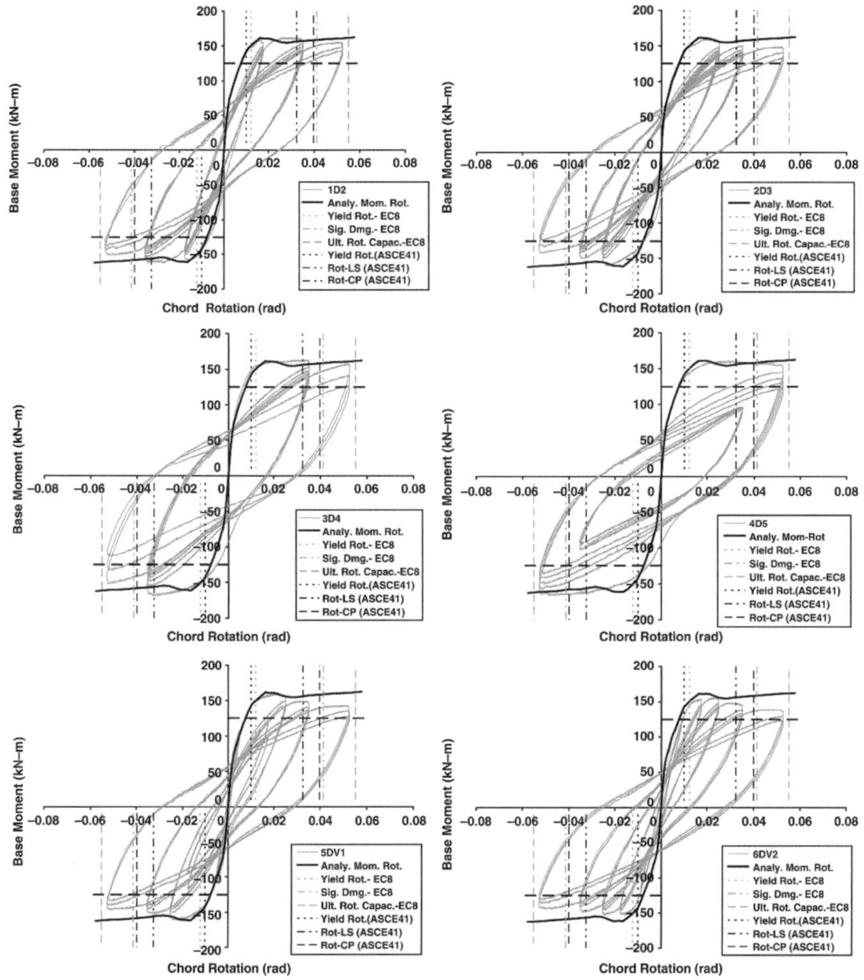

Fig. 35.3 Moment-chord rotation relations for column specimens

to plastic rotations by assuming a plastic hinge length in obtaining the analytical moment-chord rotation relations [6].

Three deformation limit states corresponding to yield rotation, significant damage and ultimate rotation capacity according to Eurocode 8, and yield rotation, life safety and collapse prevention according to ASCE/SEI 41-Update are marked respectively on each diagram in both loading directions. 80% levels of the positive and negative yield moments (0.8 My) are also indicated on the vertical axis.

The column specimens are classified as Condition (i) by ASCE/SEI 41-Update. Eurocode 8 accounts for the enhanced seismic performance of these columns with deformed longitudinal bars, low shear and axial force and proper confinement at the plastic hinge region in specifying their deformation-based performance criteria. The ultimate chord rotation values are defined in Eurocode 8 by Eq. (35.1).

$$\theta_{\mathrm{um}} = \frac{1}{\gamma_{\mathrm{el}}} 0.016.(0.3^{\nu}) \left[\frac{\max(0.01;\omega')}{\max(0.01;\omega)} f_{\mathrm{c}} \right]^{0.225} \left(\frac{L_{\mathrm{v}}}{h} \right)^{0.35} 25^{\left(\alpha \rho_{sx} \frac{f_{\mathrm{yw}}}{f_{\mathrm{c}}} \right)} (1.25^{100\rho_d}) \tag{35.1}$$

Chord rotations for significant damage limit state are taken as the ¾ of the ultimate chord rotation (near collapse) as indicated by Eurocode 8.

It is observed from Fig. 35.3 that the performance limit states proposed by Eurocode 8 and ASCE/SEI 41 are quite different. Significant damage and ultimate capacity deformation limits of Eurocode 8 are 27 and 39% larger than the life safety and collapse prevention limits of the ASCE/SEI 41, respectively, although these

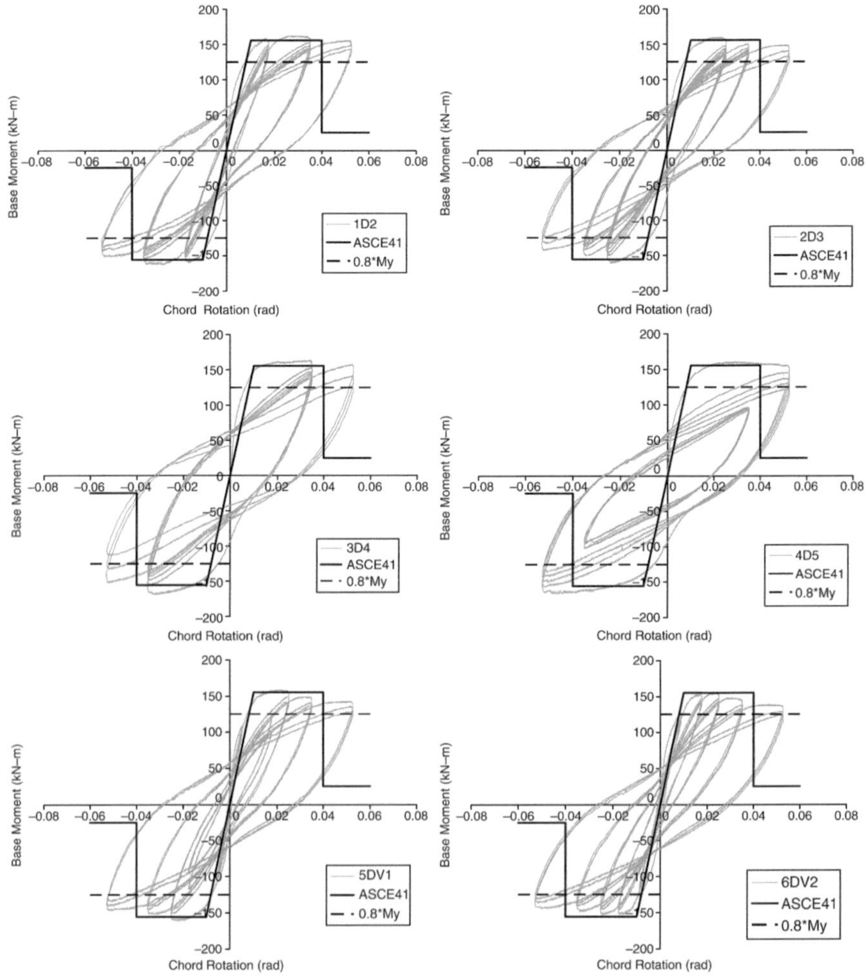

Fig. 35.4 Comparison of the responses of column specimens with the ASCE41 modelling parameters

different performance limit definitions in the two codes actually indicate similar performance levels. The results presented in Fig. 35.3 confirm the limit state predictions of Eurocode 8 meanwhile demonstrate that ASCE/SEI 41 limit state definitions are too conservative for such columns, even when the columns are subjected to severe displacement cycles. Apparently, the deformation capacities of columns conforming to modern design codes that exhibit pure flexure behavior under low axial forces are significantly larger than that permitted by ASCE/SEI 41.

Two parameters ***a*** and ***b*** are proposed in the ASCE/SEI 41 for modeling the plastic hinge behavior of flexural members where ***a*** is the plastic rotation at significant loss of plastic rotation capacity, and ***b*** is the plastic rotation at axial load failure. These two parameters are mainly employed in the nonlinear static analysis of concrete structures for constructing the capacity curves. The moment-rotation envelope relations obtained by employing the associated values of the two parameters for test specimens are calculated and compared with the results obtained from the experiments in Fig. 35.4. It can be observed from Fig. 35.4 that the plastic hinges of column specimens are capable of sustaining larger plastic deformations before significant loss of plastic rotation capacity. The modeling parameter ***a*** seems to be very conservative for defining the rotation capacity of column plastic hinges when the axial load ratio is around 0.20 whereas the suggested ***a*** values may be more reasonable at higher axial loads.

35.4 The Effect of Displacement History on Target Displacement Demand

The column specimens herein can be considered as cantilever structures carrying an assigned mass, which possess the lateral force (base shear) versus lateral drift relationship as obtained from the experiments. These relationships are presented in Fig. 35.5 for four specimens. 80% level of the yield lateral force level is also marked for each specimen on each figure.

The lateral force-lateral drift envelopes for the first positive cycles of the column structures and the design spectrum expressing seismic excitation are shown in Fig. 35.6a and b, respectively.

It is observed from Fig. 35.6a that all specimens exhibit similar envelopes until they fall below the 80% V_{max} level, regardless of the cyclic displacement histories they are enveloping. The lateral strength of all specimens falls below 80% of V_{max} approximately at the same drift ratio of 0.035, or at the column drift of 70 mm.

If the average envelope curves in Fig. 35.6a are considered as the capacity curves for the column structures, then the target drift demands of each column structure can be calculated through equivalent linearization procedure. The average envelope curves are replaced by bi-linear capacity curves by employing the equal energy approach. The capacity curves for the column structures possess similar ductility ratios of 4.9 at the drift ratio of 0.035, and similar post yield stiffness ratio of –0.048. A mass producing an initial elastic period of T_0=0.3 second is assigned to each column.

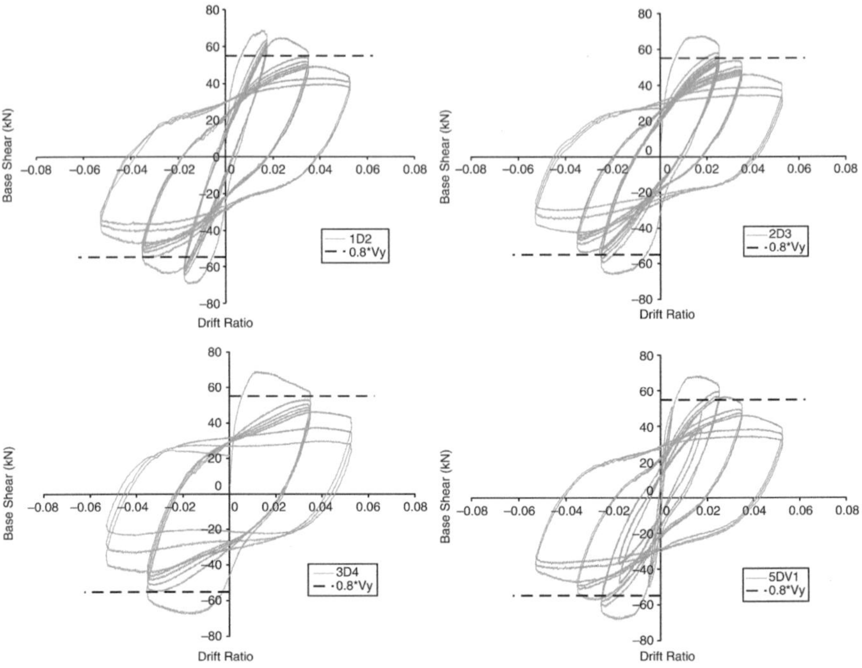

Fig. 35.5 Experimental responses of column structures

Two different approaches for equivalent linearization are employed comparatively. The first one is the improved procedure proposed in FEMA 440 where effective period and effective damping are calculated from the empirical equations based on strength degrading model, whereas they are directly calculated from the experimental data shown in Fig. 35.5 in the second approach. Effective damping in the second approach is based on the first experimental cycle in Fig. 35.5 with drift

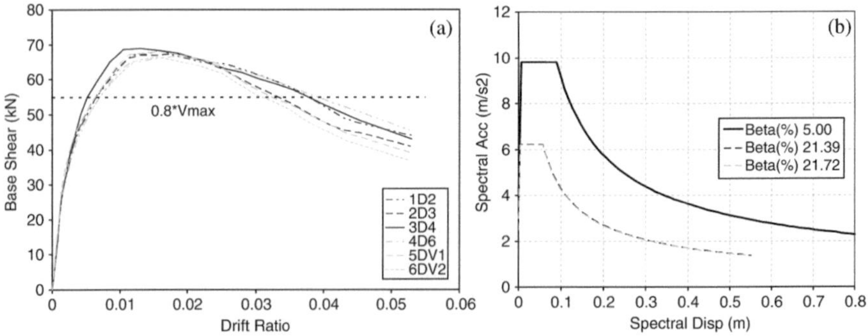

Fig. 35.6. (**a**) Envelope curves for positive base shear-drift ratio cycles of column structures, (**b**) design spectrum

Table 35.3 Target drift ratios calculated with equivalent linearization

	Experimental			Fema – 440		
Specimen	β_{eff} (%)	T_{eff}/T_0	Target drift	β_{eff} (%)	T_{eff}/T_0	Target drift
1D2	24.50	2.46	0.034			
2D3	22.49	2.55	0.037			
3D4	33.86	2.43	0.029	21.39	1.90	0.026
5DV1	23.76	2.55	0.036			
6DV2	23.69	2.60	0.037			

amplitudes of 0.035. It is assumed that each column structure reaches its target drift during its first cycle to the drift ratio of 0.035, after completing a past displacement history at lower drift values as given in Table 35.2. Hence effective damping reflects the effect of displacement history on target drift demand in the second approach. It can be noted from Fig. 35.6 that displacement history does not have a significant influence on effective (peak to peak) stiffness. The results of the first (analytical: FEMA 440) and the second (experimental) equivalent linearization approaches are summarized in Table 35.3.

The column structure 3D4 attain the drift amplitude of 0.035 in the first displacement cycle with in-cycle degradation, however it does not exhibit any cyclic degradation due to displacement history effects. Accordingly it possesses the highest effective damping value which leads to the lowest target drift ratio of the experimental approach. Target drift ratios for the other columns are higher, reflecting reduced effective damping due to cyclic degradation in energy dissipation capacity. FEMA 440 approach on the other hand does not properly reflect the effect of displacement history. Effective damping values obtained from the strength degrading model are lower whereas the effective periods are significantly lower compared to the experimental values. This leads to unconservative estimation of the target drift ratios. The target drift ratios obtained with the experimental approach for the column structures that undergo severe inelastic displacement cycles before reaching the maximum drift are 40% larger than those obtained with the FEMA 440 approach. The stiffness degrading model employed in the FEMA 440 approach perhaps do not properly represent the experimental in-cycle and cyclic degradation behavior of the columns presented herein.

35.5 Conclusions

It is observed that Eurocode 8 performance limits conform quite well to the experimental performance of columns whereas ASCE/SEI 41 performance limits appeared to be quite conservative in predicting the experimental performance of column plastic hinges designed to fail in pure flexure under moderate axial load levels. A further update might be warranted in order to reduce unnecessary conservatism.

The effect of displacement history on the target drift demands of concrete structures dominated by flexural column behavior is observed to be significant under severe earthquake ground motions. Realistic models simulating the degradation behavior of columns under severe displacement histories are required for the accurate calculation of drift demands for both static pushover analysis and nonlinear time history analysis.

References

1. ACI Committee 318 (2005) Building code requirements for structural concrete (ACI 318-05) and commentary (318R-05). American Concrete Institute, Farmington Hills, Michigan
2. ASCE/SEI 41 (2007) Seismic rehabilitation of existing buildings. American Society of Civil Engineers, Reston, VA
3. Comite Europeen de Normalisation (2005) European standard EN 1998-3 Eurocode 8: design of structures for earthquake resistance – part 3: assessment and retrofitting of buildings. Brussels
4. Elwood JK, Matamoros AB, Wallace JW, Lehman DE, Heintz JA, Mitchell AD, Moore MA, Valley MT, Lowes MT, Comartin CD, Moehle JP (2007) Update to ASCE/SEI 41 concrete provisions. Earthq Spect 23(3):493–523.
5. Iwasaki T, Kawashima K, Hasegawa K, Koyama T, Yoshida T (1987) Effect of number of loading cycles and loading velocity of reinforced concrete bridge piers. Proceedings of the 19th Joint Meeting of the US-Japan Panel on Wind and Seismic Effects, UJNR, Tsukuba, pp 225–238
6. Priestley MJN, Calvi GM, Kowalsky MJ (2007) Displacement-based seismic design of structures. IUSS Press, Foundazione EUCENTRE, Pavia, 720 pp
7. Pujol S, Sozen MA, Ramirez JA (2006) Displacement history effects on drift capacity of reinforced concrete columns. ACI Struct J 103(2):253–262.
8. Verderame GM, Fabbrocino G, Manfredi G (2009) Seismic response of RC columns with smooth reinforcement. Part II: Cyclic tests. Eng Struct 30:2289–2300.

Chapter 36
Innovative Seismic Retrofitting of RC Columns Using Advanced Composites

Dionysis Bournas and Thanasis Triantafillou

36.1 Introduction

Earthquakes worldwide have shown the seismic vulnerability of existing RC columns. Poorly detailed columns are the most critical structural elements, which may fail due to shear, compressive crushing of concrete, rebar buckling, bond at lap-splices and flexure. Seismic retrofitting of RC columns is a challenging task that may be addressed successfully today using externally bonded FRP for all the aforementioned failure mechanisms but the last one, that is flexure. FRPs, in the form of jackets with the fibers typically in the columns' circumferential direction, are quite effective in carrying shear and in providing confinement, thus increasing the shear resistance and the deformation capacity of existing RC columns. Seismic retrofitting of RC columns with FRP is summarized schematically in Fig. 36.1.

Effective strengthening of columns in flexure, often needed for instance to satisfy capacity design requirements or when existing rebars have been affected by corrosion, calls for the continuation of longitudinal reinforcement beyond the end cross sections, where moments are typically maximum; hence, placement of externally bonded FRP reinforcement is not applicable. As a result, flexural strengthening of RC columns is typically achieved today by using RC jackets or some forms of steel jackets, namely steel "cages", also followed by shotcreting. RC jackets or steel cages covered by shotcrete require intensive labor and artful detailing, they increase the dimensions and weight of columns and result in substantial obstruction of occupancy. Therefore, the implementation of a low labor and minimal obstruction flexural strengthening technique for RC columns still remains a challenging task, which is addressed in this study through the use of near-surface mounted (NSM) reinforcement.

NSM reinforcement involves cutting of grooves into the concrete cover and bonding of rebars inside the grooves through the use of an appropriate filler (typically epoxy resin or cement-based mortar). The idea of NSM reinforcement was born in

D. Bournas (✉)
EUCENTRE, Via Ferrata 1, Pavia 27100, Italy
e-mail: dionysis.bournas@eucentre.it

M.N. Fardis (ed.), *Advances in Performance-Based Earthquake Engineering*, 383
Geotechnical, Geological, and Earthquake Engineering 13, DOI 10.1007/978-90-481-8746-1_36,

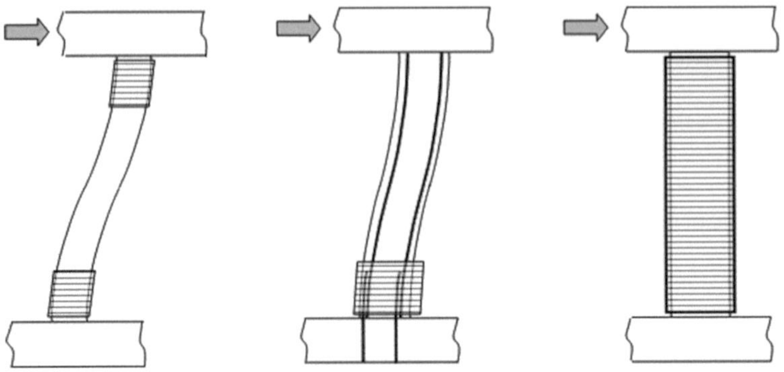

Fig. 36.1 Seismic retrofitting of RC columns with FRP: (**a**) Local jacketing to increase the deformation capacity of concrete and/or to delay rebar buckling. (**b**) Local jacketing to prevent failure at lap-splices. (**c**) Full height jacketing to increase the shear resistance

Europe for steel rebars in the late 1940s [1], but it was only recently, when more hi-tech materials, such as FRPs and high quality epoxies, became available, that the technique was given substantial attention by the research community and practicioners. Research so far on NSM reinforcement for RC structures has focused on flexural strengthening of beams or slabs with an emphasis on bond aspects (e.g. [2, 4, 7, 9, 10, 12]), on shear strengthening of RC beams [6, 8] and on flexural strengthening with prestressed NSM FRP bars [5, 11]; some of the most recent research results in these areas are reported in Triantafillou [13].

This paper presents the first systematic study on NSM-based flexural strengthening of RC columns under simulated seismic loading. The investigation addresses column strengthening with NSM carbon or glass fibers, as well as stainless steel rebars. Another innovative aspect in this study is the combination of NSM reinforcement with local jacketing, which comprised the recently developed textile-reinforced mortar (TRM) confining system, described by Triantafillou et al. [14] and Bournas et al. [3]. Details are provided below.

36.2 Experimental Investigation

36.2.1 Test Specimens and Experimental Parameters

The experimental program aimed to study the flexural strengthening of old-type non-seismically detailed RC columns with NSM reinforcement and to compare the effectiveness of different flexural strengthening schemes. A total of eleven full-scale RC column specimens with the same geometry were constructed and tested under cyclic uniaxial flexure with constant axial load (Fig. 36.2a). The specimens were flexure-dominated cantilevers with a height to the point of application of the load (shear span) of 1.6 m (half a typical story height) and a cross section of

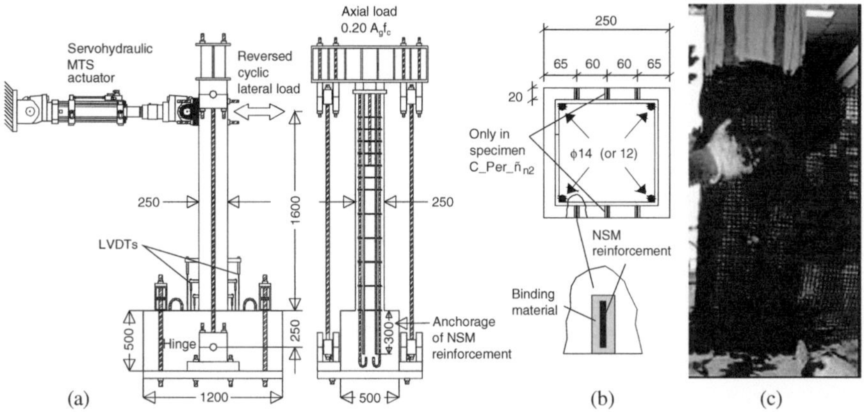

Fig. 36.2 (**a**) Schematic of test setup. (**b**) Cross section of columns. (**c**) NSM reinforcement and application of TRM jacket

250 × 250 mm. To represent old-type columns, specimens were reinforced longitudinally with four 14 mm-diameter smooth bars (except for one specimen which had 12 mm bars) and 8 mm diameter smooth stirrups, closed with 90-degree hooks at both ends, at a spacing of 200 mm. The geometry of a typical cross section is shown in Fig. 36.2b.

The specimens were designed such that the effect of a series of parameters on the flexural capacity of RC columns could be investigated. These parameters comprised: type of NSM reinforcement (CFRP strips, GFRP bars, stainless steel rebars), configuration of NSM reinforcement (CFRP strips placed with their large cross section side perpendicular or parallel to the column sides, depending on whether a proper concrete cover is available or not), amount – that is geometrical reinforcing ratio – of NSM or internal reinforcement, type of bonding agent for the NSM reinforcement (epoxy resin versus cement-based mortar) and NSM reinforcement with or without local jacketing at the member ends. A description of the specimens follows next.

One specimen was tested without retrofitting, as *Control.*

C_Per was strengthened with two CFRP strips symmetrically placed on each of two opposite sides of the column (those with highest tension/compression). The strips had a cross section of 16 × 2 mm and were placed inside 10 × 20 mm orthogonal grooves with the large cross section side *perpendicular* to the column side. This scheme is feasible only if the concrete cover is at least equal to 20 mm.

*C_Per_*ρ_{n2} was strengthened as C_Per, but with a higher geometrical reinforcing ratio for the NSM reinforcement ($\rho_n = 0.3\%$) provided by placing three strips (instead of two, corresponding to $\rho_n = 0.2\%$) on each column side.

*C_Per_*ρ_{s2} was strengthened as C_Per, but it was initially designed with a lower reinforcing ratio for the internal steel reinforcement. This specimen was reinforced with 12 mm-diameter bars ($\rho_s = 0.72\%$), whereas all others had 14 mm-diameter bars ($\rho_s = 0.98\%$).

C_Par was strengthened with two CFRP strips (with dimensions as above) symmetrically placed on each of two opposite sides of the column, but with their large cross section side *parallel* to the column side; the strips were placed inside 20 × 5 mm orthogonal grooves. This scheme is expected to have less favourable bond characteristics compared to C_Per, but it may be easily applied if the concrete cover is small.

C_Par_J had the same NSM reinforcement as C_Par and an additional confining jacket, which extended from the column base to a height of 600 mm. The aim of this jacket was mainly to protect the NSM reinforcement against premature failure due to buckling and/or debonding.

G was strengthened with two 8 mm-diameter GFRP bars symmetrically placed on each of two opposite sides of the column. The bars were placed in 20 × 20 mm square grooves.

S_R was strengthened with two 12 mm-diameter stainless steel rebars symmetrically placed on each of two opposite sides of the column, in 20 × 20 mm square grooves. As in all specimens above with NSM reinforcement, the bonding agent inside the grooves was epoxy resin.

S_M had the same NSM reinforcement as S_R, but the bonding agent inside the grooves was a cement-based mortar.

S_R_J had the same NSM reinforcement as S_R and an additional confining jacket, as used in C_Par_J.

S_M_J had the same NSM reinforcement as S_M and an additional confining jacket, as used in S_R_J.

Of crucial importance in the selection of NSM reinforcement was the requirement of equal tensile strength for each of the reinforcing elements (CFRP strips, GFRP bars, stainless steel bars). Given that all these elements are commercial products, this requirement was satisfied by proper combinations of cross section geometries and material strength data.

36.2.2 Strengthening Procedures, Test Setup and Materials

When their preparation was completed, grooves and holes were filled by injecting the bonding agent using a simple silicone gun, then the NSM reinforcement was placed into position and the bonding material in excess was removed. For the specimens receiving TRM jacketing a commercial textile with equal quantity of carbon rovings in two orthogonal directions was applied in four layers (Fig. 36.2c). Application of the mortar with this textile was made in approximately 2 mm thick layers. Of crucial importance in this method, as in the case of epoxy resins, was the application of each mortar layer while the previous one was still in a fresh state.

The columns were fixed into a heavily reinforced 0.5 m-deep base block, 1.2 × 0.5 m in plan, within which the longitudinal bars were anchored with 50 mm radius hooks at the bottom. The 14 mm-diameter longitudinal bars had a yield stress of 372 MPa, a tensile strength of 433 MPa and an ultimate strain equal to 17%; the respective values for the 12 mm-diameter bars were 330 MPa, 412 MPa and 23%.

The corresponding values for the steel used for stirrups were 351 MPa, 444 MPa and 19.5%. In order to simulate field conditions the base blocks and the columns were cast with separate batches of ready-mix concrete. The mean compressive strength of concrete on the day of testing the columns, measured on 150 × 150 mm cubes, was equal to 25.8 MPa. CFRP strips had an elastic modulus equal to 145 GPa and a tensile strength equal to 2175 MPa, whereas the respective values for GFRP bars were 65 GPa and 1490 MPa. Stainless steel bars had a conventional yield strength equal to 670 MPa, a tensile strength of 760 MPa and an ultimate strain equal to 19%. The tensile strength for each of the three NSM reinforcements (conventional yield force, in the case of stainless steel) was calculated as: 69.5 kN for the CFRP strips, 74.9 kN for the GFRP bars and 75.6 kN for the stainless steel bars.

For the specimens receiving resin adhesive for bonding of the NSM reinforcement, a commercial structural adhesive was used with a tensile strength of 30 MPa and an elastic modulus of 4.5 GPa. For the specimens receiving mortar as a binding material for bonding of the NSM reinforcement (stainless steel bars in specimens S_M and S_M_J), a commercial inorganic dry binder was used, consisting of cement and polymers at a ratio of about 8:1 by weight. The water:binder ratio in the mortar was 0.23:1 by weight, resulting in plastic consistency, good workability and high flowability. The average flexural and compressive strength values for this mortar were 6.31 MPa and 17.5 MPa, respectively.

The textile used in this study for the TRM jacketing was made with equal quantity of carbon fibers in two orthogonal directions. Each fiber roving was 3 mm wide and the clear spacing between rovings was 7 mm. The weight of carbon fibers in the textile was 348 g/m^2 and the nominal thickness of each layer (based on the equivalent smeared distribution of fibers in the circumferential direction) was 0.095 mm. The mean tensile strength and the elastic modulus of the carbon fibers (as well as of the textile, when the nominal thickness is used) were taken from data sheets equal to 3800 MPa and 225 GPa, respectively.

The columns were subjected to lateral cyclic loading through the use of a horizontally positioned MTS actuator, under a constant axial load corresponding to 20% of the members' compressive strength. Displacements and axial strains at the plastic hinge region were monitored using six rectilinear displacement transducers. The instrumentation also comprised a total of 12 strain gages for each column, which were mounted on one NSM reinforcing element per column side. Measurements from the strain gages on each NSM element were used to determine the local bond-slip relationship in the anchorage region as well as the NSM strain (equal to the fracture strain or the strain at debonding) in the section of maximum moment.

36.3 Results

The response of all columns tested is given in Fig. 36.3 in the form of load-drift ratio loops. The performance and failure mode of all tested specimens was controlled by flexure, as expected due to their design characteristics. This was an

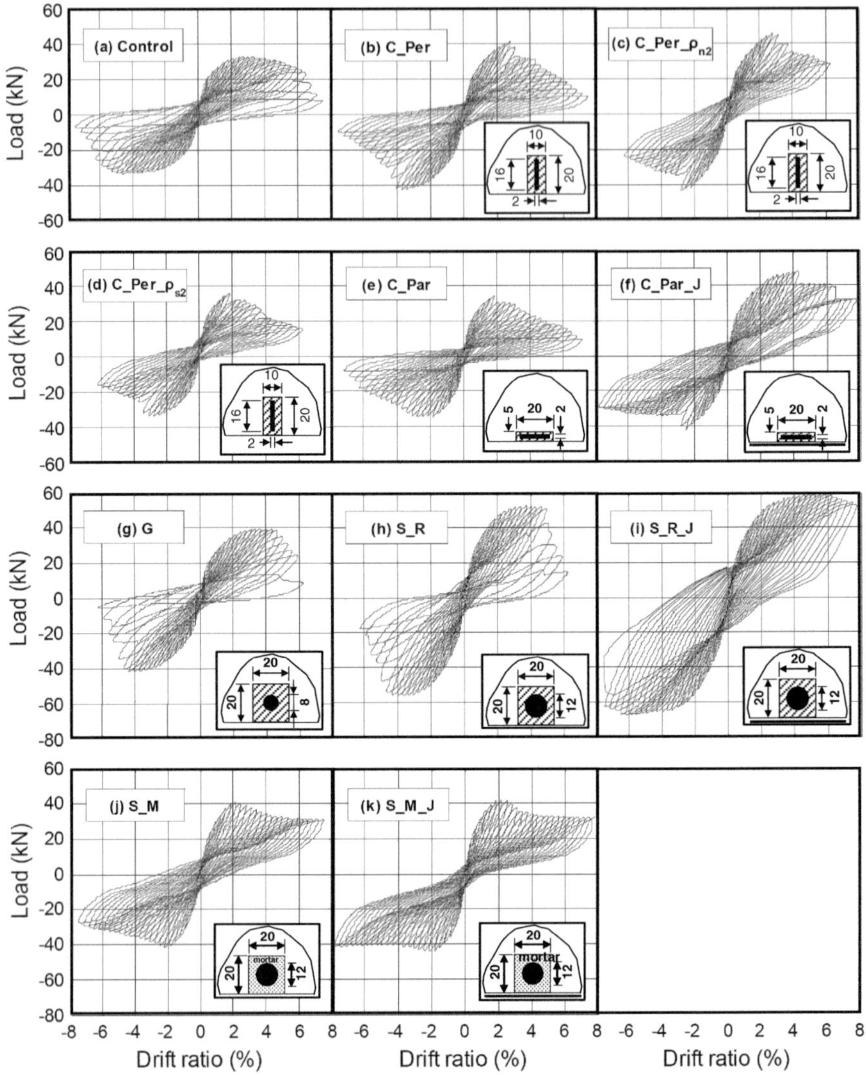

Fig. 36.3 Load versus drift ratio curves for tested specimens (the inserts illustrate NSM reinforcement configurations)

important requirement, as the main objective was to evaluate the effectiveness of NSM reinforcement as a means of flexural strengthening of RC columns. The control specimen attained a peak load of about 33 kN and a drift ratio at failure of 6.25%. After yielding of the longitudinal reinforcement, the concrete cover and a part of the core over the lower 200 mm of the column disintegrated and bar buckling initiated after the concrete cover spalled off. With only one exception (column C_Par), all strengthened specimens displayed higher (up to about 100%) flexural resistance compared to the control specimen. Flexural cracking at the column base started at the early stages of loading and the number of cracks increased and propagated with increasing drift ratios, while inclined cracks propagated in the concrete surface at both sides of the grooves as a result of high pull out forces of the NSM reinforcement for most specimens. Contrary to the unstrengthened column, the failure of the strengthened specimens was never attributed to buckling of the internal steel, as a significant portion of the total force in the compression zone was carried by the NSM reinforcement. However, buckling of the longitudinal internal bars always occurred abruptly after failure of the NSM reinforcement. The behavior of each strengthened column is described in detail below.

The observed failure mode for specimens C_Per, C_Per_ρ_{n2} and C_Per_ρ_{s2} was due to tensile fracture of the CFRP strips at the cross section of maximum moment (column base). Compared with the control specimen, the peak force increased up to about 40% and the attained drift ratio (at peak force) was approximately the same, in the order of 3%. Fracture of the NSM strips resulted in a drop of the applied force, when the mean recorded strains of CFRP at the column base were equal to 0.95%, 0.93% and 0.85% for specimens C_Per, C_Per_ρ_{n2} and C_Per_ρ_{s2} respectively. These values are nearly half the measured ultimate strain in uniaxial tests, indicating the detrimental effect of cycling on the tensile strength of CFRP strips. Partial debonding of the strips when subjected to high pull out forces in one direction of loading deprived their lateral restraint in the next loading cycle. As a consequence, the strips became vulnerable to high compressive stresses resulting in local buckling, which led to their tensile fracture at strains less than the ultimate uniaxial strain.

Specimen C_Par displayed rather poor flexural strengthening characteristics: It failed due to early debonding of the CFRP strips at a force marginally higher than the control specimen and a drift ratio of about 2%, with a mean recorded strain of the strips at peak force equal to 0.50%, that is well below their tensile capacity. Debonding of the NSM strips at such a low strain is attributed to their outward spalling due to buckling, rather than to their poor anchoring conditions and the strips' low resistance against pull out. This can be confirmed by examining the results in comparison with specimen C_Par_J, which was identical to C_Par but jacketed at the column end. In this specimen the TRM jacket provided lateral resistance to the strips against buckling, thus increasing the peak force substantially, by 46 and 26% in the push and pull direction, respectively, and the drift ratio at peak force to about 4 and 2.5% in the corresponding directions. The reduced activation of tensile strips in the pull direction as compared to the push is attributed to their debonding, a fact which is confirmed by the values of mean recorded strains at peak force equal to 1.6 and 0.85% in the push and pull direction, respectively. These values are in agreement with observations of strip tensile fracture in the push direction only.

Specimen G, strengthened with 8 mm-diameter GFRP bars, displayed some distinct behavior characteristics: At a drift ratio a little higher than 2% some of GFRP bar ribs experienced shear fracture, resulting in slippage between the bars and the epoxy adhesive inside the grooves. The mean recorded strain in the bars when this phenomenon initiated was 0.45%, well below the bars' ultimate strain. In this specimen, failure in both directions was due to buckling of the GFRP bars, at drift ratios in the order of 5%, with a mean recorded strain of GFRP equal to 1.1%; the attained degree of strengthening was about 1.20–1.25.

Specimen S_R failed when the bars buckled suddenly at a degree of strengthening more than 1.6 and a drift ratio of about 5%, with a buckled length approximately equal to 0.5 m. Its jacketed counterpart, that is specimen S_R_J, displayed an improved behavior, comprising stable hysteresis loops until large drift ratios, in the order of 8%. This specimen attained the maximum flexural resistance, which was nearly double that of the control specimen. The confinement exerted by the TRM jacket at the base of this specimen prevented buckling of the NSM bars, which fractured in the pull direction, when the tensile strain was approximately equal to 10.1%. This value is nearly half the measured ultimate strain in the uniaxial tests, indicating again, as in the case of specimens C_Per, C_Per_ρ_{n2} and C_Per_ρ_{s2}, the detrimental effect of cycling on the ultimate uniform elongation of stainless steel bars.

For specimens S_M and S_M_J with mortar-filled grooves, the relatively low strength of the mortar in the base block resulted in gradual pull out of the bars and relative slip between bars and the surrounding mortar in the anchoring region, thus limiting the force transfer into the anchoring length and reducing the utilization of the NSM bars. The damage of the mortar inside the base block increased in a stable manner as the displacement increased up to the peak resistance of the specimens, which was marked at a drift ratio of about 2% for both directions of loading, corresponding to a strengthening degree in the order of 1.25–1.30. Apart from a slight reduction of the lateral load, the post peak response of both specimens was quite stable, displaying a marginal strength degradation up to a load level defined by the residual friction between bar and mortar. This pull out resistance due to friction mechanisms resulted in a nearly rigid motion of the stainless steel bars into the anchoring region with practically the same slip along the bonded length, providing to columns S_M and S_M_J a pseudo-ductile behavior.

36.4 Discussion

All columns responded as designed and failed by flexural yielding of the internal steel, followed by failure of the NSM reinforcement. In terms of the various factors investigated in this experimental program, an examination of the results in terms of strength (average increase in the push and pull direction) but also in terms of overall response revealed the following information:

Type of NSM reinforcement (C_Per versus G versus S_R). Despite the roughly equal (monotonic) uniaxial strength of CFRP, GFRP and stainless steel bars, the latter were more effective, resulting in strength increase equal to 64%. The respective values for FRPs were lower (26% for CFRP and 22% for GFRP), due to failure of the FRP reinforcing elements at strains less than those corresponding to peak stress, as a result of cyclic loading. In terms of deformation capacity, quantified here by the drift ratio at conventional failure, stainless steel and GFRP bars outperformed CFRP strips by approximately 25%, due to the lower deformability of carbon fibers in comparison with the other two materials.

Geometrical reinforcing ratio of NSM reinforcement (C_Per versus C_Per_ρ_{n2}): Increasing the NSM reinforcing ratio by 50% (three versus two strips in each side) resulted in a nearly proportional increase in strength, that is from 26% in specimen C_Per to 35% in specimen C_Per_ ρ_{n2}. Of course, this linearity may not apply in the case of large NSM reinforcing ratios.

Geometrical reinforcing ratio of internal steel reinforcement (C_Per versus C_Per_ρ_{s2}). Through the use of cross-section analysis based analytical modelling and the rectangular stress block approach for concrete in compression (without safety factors), the control specimen of column C_Per_ρ_{s2} has a predicted strength equal to 26.15 kN (note that the same analysis predicted the experimentally obtained strength of the control column with an error of less than 5%, hence this model is considered reliable). By dividing the strength of specimen C_Per_ρ_{s2} (average value in the push and pull direction) with this value, the resulting degree of strengthening is approximately equal to 1.34. Therefore it is verified (and quantified) that the effectiveness of NSM reinforcement increases as the internal steel reinforcing ratio decreases: two NSM strips in each column side increased the strength by 34% for specimen with geometrical ratio of internal steel equal to ρ_s=0.72%, whereas the respective increase for the case of ρ_s=0.98% was only 26%.

Configuration of NSM strips (C_Per versus C_Par). In the absence of local jacketing, NSM strips placed with their large cross section side perpendicular to the column side were far more effective than those with their large cross section side parallel to the column side, due to the more favourable bond conditions. The strength increase in the former case was 26%, but only 4%, that is marginal, in the latter case.

NSM reinforcement with of without local jacketing (C_Par versus C_Par_J, S_R versus S_R_J, S_M versus S_M_J). Except for the case of mortar binder inside the grooves, which resulted in NSM debonding at the anchorage, local wrapping of the columns with TRM jackets resulted in dramatic improvements of the retrofitted specimens' response, by increasing both strength and deformation capacity. Jacketing with TRM improved the bond conditions and prevented buckling of the NSM reinforcement, thereby making the strength increase from 4 to 36% in the case of CFRP and from 64 to 90% in the case of stainless steel. In columns retrofitted with NSM bars placed inside mortar, jacketing offered a marginal increase in strength and a moderate increase in deformation capacity. Of all columns tested, the one retrofitted with the combination of stainless steel bars and TRM jacketing displayed the best response characteristics (Fig. 36.3i), with stable post peak behavior and minimal strength degradation up to large drift ratios. On the basis of the results presented herein, it seems that the combination of NSM flexural strengthening and local jacketing is a viable means for increasing strength without compromising deformation capacity, which might be the case in unjacketed columns under low axial loads. In that respect it should be noted that higher axial loads would result in a lower drift ratio, as also confirmed here by a test of a column identical to the control specimen but with a normalized axial load equal to 0.3, in which case the drift ratio at failure was 3.75% (much lower than 6.25% recorded for the case of normalized axial load equal to 0.2). Hence, the improvements in deformation capacity are expected higher as axial loads increase.

Type of bonding agent (S_R versus S_M, S_R_J versus S_M_J). Epoxy resin was a much more effective bonding agent for NSM stainless steel. For the unjacketed specimens, when mortar was used (S_M) instead of resin (S_R), the increase in strength dropped from 64 to 24%; the corresponding values for jacketed specimens were 90 and 29%. Hence, the use of mortar instead of resin reduced the effectiveness of the strengthening scheme to about 1/3, due to pullout of the NSM bars.

36.5 Conclusions

A systematic study on NSM-based flexural strengthening of RC columns under simulated seismic loading was presented in this paper. The investigation addressed column strengthening with NSM CFRP or GFRP, as well as stainless steel. Another innovative aspect in this study was the combination of NSM reinforcement with local jacketing, which comprised the recently developed textile-reinforced mortar (TRM) confining system. The design of specimens allowed for an investigation of several variables, details of which are given above. This investigation proved that NSM FRP or stainless steel reinforcement is a viable solution towards enhancing the flexural resistance of reinforced concrete columns subjected to seismic loads. With proper design, which should combine NSM reinforcement with local jacketing at column ends, it seems that column strength enhancement does not develop at the expense of low deformation capacity.

References

1. Asplund SO (1949) Strengthening bridge slabs with grouted reinforcement. ACI Struct J 20(4):397–406
2. Blaschko M, Zilch K (1999) Rehabilitation of concrete structures with strips glued into slits. Proceedings of 12th international conference on composite materials, Paris, (CD-ROM)
3. Bournas DA, Lontou PV, Papanicolaou CG, Triantafillou, TC (2007) Textile-reinforced mortar (TRM) versus FRP confinement in reinforced concrete columns. ACI Struct J 104(6): 740–748
4. Carolin A, Nordin H, Täljsten B (2001) Concrete beams strengthened with near surface mounted reinforcement of CFRP. Proceedings of international conference on FRP composites in civil engineering. The Hong Kong Polytechnic Univ, Hong Kong, II:1059–1066
5. Casadei P, Galati N, Boschetto G, Tan KY, Nanni A (2006) Strengthening of impacted prestressed concrete bridge I-girder using prestressed near surface mounted C-FRP bars. Proceedings of 2nd international fib congress. Naples, Italy, (CD-ROM, paper ID 10–76)
6. De Lorenzis L, Nanni A (2001) Shear strengthening of reinforced concrete beams with NSM fiber-reinforced polymer rods. ACI Struct J 98(1):60–68
7. De Lorenzis L, Nanni A (2002) Bond between near-surface mounted fiber-reinforced polymer rods and concrete in structural strengthening. ACI Struct J 99(2):123–132

8. Dias SJE, Barros JAO (2006) NSM CFRP laminates for the shear strengthening of T section RC beams. Proceedings of 2nd international fib congress. Naples, Italy, (CD-ROM, paper ID 10–58)
9. Harmon T, Kim YJ, Kardos J, Johnson T, Stark A (2003) Bond of surface-mounted fiber-reinforced polymer reinforcement for concrete stuctures. ACI Struct J 100(5):557–564
10. Hassan T, Rizkalla S (2003) Investigation of bond in concrete structures strengthened with near surface mounted carbon fiber reinforced polymer strips. ASCE J Comp Constr 7(3): 248–257
11. Nordin H, Täljsten B (2006) Concrete beams strengthened with prestressed near surface mounted CFRP. ASCE J Comp Constr 10(1):60–68
12. Sena-Cruz JM, Barros JAO (2004) Modeling of Bond between Near-Surface Mounted CFRP Laminate Strips and Concrete. ASCE J Comp Const 8(6):519–527
13. Triantafillou TC (ed) (2007) Fiber-reinforced polymer reinforcement for concrete structures. Proceedings of 8th international symposium on Fiber reinforced polymer reinforcement for concrete structures. University of Patras, Greece, www.frprcs8.upatras.gr
14. Triantafillou TC, Papanicolaou CG, Zissimopoulos P, Laourdekis T (2006). Concrete confinement with textile-reinforced mortar jackets. ACI Struct J 103(1):28–37

Chapter 37
Optimum Partial Strengthening for Improved Seismic Performance of Old Reinforced Concrete Buildings with Open Ground Story

Themistocles A. Antonopoulos and Stavros A. Anagnostopoulos

37.1 Introduction

Greece is one of the most seismically active countries of Europe, where in the last 3 decades several strong earthquakes have killed more than 260 people and caused large economic loss [2]. Although these consequences are relatively small compared to those from other catastrophic earthquakes around the world, they are still substantial considering that the magnitude of the quakes that caused them were on the order of $M_L = 6.0$–6.5.

A major problem of seismic safety in Greece is associated with buildings designed before 1984, i.e. the year when a major revision was introduced to the old Greek Code of 1959 for earthquake resistant design. That old Code was reflecting the state of knowledge in Greece in the 1920s and 1930s and was applied with the reinforced concrete Code of 1954. The latter, based on the German Norms of the time, was oriented towards gravity and wind loads but not earthquakes and as a result it was less demanding than codes aimed at providing earthquake resistance. Thus, buildings designed before 1984 possessed inadequate strength and very little ductility, a concept that was practically unknown then as a basic property necessary for earthquake resistant design. Another important factor that contributed to reduced seismic strengths was the fact that in the period 1956–1978, when great numbers of reinforced concrete, multistory apartment buildings were built throughout Greece, no strong earthquakes occurred that would have revealed inadequate practices and code shortcomings. This led to a lack of awareness about the aforementioned problems and, furthermore, to reduced or no attention paid to problems of quality control. As a result a large number of buildings designed and constructed in that period have inadequate seismic capacity, due, in large part, to:

T.A. Antonopoulos (✉)
Department of Civil Engineering, University of Patras, Patras GR26504, Greece
e-mail: antonopoulos@upatras.gr

M.N. Fardis (ed.), *Advances in Performance-Based Earthquake Engineering*, Geotechnical, Geological, and Earthquake Engineering 13, DOI 10.1007/978-90-481-8746-1_37,

- Absence of a clear and well defined lateral load resisting system to transfer safely all the seismic forces to the foundation of the building.
- Lack of global and local ductility, a necessary property for earthquake resistance unknown at the time. As a result such buildings behave more or less in a non-ductile manner (mostly shear failures) due to inadequate and poorly detailed transverse reinforcement. In addition low percentages of longitudinal, poorly anchored, reinforcement, lead to low member and structural strengths.
- Weak columns due to no capacity design provisions.
- Inadequate or even absent quality control in all phases of building production.

The most vulnerable class of RC buildings in the above described group, is that with an open ground story, where cars can be parked (buildings with pilotis). Compared to modern construction, buildings of this category have very low seismic capacity [3, 5, 11]. Most of the major collapses of buildings in Greece from earthquakes in the past 3 decades belong to that category [1, 7].

The goal of the paper is to investigate the feasibility of removing the soft story weakness of the above class of buildings with simple, low cost interventions in the open ground story only, so that the building remains functioning during strengthening. Low total cost and an even lower cost share for each apartment owner in a building, combined with no interruption of usage, are key factors that could make an intervention proposal of this type attractive for large scale application.

Two methods of strengthening are investigated for a 5-story building designed with the old codes: one using steel bracing in ground story bays and another using column jacketing. Since the strengthening is limited to the ground story, it will certainly not bring the building to current standards of seismic safety. Rather, it will try to optimize the results by minimizing the building vulnerability, given that stiffening and strengthening the ground story will increase the loading in the stories above. Its capacity is assessed using static pushover type of analyses, as well as non linear Response History Dynamic Analyses (RHDA) for selected real and artificial accelerograms, compatible with the design spectrum of the current code. The assessments are carried out by applying the new draft Greek retrofitting code [9], as well as the ASCE-41 provisions [4]. Subsequently, using a trial and error procedure the building is strengthened by both methods, its new capacities are assessed and conclusions are drawn.

37.2 Building Description, Strengthening Solutions and Earthquake Input

The building considered in this work is a 5-story, space frame RC building with brick infill walls. Its typical floor plan and its view are shown in Fig. 37.1 along with the two strengthened variants. A symmetric layout was selected for simplicity, being a case not quite representative of old RC buildings, which very often are quite irregular, with horizontal and vertical irregularities that can cause significant

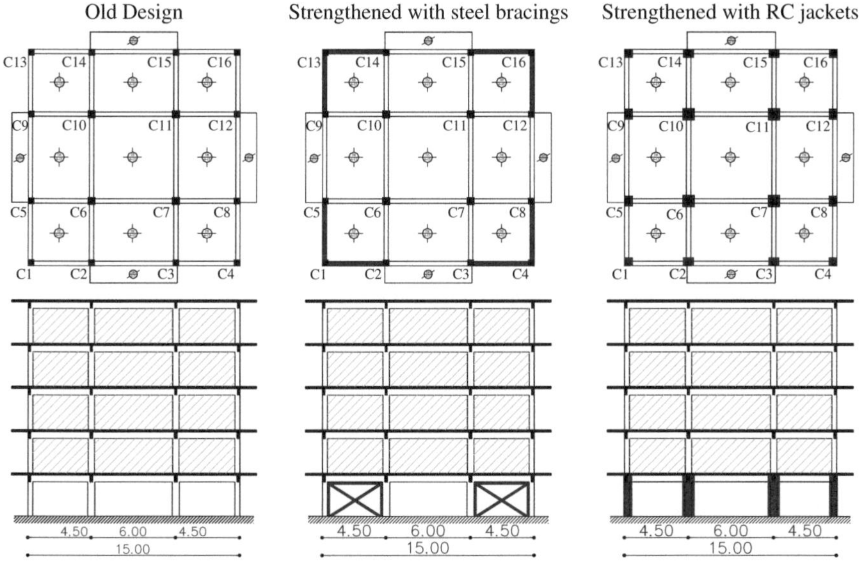

Fig. 37.1 Typical floor plan (*upper graphs*) and views (*lower graphs*) of the original building and of the strengthened variants

torsional response. Thus, this building should be considered as a good design for that period and a "best case scenario" for our study.

The building was designed according to the old Greek Codes for seismic zone II and soil class B, a combination leading to a base shear coefficient of ε=0.08. For comparison with current earthquake provisions, the pertinent design spectra of the current code and the response spectra of selected greek recordings given at the end of this chapter should be examined. B225 grade concrete (f_c = 24 MPa) and StIII steel reinforcement (f_y = 460 MPa) were assumed, materials that became almost standard in the later part of the considered period. The total gravity and live load is 12.45 kN/m^2 of floor area, where the weight of brick infills is also included. Typical beam dimensions are 0.20×0.60 m with 0.15 m slab thickness. An idea about the typical columns of those buildings can be obtained from Fig. 37.2 showing the cross section of a ground story corner column. In the same figure, a current "version" of the same column section is shown, along with the strengthened original section. All story heights were set equal to 3.0 m. Brick infill walls were assumed in all the bays, but only the exterior ones, whose thickness is about 22 cm were considered in our assessments. Interior infills, with typical thickness of about 10 cm and with door openings, were ignored through the engineering approximation that this omission was counterbalanced by ignoring any openings in the exterior infills. Compressive strength of the masonry infills was taken equal to f_k=2.5 MPa in the direction of the inclined struts while the modulus of elasticity was considered equal to 1000 f_k=2.5 GPa according to KAN.EPE. This building will be referred as "Old Design".

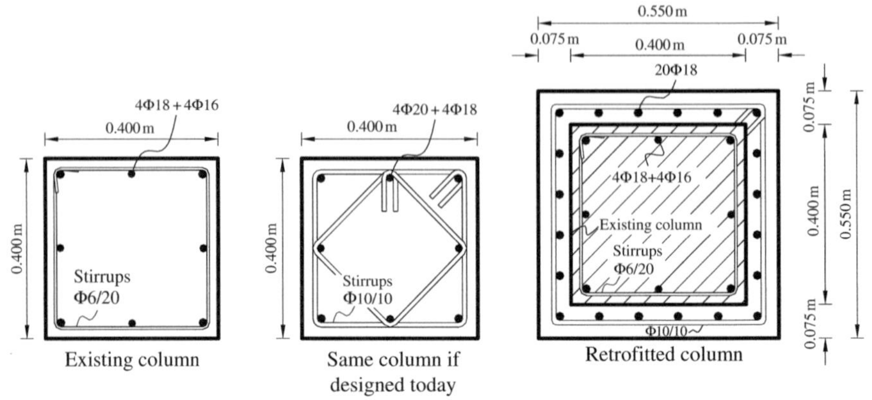

Fig. 37.2 Column sections of old design and retrofitted building

Two strengthening solutions were examined: The first using steel bracings in four bays in the building perimeter and the second using reinforced concrete jackets in some or all columns of the open ground story. For the first solution, the minimum required section (IPE 180) of X-braces satisfying the KAN.EPE slenderness requirements were used. For the second solution, the KAN.EPE and the ASCE-41 provisions were used. Due to differences between the two documents, mainly in member stiffness estimates, KAN.EPE led to jackets in all columns, while following ASCE-41, jackets in only the 4 interior columns were sufficient. The jacket thicknesses are 10 mm and 7.5 mm for the interior and the exterior (perimetric) columns, respectively. More details can be seen in Fig. 37.2.

The first 3 fundamental periods of the old design and of the strengthened variants, computed with member properties per KAN.EPE and ASCE-41, are listed in Table 37.1. It is interesting to observe the differences resulting from the different member stiffness approximations of the two documents.

Concerning the earthquake input, this was assumed to be the design spectrum of the current Greek code [8] for seismic zone II (PGA=0.24 g) and soil type B (Fig. 37.3). For the RHDA, the 3 strongest Greek recordings were used, each with its two horizontal components, whose response spectra and corresponding PGAs are also shown in Fig. 37.3. In addition, 5 semi-artificial records, labeled AR1-AR5, were generated using the program described by [10] to closely match the design spectrum (see Fig. 37.3).

Table 37.1 First three fundamental periods of the existing and the strengthened building

	Old design		Strengthened: X-Bracing		Strengthened: RC Jackets	
Mode	ASCE-41	KAN.EPE	ASCE-41	KAN.EPE	ASCE-41	KAN.EPE
$T_x = T_y$ (sec)	0.50	0.79	0.40	0.44	0.43	0.62
T_θ (sec)	0.41	0.69	0.26	0.28	0.34	0.51

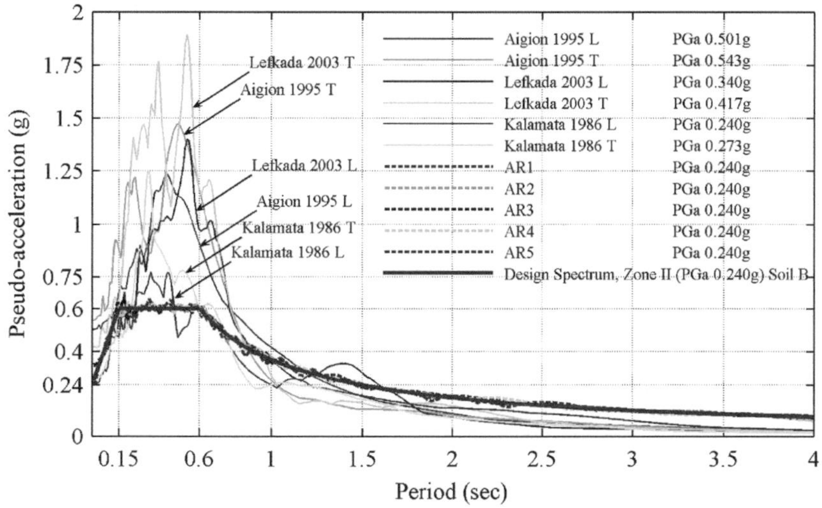

Fig. 37.3 Design spectrum for zone II (PGA=0.24 g), Soil class B, and response spectra of real and semi-artificial records

37.3 Nonlinear Static Pushover Analysis Results

Nonlinear, static pushover analyses were carried out using the Ruaumoko 3D program [6]. Figure 37.4 shows a comparison of the pushover curves for the old design and its strengthened variants with both methods, using member properties according to the Greek Retrofitting Code. The same graphs based on the ASCE-41 provisions

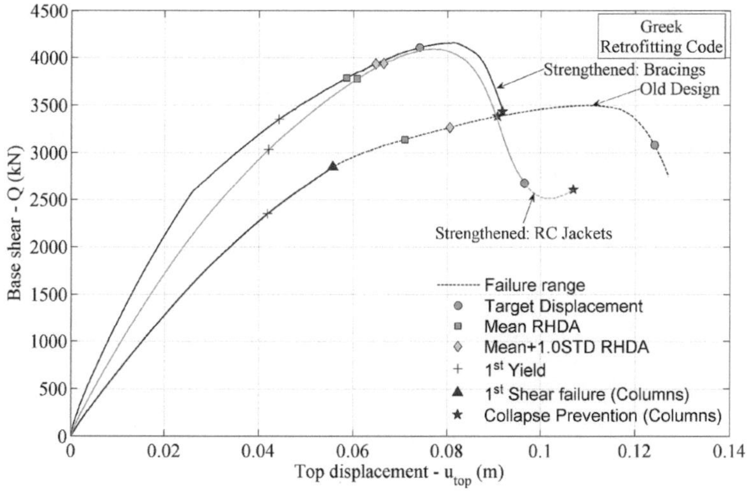

Fig. 37.4 Pushover curves of the old design and the strengthened variants – KAN.EPE properties

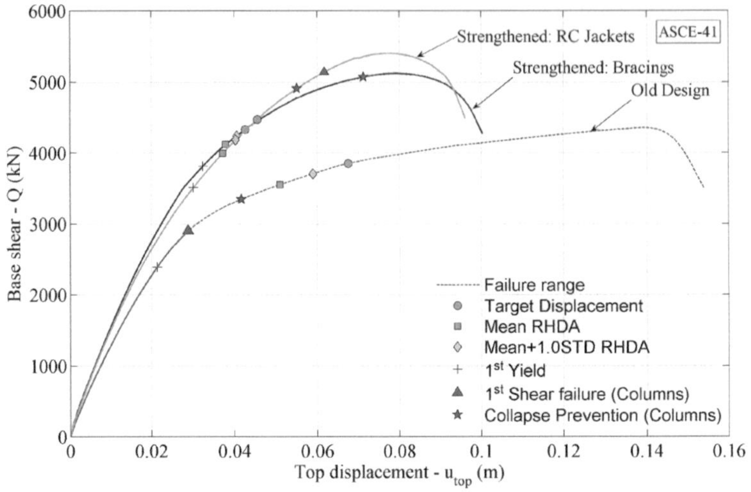

Fig. 37.5 Pushover curves of the old design and the strengthened variants – ASCE-41 properties

are given in Fig. 37.5. Performance points corresponding to first yield (first plastic hinge formation), first shear failure in columns and first exceedance of the available plastic hinge rotation capacity in columns are also given. Target displacements are also shown along with some results from RHDA i.e. the mean and mean plus one standard deviation maximum top floor displacements in order to be directly comparable to the target displacement, which is also given and marked on the graphs. The dotted line indicates failure.

Exact values of these points and corresponding ductility factors, based on the yield displacement of the bilinear approximation, as well as ductility factors of the bracings at the target and the mean displacements (from RHDA), are listed in Table 37.2.

It is no surprise that the old design cannot reach the target displacement, as determined following either of the two "codes", or the lower mean displacements computed using RHDA (with the semi- artificial motions). In fact, the graphs in both

Table 37.2 Computed values of various parameters by KAN.EPE and ASCE-41

		Pushover					RHDA			
		V_y (kN)	u_y (m)	u_{target} (m)	μ_{target}	$\mu_{bracing}$	u_{mean} (m)	$u_{mean+STD}$ (m)	$\mu_{u,mean}$	$\mu_{bracing}$
KAN.EPE	Old Design	2680	0.044	0.124	2.82	–	0.071	0.081	1.61	–
	Bracings	2850	0.025	0.074	2.96	4.96	0.058	0.065	2.32	3.93
	RC Jackets	2600	0.029	0.097	3.34	–	0.061	0.066	2.10	–
ASCE–41	Old Design	2780	0.022	0.067	3.05	–	0.051	0.059	2.32	–
	Bracings	2900	0.019	0.043	2.26	2.05	0.037	0.041	1.95	2.01
	RC Jackets	2580	0.017	0.045	2.64	–	0.037	0.040	2.18	–

Figures show premature shear failure of a column (red triangle points) and subsequent flexural failure (exceeding plastic hinge rotation capacity – red asterisk points – also indicated as "collapse prevention") both preceding the target displacement, while the shear failure preceded even the time history displacements. The more flexible building model resulting from the Greek KAN.EPE gave larger displacements defining the various performance points on the push over curves, compared to the ASCE-41 model. Their sequence, however, remains the same in both cases.

Looking now at the pushover curves of the two strengthened building variants (with bracing and with column jacketing) as determined by both the KAN.EPE and ASCE-41 models, both strengthening methods give similar curves, all higher than the curve of the old design. The descending branch of each curve indicates infill wall failures in the stories above ground.

Looking at Fig. 37.4 (KAN.EPE models), we see that both strengthening methods lead to collapse displacements (collapse prevention) larger than the corresponding target displacements. However, in the case of x-bracing this comes before the walls start failing, while the opposite happens when column jacketing is used for strengthening. The other points on the two curves are quite close.

Looking now at Fig. 37.5 we see again that both strengthening methods lead to target displacements below the corresponding collapse prevention displacements (first flexural failure of a column). However the strengthening with x-bracing produces a structure with larger collapse displacement and no shear failure in the ascending portion of the curve. Thus the data in both graphs and in Table 37.2 indicate a better performance of the building strengthened with x-bracing than the one with strengthened columns.

37.4 Nonlinear, Response History Dynamic Analysis (RHDA)

In addition to the static pushover analyses presented in the previous chapter, the old design and the two strengthened variants were evaluated using nonlinear dynamic response history analyses (RHDA). These were carried out using the 3 pairs of real Greek records and three pairs of the semi-artificial records formed from the 5 components indicated earlier. For each of the two motion sets, interstory drifts, bracing ductility factors and damage indices (DI) of the first story infill walls were computed and the average of their peaks for the 3 motion pairs in a set were obtained. The symmetry of the buildings allowed both response parameters along the x and y directions to be included in the averaging. The alternative to this would be to rotate each motion pair by 90 degrees and then have 6 sets of data for averaging. The Damage Indices of the infills, defined as ratios of the maximum computed strain from the analysis to the strain corresponding to the maximum strength of the infill, provide an indication of the effectiveness of any strengthening solution for the response of the stories above. Conventionally, values of DI between 0.5 and 1.0 represent cracked infills while values of DI slightly greater than 1.0 correspond to infills that have reached their maximum available strength and have started to degrade. Under the assumption that the state of damage of the RC members (mainly columns) in a given story is directly dependent on the interstory drift, as optimum strengthening solution is considered the one that minimizes the maximum interstory drift of the building. In this manner, the weak ground story stops being weak and its drifts are comparable to the drifts of the stories above, thus preventing the formation of a mechanism.

Figures 37.6 and 37.7 show interstory drifts and the other response parameters for the KAN.EPE and ASCE-41 models, respectively. Graphs at left are for the semi-artificial motions and at right for the real, unscaled, Greek motions. Drifts are given for the old design and for the two methods of strengthening.

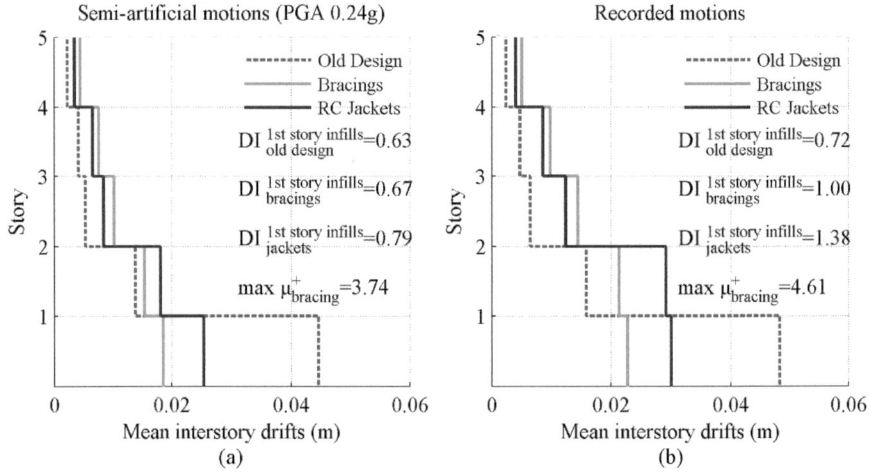

Fig. 37.6 Mean interstory drifts from RHDA – KAN.EPE properties

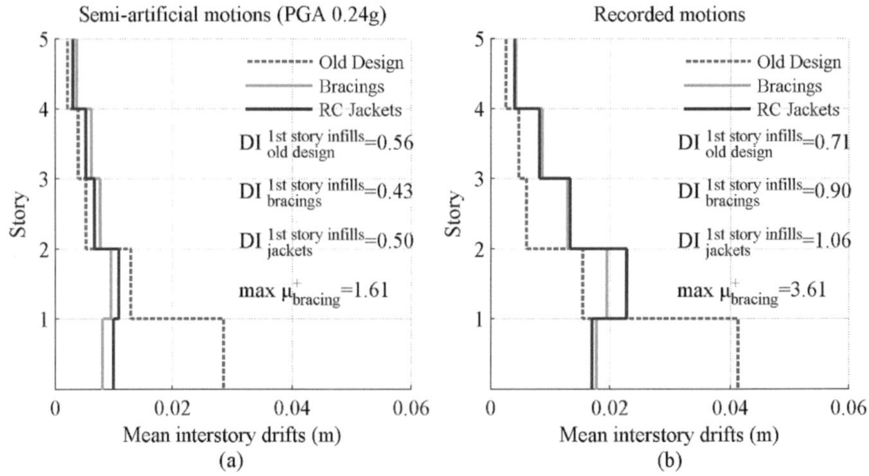

Fig. 37.7 Mean interstory drifts from RHDA – ASCE-41 properties

Both models and both sets of motions indicate that a substantial improvement of the behavior of the old design can be achieved by the two methods of strengthening. It is interesting to observe that the KAN.EPE model (Fig. 37.6) indicates a noticeable advantage of the strengthening with x-bracing over the column jacketing solution. A similar but substantially reduced advantage is seen in the ASCE-41 model (Fig. 37.7). It is also noted that the real recorded Greek motions are more penalizing than the semi-artificial ones. The latter represent the level of loading modern structures should be designed for. However, it would appear that even for these, most severe of all Greek earthquake records, the simple strengthening measures examined here could help the old design avoid collapse and reduce its damage.

37.5 Concluding Remarks

Based on the simple building examined, it appears that the use of steel bracings, properly designed and detailed, in pairs of bays on opposite sides at the building perimeter, can alleviate the soft story weakness in old buildings, designed in accordance with the old Greek codes and having open ground stories. The same may be achieved, although not as effectively, using strengthening of the ground story columns with concrete jacketing. These results were confirmed by applying both the ASCE-41 provisions and the draft Greek retrofitting code. However, substantial differences in the expected design displacements and other parameters were found with the two models, due mainly to the differences in concrete member effective stiffnesses specified in these two documents.

References

1. Anagnostopoulos S (1996) Seismic risk and earthquake damage in Greece. Paper presented at the 1996 earthquake resistant engineering structures ERES 96 Symposium, Thessaloniki, Greece
2. Anagnostopoulos S (2003) Seismic risk mitigation: Greek case study. In: Mulargia F, Geller R (eds) Earthquake science and seismic risk reduction. Kluwer Academic Publishers, London
3. Antonopoulos T, Anagnostopoulos S, Tsirlis F (2008) Evaluation of the seismic performance of old reinforced concrete buildings with an open ground storey, simple retrofitting solutions. Proceedings of the 3rd Greek conference on earthquake engineering and engineering seismology. Paper No 2061, Athens, (in Greek)
4. ASCE/SEI 41-06 (2007) Seismic rehabilitation of existing buildings. Publication by American Society of Civil Engineers, ISBN 978-0-7844-0884-1
5. Baros D, Kyrkos M, Maravas A, Anagnostopoulos S (2007) Assessment of seismic capacity of existing buildings – effects of uncertainties. In: Mucciarelli M, Herak M (eds) Increasing seismic safety by combining engineering technologies and seismological data. Dubrovnik, Croatia
6. Carr A (2005), Ruaumoko volume 3: user manual for the 3-dimensional version. Department of Civil Engineering, University of Canterbury
7. Dimitriou P, Karakostas C, Lekidis V (2000) The Athens (Greece) Earthquake of 7 September 1999: The event, its effects and the response. Proceedings of 2nd Euroconference on global

change and catastrophe risk management: earthquake risks in europe, IIASA, Laxenburg, Austria, on web site: http://www.iiasa.ac.at/Research/RMP/july2000/
8. E.A.K-2000 (Greek Seismic Design Code – 2000), OASP (Greek Organization for Seismic Planning and Protection) Greek Ministry for Environmental Planning and Public Works, Athens, Greece (In Greek): www.oasp.gr.
9. KAN.EPE. (Greek Retrofitting Code-2009) OASP (Greek Organization for Seismic Planning and Protection), Greek Ministry for Environmental Planning and Public Works, Athens, Greece. (In Greek): www.oasp.gr
10. Karabalis DL, Cokkinides GJ, Rizos DC, Mulliken JS, Chen R, (1994) An interactive computer code for generation of artificial earthquake records. In: Khozeimeh K (ed) Computing in civil engineering (ASEE). 1122–1155.
11. Repapis C, Zeris C, Vintzileou E (2006) Evaluation of the seismic performance of existing RC buildings: II. A case study for regular and irregular buildings. J Earthq Eng 10(3):429–452.

Part IV
Advanced Seismic Testing for Performance-Based Earthquake Engineering

Chapter 38
Role and Application of Testing and Computational Techniques in Seismic Engineering

Oreste Salvatore Bursi, Rosario Ceravolo, Francisco Javier Molina, and Marco Molinari

38.1 Introduction

The 2009 L'Aquila $M_W = 6.3$ earthquake caused severe damage to buildings in the city and its vicinity, the seismic motion being very strong with respect to the 10% in 50 years hazard level [5]. This event followed those of Friuli, 1976, $M_W = 6.4$- and Irpinia, 1980, $M_W = 6.9$- showing that Italy suffers from destructive earthquakes on a regular basis. Italy has a large stock of old and relatively old structures with insufficient seismic capacity. Therefore, it is crucial to accurately evaluate their existing seismic capacity – the collapse margin and the state of complete collapse at which the structure no longer can sustain gravity – and then to retrofit them accordingly. On the other hand, Performance-Based Earthquake Engineering (PBEE) is still in development and focuses on structural innovative systems aimed at enhancing functionality, operation and safety [16]: base-isolation and passive dampers are typical examples [14]. There are also notable advances in Finite Element (FE) methods [11], but still insufficient to reliably predict the behaviour of structures at collapse. All such research and development should however be checked against actual performance, before being transferred with confidence to actual design and construction practice. Therefore, the need for actual data obtained by experimentation is deemed to be urgent for the advancement of PBEE, especially for those issues that are pertinent to collapse, rate-of-loading, realistic scale and structural systems where actual data are scarce.

A taxonomy of modern structural testing techniques identifies the following: (i) Pseudo-Dynamic (PsD) testing, conducted with inertia effects of a structural system numerically simulated in the computer and displacements applied by hydraulic actuators [12]; (ii) Real-time testing, where inertia effects are physically developed in the test structure; dynamic are applied either by actuators or shake tables; (iii) Real-time testing with dynamic substructuring, which melds physical and

O.S. Bursi (✉)
DIMS, Università di Trento, via Mesiano, 77 38126 Trento, Italy
e-mail: oreste.bursi@ing.unitn.it

M.N. Fardis (ed.), *Advances in Performance-Based Earthquake Engineering*, Geotechnical, Geological, and Earthquake Engineering 13, DOI 10.1007/978-90-481-8746-1_38,

numerical substructures to predict the seismic behaviour of complex specimens [1]; (iv) Structural Behaviour Monitoring (SBM), that employs vibration tests and limited sensors to collect mission-critical data only from the part of a structure that may potentially experience damage; the rest can be numerically modelled [20].

In order to improve the damage evaluation of structures at different limit states, it is natural to combine some of the above techniques. This is exactly the topic of the paper. Section 38.2 presents the main characteristics of a 3D prototype steel-concrete composite structure and the use of both continuous PsD and SBM techniques. The 3D and 2D FE model of the structure and its joints are described in Section 38.3, together with modal updating techniques and the corresponding global damage index results. In Section 38.4, linear and non-linear identification techniques are adopted to characterize dynamic properties of this structure as well as hysteretic displacement-restoring force relationships of each storey. Besides, it is shown how experimental relationships combined with instantaneous modal properties can be presented in a demand/capacity spectrum format.

38.2 3D Structure and Test Programme

38.2.1 The 3D Test Structure

The full scale, 3D steel-concrete composite structure under test consists of three identical Moment Resisting (MR) frames in Fig. 38.1a, and b, aligned in the longitudinal direction. X-shaped braces were placed in the transverse direction. Steel-concrete, full shear connection composite beams and partially-encased steel-concrete composite columns were employed. High-ductile partial-strength composite beam-to-column joints were designed to provide a plastic joint rotation of 35 mrad at a residual strength of at least 80 per cent of the maximum value [6]. A detailed description of the structure, including the design methods and structural performance data, can be found in [2, 3].

Figure 38.1a, b show the location of the excitation forces: Stepped Sinusoidal Tests (SST) and Shock Hammer Tests (SHT) were performed in the three Phases outlined in Table 38.1. Three different accelerometer configurations were employed in each test: (i) six accelerometers were placed on the structure, one for each of its global degrees of freedom (Fig. 38.1a, and b); (ii) local configurations, termed B and C in Fig. 38.1c, and d, were used to characterise the local dynamic behaviour of interior and exterior beam-column joints, respectively, where damage was expected.

38.2.2 Dynamic, Pseudo-Dynamic and Cyclic tests

To perform a damage estimation, three vibration tests were performed at different damage levels: Phase I for the identification of the intact structure; Phase II for

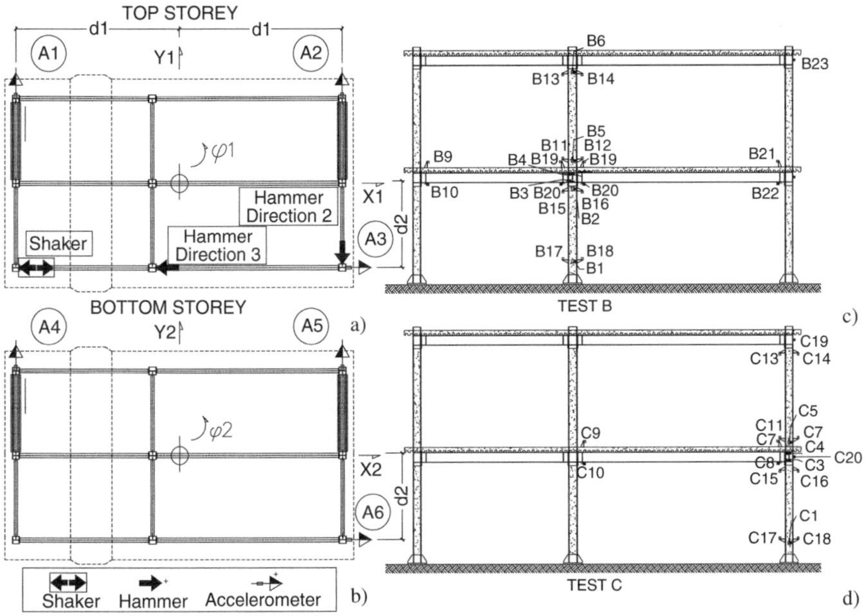

Fig. 38.1 Location of forcing devices and accelerometers: global configuration A at the top (**a**) and bottom (**b**) storeys; local configurations B for interior (**c**) and C for exterior members (**d**)

Table 38.1 Summary of the test programme and performance objectives

Test	Vibration test	PsD test PGA [g]	Performance objective
I	Phase I		Identification at the Undamaged State
1		0.10	Pseudo-Elastic State
2		0.25	Serviceability Limit State (SLS)
3		1.40	Life Safe Limit State (LS LS)
II	Phase II		Identification at the LS LS
4		1.80	Collapse Onset Limit State (CO LS)
5		Cyclic	Max top displacement equal to 300 mm
III	Phase III		Identification beyond the CO LS

the structural identification at the Life Safe Limit State (LS LS); Phase III for the identification beyond the Collapse Onset Limit State (CO LS), as outlined in Table 38.1. To investigate the structural performance at different PGA levels, four PsD tests were carried out as detailed in both the test programme and performance objectives summarized in Table 38.1. Moreover, a final cyclic test beyond the CO LS was performed.

38.3 Identification and Damage Evaluation Under Dynamic and Cyclic Loadings

38.3.1 Linear Identification of Structural Modal Properties

A linear identification of the intact structure was obtained from shock test signals acquired during Phase I of Table 38.1. The numerical procedure consists of an Eigenvalue Realization Algorithm (ERA) applied in classical free decay conditions, which supplied the first six natural frequencies of the structures. The same results, at least for modal frequencies and shapes, were obtained from ambient vibration signals through a Stochastic Subspace technique [18]. Damping values were small and rather scattered, especially when evaluated from ambient vibration signals (see Table 38.2). Experimental modal data obtained at different Phases and not reported here for brevity can only give a qualitative detection of damage [15].

38.3.2 Model Updating Methodology and Damage Evaluation

Unlike the linear identification technique of Section 38.3.1, the FE-based nonlinear Modal Updating (MU) technique allows damage location and quantification. Therefore, in order to complement global load-displacement data provided by PsD and cyclic tests, redundant measurements provided by accelerometers or static sensors were used in conjunction with a MU technique. With this objective in mind, the 3D and 2D FE models of the benchmark structure in Fig. 38.2 were developed and implemented (see [15] for details). The 3D model consists of three identical parallel frames (Fig. 38.2a), connected through master-slave numerical constraints. Braces were modelled as truss elements; beams and columns as beam-column elements; base joints and beam-column connections by rotational springs. Beam-column panels were modelled by a mechanical idealization involving four rigid bars connected by pins and rotational springs. The 2D FE model of the interior frame adopted for cyclic MU (Fig. 38.2b) was appropriate to update the structure under cyclic loading in the X-direction only. The eigen-sensitivity based approach to FE MU was followed for dynamic identification. Discrepancies between numerical and experimental eigenvalues and eigenvectors were minimized by adjusting unknown model

Table 38.2 Synthesis of experimental modal data extracted for Phase I

Mode	Frequency [Hz]	Mean damping ratio (%)	Description
1	3.38	0.26	1st translational in X-dir.
2	5.10	0.91	1st translational in Y-dir.
3	6.87	2.58	1st torsional
4	11.19	2.2	2nd translational in X-dir.
5	15.34	1.75	2nd translational in Y-dir.
6	22.31	2.83	2nd torsional

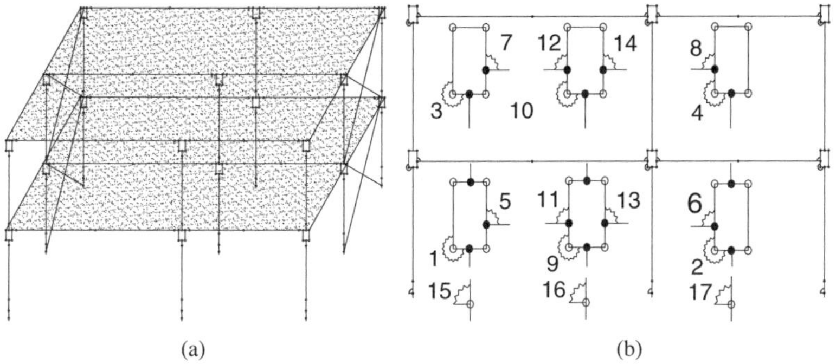

Fig. 38.2 FE baseline models of the benchmark structure: (**a**) 3D model for dynamic MU; (**b**) 2D model of the interior frame for static MU

parameters for the stiffness of some elements in the FE model. Linearization of the problem gives the over-determined system:

$$\mathbf{\Delta}_t = \mathbf{S}_t \mathbf{h}_t \tag{38.1}$$

where $\mathbf{\Delta}$ is the difference between experimental and numerical eigendata, $\mathbf{S}$ is the sensitivity matrix and $\mathbf{h}$ the increment of non-dimensional parameters applied to sensitive FE quantities: base and connection springs, column and brace stiffness. The optimization problem was solved via the Powell's Dog-Leg (DL) technique, by minimizing the scalar objective function $F(\mathbf{p})$, given by

$$F(\mathbf{p}) = \mathbf{\Delta}(\mathbf{p})^{\mathrm{T}} \mathbf{\Delta}(\mathbf{p}) \tag{38.2}$$

As a result of the structural identification in the different test phases of Table 38.1, both local D_{L} and global D_{G} damage indices, reported in Table 38.3, were computed. These values allow a better evaluation of the structure. With regard to the identification under cyclic loading, a secant stiffness approach combined with the Dog-Leg optimisation technique was implemented [17]. As a result, moment-rotation diagrams of the joint components, as shown in Fig. 38.3, can be estimated. The strength and stiffness degradation are evident.

Table 38.3 Global damage indices D_{G} of the structure between of Phases I and II and II and III

Method	$D_{\mathrm{G, I\text{-}II}}$	$D_{\mathrm{G, I\text{-}III}}$
Based on three experimental Modes	0.4652	0.6274
Based on six experimental Modes	0.4650	0.6274
Based on the updated flexibility matrix	0.5212	0.6669

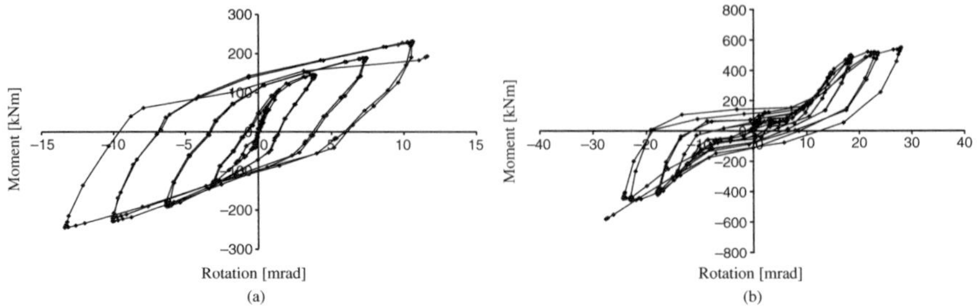

Fig. 38.3 Identified joint components under cyclic loading: (**a**) shear panel spring No. 3 in Fig. 38.2; column base joint spring No 16 in Fig. 38.2

38.4 Non-Linear Identification Techniques Under Pseudo-Dynamic Loadings

38.4.1 Structural Dynamic Characteristics from Spatial Model Identification

The technique presented herein allows identification of instantaneous natural frequencies and damping ratios from nonlinear force/displacements during PsD tests. Spatial Model Identification [13] exploits response histories of displacements and restoring forces, either experimental or numerical, to identify the stiffness and damping matrices of an equivalent linear model with a least-squares minimization method: for known theoretical mass matrix, the complex modes and associated frequencies and damping ratios are computed. The identification is repeated at subsequent time instants, with constant time interval length covering several oscillation periods (100 time increments, each of 0.01 s). In this case, the structure in Fig. 38.1 has been considered as a 2-DoF system in agreement with the PsD assumptions. For brevity, attention is limited to the 1st translational mode, see Table 38.1; additional information may be found in [4]. The evolution of the identified instantaneous frequency and damping ratio in the different PsD tests is shown in Fig. 38.4. Highly non-linear response and significant reduction in the natural frequencies are observed under stronger earthquakes.

38.4.2 Structural Dynamic Characteristics Provided via a Hysteretic Model

Instantaneous estimates of structural dynamic parameters in PsD tests are very helpful. However, when structures experience inelastic phenomena in seismic engineering, it is much more effective to deal with hysteretic models which offer a physical and compact representation of hysteresis [10, 19]. The Bouc-Wen model was adopted here, because it is a continuous model. For a SDoF system it gives:

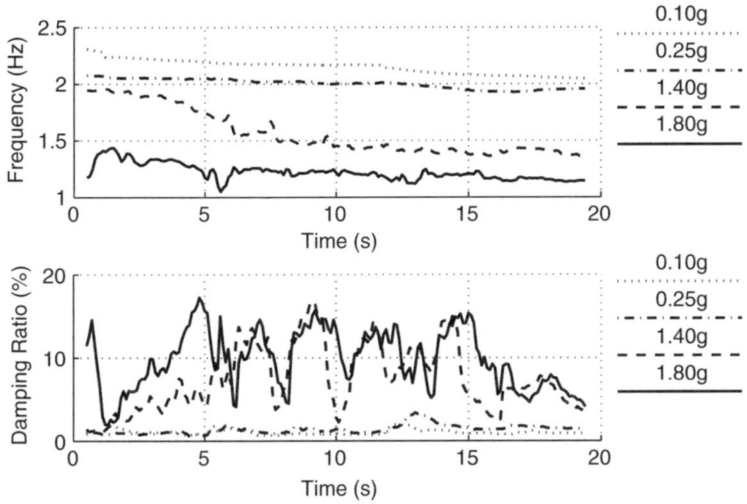

Fig. 38.4 Spatial-Model identified instantaneous frequency and damping for the first mode from PsD test results at different PGA levels

$$\begin{cases} m\ddot{x} + r = u \\ \dot{r} = A\dot{x} - (\beta \cdot sign\,(r\dot{x}) + \gamma)\,|r|^n\,\dot{x} \end{cases} \tag{38.3}$$

where $\dot{x}$ and $\ddot{x}$ are velocity and acceleration, respectively, u is the external excitation and r the hysteretic restoring force. A, β, n>0 and $0 \leq \gamma \leq \beta$ are model parameters [19]. By the use of parametric identification, two options are considered here: (i) the direct identification of the Bouc-Wen parameters, as the restoring force r is available during a PsD test (this technique is applied in this section); (ii) a more general case, where polynomial fitting is first performed to characterize the restoring force and then the hysteretic model parameters are identified (this case is treated in Section 38.4.3).

Modelling the structure as a two-DoF system, the Bouc-Wen model entails:

$$\begin{Bmatrix} \dot{f}_1 \\ \dot{f}_2 \end{Bmatrix} = [K]\{\dot{x}\} - \{f_{BW}\} = \begin{bmatrix} k_{11} & k_{12} \\ k_{21} & k_{22} \end{bmatrix} \cdot \begin{Bmatrix} \dot{x}_1 \\ \dot{x}_2 \end{Bmatrix} - \begin{Bmatrix} \beta_2 f_2\,|\dot{x}_2 - \dot{x}_1| - \beta_1\,(f_1 + f_2)\,|\dot{x}_1| \\ \beta_2 f_2\,|\dot{x}_2 - \dot{x}_1| \end{Bmatrix} \tag{38.4}$$

where [K] is the matrix of tangent stiffness, β_{i} are the Bouc-Wen parameters and f_i restoring forces applied to the two composite slabs. As a result, Bouc-Wen model parameters can be determined by solving for each time step the following non-linear least-square problem [8]:

$$F_{ob}\,(k_{11}, k_{12}, k_{21}, k_{22}, \beta_1, \beta_2) = \left\| \sum_i \dot{f}_{i,measured} - \dot{f}_i \right\| \tag{38.5}$$

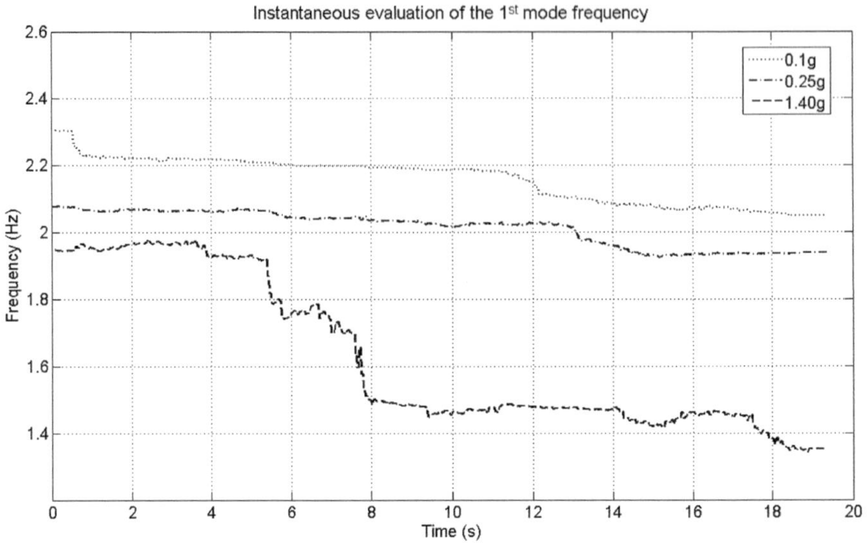

Fig. 38.5 Evolution of the first mode frequency evaluated from Bouc-Wen elastic coefficients for different PGAs

Equation (38.5) was applied on a window length of 101 samples at a sampling rate Δt=0.002 s. For brevity, Fig. 38.5 reports only the evolution of the first modal frequency, which is comparable to results shown in Fig. 38.4. The 1.8g test results are not plotted, because the model should be enriched to take pinching into account.

38.4.3 Identification of a Hysteretic Model via a Polynomial Form

This method is more general than that of the Section 38.4.2, because restoring forces are considered as unknown [7]. For a 2 DoF non-linear hysteretic system the following polynomial approximation is assumed:

$$\begin{bmatrix} m_1 & 0 & 0 & 0 \\ 0 & m_2 & 0 & 0 \\ 0 & 0 & 0 & 0 \\ 0 & 0 & 0 & 0 \end{bmatrix} \begin{Bmatrix} \ddot{x}_1 \\ \ddot{x}_2 \\ \ddot{r}_{10} \\ \ddot{r}_{21} \end{Bmatrix} + \begin{bmatrix} 0 & 0 & 0 & 0 \\ 0 & 0 & 0 & 0 \\ -\alpha_1 & 0 & 1 & 0 \\ \beta_1 & -\beta_1 & 0 & 1 \end{bmatrix} \begin{Bmatrix} \dot{x}_1 \\ \dot{x}_2 \\ \dot{r}_{10} \\ \dot{r}_{21} \end{Bmatrix} + \begin{bmatrix} 0 & 0 & 1 & -1 \\ 0 & 0 & 0 & 1 \\ 0 & 0 & \alpha_2 & 0 \\ 0 & 0 & 0 & \beta_2 \end{bmatrix} \begin{Bmatrix} x_1 \\ x_2 \\ r_{10} \\ r_{21} \end{Bmatrix} =$$

$$= \begin{Bmatrix} u_1 \\ u_2 \\ \alpha_3 \dot{x}_1^3 + \alpha_4 \dot{x}_1^2 r_{10} + \alpha_5 \dot{x}_1 r_{10}^2 + \alpha_6 r_{10}^3 \\ \beta_3 (\dot{x}_2 - \dot{x}_1)^3 + \beta_4 (\dot{x}_2 - \dot{x}_1)^2 r_{21} + \beta_5 (\dot{x}_2 - \dot{x}_1) r_{21}^2 + \beta_6 r_{21}^3 \end{Bmatrix} \tag{38.6}$$

where m_i and $\ddot{x}_i$ are the mass and the acceleration of the DoF $i \geq 1$ relative to the ground, respectively; $\dot{x}_i - \dot{x}_j$ is the difference of velocities between two adjacent

DoFs, i.e. $j=i-1$; u_i the external force acting on the mass m_i; r_{ij} is the internal restoring force between DoFs i and $j=i-1$. $\alpha_1, \alpha_2 \ldots \alpha_6$ and $\beta_1, \beta_2 \ldots \beta_6$ are coefficients of the polynomial approximations. If N_m defines the number of data, i.e. acquired displacements, velocities or accelerations, then the vector of the instantaneous estimators **p** can be determined by minimizing, instant-by-instant in a certain interval the following objective function (details are given in [8]):

$$F_{ob}\left(n^*, \mathbf{p}\right) = \sum_{j=1}^{M} \sum_{m=0}^{M-1} \left| \left|STFT_j\left(n^*, m\right)\right|^2 - \left|STFT_{V,j}\left(n^*, m\right)\right|^2 \right| \tag{38.7}$$

where $STFT_j(n^*, m)$ and $STFT_{V,j}(n^*, m)$ are values of the Short Time Fourier Transform (*STFT*) at the discrete time instant $n{*}\Delta t$ and at the frequency $1/(m\Delta t)$, of the *j*th system response, respectively. They were measured and calculated through the polynomial system for a given configuration of the parameters **p** – see Eq. (38.6). This procedure was applied for the non-linear identification of the benchmark structure of Fig. 38.1 in the PsD test at a 0.1*g* PGA. The identification of PsD test data entails some advantages, indeed: it is possible to neglect rate-dependent damping effects, since displacements are applied in a quasi-static fashion. The fitting in the time-frequency domain used a window length of 101 samples. A 5th degree polynomial representation was used in Eq. (38.6) corresponding to three terms, since only odd polynomial terms were retained. Details are given in [7]. The force-interstorey drift response of the top storey, representing the main result of the non-linear identification technique, is shown in Fig. 38.6a. It clearly shows a hysteretic behaviour activated even at relatively low ground motion amplitudes, owing to the cracking of the concrete slab around beam-to-column joints [4]. Measured and estimated time-histories of the relative restoring force are shown in Fig. 38.6b, demonstrating the high quality of the identification process.

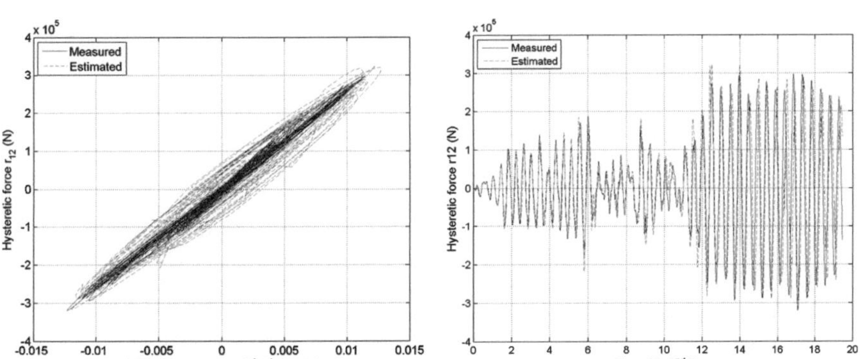

Fig. 38.6 Measured and identified responses at the top storey: (**a**) hysteretic force-interstorey drift response; (**b**) time-history of the restoring force

38.4.4 Data in a Design Format from PsD Tests

A proper re-evaluation of dynamic characteristics from PsD test data – see Section 38.4.1 – can provide information on seismic demands of steel-concrete composite buildings along the line of the Capacity Spectrum Method [9]. In this respect an automatic procedure was implemented in [14], where the modulus of the Hilbert

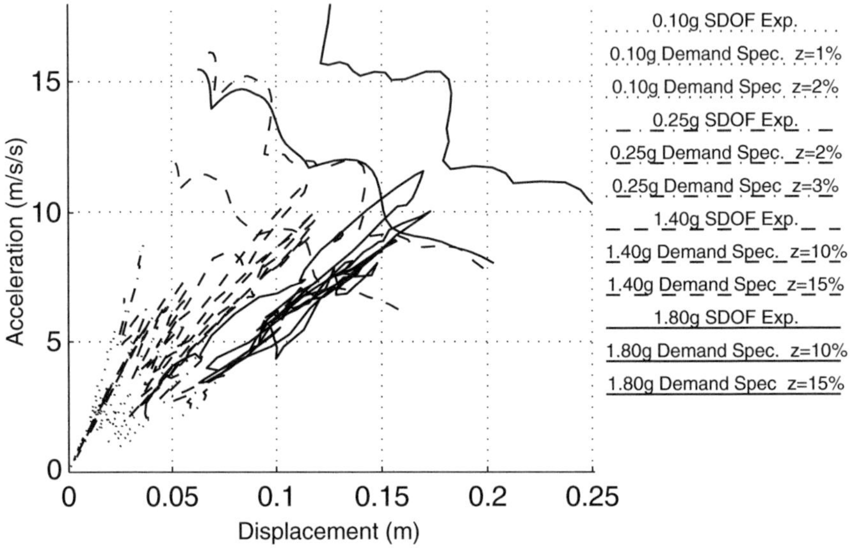

Fig. 38.7 SDoF-equivalent pseudo acceleration-displacement representation of identified values of oscillatory amplitude and frequency. Matching with linear demand spectra

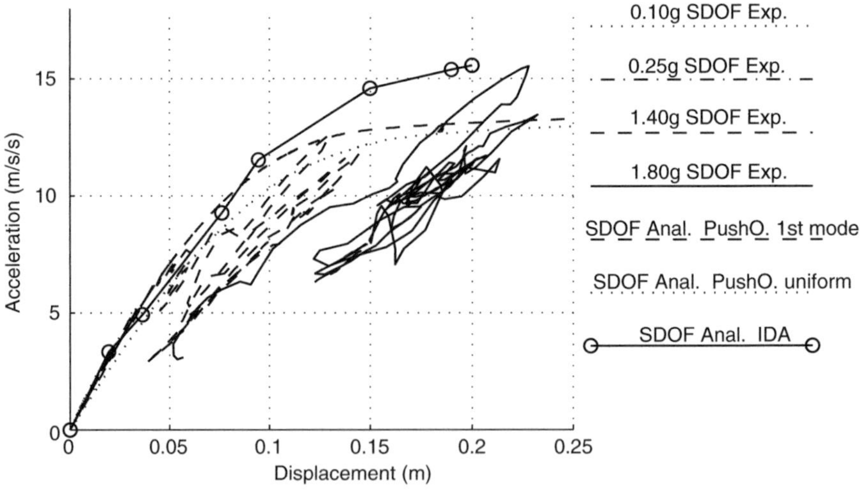

Fig. 38.8 SDoF-equivalent pseudo acceleration-displacement representation of identified values of total amplitude and frequency. Comparison with analytic pushover and IDA curves

transform of the top storey displacement, considered as *oscillatory amplitude*, was computed at each PGA level, together with residual displacements i.e. permanent deformations. They were added together to form the *total amplitude*. Hence, the relationship between frequency and amplitude can be represented as a capacity curve in an acceleration-displacement diagram of an equivalent SDoF system [9]. This is shown in Fig. 38.7, where the curves of the equivalent SDoF system were obtained by scaling the modal participation factor with a constant mode shape of components: [0.454 1] [3]. The linear response spectrum for each earthquake is represented as a pair of thinner lines, with the same line-type of the response to that earthquake, corresponding to damping ratios of 1 and 2% (at 0.1g PGA), 2 and 3% (at 0.25g PGA) and 10 and 15% (1.4 and 1.8g PGA) in agreement with Fig. 38.4. One can observe both the demand and the capacity spectra for this structure.

A similar procedure was followed to compute non-linear responses depicted in Fig. 38.8, considering *total amplitudes*. Three different capacity relationships provided by FE analyses [4] were plotted: (i) a pushover analysis based on the [0.454 1] first mode shape; (ii) a pushover analysis based on a uniform shape; (iii) an Incremental Dynamic Analysis (IDA). Results of the pushover analyses match the experimental capacity curve up to the 1.4g PGA, whilst experimental results for the 1.8g PGA earthquake are better tracked by the IDA curve.

38.5 Conclusions

An evaluation of structural performances in terms of forces and deformations has been presented and applied to a partial strength 3D two-storey moment-resisting steel-concrete composite structure. The framed structure was subjected to different pseudo-dynamic tests, followed by a final cyclic test and alternated with vibration tests using a limited number of sensors to collect mission-critical data only from beam-column and column base joints that mainly experience damage. Apart of standard modal extraction tools used to identify frequency and mode shapes in the linear regime, different automatic and semi-automatic identification numerical tools were used, owing to the dependence between dynamic properties and deformation amplitude. Results allow assessing structural damage for different limit states, as well as the present PsD results in a practical demand/capacity spectrum format. Melting of experimental and numerical identification techniques in association to new sensor developments hold great promise for the future.

References

1. Blakeborough A, Williams MS, Darby AP, Williams DM (2001) The development of real–time substructure testing. Phil Trans R Soc Lond A 359:1869–1891
2. Braconi A, Bursi OS, Fabbrocino G, Salvatore W, Tremblay R (2008). Seism. performance of a 3D full-scale high-ductility steel-concrete composite MR structure – Part I: Design and testing procedure. Earth Eng Str Dyn 37(14):1609–1634.

3. Braconi A, Bursi OS, Fabbrocino G, Salvatore W, Tremblay R (2008) Seismic performance of a 3D full-scale high-ductile steel-concrete composite MR frame – Part II: Test results and analytical validation. Earth Eng Str Dyn 37(14):1635–1655
4. Bursi OS, Caramelli S, Fabbrocino G, Molina J, Taucer F, Salvatore W, Zandonini R, (2004) 3D full scale seismic testing of a steel-concrete composite building at ELSA. Eur Report, EUR 21299 EN. JRC, CE, Ispra, Italy
5. Bursi OS, Dusatti T, Pucinotti R (2009), A Reconnaissance report. The April 6, 2009, L'Aquila Earthquake. Italy, Report, http://www.reluis.it
6. CEN (2005) EN 1998-1:2005, Eurocode 8: design of structures for earthquake resistance. Part 1: general rules, seismic actions and rules for buildings. European Committee for Standardization, Brussels.
7. Bursi OS, Ceravolo R, Demarie G, Erlicher S, Molinari M, Zanotti Fragonara L (2009) Id. of the Damage Evolution in a Benchmark Steel-Concrete Composite Structure during PsD Testing, In Proceedings of compdyn 2009, Rhodes, Paper n. 324
8. Ceravolo R, Demarie GV, Erlicher S (2007) Instantaneous identification of Bouc-Wen-type Hyst. Systems from seismic response data. Key Eng Mat 347:331–338.
9. Fajfar, P. (1999) Capacity spectrum method based on inelastic demand spectra Earth Eng Str Dyn 28, 979–993.
10. Iwan WD (1966) A distributed-element model for hysteretic and its steady-state dynamic response. J Appl Mech 33:893–900.
11. Lavan O, Sivaselvan MV, Reinhorn AM, Dargush GF (2009), Progressive collapse analysis through strength degradation and fracture in the mixed Lagrangian formulation. Earth Eng Str Dyn early view
12. Magonette G (2001) Development and application of large–scale continuous pseudo–dynamic testing techniques. Phil Trans R Soc London A, 359:1771–1799.
13. Molina FJ, Pegon P, Verzeletti G (1999) Time-domain identification from seismic pseudo-dynamic test results on civil engineering specimens. Proceeding of the 2nd international conference on identification in engineering systems, University of Wales Swansea
14. Molina FJ, Sorace S, Terenzi G, Magonette G, Viaccoz B (2004) Seismic tests on reinforced concrete and steel frames retrofitted with dissipative braces. Earthq Eng. Struct Dyn 33:1373–1394
15. Molinari M, Savadkoohi A, Bursi OS, Friswell MI, Zonta D (2009) Damage identification of a 3-D full scale steel-concrete composite structure with partial strength joints at different pseudo-dynamic load levels. Earthq Eng Struct Dyn 8:1219–1236
16. Porter KA (2003) An overview of PEER's performance-based earthquake engineering methodology. ICASP9, San Francisco, July 6–9, 2003
17. Savadkoohi TA (2008) Inverse modelling of a steel-concrete composite moment resisting structure subjected to severe cyclic loads and forced vibration tests. Ph.D. Thesis, University of Trento, Italy
18. Van Overschee P, De Moor B (1996). Subspace identification for linear systems: theory – implementation – applications. Kluwer, Dordrecht
19. Wen YK (1976) Method for random vibration of hysteretic systems. J Eng Mech ASCE 102:249–263
20. Xu B, Wu Z, Chen G, Yokoyama K (2004) Direct identification of structural parameters from dynamic responses with neural networks. Eng Appl Artif Int 17(8):931–943.

Chapter 39
Reliability Assessment in Pseudo-Dynamic and Dynamic Tests

Francisco Javier Molina, Georges Magonette, and Pierre Pegon

39.1 Introduction

Control errors are responsible for the largest aberrations in the obtained response of shaking-table (ST), pseudo-dynamic (PsD) and hybrid tests. However, the consequences of those errors cannot be assessed just by looking at the magnitude of them, since they have cumulative effects that depend on many interacting aspects. This has been observed by some authors who have also proposed techniques for its assessment in PsD tests [5, 8, 10, 11] and in ST tests [6]. This paper presents a general philosophy for the definition and assessment of the reliability of these tests regarding control errors, based on the comparison of the obtained experimental eigenfrequencies and damping ratios with the ones of the ideal prototype structure. Those frequencies and damping ratios are estimated by means of linear models that approximate both the experimental system and the prototype one and are always identified exclusively from the results of the test in question, assuming that all excitations really acting on the specimen during the test are properly measured. A different way to apply this methodology to pseudo-dynamic and ST tests is proposed. In the case of the pseudo-dynamic tests the linear model to identify is a spatial model where the unknowns are the coefficients of the stiffness and damping matrices that define the eigenvalue problem. In the case of the ST test, a linear filter model is identified, which poles determine the frequencies and damping ratios as well.

39.2 General Approach for Definition and Assessment of Testing Reliability

We will consider that the testing set-up models a prototype structure under certain conditions, ideally represented by the system of Fig. 39.1 (a), where $i(t)$ is the specified input (excitation) as a function of time t, $o(t)$ is the corresponding output

F.J. Molina (✉)
European Laboratory for Structural Assessment (ELSA), European Commission – JRC, Via E. Fermi, 2749 21027 Ispra (VA), Italy
e-mail: francisco.molina@jrc.ec.europa.eu

M.N. Fardis (ed.), *Advances in Performance-Based Earthquake Engineering*, Geotechnical, Geological, and Earthquake Engineering 13, DOI 10.1007/978-90-481-8746-1_39,

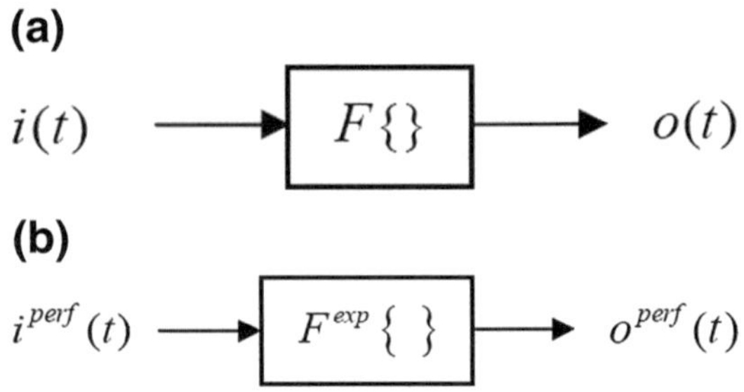

Fig. 39.1 Ideal representation of prototype (**a**) and experimental (**b**) idealised systems

(response) and $F\{\}$ is ideally the functional operator of the prototype system that produces the output function when applied to the input function. That is to say, for the prototype system:

$$o(t) = F\{i(t)\} \tag{39.1}$$

The performed response during the test, $o^{\text{perf}}(t)$, will differ from the "ideal" (prototype) one, so that, for the experimental model of Fig. 39.1 (b), Eq. (39.1) is transformed to:

$$o^{\text{perf}}(t) = F^{\text{exp}}\left\{i^{\text{perf}}(t)\right\} \tag{39.2}$$

where the operator of the prototype system has been substituted by the one of the experimental system. Even though the test can be performed in a different scale for the time or other magnitudes, the variables in Fig. 39.1 and Eqs. (39.1) and (39.2) all refer to the original prototype scale.

The reliability of the test will be assessed by the fidelity in the reproduction of the ideal response by the performed response:

$$o^{\text{perf}}(t) \approx o(t) \quad ? \tag{39.3}$$

or, equivalently, by the fidelity in the reproduction of the ideal operator of the prototype system by the experimental system operator:

$$F^{\text{exp}}\{\,\} \approx F\{\,\} \quad ? \tag{39.4}$$

which may give a more general assessment independently of the particular input of a single test. In fact, as we will see in the case of the ST test, also the input signal may be modified by the testing system and the question:

$$i^{\text{perf}}(t) \approx i(t) \quad ? \tag{39.5}$$

may arise. So the latter comparison, Eq. (39.4), is wiser than the former one, Eq. (39.3).

When a test is executed, the performed response $o^{\mathrm{perf}}(t)$ is part of the test results, but normally the other entities appearing in Eqs (39.3) and (39.4) are not known and the reliability of the test is not assessed. In the following sections, methods are proposed for assessing the test reliability based on the estimation and comparison of the functional operators for the prototype and for the experimental systems by using linear-equivalent models identified from the measurements. Then, instead of a direct use of Eq. (39.4), the reliability of the test will be assessed by comparing some characteristic values, which are the eigenfrequencies and damping ratios, for both systems:

$$\omega^{\exp} \approx \omega \quad ?; \; \zeta^{\exp} \approx \zeta \quad ? \tag{39.6}$$

It is worth mentioning that a similar technique is used for example when analysing the accuracy of discrete-time integration methods for the equation of motion, as compared to the continuous ideal solution. As done also there, those characteristic values are obtained for linear systems expected to suffer the same kind of error consequences as the real non-linear structures [3].

In the case of a seismic problem for a N-DoF structure, the input is the specified accelerogram $\mathbf{a_g}(t)$ and the output would be the relative displacements $\mathbf{d}(t)$. The representation in Fig. 39.1(left) for the prototype structure, is transformed into the one in Fig. 39.2.

The functional operator in Fig. 39.2 is defined as the one that solves the equation of motion:

$$\mathbf{M}\ddot{\mathbf{d}}(t) + \mathbf{r}(\mathbf{d}(t), \dot{\mathbf{d}}(t)) = -\mathbf{MJa_g}(t) \tag{39.7}$$

where $\mathbf{M}$ is the mass matrix that multiplies the relative accelerations, $\mathbf{r}$ are the restoring forces that depend non-linearly on the history of displacements and velocities and $\mathbf{J}$ is the influence matrix.

We can write here the respective formulas for a linear-equivalent system for a seismic problem of a prototype structure where the restoring forces can be expressed as:

$$\mathbf{r}(t) = \mathbf{C}\dot{\mathbf{d}}(t) + \mathbf{K}\mathbf{d}(t) \tag{39.8}$$

In this case, the equation of motion, Eq. (39.7), would be transformed to:

$$\mathbf{M}\ddot{\mathbf{d}}(t) + \mathbf{C}\dot{\mathbf{d}}(t) + \mathbf{K}\mathbf{d}(t) = -\mathbf{MJa_g}(t) \tag{39.9}$$

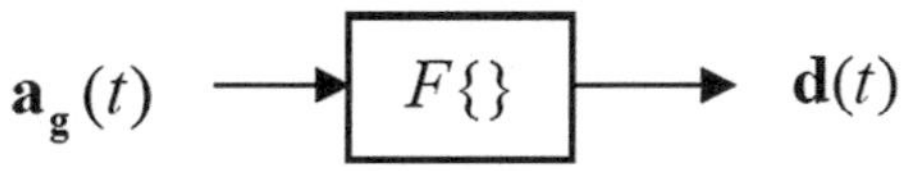

Fig. 39.2 Ideal representation of prototype system for a seismic problem

or, in the Laplace domain:

$$\mathbf{M}s^2\mathbf{D}(s) + \mathbf{R}(s) = \mathbf{M}s^2\mathbf{D}(s) + \mathbf{C}s\mathbf{D}(s) + \mathbf{K}\mathbf{D}(s) = -\mathbf{M}\mathbf{J}\mathbf{A}_{\mathbf{g}}(s) \tag{39.10}$$

which can also be rewritten as

$$\mathbf{D}(s) = \left[\mathbf{M}s^2\right]^{-1}\left[-\mathbf{M}\mathbf{J}\mathbf{A}_{\mathbf{g}}(s) - \mathbf{R}(s)\right] \tag{39.11}$$

corresponding to the block diagram of Fig. 39.3.

Working with linear systems in the Laplace domain, the application of the operator to give the solution consists of a multiplication of the excitation by the transfer function between the input and the output:

$$\mathbf{D}(s) = \mathbf{F}(s)\mathbf{A}_{\mathbf{g}}(s) \tag{39.12}$$

which would substitute Eq. (39.1) and where, for this particular case of a seismic problem, the transfer function, as derived from Eq. (39.10), is:

$$\mathbf{F}(s) = -\left[\mathbf{M}s^2 + \mathbf{C}s + \mathbf{K}\right]^{-1}\mathbf{M}\mathbf{J} \tag{39.13}$$

The poles of this transfer function can be obtained by solving the eigenvalue problem [1]:

$$s\begin{bmatrix} \mathbf{C} & \mathbf{M} \\ \mathbf{M} & \mathbf{0} \end{bmatrix}\varphi + \begin{bmatrix} \mathbf{K} & \mathbf{0} \\ \mathbf{0} & -\mathbf{M} \end{bmatrix}\varphi = \mathbf{0} \tag{39.14}$$

whose complex conjugate eigenvalue couples can then be expressed in the form

$$s_i, s_i^* = \omega_i\left(-\zeta_i \pm j\sqrt{1-\zeta_i^2}\right) \tag{39.15}$$

where ω_i is the natural frequency and ζ_i the damping ratio of the ith mode.

In order to estimate linear-equivalent natural frequencies and damping ratios for a non-linear seismic problem, we first estimate linear-equivalent matrices of stiffness, damping and mass by using the measurements done during the test and then

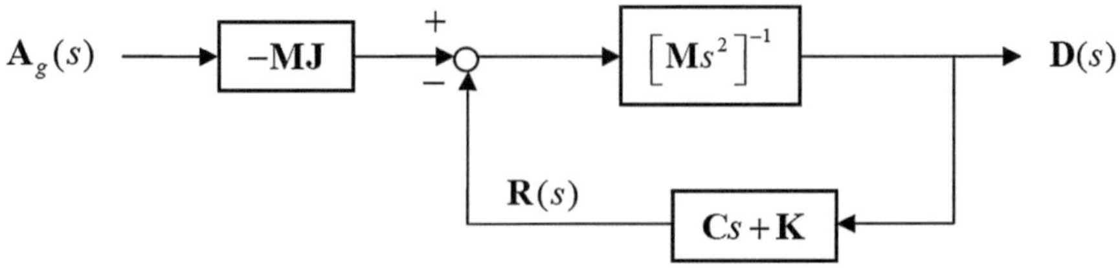

Fig. 39.3 Prototype system for a linear seismic problem

solve the eigenvalue problem Eq. (39.14) for those matrices associated to the prototype structure. A similar approach is followed for the matrices and eigenvalues associated to the experimental system that includes the distortion introduced by the testing method. Depending on the different testing techniques dealt in the following sections, the formulation of the respective matrices and operator will be seen on a case by case basis.

39.3 Pseudo-Dynamic Test

In this section we consider the case of a PsD test on a global specimen (i.e., one that contains all the elements of the system) and we analyse the discrepancies in the response introduced exclusively by the control errors. Thus, we do not consider here other aberrations, such as the ones introduced by the application of a discrete number of DoFs (lumped masses) or by neglecting the strain-rate effect, which can be ignored if the structure and its materials are appropriate for a PsD test [2]. Also the alterations in the response introduced by the discrete-time integration method will not be considered. In fact, the latter alterations are completely negligible when using the continuous PsD technique that employs extremely small integration time steps [9].

Within such assumptions, the linear-equivalent system of the prototype structure can be represented as in the diagram of Fig. 39.3, while the one of the PsD experiment can be represented as in Fig. 39.4. In both diagrams, the time and the Laplace variable are in the scale of the prototype in order to facilitate the comparison, independently of the fact that the PsD test is done slower than reality. In fact, as it is well known, the testing speed used affects the reliability of the test results [4].

According to the definition, Eq. (39.2), the solution of the PsD equation of motion will be called performed displacement:

$$\mathbf{D}^{\mathrm{perf}}(s) = \mathbf{F}^{\mathrm{exp}}(s)\mathbf{A_g}(s) \tag{39.16}$$

in order to distinguish it from the one of the prototype system, Eq. (39.12). However, the physical displacements that are measured during the test are:

$$\mathbf{D}^{\mathrm{meas}}(s) = \mathbf{H}(s)\mathbf{D}^{\mathrm{perf}}(s) \tag{39.17}$$

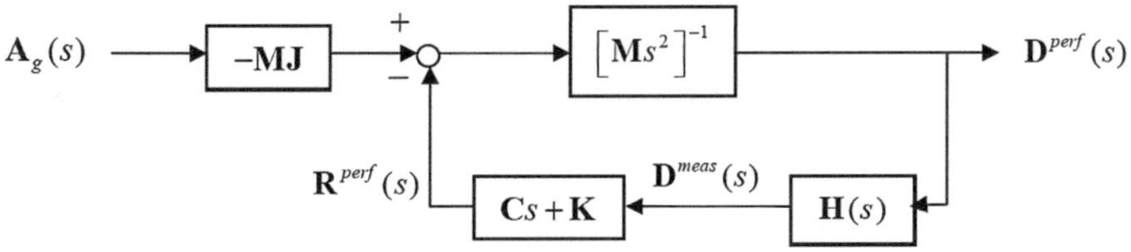

Fig. 39.4 Linear-equivalent experimental system for a PsD test

where $\mathbf{H}(s)$ is the transfer function of the control system that imposes the displacements to the specimen, normally by hydraulic actuators. This transfer function depends on the controller type and parameters, on the actuators, on the specimen and also on the testing speed [5]. Because the physical displacements of the specimen are different from those obtained from the solution, the physical restoring forces measured and introduced in the PsD equation obey the relationship:

$$\mathbf{R}^{\text{perf}}(s) = \mathbf{C}s\mathbf{D}^{\text{meas}}(s) + \mathbf{K}\mathbf{D}^{\text{meas}}(s) = [\mathbf{C}\mathbf{H}(s)s + \mathbf{K}\mathbf{H}(s)]\,\mathbf{D}^{\text{perf}}(s) \tag{39.18}$$

and the PsD equation of motion is written as:

$$\mathbf{M}s^2\mathbf{D}^{\text{perf}}(s) + \mathbf{R}^{\text{perf}}(s) = \left[\mathbf{M}s^2 + \mathbf{C}\mathbf{H}(s)s + \mathbf{K}\mathbf{H}(s)\right]\mathbf{D}^{\text{perf}}(s) = -\mathbf{M}\mathbf{J}\mathbf{A}_{\mathbf{g}}(s) \tag{39.19}$$

which corresponds to a relationship of the whole experimental system, Eq. (39.16), with the experimental transfer function:

$$\mathbf{F}^{exp}(s) = -\left[\mathbf{M}s^2 + \mathbf{C}\mathbf{H}(s)s + \mathbf{K}\mathbf{H}(s)\right]^{-1}\mathbf{M}\mathbf{J} \tag{39.20}$$

In order to simplify this formulation, we will substitute Eq. (39.18) by the approximation:

$$\mathbf{R}^{\text{perf}}(s) = \left[\mathbf{C}^{\text{perf}}s + \mathbf{K}^{\text{perf}}\right]\mathbf{D}^{\text{perf}}(s) \tag{39.21}$$

where, by definition, the performed stiffness and damping matrices are:

$$\mathbf{K}^{\text{perf}} = \lim_{s\to 0}\,[\mathbf{C}s + \mathbf{K}]\,\mathbf{H}(s) \tag{39.22}$$

and

$$\mathbf{C}^{\text{perf}} = \lim_{s\to 0}\frac{[\mathbf{C}s + \mathbf{K}]\,\mathbf{H}(s) - \mathbf{K}^{\text{perf}}}{s} \tag{39.23}$$

Then, approximately:

$$\mathbf{M}s^2\mathbf{D}^{\text{perf}}(s) + \mathbf{R}^{\text{perf}}(s) = \left[\mathbf{M}s^2 + \mathbf{C}^{\text{perf}}s + \mathbf{K}^{\text{perf}}\right]\mathbf{D}^{\text{perf}}(s) = -\mathbf{M}\mathbf{J}\mathbf{A}_{\mathbf{g}}(s) \tag{39.24}$$

Equation (39.24) corresponds to the experimental system represented in the diagram of Fig. 39.5, which is formally identical to the prototype system of Fig. 39.3 by substituting the prototype damping and stiffness matrices by the performed ones.

Also analogously to Eq. (39.13), the transfer function that solves Eq. (39.24) is:

$$\mathbf{F}^{\text{exp}}(s) = -\left[\mathbf{M}s^2 + \mathbf{C}^{\text{perf}}s + \mathbf{K}^{\text{perf}}\right]^{-1}\mathbf{M}\mathbf{J} \tag{39.25}$$

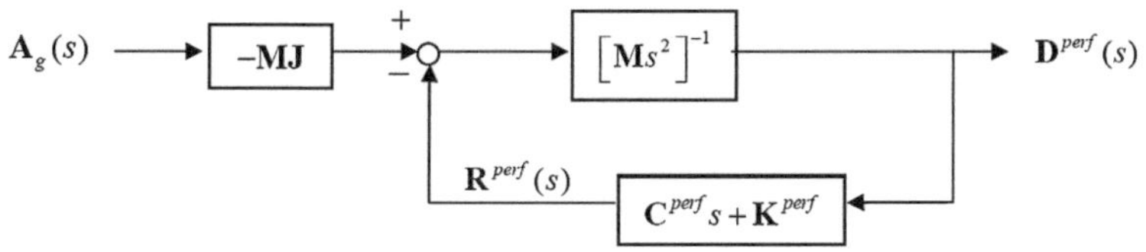

Fig. 39.5 Approximation of the experimental system based on the performed damping and stiffness matrices

As introduced in the previous section, the assessment of the test reliability will be done by comparing the characteristics of this experimental transfer function, Eq. (39.25), with the ones of the prototype transfer function, Eq. (39.13). The method proposed for this consists of estimating the matrices entering in these formulae by using the results of the performed test without introducing any analytical model of the structure or the control system. In fact, just using the variables that are available during the test (Fig. 39.4), we establish the relationship:

$$\mathbf{R}^{\text{perf}}(s) = [\mathbf{C}s + \mathbf{K}]\,\mathbf{D}^{\text{meas}}(s) \tag{39.26}$$

which is then used in the time domain as:

$$\mathbf{r}^{\text{perf}}(t) = \mathbf{C}\dot{d}^{\text{meas}}(t) + \mathbf{K}d^{\text{meas}}(t) \tag{39.27}$$

in order to identify the damping and stiffness matrices of the prototype **C** and **K** (spatial model method by Molina et al. [7]). The eigenvalue problem Eq. (39.14) is applied to such identified matrices in order to solve the frequencies and damping ratios of the prototype:

$$s_i, s_i^* = \omega_i\left(-\zeta_i \pm j\sqrt{1-\zeta_i^2}\right) \tag{39.28}$$

Then, using always only variables accessible during the test, from the established relationship Eq. (39.21), using again the time domain spatial model identification, the performed damping and stiffness matrices of the experimental model $\mathbf{C}^{\text{perf}}$ and $\mathbf{K}^{\text{perf}}$ are identified from:

$$\mathbf{r}^{\text{perf}}(t) = \mathbf{C}^{\text{perf}}\dot{d}^{\text{perf}}(t) + \mathbf{K}^{\text{perf}}d^{\text{perf}}(t) \tag{39.29}$$

With them the eigenvalue problem:

$$s\begin{bmatrix}\mathbf{C}^{\text{perf}} & \mathbf{M}\\ \mathbf{M} & \mathbf{0}\end{bmatrix}\boldsymbol{\varphi} + \begin{bmatrix}\mathbf{K}^{\text{perf}} & \mathbf{0}\\ \mathbf{0} & -\mathbf{M}\end{bmatrix}\boldsymbol{\varphi} = \mathbf{0} \tag{39.30}$$

is formulated. Its solution:

$$s_i^{\text{exp}}, s_i^{\text{exp}*} = \omega_i^{\text{exp}}\left(-\zeta_i^{\text{exp}} \pm j\sqrt{1-\zeta_i^{\text{exp}2}}\right) \tag{39.31}$$

gives the experimental frequency and damping values. Finally, the reliability of the experiment is assessed by comparing these experimental values, Eq. (39.31), with those estimated for the prototype, Eq. (39.28).

An example of application of the proposed assessment is given by Molina and Géradin [4].

39.4 Shaking-Table Test

As in the PsD case above, we identify what the consequences of the control errors in a ST test are on the response of the tested structure and how it can be compared with the ideal response of the prototype. Firstly, we define the linear approximation of the prototype system as the one of the diagram in Fig. 39.3 or, in a more simplified form, as the one in Fig. 39.6, where the prototype transfer function of the structure $\mathbf{F}(s)$ is given by Eq. (39.13).

Regarding the experimental system, we use the diagram in Fig. 39.7, where it is assumed that the structure's transfer function still corresponds to the one of the prototype, but the specified excitation has been modified by the testing apparatus before it physically arrives at the structure. The functions appearing in this figure are explained below.

Because the ST test is done at real time, the control errors are severe and, in order to compensate for them, the specified accelerogram is numerically prefiltered in order to produce a modified target function to be sent to the controller:

$$\mathbf{A}_g^{\mathrm{targ}}(s) = \mathbf{H}^{\mathrm{comp}}(s)\mathbf{A}_g(s) \tag{39.32}$$

This compensation is typically done off-line before carrying out the test and is based on measurements done at preliminary calibration tests for the current testing configuration but at a lower level of excitation, to avoid damage to the specimen in this calibration phase. The specified acceleration may have components only in some of the DoFs at which the ST responds (typically in translations). Those will be called object DoFs. Once the target signal is introduced in the controller, the table will respond with acceleration at the object DoFs, called here the performed ground acceleration:

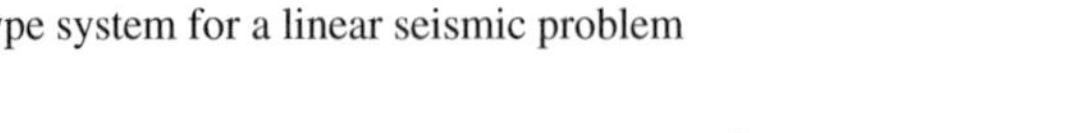

Fig. 39.6 Prototype system for a linear seismic problem

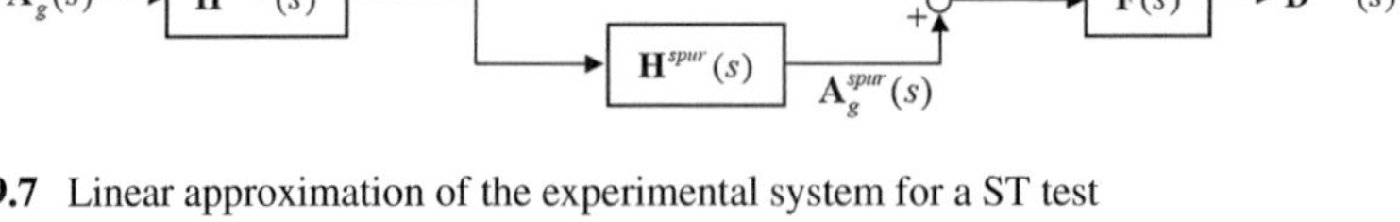

Fig. 39.7 Linear approximation of the experimental system for a ST test

$$\mathbf{A}_g^{\mathrm{perf}}(s) = \mathbf{H}^{\mathrm{obj}}(s)\mathbf{A}_g^{\mathrm{targ}}(s) \tag{39.33}$$

but also at the remaining DoFs (typically rotations) as a spurious response:

$$\mathbf{A}_g^{\mathrm{spur}}(s) = \mathbf{H}^{\mathrm{spur}}(s)\mathbf{A}_g^{\mathrm{targ}}(s) \tag{39.34}$$

Note that this spurious response exists also in STs whose nominal DoFs coincide with the object ones, because the guiding systems that restrict the spurious response are never perfect. Moreover, this spurious response may also include internal vibration modes of the platform itself, if they are able to modify the response of the test structure.

The combination of the performed ground acceleration and the spurious one gives the effective excitation physically arriving at the structure [6]:

$$\mathbf{A}_g^{\mathrm{eff}}(s) = \mathbf{A}_g^{\mathrm{perf}}(s) + \mathbf{A}_g^{\mathrm{spur}}(s) \tag{39.35}$$

which, according to Fig. 39.7, is causing the obtained response:

$$\mathbf{D}^{\mathrm{perf}}(s) = \mathbf{F}(s)\mathbf{A}_g^{\mathrm{eff}}(s) \tag{39.36}$$

However, usually the existence of the spurious excitation is neglected in the test results and the structural response is assumed to be due only to the performed ground acceleration at the object DoFs, i.e.,

$$\mathbf{D}^{\mathrm{perf}}(s) = \mathbf{F}^{\mathrm{exp}}(s)\mathbf{A}_g^{\mathrm{perf}}(s) \tag{39.37}$$

where the prototype transfer function of the structure has been substituted by the experimental one. In fact, within the linear approximation, according to Eq. (39.33):

$$\mathbf{A}_g^{\mathrm{targ}}(s) = \left[\mathbf{H}^{\mathrm{obj}}(s)\right]^{-1} \mathbf{A}_g^{\mathrm{perf}}(s) \tag{39.38}$$

and, using Eq. (39.34):

$$\mathbf{A}_g^{\mathrm{spur}}(s) = \mathbf{H}^{\mathrm{spur}}(s)\left[\mathbf{H}^{\mathrm{obj}}(s)\right]^{-1} \mathbf{A}_g^{\mathrm{perf}}(s) \tag{39.39}$$

so that, introducing Eq. (39.39) in Eqs. (39.35) and (39.36):

$$\mathbf{D}^{\mathrm{perf}}(s) = \mathbf{F}(s)\left[\mathbf{I} + \mathbf{H}^{\mathrm{spur}}(s)\left[\mathbf{H}^{\mathrm{obj}}(s)\right]^{-1}\right]\mathbf{A}_g^{\mathrm{perf}}(s) \tag{39.40}$$

By comparing with Eq. (39.37) we obtain:

$$\mathbf{F}^{\mathrm{exp}}(s) = \mathbf{F}(s)\left[\mathbf{I} + \mathbf{H}^{\mathrm{spur}}(s)\left[\mathbf{H}^{\mathrm{obj}}(s)\right]^{-1}\right] \tag{39.41}$$

where $\mathbf{I}$ is the identity matrix.

In order to derive the characteristics of the transfer functions of the prototype and the experimental systems from just the experimental measurements, we propose the filter model [7], which models the system as a linear filter with known input and output and whose coefficient matrices in the discrete-time domain can be identified by least squares. Then, from those matrices of the filter, the eigenvalues and corresponding poles of the continuous free solution are calculated and from them the associated frequencies and damping ratios.

Thus, it is necessary that, during the experiment, apart from the performed table acceleration at the object DoFs, all the other spurious components are accurately measured and combined in a global effective table acceleration vector, Eq. (39.35), arriving at the structure. According to Eq. (39.36), by applying the filter model with the effective acceleration as input and the measured performed displacement as output, the eigenfrequencies and damping ratios of the prototype system are identified as:

$$s_i, s_i^* = \omega_i \left(-\zeta_i \pm j\sqrt{1 - \zeta_i^2} \right) \tag{39.42}$$

Analogously, according to Eq. (39.37), by applying the filter model with the performed acceleration as input and the measured performed displacement as output, the eigenfrequencies and damping ratios of the experimental system are identified as:

$$s_i^{\exp}, s_i^{\exp^*} = \omega_i^{\exp} \left(-\zeta_i^{\exp} \pm j\sqrt{1 - \zeta_i^{\exp^2}} \right) \tag{39.43}$$

As in the previous section, the reliability of the experiment is assessed by comparing these two sets of frequencies and damping ratios.

An example of the application of this approach is given by Molina et al. [6].

39.5 Conclusions

This paper is a contribution to the definition of a general approach for the assessment of reliability in PsD and ST tests regarding the disturbances of the ideal response of the specimens due to the presence of control errors. The assessment is done through the comparison of the experimental response eigenfrequencies and damping ratios to the ones of the prototype structure. All the required parameters are estimated using exclusively experimental information from the test.

Acknowledgement The work presented in this paper was funded by the European Commission through the SAFECONSTRUCT institutional action n. 32003 of the IPSC, JRC.

References

1. Ewins DJ (1984) Model testing: theory and practice. Research Studies Press, Taunton
2. Gutierrez E, Magonette G, Verzeletti G (1993) Experimental studies of loading rate effects on reinforced concrete columns. ASCE J Eng Mech 119:887–904
3. Hilber HM, Hughes TJR, Taylor RL (1977) Improved numerical dissipation for time integration algorithms in structural dynamics. Earthq Eng Struct Dyn 5:283–292
4. Molina FJ, Géradin M (2007) Earthquake Engineering experimental Research at JRC-ELSA. In: Ibrahimbegovic A, Kozar I (eds) NATO workshop 2006. Extreme man-made and natural hazards in dynamics of structures. Springer, New York, pp 311–351
5. Molina FJ, Magonette G, Pegon P (2002) Assessment of systematic experimental errors in pseudo-dynamic tests. 12th European conference on earthquake engineering, London, Paper 525
6. Molina FJ, Magonette G, Viaccoz B, Geradin M (2008) Apparent damping induced by spurious pitching in shaking-table testing. Earthq Eng Struct Dyn 37(1):103–119
7. Molina FJ, Pegon P, Verzeletti G (1999) Time-domain identification from seismic pseudodynamic test results on civil engineering specimens. 2nd international conference on identification in engineering systems, University of Wales, Swansea
8. Mosqueda G, Stojadinovic B, Mahin SA (2007) Real-time error monitoring for hybrid simulation. Part I: methodology and experimental verification. J Struct Engrg 133:100–1108
9. Pegon P, Molina FJ, Magonette G (2008) Continuous pseudo-dynamic testing at ELSA. In: Saouma VE, Sivaselvan MV (eds) Hybrid simulation; theory, implementation and applications. Taylor & Francis, Balkema, pp 79–88
10. Shing PSB, Mahin SA (1987) Cumulative experimental errors in pseudodynamic tests. Earthq Eng Struct Dyn 15:409–424
11. Thewalt C, Roman M (1994) Performance parameters for pseudodynamic tests. ASCE J Struct Eng 120:9

Chapter 40
Dynamic Interaction Between the Shaking Table and the Specimen During Seismic Tests

Alain Le Maoult, Jean-Claude Queval, and Rogerio Bairrao

40.1 Introduction

The optimisation of powerful experimental tools, such as shaking tables, is one of the main fields of research and progress in the European community of experimental earthquake engineering [1, 2, 5]: control systems, hybrid tests, high-speed data transfer, sub-structuring, etc. The improvement of shaking table technologies is of paramount importance for the reduction of the seismic vulnerability of the building stock and the mitigation of the consequences of future, inevitable earthquakes, through the improvement of the technology for earthquake-resistant construction.

This paper focuses on the boundary conditions between the test structure and the platform of a shaking table. These boundary conditions are a major parameter for the design and numerical analyses of shaking table tests. In shaking table studies, upmost attention is paid to the design of the fixing and anchorage of the test structure. All analyses are made assuming a completely rigid shaking table (both its actuators and the platform). However, the interaction between the shaking table and the structure has been clearly observed since long. Blondet and Esparza [3] studied the interaction between actuators and shaking table, depending on the control tuning. In CEA Saclay, decrease of massive test structure frequencies compared to calculations have been observed in the past 15 years, notably in projects CASSBA, CAMUS 1–4, CAMUS 2000 [4] and more recently SMART. The global stiffness that the "Azalée" shaking table in CEA Saclay should have to explain this decrease has been calculated after the test since long.

Recently, the validity domain of the rigid platform hypothesis has been studied for the "Azalée" shaking table of the CEA Saclay laboratory, one of the largest shaking tables in Europe with a quite standard design of its platform. The first part of the paper describes and validates the Finite Element (FE) model of the platform of the shaking table. The second part presents the test specimen used to check

A. Le Maoult (✉)
CEA/DEN/SEMT/EMSI, 91191 Gif sur Yvette, France
e-mail: alain.lemaoult@cea.fr

M.N. Fardis (ed.), *Advances in Performance-Based Earthquake Engineering*, Geotechnical, Geological, and Earthquake Engineering 13, DOI 10.1007/978-90-481-8746-1_40,

the interaction, while the third part deals with and concludes about the interaction between the shaking table and the test structure.

40.2 Description and Validation of the Platform FE Model of the "Azalée" Shaking Table in CEA (Saclay)

The Azalée platform is a 6 m square plate, 2 m deep Figs. 40.1 and 40.2. It is made of 36 welded aluminium "boxes" and 4 lateral anchorages for the horizontal actuators. The mass of the platform is 23.6 tonnes and its maximum upload mass is 100 tonnes.

Simple analytical models were employed initially, using common springs and plates. However, that kind of models cannot reproduce the more complex details of the "Azalée" geometry. So, a more detailed linear FE model (Fig. 40.3), using thin

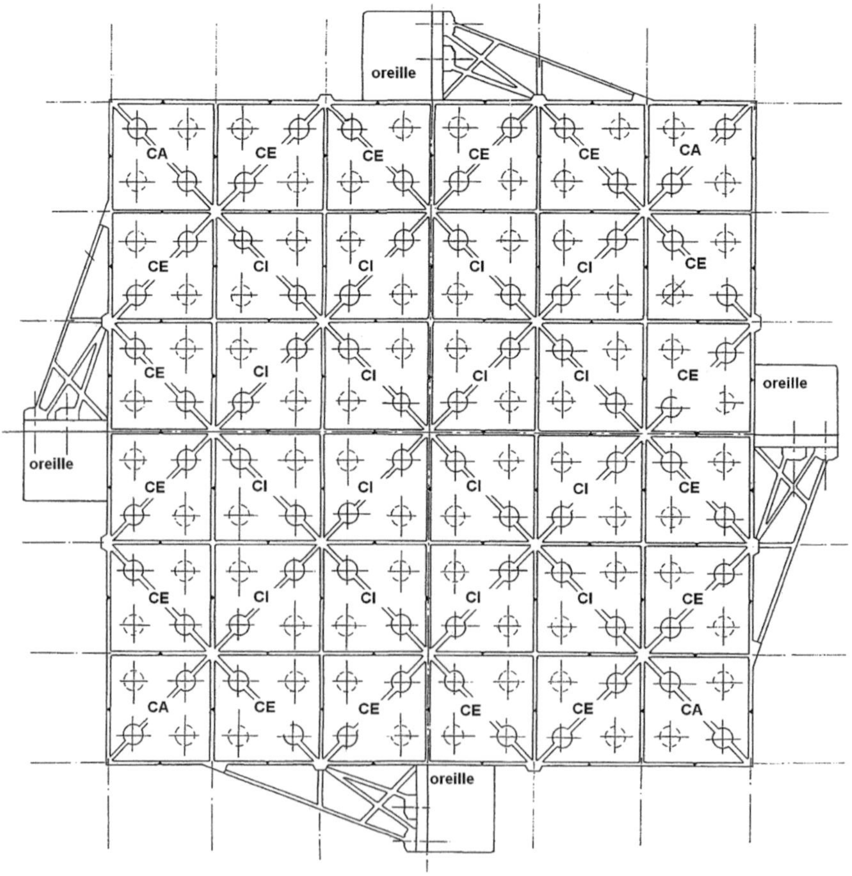

Fig. 40.1 Plant view of the Azalée platform

Fig. 40.2 Azalée platform

shell elements has been constructed and appropriately simplified (Fig. 40.4) in order to be easily implemented on any FE software.

Two boundary conditions were then analysed:

- In the first one, the platform was considered supported by four air cushions (64 springs in the model). This configuration is used in the next section to validate the platform model.
- In the second one, the platform was considered fixed to eight rigid actuators (86 unidirectional restraints).

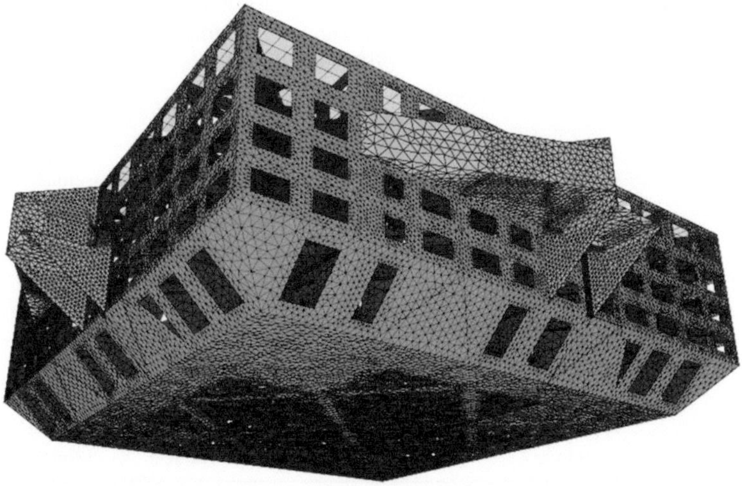

Fig. 40.3 Detailed CASTEM model of Azalée platform

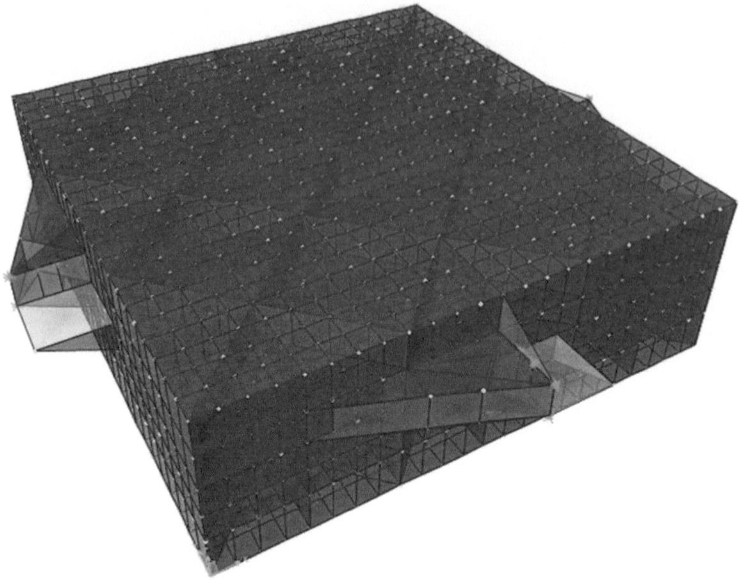

Fig. 40.4 Simplified FE model

The validation of the linear FE platform model was done using as reference two experiments performed in 1989 and 2000. For these tests for the determination of frequencies and modes the platform was simply supported by four air cushions (having stiffness: $K_{Z\_cus} = 908$ kN/m), without any actuators. Table 40.1 shows a comparison between the calculated and the measured frequencies for each mode.

The platform was then analysed under its usual condition: eight rigid actuators (assuming perfect control compensation of the oil column stiffness), represented by 86 unidirectional restraints. The frequencies for this situation are given in Table 40.2. Note that these modes are different from the ones in the previous configuration. The new boundary conditions have changed mainly the behaviour of the

Table 40.1 Experimental and calculated frequencies (Hz) with air cushions

	Mode 1	Mode 2	Mode 3	Mode 4	Mode 5	Mode 6
Experimental	80.5	95.7	116.2	129.2	129.3	151.1
Calculated	80.0	95.0	117.0	134.5	134.5	162.2

Table 40.2 Calculated frequencies (Hz) of model with 8 rigid actuators

Mode 1	Mode 2	Mode 3	Mode 4	Mode 5	Mode 6
54.3	54.3	65.7	73.5	73.5	99.0

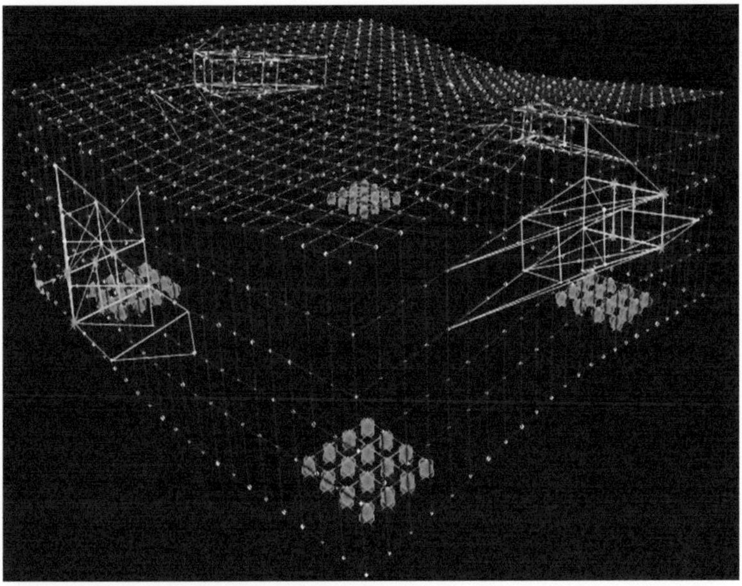

Fig. 40.5 Platform on the actuators; mode 6

platform. Five modes appeared between 54 and 74 Hz. The first two are generated by the deformation of the anchorage of the horizontal actuators to the platform. The others are due to the global flexure of the platform (see Fig. 40.5 for an example).

40.3 The Specimens

The evaluation of the platform with the specimen on top is done through a study of the reduction of the specimen frequency between the configurations "on a rigid base" and "on the shaking table platform". This section presents the specimen used to check the interaction between the platform and the test structure.

Two types of specimens have been considered and modelled with linear FEs: a simple specimen and more complex and realistic structure, representative of a real test.

A simple stick specimen and its model can be used as a basis for a parametric study.

The simple stick specimen is a beam fixed on a very rigid square plate, with no local deformation at the bottom (Fig. 40.6). Several values of the mass and stiffness of the beam have been used to evaluate the frequency reduction between the configurations "on a rigid base" and "on the shaking table platform". The beam should allow the simulation of lateral (flexural) and vertical modes of a mock-up, without simulation of any torsional mode. For example, one with hollow square section,

Fig. 40.6 Azalée platform model with the simple stick model

2 m wide and 100 mm thick. It was checked that this model can correctly simulate flexure of the CAMUS mock-up, considering these beam parameters:

Stiffness = 780.6 MN/m.
Mass = 36 tonnes.
Height = 3.4 m.

More complex and realistic structures, representative of real tests, allow the analysis of different boundary conditions of the specimen on the platform (influence of local deformations).

The first test structure is an asymmetric reinforced concrete specimen (Fig. 40.7) tested at the CEA (Saclay) lab within the SMART project in 2008. Its mass is 45 tonnes, and its main dimensions are a 3.6 m height and a plan area of 2.5 m × 3 m.

Special attention has been paid to the boundary conditions between the specimen and the platform. To evaluate their influence, another configuration has been tested, with a very rigid square plate without local deformation at the base of the specimen.

The second test structure is a reinforced concrete frame with 2 storeys (Fig. 40.7). It was tested in the CEA lab in 2004, within the ECOLEADER project. Its mass is 33 tonnes and its main dimensions are a height of 7 m and a plan area of 2.5 m × 3 m.

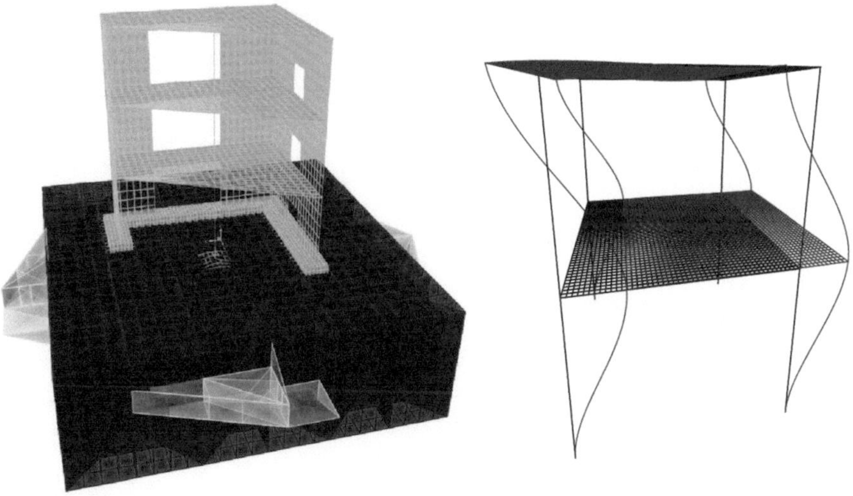

Fig. 40.7 More complex and realistic structures *Left*: SMART specimen, *Right*: ECOLEADER specimen

40.4 Analyses

The evaluation is carried out by studying the frequency reduction of the specimen between the configurations "on a rigid base" and "on the shaking table platform". This was done for each lateral (flexural) and vertical mode of the specimen; generally the most important ones.

Two charts have been constructed to analyse these results:

- A first chart (Fig. 40.8) for the specimen's lateral (flexural) and vertical modes without local effects: from the modal mass and the centre of mass of the specimen (flexural moment), one reads the frequency decrease between the two configurations "on a rigid base" and "on the platform".
- Another chart (Fig. 40.9) for the specimen's vertical modes without local effects: from the modal mass and the centre of mass of the specimen one reads the frequency reduction between the configuration "on a rigid base" and "on the platform".

The experimental results of the mock-ups used for the projects SMART and CAMUS and of the ECOLEADER frame are indicated on those abacuses too.

Radial lines give possible "iso-frequency" curves (of the specimen on a rigid base).

These simple charts are a first approximation allowing a quick evaluation of the frequency decrease with respect to a specimen fixed to the Azalée platform.

Most of these comparisons with experimental results were made for the lateral (flexural) modes. For the CAMUS, ECOLEADER and SMART specimens on a

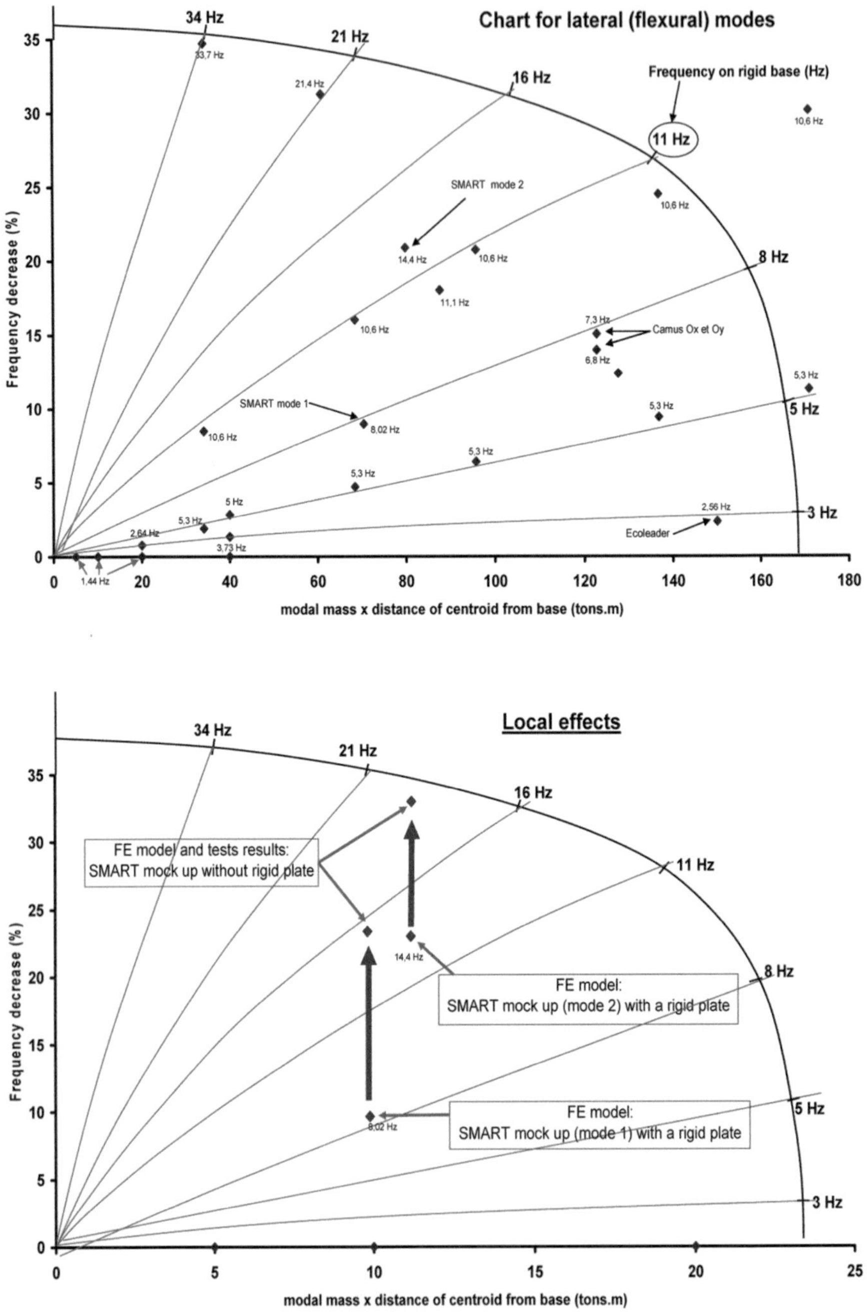

Fig. 40.8 Chart for lateral (flexural) modes, giving the frequency decrease with respect to the case "rigid base" from the modal mass times the distance of the centre of mass of the specimen from the base (top) without local effects; (bottom) SMART specimen, with and without local effects

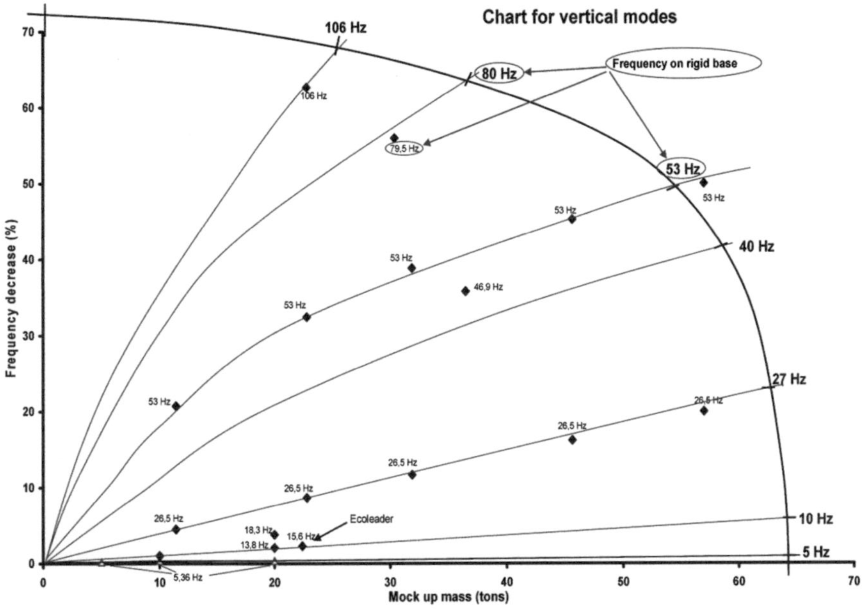

Fig. 40.9 Chart for vertical modes, giving the frequency decrease with respect to the "rigid base" without local effects case from the modal mass of the specimen

rigid base, the chart prediction of the frequency decrease is very accurate. That means that the actuators were indeed rigid and the frequency reduction is not due to them.

In Fig. 40.8 (bottom), four points for the SMART specimen are shown: two of them for the specimen on a rigid platform and two other for the specimen directly fixed to the platform of the Azalée shaking table. The comparison between the configurations "with a rigid base" and "without a rigid base" gives an estimation of the decrease due to the boundary conditions between the specimen and the platform:

- From 9% to 23% for the first mode; for this mode, more than half of the decrease is due to local deformations between the specimen and the platform.
- From 21% to 33% for the second mode; more than 30% of which is due to the same local deformations.

When a very large decrease is expected, it is recommended and possible to perform a numerical analysis with the specimen on the simplified model of the Azalée platform.

40.5 Conclusions

In shaking table studies considerable attention is paid to the design of the foundation and the anchorage of the specimen. All analyses consider completely rigid shaking table (actuators and platform). However, since long interaction between the shaking table and massive specimens has been clearly observed, namely a decrease in frequency of the specimen "on the shaking table" with respect to that "on a rigid base". This paper investigates the validity domain of the rigid "Azalée" shaking table at the CEA (Saclay) laboratory.

The paper demonstrated that, for the large "Azalée" shaking table, most of the interaction is due to the platform deformation during the test. Two charts giving the reduction in frequency as a function of specimen parameters have been obtained for lateral (flexural) and vertical modes, using a simplified specimen model, without local deformation between the specimen and the platform. They have been validated in three large experimental tests performed in the past 10 years, within the CAMUS, ECOLEADER and SMART projects. Using them it is easy to estimate the minimum decrease in frequency of a specimen fixed on the platform. If this first estimation shows significant interaction, a more detailed study is needed. A simplified FE model of the platform is available for a definite evaluation of the interaction. A second evaluation takes into account the local deformations between the specimen and the platform. For large specimens these local deformations can reduce the frequency by almost 50%.

Evaluations of the interaction were made after past "Azalée" shaking table tests, using simplified models. This paper will help experimental and numerical researchers to better take care, during the design of the test, of the boundary conditions between the platform of the shaking table and the specimen. More detailed comparisons to numerical analyses of large shaking table tests will now be possible for the Azalée table.

References

1. Bairrao R (2008) Shaking table testing. In: Bursi OS, Wagg DJ (eds) Modern testing techniques for structural systems. Springer, New York, NY, pp 165/196. ISBN-978-3-211-09444-0
2. Bairrao R, Falcao MJ, Carydis P, Mouzakis H, Karapitta L, Queval JC (2006) Performance benchmark of three major european shaking tables. 13th european conference on earthquake engineering, Geneva, paper 110
3. Blondet M, Esparza C (1988) Analysis of shaking table-structure interaction effects during seismic simulation tests. Earthq Eng Struct Dyn 16:473/490
4. Combescure D, Ragueneau F (2002). CAMUS2000 Benchmark. Experimental results and specifications to the participants. Report CEA/SEMT/EMSI/RT/02-067/A
5. Taucer F (2005) Recent advances and future needs in experimental earthquake engineering. In: Severn R, Bairrao R (General eds) CASCADE – cooperative advancements in seismic and dynamic experiments. Report No. 7, LNEC. ISBN-972-49-1971-4.

Chapter 41
Frameworks for Internet Online Hybrid Test

Peng Pan and Masayoshi Nakashima

41.1 Introduction

Online hybrid testing, particularly when combined with substructuring techniques, is able to conduct large-scale tests. An extension of this technique is to combine multiple loading tests conducted in remote locations and to integrate the tests with sophisticated numerical analysis codes. The idea is very appealing, as it will significantly increase the capacity and versatility of the conventional online hybrid tests. The concept, referred as "Internet online hybrid test" hereafter, has been addressed over the past few years [2–4, 9]. Several notable systems were developed and demonstrated by actual applications [1, 7, 8, 10, 12, 14]. In this paper, three frameworks for the Internet online hybrid tests, designated as the host-station, dual-model, and peer-to-peer frameworks, respectively, are presented. The host-station framework incorporates a user-defined experimental element into the numerical model for the entire structure. The dual-model framework solves the dynamics of the entire structure separately from the static tests and analyses for the substructures. The peer-to-peer framework is, which divides the entire structure into encapsulated substructures with a generalized interface, more extensible. Multiple tested substructures and numerical substructures using various structural analyses codes can be accommodated within the single framework, simply interfaced with the boundary displacements and forces. Coordinator programs are developed to keep the boundaries among all substructures compatible and equilibrated. A series of online hybrid tests are introduced. The flexibility, extensibility and accuracy of the developed online hybrid test systems using the three frameworks are demonstrated.

41.2 Host-Station Framework

The host-station framework is illustrated in Fig. 41.1. The system consists of the host, stations, and data exchange interfaces. The simulated structure is divided into

---

P. Pan (✉)
Department of Civil Engineering, Tsinghua University, Haidian District, Beijing 100084, China
e-mail: panpeng@tsinghua.edu.cn

M.N. Fardis (ed.), *Advances in Performance-Based Earthquake Engineering*, Geotechnical, Geological, and Earthquake Engineering 13, DOI 10.1007/978-90-481-8746-1_41,

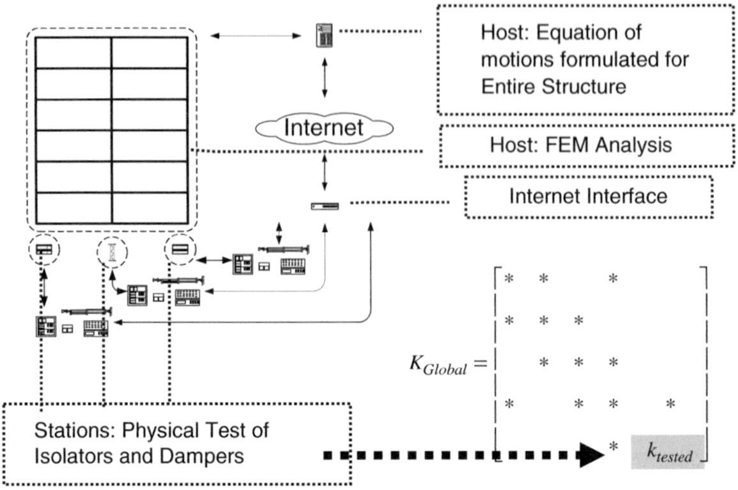

Fig. 41.1 Host-station framework

the computed parts and the tested parts through substructuring techniques. The host analyzes the computed parts, collects information from the stations and simulates the response of the entire structure. Each station performs a physical test in accordance with the (displacement) command transferred from the host and creates information about the behavior of each substructure. The data exchange interfaces are in charge of communication between the host and the stations. The host and the stations are geographically distributed, but they do the test collaboratively by carefully exchanging data. The details about the framework are presented in [5].

41.3 Dual-Model Framework

The basic idea of the dual-model framework is to solve the dynamics of the structure separately from its static tests and analyses, i.e., to formulate and solve the equations of motion of the entire structure using the restoring forces obtained from the static substructures. Figure 41.2 illustrates the test scheme using a three-story steel braced frame structure as an example. It is reasonable that the dynamics of the structure are simplified into the equations of motion of a three degree-of-freedom (DOF) system, where each DOF corresponds to one story level. The static behavior, i.e., the restoring forces of respective DOFs, is obtained from two substructures: the conventional frame simulated numerically by a finite element program, and the braces tested physically. The dynamic model uses the information from previous steps to predict the displacement vector for the current step and sends the next displacements to the substructures for both the physical loading and the static model. The restoring forces corresponding to these displacements are then collected and sent back to the dynamic model. The global restoring force vector is formed, each component of

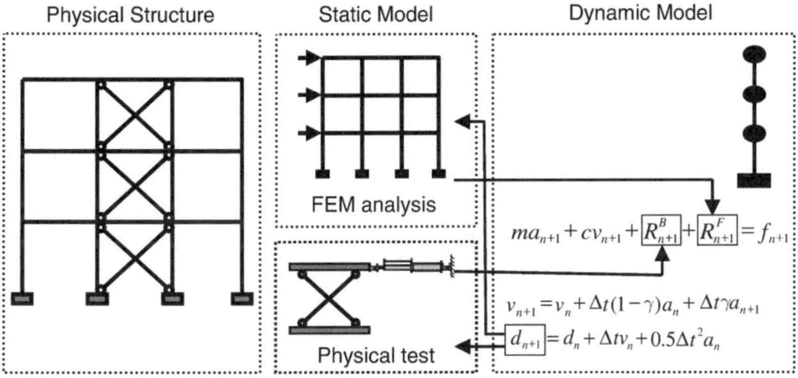

Fig. 41.2 Dual-model framework

which is associated with one dynamic DOF. Finally, the dynamic model updates the state variables for the next step simulation. In the dynamic model, the boundaries between substructures are always associated with the dynamic DOFs and the equilibrium at the boundary is satisfied explicitly by solving the equations of motion. The details about the framework are given in [11].

41.4 Peer-to-Peer Framework

The peer-to-peer framework is more extensible. As shown in Fig. 41.3, the simulated structure is divided into several substructures. The peer-to-peer system treats all substructures equally as independent subsystems, regardless of their being tested

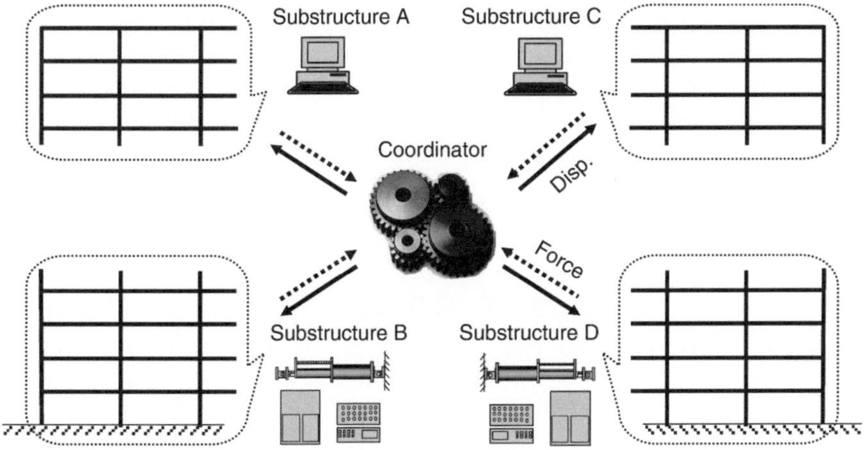

Fig. 41.3 Peer-to-peer framework

or analyzed. The equations of motion are not formulated for the entire structure, but for each substructure separately. Interactions among substructures are considered, so that compatibility and equilibrium are satisfied at the boundaries. Figure 41.4 shows a case in which the simulated structure is divided into two substructures. One substructure, called "Analysis Substructure" is analyzed numerically, and the other, termed "Test Substructure," is tested physically. A program called "Coordinator" is developed to satisfy equilibrium and compatibility at the boundaries between the two substructures. First, Coordinator determines the displacements at the boundary and sends the displacements to the two substructures. The boundary displacements are taken to be identical in the two substructures, so that comparability is satisfied all the time. For the analysis substructure, the boundary displacements are taken as the external load and the reaction forces at the boundary are calculated by solving the equations of motion. For the test substructure, the boundary displacements are imposed on the test specimen and the reaction forces are measured directly. These reaction forces at the boundary are sent to Coordinator and Coordinator checks the equilibrium at the boundary. If the equilibrium is satisfied, the simulation of the current step is completed. Otherwise, Coordinator specifies new boundary displacements in reference to the unbalanced forces and repeats the procedure, until the boundary equilibrium is satisfied. In the above procedure, the equations of motion are not formulated for the entire structure, but for respective substructures separately, to ensure their independence. In each substructure, only the boundary

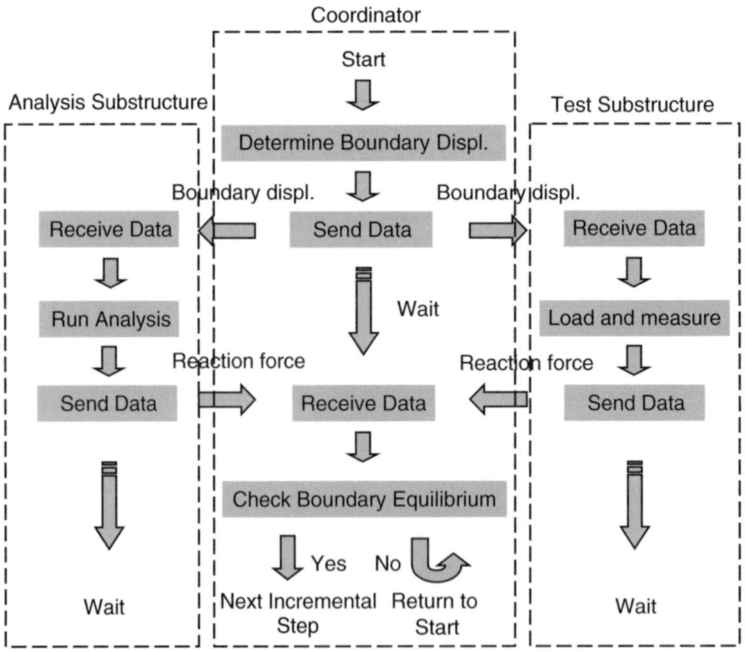

Fig. 41.4 Workflow for peer-to-peer framework

displacements and corresponding reaction forces are treated as I/O interfaces, while the rest is encapsulated within the substructure. Such encapsulation makes it feasible to take the substructure as a black box and use existing analysis tools. A "Coordinator" equipped with an iterative algorithm based on the quasi-Newton iterations is developed to achieve compatibility and equilibrium at boundaries. A test procedure, featuring two rounds of quasi-Newton iterations and using assumed elastic stiffness, is adopted to avoid iteration for the substructure being tested physically. The details about the framework are given in [6].

41.5 Test Using Host-Station Framework

A base-isolated structure shown in Fig. 41.5(a) was tested. It was an eight-story, two-span steel moment-resisting frame isolated by rubber bearings and treated as a planar structure. As shown in Fig. 41.5(b), the substructuring technique

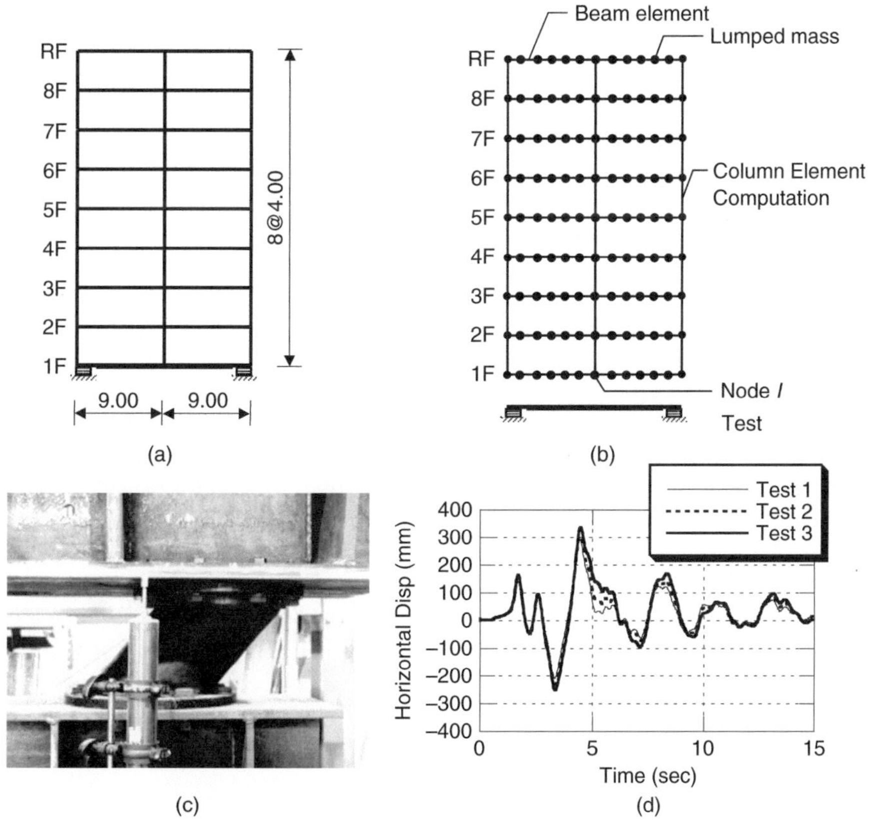

Fig. 41.5 Demonstration tests using host-station framework: (**a**) base-isolated structure; (**b**) model; (**c**) test specimen; (**d**) test results

was employed in the test. The superstructure was modeled numerically and the base-isolation layer was tested physically.

The superstructure was modeled by lumped masses, beam elements and column elements. In the model, each beam was divided into six beam elements along the beam axis and an associated lumped mass was assigned at each node. This discretization was adopted to trace the beam deflections more accurately. A concentric plastic hinge was assigned to each end of the beam and column elements to allow for member plastification. Interaction between the axial force and moment was taken into account for yielding of the plastic hinges. Both the material and geometric nonlinearities were taken into account.

In this demonstration test, the base-isolation layer was taken to move only horizontally, with the assumption that the base-isolators were infinitely stiff in the vertical direction. Then, the degree of freedom shared by the superstructure and the base-isolation layer was reduced to one [designated as Node I in Fig. 41.5(b)], i.e., the horizontal displacement and force of the base-isolation layer. The test specimen is given in Fig. 41.5(c). Three physical Internet online tests, i.e. Tests, 1, 2, and 3, were carried out. Reasonable test results shown in Fig. 41.5(d) were obtained. Details about the tests are given in [5].

41.6 Test Using Dual-Model Frame

The effectiveness of the Internet online hybrid test adopting the dual-model framework was calibrated by a test applied to the same eight-story base-isolated structure described above, but the simulated response involved collision of the base-isolation layer with the surrounding retaining walls (Fig. 41.6(a)). As shown in Fig. 41.6(b), the superstructure was assigned as the numerical part and a general-purpose FEM code, ABAQUS, was used to model the superstructure. A total of 240 beam elements (with 675 DOFs) were used, so that inelastic behavior of the beams and columns particularly during the collision would be simulated accurately. The base-isolation layer and the surrounding retaining walls were tested physically, and the restoring force characteristics of the rubber bearings and collision and inelastic behavior of the retaining walls were obtained from the test. The specimen is given in Fig. 41.6(c). The test was successfully conducted, and the obtained responses were reasonable. The corresponding hysteresis curves of the base isolation layer are plotted in Fig. 41.6(d). It reveals that at each collision the resisting force increases significantly. Details are given in [11].

41.7 Test Using Peer-to-Peer Framework

The peer-to-peer framework was applied to investigate the seismic response of a steel reinforced concrete (SRC) structure with a steel braced tower built on top, as shown in Fig. 41.7(a). The entire structure was divided into three substructures,

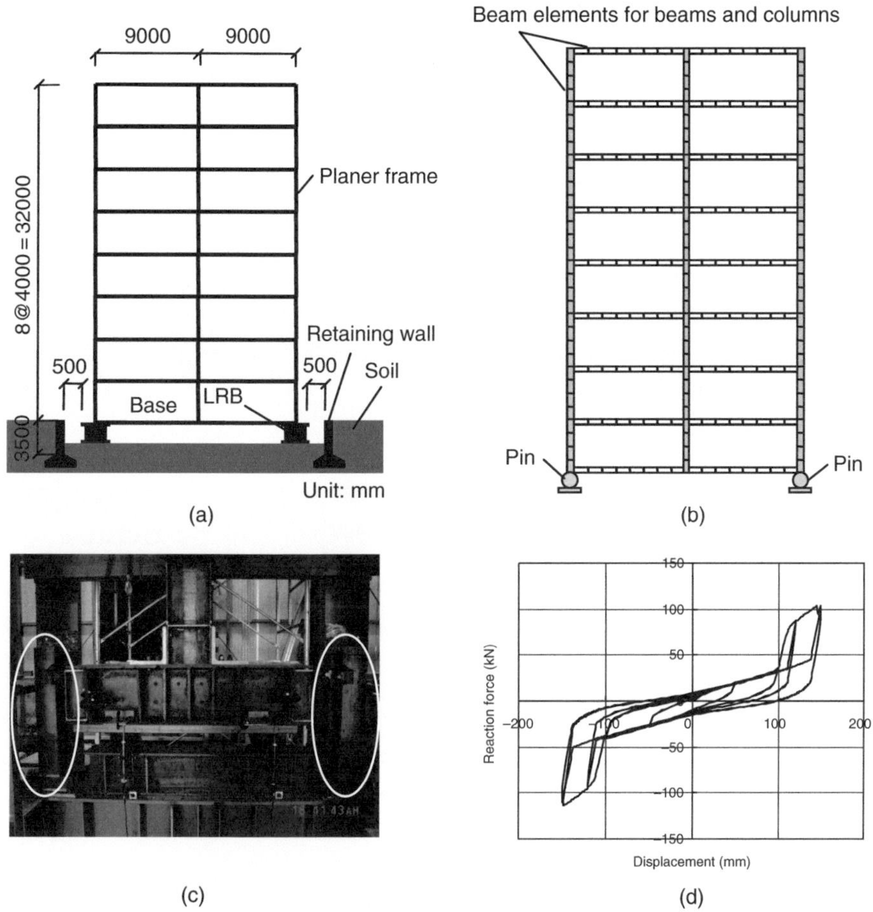

Fig. 41.6 Demonstration test using dual-model framework: (**a**) base-isolated structure with retaining wall; (**b**) superstructure model; (**c**) test specimen; (**d**) test results

namely, the SRC frame, the first story of the tower and the upper part of the tower. The static coordinator was employed to handle the horizontal displacements among the three substructures.

The first story of the tower was tested physically, using a scaled model that maintained the similitude, while the other two substructures were treated numerically. OpenSEES was used to simulate the SRC frame, as this analysis code is equipped with an excellent fiber-formulated element that is particularly suitable for composite members. A general-purpose finite element program, ABAQUS, was used to simulate the upper part of the tower, because of its advantage in handling strong geometric nonlinearities. Each model contained a few hundreds DOFs.

The setup for the test substructure is shown in Fig. 41.7(b). The vertical jack adopted force control to provide a constant gravity of the tower, while the horizontal jack was used in displacement control to realize the target boundary displacement

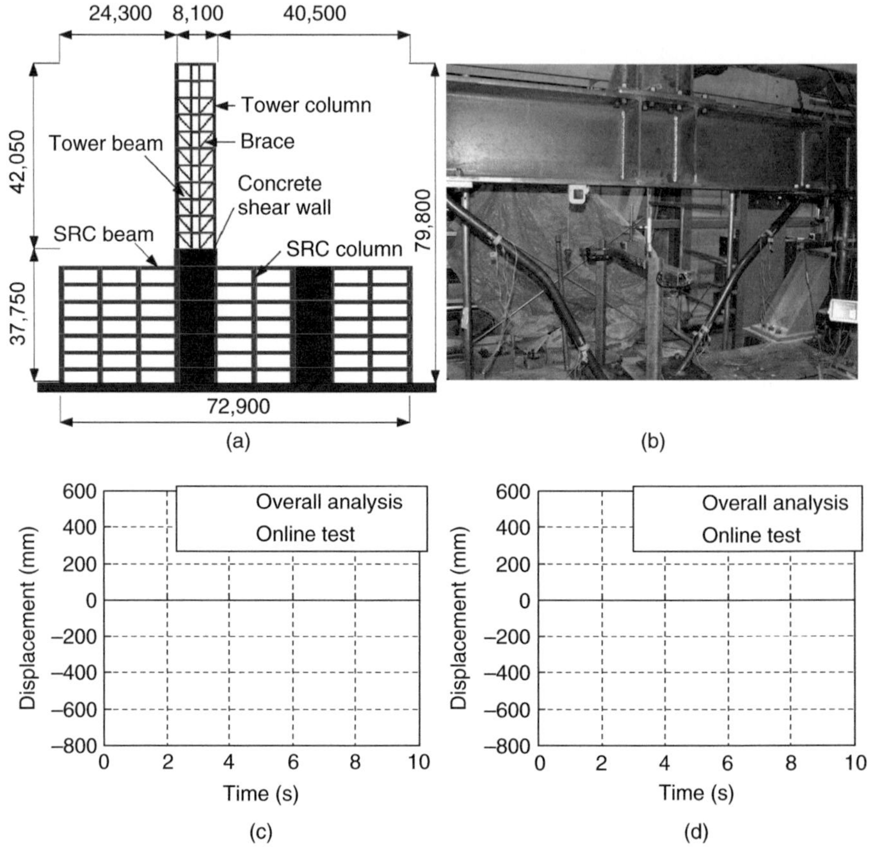

Fig. 41.7 Demonstration test using peer-to-peer framework: (**a**) Prototype structure; (**b**) Photo of specimen; (**c**) Response of roof of SRC frame; (**d**) Response of first story of tower

received from the coordinator. All substructures were encapsulated by means of the generalized interface, with boundary displacements and reaction forces exchanged by the input and output files. The restart option of each finite element program was employed, enabling the simulation to advance step by step.

All three substructures exhibited considerable inelastic behavior. In particular the test substructure sustained the largest deformation and the braces buckled seriously. In order to validate the interface implementation, the seismic responses obtained from the online hybrid test were compared with those of the overall numerical simulation. The displacements at the substructure interfaces are compared in Fig. 41.7(c) and (d) for the top of the SRC frame and the specimen, respectively. For both interfaces, the displacements are very close to each other. The discrepancy observed in Fig. 41.7(d) is due to the difference in hysteretic behavior between the numerical model (assumed behavior) and the physical test (actual behavior). This discrepancy had a significant influence on the tower behavior, but had little effect on the SRC frame, because the frame was much heavier than the tower.

This application confirms that: (1) the generalized interface is able to encapsulate multiple numerical substructures using different analysis codes; and (2) the static coordinator employing the two-round quasi-Newton procedure is capable of reproducing the seismic response of complex structures.

Details about this test are given in [13].

41.8 Conclusion

Three frameworks for the online hybrid test, named host-station, dual-model, and peer-to-peer, respectively, are developed. Major conclusions are as follows:

(1) The three online hybrid test frameworks are extensible to host multiple geographically distributed tested and numerical substructures.
(2) The series of online hybrid tests conducted demonstrate that the online hybrid test frameworks are flexible, extensible and accurate.

Acknowledgments Financial support by the National Science Foundation of China under Grant No. 50808107 is gratefully acknowledged. The first author gratefully acknowledges the Japan Society for the Promotion of Science (JSPS) for giving him a financial support for most of the research reported in this paper. The authors are grateful to the following individuals for their assistance in developing the system and conducting the tests: T. Wang of Institute of Engineering Mechanics, China, H. Tomofuji of Okumura Corporation, Japan, N. Yoshitake of Ministry of Land, Infrastructure, Transport and Tourism of Japan, J. McCormick of University of Michigan, USA and Y. Murata of Nippon Steel Engineering, Japan. Any errors in this paper are those of the authors alone.

References

1. Mosqueda G, Stojadinovic B, Hanley J, Sivaselvan M, Reinhorn MA (2008) Hybrid seismic response simulation on a geographically distributed bridge model. J Struct Eng 134(4):535–543
2. NSF (2000a) Network for earthquake engineering simulation (NEES): earthquake engineering research equipment. Program solicitation. Report NSC00-6, National Science Foundation, USA
3. NSF (2000b) Network for earthquake engineering simulation (NEES): system integration, program solicitation. Report NSC00-7, National Science Foundation, USA
4. NSF (2001) Network for earthquake engineering simulation (NEES): consortium development, program solicitation. Report NSC 01-56, National Science Foundation, USA
5. Pan P, Tada M, Nakashima M (2005) Online hybrid test by internet linkage of distributed test and analysis domains. Earthq Eng Struct Dyn 34:1407–1425
6. Pan P, Tomofuji H, Wang T, Nakashima M, Ohsaki M, Mosalam KM (2006) Development of peer-to-peer (P2P) Internet online hybrid test system. Earthq Eng Struct Dyn 35:867–890
7. Sugiura K, Nagata N, Suzuka Y, Watanabe E (1998) Internet related structural testing. Proceedings of the Eighth KKNN Seminar on Civil Engineering, Singapore, 219–224
8. Takahashi Y, Fenves GL (2006) Software framework for distributed experimental-computational simulation of structural systems. Earthq Eng Struct Dyn 35:267–291

9. Tsai K et al (2003) Network platform for structural experiment and analysis. Report NCREE-03-021, National Center for Research on Earthquake Engineering, Taiwan
10. Watanabe E, Yun C, Sugiura K, Park D, Nagata K (2001) Online interactive testing between KAIST and Kyoto University. Proceedings of the Fourteenth KKNN Symposium on Civil Engineering, Kyoto, Japan, pp 369–374.
11. Wang T, Nakashima M, Pan P (2006) On-line hybrid test combining with general-purpose finite element software. Earthq Eng Struct Dyn 35:1471–1488
12. Wang K, Tsai K, Wang S, Cheng W, Yang Y (2003) Networked hybrid test frameworks and examples. The Fifth Seminar on Earthquake Engineering for Building Structures (SEEBUS), Kyoto, Japan, 81–90.
13. Wang T, Yoshitake N, Pan P, Lee TH, Nakashima M (2008) Numerical characteristics of peer-to-peer (P2P) Internet online hybrid test system and its application to seismic simulation of SRC structure. Earthq Eng Struct Dyn 37:265–282
14. Yang YS, Hsieh, SH, Tsai KC, Wang SJ, Wang KJ, Cheng WC, Hsu CW (2007) ISEE: Internet-based simulation for earthquake engineering, Part I: Database approach. Earthq Eng Struct Dyn 36:2291–2306

Chapter 42
Large Scale Seismic Testing of Steel-Framed Structures at NCREE

Keh-Chyuan Tsai, Chao-Hsien Lee, Ching-Yi Tsai, Chih-Han Lin, Po-Chien Hsiao, Min-Lang Lin, Yuan-Tao Weng, Ker-Chun Lin, and Jui-Liang Lin

42.1 Introduction

The seismic performance of structures can be most realistically evaluated through large-scale experiments. As a result, the demand of large-scale structural tests has significantly increased in recent years. Nevertheless, due to the limited capacity of each individual laboratory, it has been found difficult and not very cost-effective to carry out such large-scale structural tests by one single research institute. Thus, the hybrid experiments collaboratively conducted by several research institutes have become the trend of large-scale structural tests in modern research laboratories. Recently, several large-scale international-cooperative hybrid tests on frame structures have been conducted at the National Center for Research on Earthquake Engineering (NCREE) in Taiwan. In this paper, four large scale frame tests are introduced. The first one is a hybrid test of full-scale three-storey three-bay buckling restrained braced frame (BRBF). This 3-storey BRBF was the largest steel and concrete composite frame ever tested. Several kinds of BRBs and beam-to-column connections were investigated. The second one is a Taiwan-US collaborative substructure pseudo-dynamic test (SPDT) in 2005, on a full-scale two-storey BRBF subjected to the bi-directional seismic ground motions. The main objective of this test was to verify the design of the gusset plates for BRBFs under bi-directional ground excitations. The third one is the cyclic testing of four full-scale two-storey steel plate shear walls (SPSWs) tested in 2007. The main issues of this test were to verify the proposed capacity design for boundary elements of SPSW frames and the advantages of adding restrainers to SPSW frames. The fourth project is the cyclic tests of three full-scale two-storey steel special concentrically braced frames (SCBFs), and this project is a collaboration of NCREE and several US universities in 2008. The objective of these tests was to investigate the performance of SCBFs using 2t-linear clearance and 8t-elliptical clearance between brace ends and gusset plate corners.

K.-C. Tsai (✉)
National Taiwan University, Taipei, Taiwan; National Center for Research on Earthquake Engineering, 200, Sec.3, Xinhai Rd., Taipei 10668, Taiwan
e-mail: kctsai@ncree.org

M.N. Fardis (ed.), *Advances in Performance-Based Earthquake Engineering*, 451
Geotechnical, Geological, and Earthquake Engineering 13, DOI 10.1007/978-90-481-8746-1_42,

42.2 Pseudo-Dynamic Testing of 3-Storey CFT/BRB Frame

42.2.1 Information on the Specimen

This was a collaborative research among Taiwan, Japan and USA [7, 8]. Measuring 12 m tall and 21 m long, the specimen had square and circular CFT columns as the two exterior and interior columns, respectively (Fig. 42.1). In this CFT/BRB frame, only the two exterior beam-to-column connections in each floor were moment connections. All other beam-to-column connections were assumed not to transfer any bending moment. The BRBs were installed in the central bay. Square CFT columns were chosen for the two exterior columns and the two columns at the center were circular CFTs. The material of all beams and columns was A572 Gr.50. The strength of the infill concrete in CFT columns was 35 MPa. Three types of BRBs were adopted for the three different floors. The two single-cored unbonded braces (UBs), each consisting of a steel flat plate in the core, were donated by Nippon Steel Company and installed at the 2nd storey. The two BRBs installed in the 3rd storey were double-core, with cement mortar infilled in two rectangular tubes. The two all-metal BRBs in the 1st storey were also double-core but fabricated with detachable features [9]. The elevation and the floor framing plan of the specimen are shown in Fig. 42.2.

Fig. 42.1 Three storey CFT/BRB frame

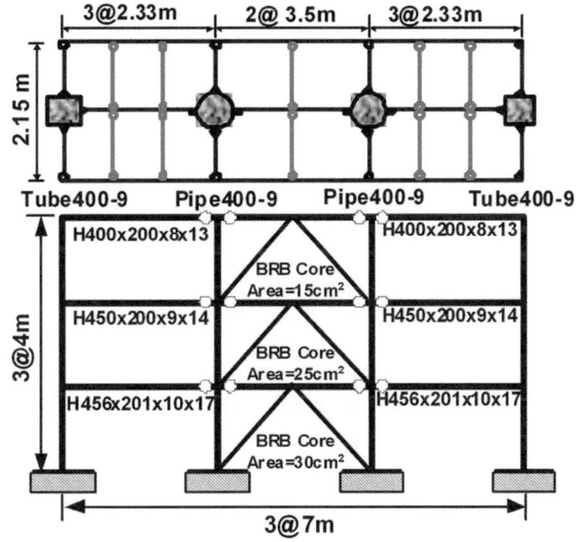

Fig. 42.2 Floor framing plan and elevation of the CFT/BRB Frame

42.2.2 Experimental Program

Two earthquake records were used in this study: TCU082EW, from the 1999 ChiChi earthquake, and LP89g04NS, from the 1989 Loma Prieta earthquake. An unexpected buckling of the gusset plate in the first storey interrupted the fist loading event. The test stopped and the stiffeners were added at the free edges of all the gusset plates. The test resumed using the same ground motions. After all PDTs were conducted, no fracture found in any of the BRBs. Thus, cyclic increasing uniform storey drifts were imposed until failure of the BRBs. Since the scheduled PDT and cyclic tests were completed with failures only in bracing components including the BRBs, UBs and the gusset plates, it was decided to conduct the Phase II tests, after repairing or replacing the damaged components. All the key analytical predictions and the experimental responses were broadcast from a website: http://cft-brbf.ncree.gov.tw.

42.2.3 Internet-Based Simulation on Earthquake Engineering System

In order to efficiently carry out PDTs in single or multiple laboratories, the internet-based simulation on earthquake engineering (ISEE) system has been developed by NCREE [12]. The ISEE system provides a platform integrating a number of different laboratories through the internet, to jointly conduct a single structural experiment. In this system, there is a command generation module (CGM) responsible for computing the next time-step displacement and passing this information to the platform for the networked structural experiments (PNSE) server. The facility control module (FCM) transforms this information received from the PNSE server into the real displacement of each actuator mounted on the specimen. The actuators push or pull the specimen following the commands received from the FCM. There are sensors and data loggers for recording all the testing results and transmitting the results to the PNSE server. Then, the measured restoring force is passed to the CGM for computing the next time-step displacement. All the testing results passed to the PNSE server are recorded in a database. Some of the data recorded in the database are broadcast on-line, enabling the data viewers to see the test results simultaneously. Videos record the whole test process and broadcast on-line via the video stream server.

42.3 Substructure PDT of Two-Storey BRB Frame Subjected to Bi-Directional Earthquake Loads

42.3.1 Information on the Specimen

This was a Taiwan-USA collaborative research. The 2-storey prototype building was located at Chiayi City on hard rock and was designed according to the storey force distribution prescribed in the 2002 Taiwan Seismic Building Specifications [1]. The design dead load (DL) of the floor was 6.89 kN/m^2, and the design live load (LL) was 2.45 kN/m^2 for each floor. The corresponding design base shears were about 20% of the building weight in both directions. The specimen was 8 m wide and 8 m tall. A 2.28 m wide, 150 mm thick concrete slab was used over the steel beams. The BRBF resisted a substantial proportion of the lateral forces: the cross-sectional areas of the BRB steel cores were 5000 mm^2 and 3300 mm^2 in the first and second floors, respectively. The details of the design procedures can be found in [11]. All members of the frame were A572 Gr50. The beam-to-column joints of the BRB frame were pin-connected. Welded connections were used for the first storey BRB-to-gusset connections and bolted connections with 10–24 mm A490 bolts for those in the second storey. The fundamental periods of the entire building system were 0.69 s and 0.57 s in the longitudinal (MRF, denoted as X-direction) and transverse (BRBF+MRF, denoted as Y-direction) directions, respectively.

42.3.2 Experimental Program and Results

The specimen was tested under bi-directional pseudo-dynamic loads. The bi-directional ground motion records (TCU076 and CHY024) scaled to three different hazard levels (2%, 10% or 50% exceedance probability in 50 years), were chosen to investigate both the in-plane and out-of-plane inelastic deformational demands. In order to reduce the residual deformations accumulating after each event, the direction of the ground motions were reversed in the subsequent earthquake load event. All the key analytical predictions and substructure pseudo-dynamic test (SPDT) results were broadcast from a website: http://substructure- brbf.ncree.org.tw. Test results demonstrated the energy dissipation ability of the specimen. They also confirm that adding stiffeners at the gusset edges can effectively increase the compressive stability of the gusset. The peak experimental inter-storey drifts reached 1.5% and 2.2% in the longitudinal and the transverse directions respectively in the 2% in 50 years event.

42.4 Cyclic tests of Four Full-Scale Two-Storey Steel Plate Shear Wall Frames

42.4.1 Restrainer Effects on Boundary Elements of SPSW Frames

Previous studies [3, 6] have shown that the out-of-plane deformation and the noise induced by steel plate buckling of SPSW frames can be significantly reduced by using restrainers. This confirmed the serviceability of restrained SPSW (R-SPSW) frames to be better than that of conventional SPSW frames. Two other issues about R-SPSW were verified in the present test. The first issue is to verify the proposed capacity design method for the boundary elements of SPSW frames [2]. The SPSW would not develop its full energy dissipation capacity if plastic hinge form in-between the ends of boundary elements. Thus, the aim of the proposed capacity design method is to ensure that the plastic hinges form at the ends of boundary elements. The second issue is to verify the effect of restrainers on reducing the column flexural demands and the beam axial force demands [2]. Thus, the size of boundary elements of SPSW frames can be significantly reduced by using restrainers. That way, the total steel weight of R-SPSW frames is less than in conventional SPSW frames.

42.4.2 Information on Specimen and Experimental Program

Four 2-storey 2.14 m wide by 6.5 m tall SPSW were constructed and cyclically tested to a roof drift of 5% at NCREE. The setup of the SPSW test system is shown in Fig. 42.3. Low yield strength steel plates of 2.6 mm were adopted for all four specimens. Two out of four specimens were constructed with horizontal restrainers. The key parameter of this series of tests is the size of the boundary elements of the specimens and the use of restrainers. All the boundary elements were of A572 GR 50 steel. A lateral support system was constructed to prevent the out-of-plane instability of the SPSW specimens. Special lateral supports were provided at the beam-to-column joints of the middle beams and the top beams, so that the unbraced length of the columns was equal to the typical height of one storey.

The four specimens are denoted as Specimen N, RS, S and CY, respectively. Specimen N is normal, in the sense that its columns complied with the proposed capacity design method. Specimen RS is an R-SPSW, designed considering proper capacity design. Because of the effectiveness of the restrainers, the size of boundary

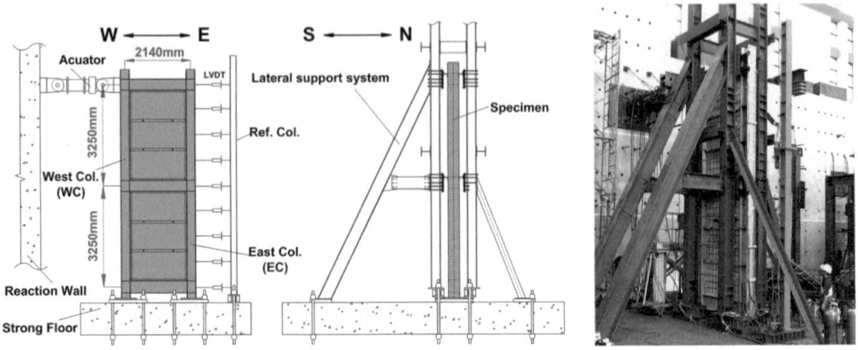

Fig. 42.3 Test setup

elements in Specimen RS is smaller than in Specimen N. Specimen S, without restrainers, has the same boundary elements as those of Specimen RS. Except the top beam, the other boundary elements of Specimen CY are as in Specimen RS. However, its top beam member size was intentionally chosen bigger than that of Specimen RS, to allow yielding (CY) at the column top. Moreover, the reduced section details were adopted at the column top of Specimen CY. The flexural demand-to-capacity ratio (DCR) of the column is lower than 1.0 for specimens N, RS and CY. This suggests that the 1F columns in the three specimens have been well-designed. The fact that the ratio exceeds 1.0 for Specimen S implies that the column strength is insufficient.

42.4.3 Key Experimental Observations and Results

In general, all specimens remained elastic up to a drift of 0.3%. Fairly nonlinear behaviour was observed due to yielding of the panel at a roof drift level of about 0.4%. Hysteretic loops notably widened at a roof drift level of 0.75%. The boundary elements slightly yielded at this stage. Plastic hinges were observed at the boundary element ends at a roof drift level of about 1%. As the roof drifts exceeded 2.5%, flange or web local buckling occurred gradually at the plastic hinge zones. Nevertheless, the load-carrying capacity did not decrease significantly. At the end of the test, it was found that a plastic hinge developed at the 1F column bottom ends, as recognized from the flaking of the whitewashes. A plastic hinge was evidently formed above the bottom end of the column of Specimen S. The local buckling of the web at middle beam RBS of Specimen S was initiated at about a 2% roof drift. For other specimens, slight local buckling of the web of middle beams occurred at a 3% roof drift. The middle beam sizes in Specimens S, RS and CY are the same. The local bucking of the beam web in Specimen CY and RS occurred at a lager roof drift level (3%). It appears that the restrainers have reduced the middle beam axial force. The presence of beam axial forces was confirmed, as well as the effectiveness of the restrainers in reducing the axial forces. Test results verified that the proposed capacity design method is effective in causing the plastic hinge to form in the compressed 1F boundary column. Furthermore, the effectiveness of the restrainers in reducing the column flexural demands was confirmed. The total weight of Specimen RS, including the restrainers (about 2.7 tons) is less than that of Specimen N (about 3.3 tons), suggesting that the restrained SPSW could potentially save more material than a typical SPSW.

42.5 Cyclic Tests of Three Full-Scale Two-Storey Concentrically Braced Frames

42.5.1 Information on the Specimen

This study was a collaborative research of NCREE and several universities in the US [4, 5, 10]. The specimen is a single bay, two-storey special concentrically-braced frame (SCBF) in X-brace configuration. The width and height of the frame are 6.7 m and 6.67 m, respectively. The specimen is the first of a series of large scale SCBF tests being conducted in the NCREE Laboratory. All beams and columns are of A572 GR 50 steel. There were three tests in this study. The main differences among them were the brace types (hollow structural or wide-flange section) and the gusset plate connection designs. During the three tests, there was no evident fracture of beams or columns. Thus, damaged braces and gussets were replaced at the end of each test. In the Phase I test, four A500 grade steel tubes were used for braces and the 8t-clearance gusset design detail [5] was adopted. Wide-flange braces and the 8t-clearance gusset detail were adopted in the Phase II test. Finally, in the Phase III test, the A500 steel tube braces and the 2t-clearance gusset detail were used. The Phase I specimen is shown in Fig. 42.4. The SCBF specimen was tested under cyclically increasing roof displacements. The key experimental responses were broadcast from a website: http://exp.ncree.org/cbf during each test.

42.5.2 Key Experimental and Analytical Results

Results of these three tests confirm that the two-storey X-shape steel SCBFs all have rather good energy dissipation characteristics (Fig. 42.5) up to a storey drift of about 3% under the cyclically increasing roof displacements. Quite evident brace local buckling and out-of-plane displacements were observed during each test. Tests confirm that both the 2t-linear and the 8t-elliptical clearance designs for out-of-plane deformations of gussets provide satisfactory ductility for the seismic steel SCBFs. Hollow structural sections braces fractured at a storey drift smaller than in the test of the SCBF using wide flange braces.

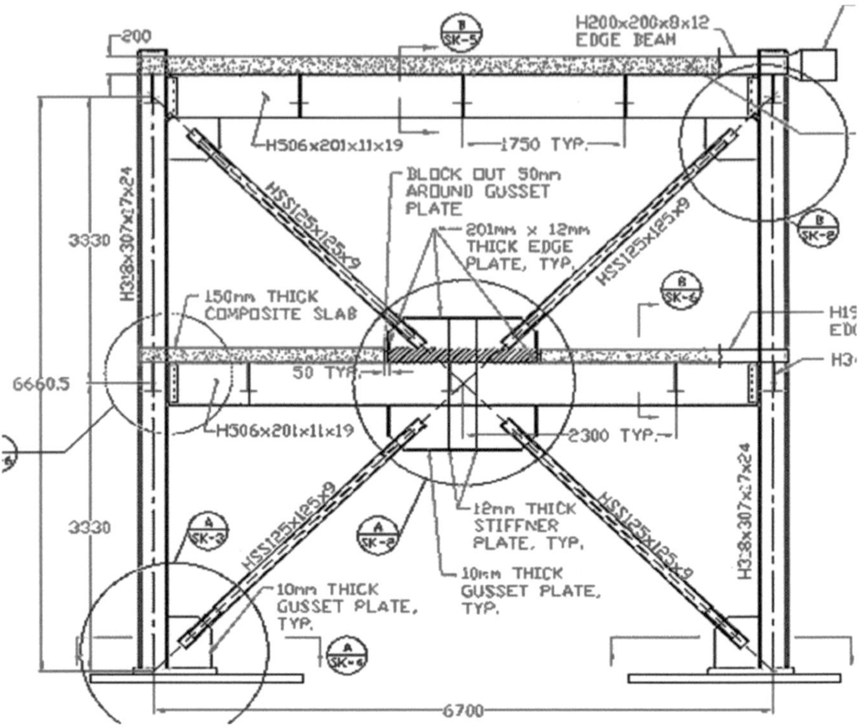

Fig. 42.4 Elevation of the SCBF specimen

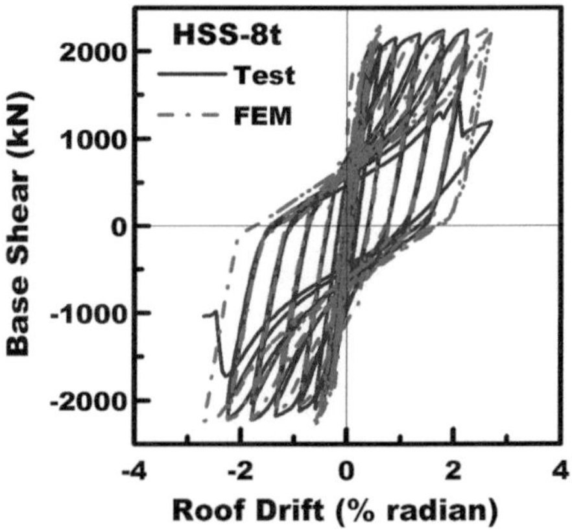

Fig. 42.5 Hysteretic loops of the SCBF specimen

42.6 Conclusions

Based on these test results, conclusions and recommendations are made as follows:

(1) Stiffeners added along the free edges of the gusset plate are effective in preventing out-of-plane instability of the brace-to-gusset connections. However, they also increase rotational stiffness and introduce flexural demands on the BRBs. Further research is required to study the BRB end connections.
(2) Based on the two-storey BRBF test results, the gusset plates for the BRBs have been found to sustain bi-directional earthquake force and deformational demands without fracture or buckling, when properly designed and detailed.
(3) Test results of the two-storey SPSW frames verify the proposed capacity design method ensuring that plastic hinges form at the ends of boundary elements. Besides reducing the out-of-plane deformation of the steel plate, the restrainers have been also confirmed by the tests to effectively reduce the flexural demands on boundary columns and the axial force demands on boundary beams. That is, restrained SPSW frames can be more economically designed with improved serviceability, compared to SPSW frames without restrainers.
(4) Results of three tests confirm that the two-storey X-shape steel SCBFs have rather good energy dissipation capacity up to a storey drift of about 3%, under the cyclically increasing roof lateral displacements.
(5) Test results confirm that both the 2t-linear and the 8t-elliptical clearance details for out-of-plane deformations of the gussets provide satisfactory ductility for seismic resistant design of SCBFs.

References

1. ABRI (2002) Seismic force requirements for building structures. Architectural Building Research Institute. Taipei, Taiwan
2. Li CH, Tsai KC, Lin CH (2009) Cyclic tests of four two-storey narrow steel plate shear walls. Proceedings of 5th International Symposium on Steel Structures, Hotel Prima Seoul, Korea, Mar 12–13, 2009
3. Lin CH, Tsai KC, Lin YC, Wang KJ, Qu B, Bruneau M (2007) Full scale steel plate shearwall: NCREE/MCEER Phase I tests. Proceeding of the 9th Canadian Conference on Earthquake Engineering, Ottawa, Ontario, Canada, June 26–29, pp. 170–179
4. Lin CH, Wei CY, Tsai KC, Powell J, Clark K, Roeder C (2009) Cyclic tests of a full-scale 2-story steel concentrically braced frame. Proceedings of 5th International Symposium on Steel Structures, Hotel Prima Seoul, Korea, March 12–13, 2009
5. Powell J, Clark K, Tsai KC, Roeder C, Lehman D (2008) Test of a full scale concentrically braced frame with multi-story X-bracing. Proceeding, Structures Congress 2008, Vancouver, Apr 24–26, 2008
6. Qu B, Bruneau M, Lin CH, Tsai KC (2008) Testing of full-scale two-storey steel plate shear wall with reduced beam section connections and composite floors. J Struct Eng ASCE 134:364–373
7. Tsai KC, Hsiao PC, Wang KJ, Weng YT, Lin ML, Lin KC, Chen CH, Lai JW, Lin SL (2008) Pseudo-dynamic tests of a full-scale CFT/BRB frame—Part I: Specimen design, experiment and analysis. J Earthq Eng Struct Dyn 37:1081–1098

8. Tsai KC, Hsiao PC (2008) Pseudo-dynamic tests of a full-scale CFT/BRB frame—Part II: Seismic performance of buckling-restrained braces and connections. J Earthq Eng Struct Dyn 37:1099–1115
9. Tsai KC, Lin SL (2003) A study of all metal and detachable buckling restrained braces. Center for Earthquake Engineering Research, National Taiwan University, Report No. CEER/R92-03
10. Tsai CY, Tsai KC, Lin ML, Chou CC (2009) Numerical responses of a full scale steel concentrically braced frame. Proceedings of 5th International Symposium on Steel Structures, Seoul, Korea, Mar 12–13, 2009
11. Tsai KC, Weng YT, Wang KJ, Tsai CY, Lai JW (2006) Bi-directional sub-structural pseudo-dynamic testing of a full-scale 2-story BRBF, part 1: seismic design, analytical and experimental performance assessment. 8NCEE-001097, 100th Anniversary Earthquake Conference, San Francisco
12. Wang KJ, Tsai KC, Wang SJ, Cheng WC, Yang YS (2007) ISEE: internet-based simulation for earthquake engineering-part II: the application protocol approach. J Earthq Eng Struct Dyn 36:2307–2323

Chapter 43
Large Scale Shaking Table Tests for High-Rise Buildings: New Projects of E-Defense

Takuya Nagae, Kouichi Kajiwara, Takahito Inoue, and Masayoshi Nakashima

43.1 Introduction

43.1.1 Seismic Vulnerability of Japan

The 1995 Hyogoken-Nanbu (Kobe) earthquake caused devastating damage to buildings and infrastructures in Kobe and its vicinity [1, 8–10]. This earthquake improved our knowledge and understanding in many aspects, such as structural, economic, societal, cultural and human ones.

It is known that Japan is destined to suffer from large ocean-ridge earthquakes on a periodical basis. Figure 43.1 shows a map of Japan and an ocean ridge, called the Nankai trough, running deep along the Pacific Coast of Japan. The trough is divided in three regions: Tokai, Tonankai and Nankai, from east to west. For many centuries, slips and ruptures along the three regions have been occurring at an interval of 100–150 years. From the pattern of the previous earthquakes, Japan is most likely to be hit by the next large earthquake by the middle of this century. In 2005, the Council of National Disaster Mitigation, chaired by the Prime Minister of Japan, disclosed a damage estimate that Japan would sustain if the Nankai trough were to rupture again [3]. According to this estimate, should all three regions rupture together, about forty million people, i.e., one third of the entire population of Japan, would be affected; about one million houses and buildings would collapse; about twenty-five thousand people might lose their life and the direct economic loss would amass to close to one trillion US dollars. These estimates are at least ten times greater than those observed in the 1995 Kobe earthquake. Another very serious piece of data was disclosed recently. Twenty-two percent of large companies listed at the Tokyo stock market are headquartered in high-rise buildings of downtown Tokyo; their sales accumulate to about 30% of the total Japanese sales. Should the Tokyo metropolitan region be

T. Nagae (✉)
Hyogo Earthquake Engineering Research Center, National Research Institute for Earth Science and Disaster Prevention, Miki, Hyogo 673-0515, Japan
e-mail: nagae@bosai.go.jp

M.N. Fardis (ed.), *Advances in Performance-Based Earthquake Engineering*, Geotechnical, Geological, and Earthquake Engineering 13, DOI 10.1007/978-90-481-8746-1_43,

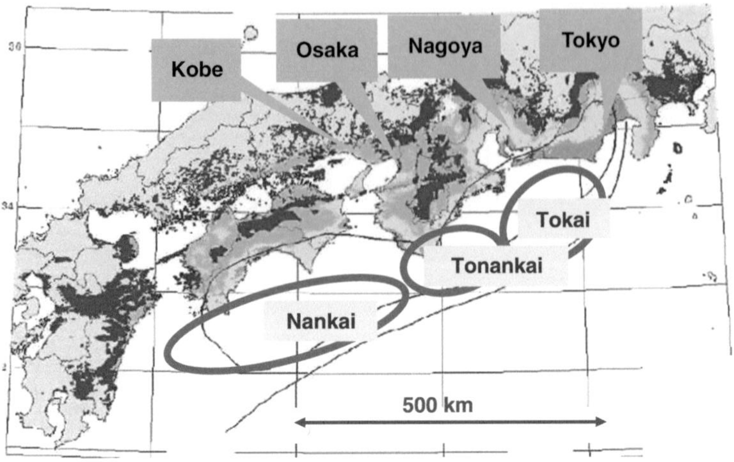

Fig. 43.1 Ocean-ridge earthquakes anticipated in Japan

hit by a large shaking, the Japanese economy would be affected very seriously. As these statistics clearly indicate, the earthquake disaster was, is, and will remain the most critical national problem in Japan.

43.1.2 Establishment and Activities of E-Defense

Stimulated by the 1995 Kobe earthquake, the Government of Japan decided to establish a large experimental facility for the advancement of earthquake engineering. Along this effort, the National Research Institute for Earth Science and Disaster Prevention (NIED) administered the construction of a shaking table facility, known as E-Defense [5]. E-Defense was completed in March 2005; its operation started in April 2005.

The E-Defense table has five actuators in each horizontal direction and is supported by fourteen vertical actuators installed underneath the table (Fig. 43.2). The

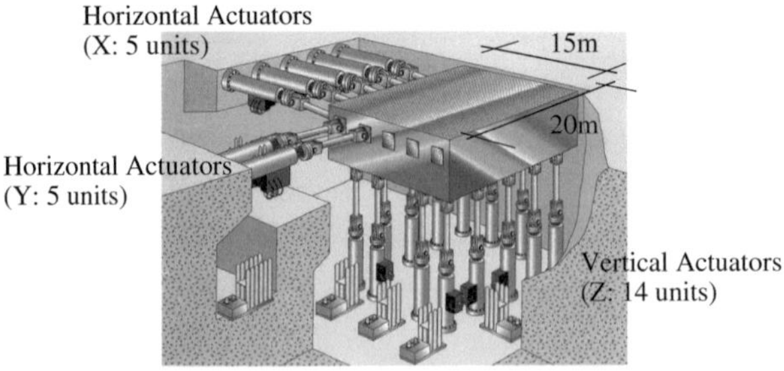

Fig. 43.2 Shaking table in E-Defense

table is 20 m by 15 m in plan and can accommodate specimens with weight of up to 12 MN (1,200 metric tons). The unique feature of the table is that it can produce shaking with a velocity of up to 2 m/s and displacements of up to 1 m in the two horizontal directions simultaneously.

Between 2005 and 2008 E-Defense conducted various shaking table tests of different types of structures. Two examples [11, 12] are the following:

1. A pair of thirty-year-old houses were tested on the table side by side. The retrofitted house was able to endure the JR Takatori motion, while the as built house collapsed. The test became a perfect appeal to the public for the importance of seismic retrofit.
2. A full scale four-story steel moment frame was tested. It had two-bays-by-one-bay in plan and weighed about 5,000 kN (500 metric tons). Under the JR Takatori motion the test structure collapsed owing to the large story drift of the first story. Numerical analyses to trace the behavior of the structure were carried out as well.

43.2 Tests for High-Rise Buildings

Large ocean-ride earthquakes are likely to occur in Japan. One serious concern about such earthquakes is long-period, long-duration shaking [6, 7]. Such shaking may produce very large floor response of high-rise buildings, characterized by large velocities and displacements. At the same time, the structure may sustain a number of cyclic inelastic deformations.

Most of Japanese high-rise buildings built in the past thirty years are in the 80–100 m height range [2, 4]. Figure 43.3 shows a generic building and the size of the E-Defense shaking table facility. The height of the building is about four times what the facility can accommodate. Thus, for a series of high-rise building tests, new substructure test methods were employed that can apply the seismic response of a high-rise building to a test specimen.

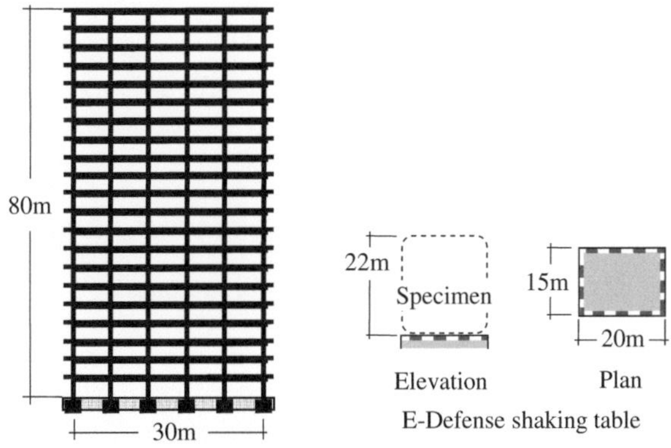

Fig. 43.3 A generic high-rise building and E-Defense shaking table

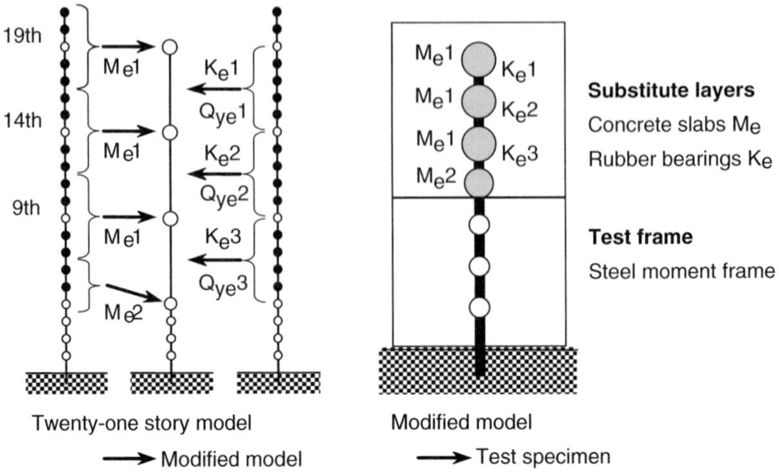

(a) Development of modified model

(1) Shop welded connection (3) Steel moment frame

(2) Field welded connection (4) Test specimen

(b) Construction of test specimen

Fig. 43.4 Development of test method for structures

43.3 Seismic Resistance Capacity

A number of existing high-rise buildings have never been assessed in terms of structural damage due to long period ground motions. The special setup shown in Fig. 43.4 was developed to reproduce seismic responses of high-rise buildings. Under severe cyclic loading, structural components in such buildings, particularly beam-to-column connections, may be damaged seriously as a result of a number of cyclic deformations. In the setup, a multi-layer mass-spring-damper system was placed on top of the physical steel moment frame that represented the details of beam-to-column connections. Figure 43.4(a) shows the procedure for developing the modified model from the prototype model having 21 degrees of freedom. The whole test structure was designed in reference to the modified model. The test structure has two-bays-by-one-bay in plan and weighed about 11,000 kN (1,100 metric tons). The addition of the mass-spring-damper system made it possible to match the vibration characteristics, such as the natural periods, between the prototype and test structure.

The test specimen was prepared according to a normal procedure of construction (Fig. 43.4b). In the longitudinal direction, a built-up wide flange section H600 was arranged, with the shop weld connection detail. In the transverse direction, H800 was employed with the field weld connection detail. The details of beams and beam-to-column connections were chosen based on past design practice.

Synthesized ocean-ridge motions were used as input to the test structure. The frame was subjected to a number of cyclic deformations in these motions. Figure 43.5 shows the time histories of interstory drift angle in the steel moment frame. The cumulative inelastic deformation finally caused fractures at beam-to-column connections. The bottom flange fracture was attributed to the amplified strains due to the composite effect of the RC floor slabs as well as the large cumulative inelastic deformations. The tests provided a set of unprecedented data on the seismic behavior of high-rise buildings.

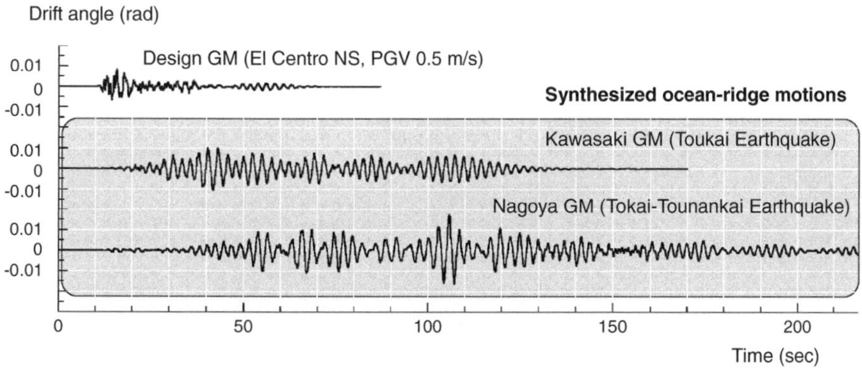

Fig. 43.5 Deformations applied to the steel moment frame

43.4 Safety of Rooms

The response of furniture placed in high-rise buildings was studied. The test for furniture was set focusing on the disaster prevention enlightenment. Long-period, long-duration shaking may hit high-rise buildings and produce very large floor response. Such response would cause serious damage to nonstructural elements, furniture and other building contents, particularly in upper floors. To reproduce such large floor response, the special test setup shown in Fig. 43.6 was developed. The test structure was treated as a rigid frame. A two-layer mass-spring system consisting of a concrete slab and rubber bearings were inserted between the rigid frame and the shaking table. This system served as an amplifier of the table motion to the level of floor response expected in upper floors of high-rise buildings. The input motion to the table had to be carefully adjusted, so that the motion would reproduce the desired floor response on the rigid frame placed on the table.

Various types of realistic rooms were set up on the specimen's floors. In the test, the maximum floor response of 1.3 m in displacement and 2.4 m/s in velocity were achieved. Figure 43.7 shows some notable test results. The unprepared rooms suffered significant damage to the contents, while rooms prepared with special tools had very little damage. From this test, the critical need of clamping the furniture against overturning and sliding became very evident.

The concept of disaster prevention enlightenment is to disseminate the unprecedented aspects to the public as shown in Fig. 43.8. Contrast between prepared rooms and unprepared rooms strongly highlight the need of preparations. Such data become a very powerful tool to enlighten the public in disaster prevention. Video files are edited to be easily used in schools, related conferences and so on. The files are now open at the NIED web site.

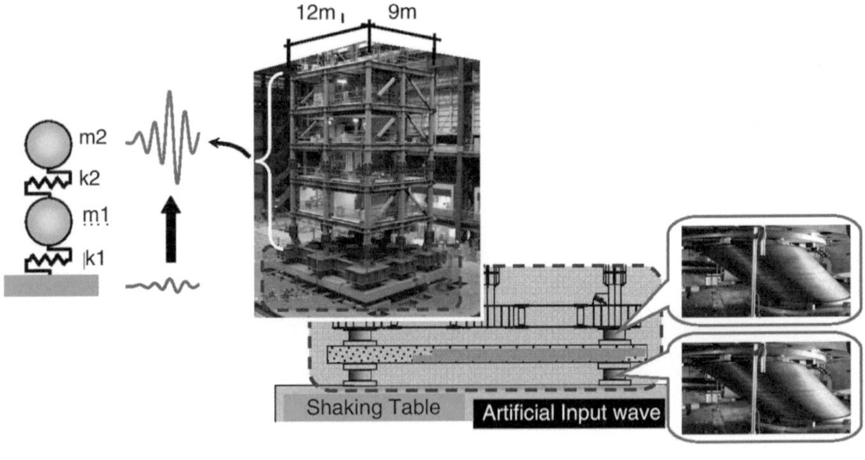

Fig. 43.6 Development of test method for rooms

(a) Devastation of unprepared office

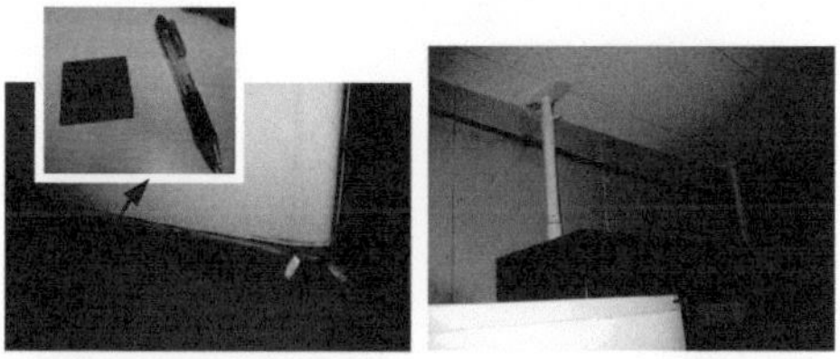

Replaceable adhesive mat Compressive vertical poles

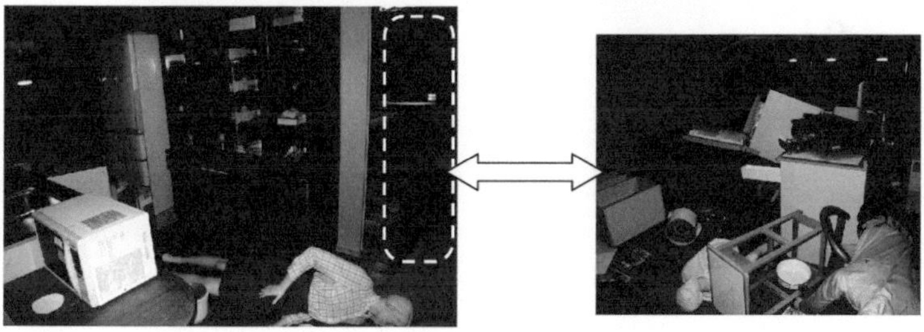

(b) Effects of preparation

Fig. 43.7 Results of room tests

43.5 Summary

Stimulated by the 1995 Kobe earthquake, the Government of Japan decided to establish a large experimental facility for the advancement of earthquake engineering. Along this effort, the National Research Institute for Earth Science and Disaster Prevention (NIED) administered the construction of a shaking table facility, known as E-Defense. E-Defense has already conducted a series of shaking table tests of various types of structures.

The pattern of previous earthquakes suggests that Japan is most likely to be hit by a next large ocean-ride earthquake by the middle of this century. One serious concern about such events is long-period, long-duration shaking that may hit large cities. The long-period ground motion may induce to hundreds of high-rise buildings very large floor response, characterized by large velocities and displacements. The structures may also sustain a number of cyclic inelastic deformations. In new projects of E-Defense, substructure test methods were employed for the large-scale

(a) Downloadable movies

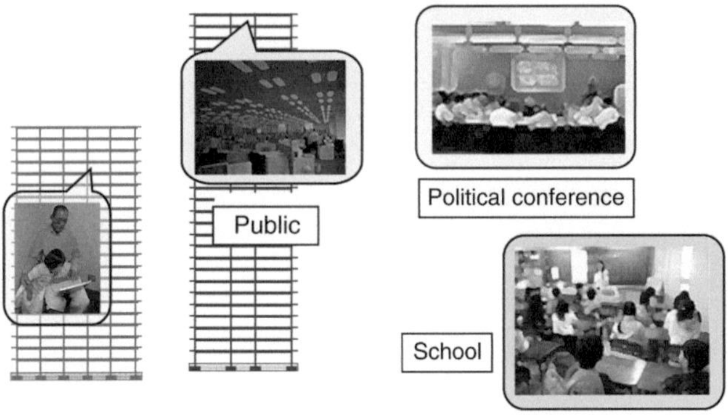

(b) Disaster prevention enlightenment

Fig. 43.8 Powerful tool opened at the web site of NIED (http://www.bosai.go.jp/hyogo/ehyogo/movie.html)

tests of high-rise buildings. Focusing on the structural performance, a steel moment frame having real connection details was tested with such a test system. In the long period ground motions, a number of cyclic deformations were applied to the test frame. The capacities of the beam to column connections were identified in term of cumulative inelastic deformation. Another shaking table test focused on the safety of rooms. The contrast between prepared and unprepared rooms was physically produced and successfully recorded in the videos. The edited video files are now open at the NIED web site. Such data begin to practically contribute to the seismic safety of Japan's society.

A series of tests provided a set of unprecedented data on the seismic behavior of high-rise buildings. According to these tests, existing buildings should be checked immediately. If the capacity to resist long-period ground motions is lacking, such buildings need to be retrofitted appropriately. As part of the new projects, structural tests focusing on seismic retrofitting of high-rise buildings are held in autumn of 2009 (Fig. 43.9).

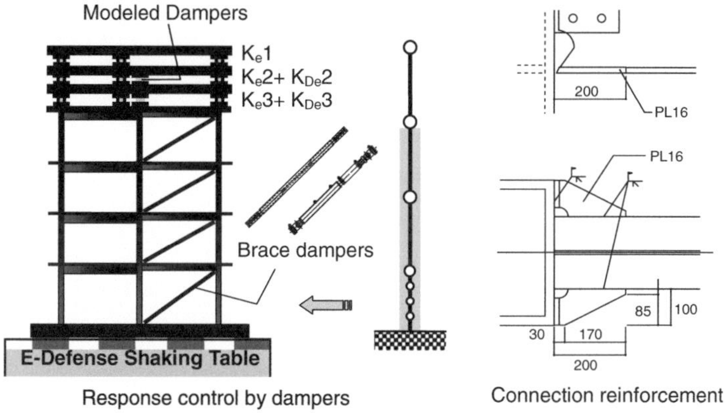

Fig. 43.9 Seismic retrofit tests coming in autumn of 2009

References

1. Architectural Institute of Japan (1995) Preliminary reconnaissance report of the 1995 Hyogoken-Nanbu earthquake. Architectural Institute of Japan, Tokyo, 216 pp
2. Building Research Institute (2005). Research report on the influence of seismic retrofit technology to buildings subjected to long-period ground motion. Tokyo, Japan (in Japanese)
3. Council of National Disaster Mitigation (2004). http://www.bousai.go.jp/jishin/chubou.html (in Japanese)
4. Fukushima T, Ichimura S, Teramoto T (1999) The basic feature history of high rise building. Architectural Institute of Japan, Tokyo, pp 307–308 (in Japanese)
5. Hyogo Earthquake Engineering Research Center, National Research Institute for Earth Science and Disaster Prevention (2005). http://www.bosai.go.jp/hyogo/ehyogo/index.html
6. Kamae K, Kawabe H, Irikura K (2004) Strong ground motion prediction for huge subduction earthquakes using a characterized source model and several simulation techniques. Proceedings of the13th world conference on earthquake engineering, Vancouver, Canada, pp 655–666
7. Kawabe H, Kaname K, Irikura K (2008). Damage prediction of long-period structures during subduction earthquakes-Part 1: Long-period ground motion prediction in the Osaka basin for future Nankai Earthquakes. Proceedings of the 14th world conference on earthquake engineering, Beijing, China
8. Kinki Branch of the Architectural Institute of Japan (1995). Reconnaissance Report on damage to steel building structure observed from the 1995 Hyogoken-Nanbu Earthquake, Steel Committee, Osaka (in Japanese with attached abridged English version)
9. Nakashima M (2001) Appendix C. Overview of damage to steel building structures observed in the 1995 Kobe Earthquake, Past Performance of Steel Moment-Frame Buildings in Earthquakes, Federal Emergency Management Agency, Report FEMA-355E, C-1-C24
10. Nakashima M, Inoue K, Tada M (1998a). Classification of damage to steel buildings observed in the 1995 Hyogoken-Nanbu Earthquake. Eng Struct 20:4–6, 271–281
11. Nakamura I, Shimizu H, Minowa C, Sakamoto I, Suzuki Y (2008) E-Defense experiments on full-scale wooden houses. Proceedings of the 14th World conference on earthquake engineering, Beijing, China
12. Suita K, Yamada S, Tada M, Kasai K, Matsuoka Y, Shimada Y (2008) Collapse experiment on 4-story steel moment frame: part 2 detail of collapse behavior. Proceedings of the 14th world conference on earthquake engineering, Beijing, China

Chapter 44
Verification Through Shaking Table Testing of EC8-Based Assessment Approaches Applied to a Building Designed for Gravity-Loads

Alberto Pavese and Igor Lanese

44.1 Introduction

The last decades have seen an increased interest in the seismic performance of existing buildings designed for gravity loads. This is especially so in Europe, particularly after catastrophic events like Friuli 1976 and Irpinia 1980 in Italy, Izmit 1999 in Turkey and recently Abruzzo 2009, Italy. Tackling this problem is not easy, for several reasons. First of all, it is necessary to address a large population of structures with different characteristics and built with different techniques. Second, there is often lack of information on the geometry, the materials properties, the reinforcement, etc.. Finally, the past history (earthquakes, time-dependent deformations, etc.) of an existing building may have reduced its seismic resistance without this being evident without accurate analyses.

Seismic risk assessment of existing buildings is a very real problem, as buildings built before the introduction of seismic codes represent, according to recent estimates, about 40% of the total in the south of Europe.

The *NEARB* project [17], Numerical and Experimental Assessment of Recommendations inherent Existing Buildings included in OPCM 3274 (the trial version [15] of the new Italian seismic regulations, the new Italian code having been published in 2008 [7]), presented in this paper, addresses the topic of the evaluation of the assessment procedures of the seismic vulnerability of the existing structures. The evaluation has been conducted by means of linear and nonlinear analysis and a shake table testing campaign performed at the TREES Lab of Eucentre on a 1:2 scale 3-storey RC frame building with masonry infills, representative of a typical design in Europe of the 1950–1960s (Fig. 44.1). The specimen is similar in many respects to that studied at full scale in the SPEAR project (Seismic Performance Assessment and Rehabilitation of Existing Buildings, [8])

A. Pavese (✉)
Department of Structural Mechanics, Università degli Studi di Pavia,
Via A. Ferrata, No. 1, Pavia 27100, Italy
e-mail: alberto.pavese@eucentre.it

M.N. Fardis (ed.), *Advances in Performance-Based Earthquake Engineering*,
Geotechnical, Geological, and Earthquake Engineering 13, DOI 10.1007/978-90-481-8746-1_44,

Fig. 44.1 Specimen on the shake table of Eucentre

and tested at the ELSA laboratory of the Joint Research Centre in Ispra (Italy) via the pseudo-dynamic testing technique.

44.2 Description of the Structure

44.2.1 General Description

The building has been designed according to the Greek code in force between 1954 and 1984, with typical construction practices of the early 1960s. It is asymmetric and torsionally imbalanced (cf. the eccentricity between centre of mass CM and centre of stiffness CR in Fig. 44.2); two beams are not directly supported by columns and one beam-column joint (at column C2) has large eccentricity.

The cross section dimensions are 125 $\times$ 250 mm for the beams and 125$\times$125 mm for the columns, except C2 (125 $\times$ 375 mm). The concrete has characteristic compressive strength f_{ck} = 25 MPa; the reinforcement is of smooth rebars with yield strength f_y = 370 MPa and 180°-hooks. Vertical bars of the columns are of 6 mm dia. and stirrups are of 3 mm dia. at 70 mm centres. There are no stirrups in beam-column joints. Further details can be found at http://www.eucentre.it/pe1.

44.2.2 Problems Deriving from the Gravity Load Design

The structure has several features typical of the design for gravity loads [14]. The columns are slender, have low flexural resistance and are weaker than the beams (*weak column/strong beam* design). The stirrups in beams or columns are designed for the shear forces due to gravity loads alone and their spacing is too large for good

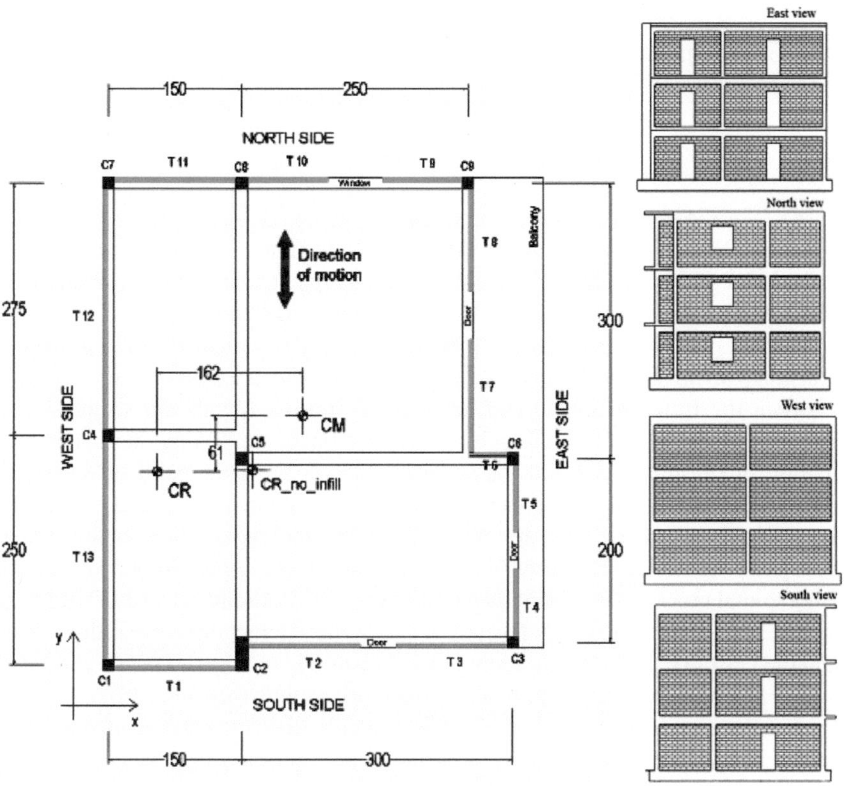

Fig. 44.2 Framing plan of the building and four views in elevation (dimensions in cm)

confinement; so they provide low ductility in terms of curvature capacity. Moreover, the floor plan irregularity produces torsional effects and larger displacements at the more flexible sides of the structure (Fig. 44.2).

44.3 Shake Table Tests

The building has been tested on the shake table using the accelerogram of Montenegro 1979, Herceg-Novi station (the same used in the test carried out on the full scale building at ELSA in the SPEAR project) in the longitudinal direction (Y axis, Fig. 44.2), scaled to different levels of PGA to simulate minor, moderate and severe earthquakes.

For a representative simulation of the full scale structure the time axis of the accelerogram has been scaled by a factor $\sqrt{\lambda} = \sqrt{2}$ and further additional masses have been added to re-tune the modal parameters of the structure [11]. The building has been fully instrumented using two main acquisition systems: one with traditional transducers commonly adopted in such tests, such as accelerometers, LVDTs, potentiometers and strain gauges; the other is an optical video-acquisition system composed of high resolution cameras and frame grabber devices.

44.3.1 Observed Damage

Test #1 – PGA=0.08g: After the first test, no visible damage was observed in the RC frame or in the masonry infills. After the test the natural frequencies were found to be reduced by about 5% [2] (Table 44.1), as the stiffness slightly decreased as result of a partial separation of the panels from the surrounding frame.

Test #2 – PGA=0.30g: In the second test the RC structure suffered a limited cracking, in flexure at the ends of the columns as well as due to the interaction between the frame members and the masonry infills. At the end sections of the columns of the first storey through cracks developed, suggesting the incipient formation of plastic hinges and the typical effect of the poor bond of the smooth reinforcing bars inside the joint). At the top of Columns C1 and C7 of the first floor, a horizontal crack was observed in the joint panel region showing the incipient occurrence of a so-called "shear-hinge" [12], as shown in Fig. 44.3c. Additionally, a horizontal crack developed at about mid-height of Columns C1 and C7 of the first storey, mainly at the external face, as result of the interactionbetween the columns and the masonry infills (shear-sliding mechanism, Fig. 44.3(a)). The masonry infills suffered severe cracking, mainly along the mortar joints (Fig. 44.3(b)), which indicates a shear-sliding mechanism. On the East side (the furthest away from the CR, see Fig. 44.2), the masonry infills at Levels 1 and 2 suffered significant damage; the

Table 44.1 Fundamental frequency of the system in different configurations

Test/configuration	First natural frequency (Hz)
Pre-test bare frame	2.3
Pre-test infilled frame	5
(1) Test #1 – PGA = 0.08*g*	4.7
(2) Test #2 – PGA = 0.30*g*	2
(3) Test #1 – PGA = 0.54*g*	2

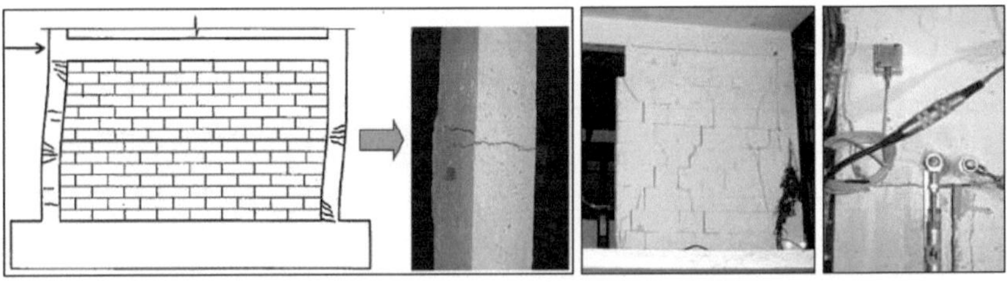

(a) (b) (c)

Fig. 44.3 (**a**) Shear-sliding mechanism of the panel and cracking at mid-height of column C1; (**b**) damage in masonry infill; (**c**) damage at the top of column C1 at the 1st storey largest panel at Level 2 collapsed, owing to the lack of a column at one end of the panel and the presence of an opening.

Test #3 – PGA=0.54 g: The final test increased the damage to the square columns (125×125 mm) developed during the previous tests. A plastic hinge (mainly with a single crack) developed in all of them. Column C2 (125×375 mm) suffered similar damage to the other columns, but more clear and visible (see Fig. 44.4a and c). All beams stayed essentially in the elastic range, because their flexural strength was significantly higher than those of the columns. The damage in the masonry infills also increased (Fig. 44.5) and the cracking patterns developed in the previous tests increased in extent and degree. The damage in the exterior beam-columns joints of Columns C1 and C7 at Levels 1 and 2 was significant. The failure mode is illustrated in Fig. 44.6, where it can be observed that a different cracking pattern and damage occurred at the top and bottom faces of the joint, as described in the following section.

Fig. 44.4 Damage after Test#3 (PGA = 0.54g); (**a**) longitudinal section of Column C2; (**b**) East view; (**c**) Spalling of the concrete cover and rebar buckling in Column C2 at 2nd storey

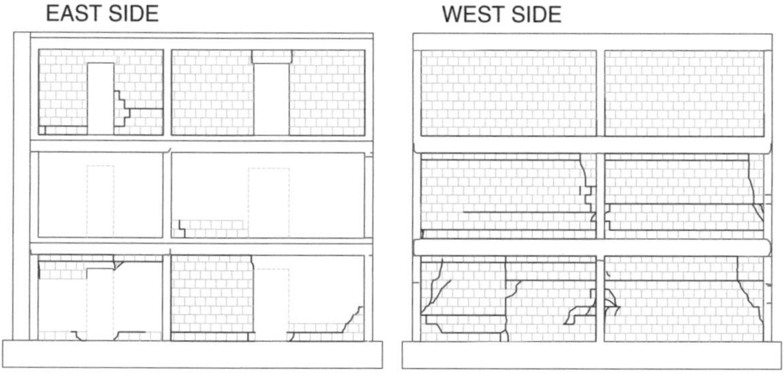

Fig. 44.5 Damage of masonry infills after Test #3 (PGA = 0.54g)

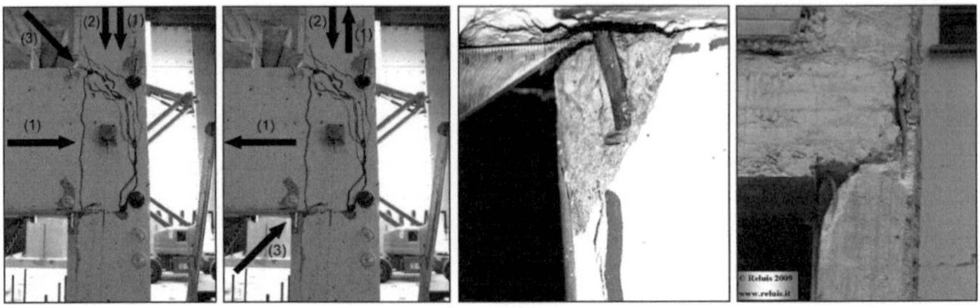

(a) (b) (c) (d)

Fig. 44.6 (**a**–c): Damage of the beam-column joint of Column C1 at the 1st storey: (**a**) forces under positive loading; (**b**) forces under negative loading; (**c**) concrete spalling and rebar buckling; (**d**) Pettino (Aq), damage of a beam-column joint in the Aquila 2009 earthquake

44.3.2 Frame-Panel Interaction

Field experience and analytical/experimental results have shown that the non-structural infills built in contact with the frame have an important influence on the seismic response of the structure. Such elements may produce both beneficial and adverse effects [1,4,5,9,16]. In general the infill panels can:

- reduce the seismic deformation demand due to their in-plane stiffness;
- increase the lateral force resistance;
- enhance the global energy dissipation capacity with their hysteresis.

Often they play a fundamental role in preventing the global collapse, particularly in non-seismic designed structures.

On the other hand they:

- stiffen the structure, possibly attracting higher inertial forces;
- may produce a soft storey or plan irregularity;
- may cause brittle failure of frame members, especially columns, due to local effects.

These effects have been observed on the tested building: the panels have in fact stiffened the RC frame (see Table 44.1) and increased the eccentricity between CR and CM (see Fig. 44.2) both in the initial configuration (i.e., the bare frame relative to the infilled one) and even after the stronger shakes despite the collapse of those located on the east side. They have caused cracking at mid-height of some columns due to shear-sliding failure and triggered failure of some beam-column joints (Fig. 44.6), due to lack of stirrups in the beam-column joints, a typical detail in old construction practice. Similar joint failures have been observed in several buildings of l'Aquila after the 2009 earthquake (Fig. 44.6d).

Such failure modes of joints can be explained considering the internal forces in the RC members and the effect of the masonry infill, acting as a diagonal strut.

When the structure responds from left to right (Fig. 44.6a), the masonry infill transfer compressive forces at the top corner of the joint (represented by arrow 3)); the compressive axial load in the column due to gravity load (arrow 2) is increased by the axial force induced by the earthquake (arrow 1). Therefore the shear forces are transferred mainly through the concrete, inducing the inclined crack (approximately at 45°) at the top of the joint. When the structure responds from right to left (Fig. 44.6b), the masonry infill is in contact with the frame at the bottom corner of the joint (arrow 3) while the earthquake induces a tensile axial force (arrow 1). Therefore, the total axial force in the column is reduced (compared to the compressive gravity load (arrow 2)) and may even result in a tensile axial load. Under these conditions, a horizontal crack forms at the bottom face of the joint and the shear forces activate the dowel mechanism of the longitudinal reinforcement of the column (Fig. 44.6c).

The shear damage in these nodes has reduced the interstorey drift demand, postponing the soft-storey mechanism. However, this positive consequence may lead to a sudden reduction of the strength of the columns with the loss of their bearing capacity [12].

In the end the external infills have given a notable contribution to the overall resistance of the structure [12]. In fact the specimen was able to carry without failing (but with a severe damage) a level of PGA equal to $0.54g$, about 3 times larger than the capacity of the bare frame tested at the ELSA laboratory in Ispra on the ($0.20g$, pseudo-dynamic testing). Recall that the acceleration does not change due to scaling [11].

44.4 EC8-Based Assessment Approaches

The recent seismic codes, such as the Italian Code "Norme Tecniche per le Costruzioni" [7] and EC8 [3], treat in detail the assessment of existing buildings. The first step concerns knowledge of the structural system, through a historical analysis, a geometric survey and mechanical characterization of materials. Based on the accuracy of the collected details, the codes define a corresponding knowledge level which implies a confidence factor (c.f.) to divide the estimated mean values of the material properties. For the examined case study the c.f. has been taken equal to 1.0, as a consequence of the refined knowledge of the geometry and material properties. Such refined knowledge has allowed the application of both linear and non-linear approaches as described in the following.

44.4.1 Linear Model

A linear elastic model has been implemented by using the open-source code *OpenSees* (http://opensees.berkeley.edu/index.php); both the RC frame and the infill panels have been modelled. An eigenvalue analysis has been performed to identify

Table 44.2 Modal quantities from the numerical model

Mode	T (s)	Freq. (Hz)	Modal mass X (%)	Modal mass Y (%)
1	0.36	2.81	82.58	1.91
2	0.29	3.44	5.83	63.18
3	0.23	4.28	1.98	21.21
4	0.14	9.19	7.45	0.20
5	0.11	7.17	0.59	5.66

the dynamic response of the structure and validate the linear model by comparing modes and frequencies with those of the specimen. In Table 44.2 the modal parameters of the most representative modes of the model are shown. This model already accounts for the reduction of the inertia of the RC elements for the verification at the ultimate limit states. So the stiffness decreases and the period increases with respect to the experimental ones in the undamaged configuration.

From the results of the eigenvalue analysis in the direction of the seismic input of the shake table (Y), it is clear that the 1st mode alone does not suffice. It is necessary to consider higher mode. A modal response spectrum analysis has been performed, combining the contributions of each mode with a SRSS combination. The behaviour factor (q) used to scale the elastic spectrum has been chosen equal to 2, according to the code provisions.

44.4.2 Non-linear Model

A non-linear model of the structure has been implemented in order to compare the accuracy of a more refined type of analysis with the previous prediction and with the experimental results. In order to have a compromise between accuracy of the analysis and computational effort, a non-linear static analysis (*pushover*) has been performed. As it is required both in EC8 and NTC, two force distributions have been applied to in the positive and negative directions along the Y axis: a "uniform" and a "1 st mode" pattern.

44.4.3 Verifications and Numerical Results

The structure is a weak column/strong beam design. Moreover, the RC slabs increase the flexural capacity of the beams. By comparing the capacity of the vertical and horizontal elements suggests problems in the columns. In the vertical elements the flexural capacity is reached before the shear capacity (assuming a double-bending deformed shape). This implies that the failure mechanism will be ductile. So the verification, according to the code provisions, will be done in terms of deformations (i.e. chord rotation). In Fig. 44.7, for each column, the experimental

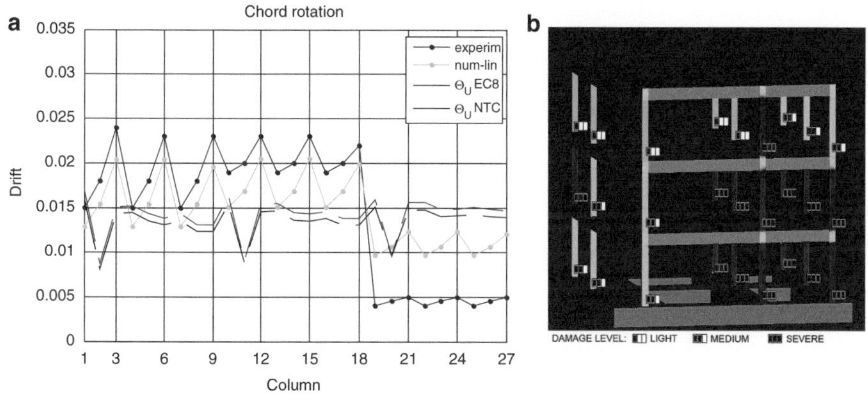

Fig. 44.7 (**a**) Chord rotation, experimental, numerical (linear model) and code provisions; (**b**) predicted damage of the columns

drift, the numerical one and the ultimate drift computed with the formulas suggested by the codes [3, 7] are shown.

The prediction of the linear model for what concerns the first two floors is not far from the test results; in particular the numerical drifts are smaller than the real experimental ones. This can be explained taking into account that in general for the Ultimate Limit States (ULS) the moment of inertia of the elements which will likely suffer cracking is reduced; this concept has not been applied to the infill panels which will be certainly affected by a stiffness degradation at the ULS. This issue is not easy to tackle, starting from the model of an infill panel [6] which has several inherent uncertainties. For this reason further parametric analysis are required to account in a suitable way for the stiffness degradation of the panels.

The prediction obtained from the non-linear model is shown in Fig. 44.8. A soft storey mechanism is evident at the first floor, for all the force distributions and the

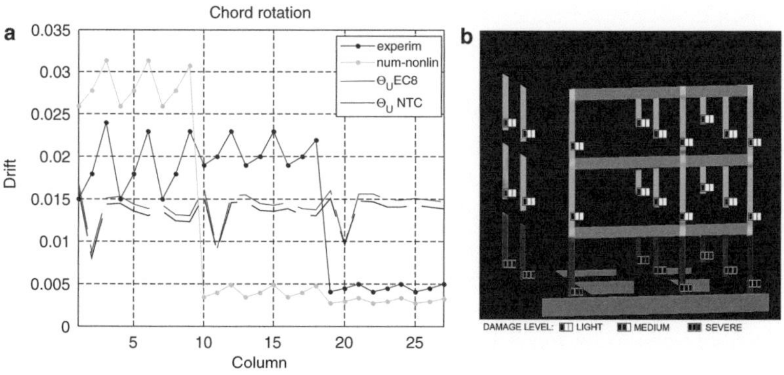

Fig. 44.8 (**a**) Chord rotation, experimental, numerical (non-linear model) and code provisions; (**b**) predicted damage of the columns

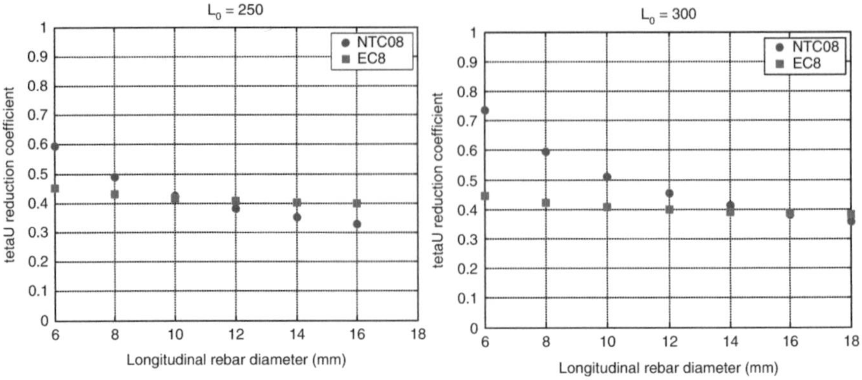

Fig. 44.9 Chord rotation reduction coefficient for several diameters and 2 overlapping lengths in EC8 and NTC

considered directions. The drift of all the elements of the 1st floor is overestimated, whereas at the 2nd and 3rd floor is underestimated.

Considering the elements of the 1st floor, all the numerical results as well as the experimental ones give a chord rotation greater than the ultimate value provided by the codes. This is also because the ultimate chord rotation capacity is reduced by a coefficient which accounts for the use of smooth bars with lapping in the plastic hinge region and the lack of seismic detailing [3, 7]. Such a coefficient is computed in a slightly different way in EC8 and NTC; its value is a function of the diameter of the longitudinal bars and the lap length. The trend is sketched in Fig. 44.9.

Looking at the damage in the columns after the tests, this reduction coefficient seems to be too conservative; in fact the codes suggest to assess the chord rotation capacity based on experimental tests when possible. Moreover, in a displacement-based approach the slippage of the smooth bars might be considered as a contribution which increases the drift capacity [13].

44.5 Conclusions and Further Development

This experimental campaign has given interesting results to understand the real seismic behaviour of this typology of existing buildings. It is worth noting that it is difficult to obtain a realistic simulation of these dynamic problems in terms of forces and consequently get the real failure mechanism without a dynamic test of the entire structure.

For what concerns the code provisions, the value of the ultimate drift suggested as reduced to account for the use of smooth bars with lapping and the lack of seismic details seems to be too conservative with respect to the experimental results. About the numerical analysis, both linear and nonlinear models have been used; the prediction of the linear model is not far from the experimental results, whereas the nonlinear one predicted a soft storey mechanism at the first floor that didn't occur during the tests. These results also depend on the intrinsic uncertainties inside the models, which are particularly increased by the infill panels. The beam-column joints have not been modeled in such a way to account for a possible strength reduction and failure of them, which indeed happened in some joints in Tests #2 and #3. A more refined model of the joints will be necessary to better fit the experimental results.

Acknowledgments Part of the current work has been carried out under the financial auspices of the Italian Civil Protection, within the framework of the Executive Project 2005–2008 (Research Project #1) and of the Executive Project 2008–2011 (Project e3). Such support is gratefully acknowledged by the authors. The authors would also like to thank the Cariplo Foundation, for the contribution within the project "Use of innovative materials for strengthening and reparation of R.C. structures in high seismicity areas" and Consorzio ReLUIS for the contribution within the project "ReLUIS Linea 1".

Finally, a special thank is due to the EUCENTRE lab staff, whose support has been crucial in the implementation and interpretation of the whole test campaign.

References

1. Calvi GM, Bolognini D, (2001) Seismic response of reinforced concrete frames infilled with weakly reinforced masonry panels. J Earthq Eng 5(2):153–185
2. Casarotti C, Peloso S, Lunghi F (2009) NEARB-3274: System identification of a torsionally-coupled building. XIII ANIDIS National Conference Proceedings to be published.
3. CEN (2005) EN 1998-3:2005 – Eurocode 8 Design of structures for earthquake resistance – Part 3: Assessment and retrofitting of buildings. European Committee for Standardisation, Brussels
4. Crisafulli FJ (1997) Seismic behavior of reinforced concrete structures with masonry infills. PhD Thesis, Department of Civil Engineering, University of Canterbury, New Zealand, http://ir.canterbury.ac.nz/handle/10092/1221
5. Crisafulli FJ, Carr AJ, Park R (2005) Experimental response of framed masonry structures designed with new reinforcing details. Bull NZ Soc Earthq Eng 38(1):19–32
6. Decanini LD, Bertoldi SH, Gavarini C (1993) Telai tamponati soggetti ad azione sismica, un modello semplificato: confronto sperimentale e numerico. Atti del 6° convegno nazionale. ANIDIS 2:815–824, Perugia 13–15
7. Decreto Ministeriale 14/01/2008, Gazzetta Ufficiale n.29 del 4/2/2008, supplemento ordinario n. 30, Norme Tecniche per le Costruzioni
8. Fardis M, Negro P (2005) Seismic assessment and rehabilitation of existing buildings. Proceeding of SPEAR international workshop, Ispra, Italy.
9. Kakaletsis DJ, Karayannis CG (2008) Influence of masonry strength and openings on infilled R7C frames under cycling loading. J Earthq Eng 12:197–221
10. Lanese I (2007) Valutazione numerico-sperimentale del comportamento sismico di un edificio esistente progettato per soli carchi verticali. Tesi di Laurea, Università degli Studi di Pavia, Italy
11. Lanese I, Marazzi F, Nascimbene R (2008) Il cambiamento di scala delle strutture per la verifica sismica su tavola vibrante di un telaio in c.a. progettato per carichi verticali. Ingegneria Sismica 4:13–25
12. Magenes G, Pampanin S (2004) Seismic response of gravity-load design frames with masonry infills, 13th World Conference on Earthquake Engineering, Vancouver, BC, Canada
13. Magenes L (2008) Confronto tra un metodo di valutazione strutturale agli spostamenti e i metodi alle forze proposti dall'OPCM 3274. Tesi di Laurea, Università degli Studi di Pavia, Italy

14. Masi A (2003) Seismic vulnerability assessment of gravity load designed R/C frames. Bull Earthq Eng 1:371–395
15. Ordinanza n. 3274 del Presidente del Consiglio dei Ministri, Gazzetta Ufficiale 5 marzo 2003, Supplemento Ordinario n. 72 della G.U. n. 105 del 8.05.2003, *Primi elementi in materia di criteri generali per la classificazione sismica del territorio nazionale e di normative tecniche per le costruzioni in zona sismica*
16. Paulay T, Priestley MJN (1992) Seismic design of reinforced concrete and masonry buildings. Wiley, New York, NY
17. Pavese A, Lanese I, Scovenna MP, Franzolin R, Lunghi F, Airouche H, Casarotti C, Peloso S (2008) NEARB project: numerical and experimental assessment recommendations inherent existing RC buildings included in OPCM 3274. Final research report

Index

Geotechnical, Geological, and Earthquake Engineering 13, DOI 10.1007/978-90-481-8746-1,

P

R

S

MIX
Papier aus verantwortungsvollen Quellen
Paper from responsible sources
FSC® C105338
FSC
www.fsc.org

If you have any concerns about our products,
you can contact us on
ProductSafety@springernature.com

In case Publisher is established outside the EU,
the EU authorized representative is:
Springer Nature Customer Service Center GmbH
Europaplatz 3, 69115 Heidelberg, Germany

Printed by Libri Plureos GmbH
in Hamburg, Germany